Microbiology and Biochemistry of Cheese and Fermented Milk

Microbiology and Biochemistry of Cheese and Fermented Milk

Second edition

Edited by

B.A. LAW
Research Consultant
Burghfield Common
Berkshire, UK

BLACKIE ACADEMIC & PROFESSIONAL
An Imprint of Chapman & Hall

London · Weinheim · New York · Tokyo · Melbourne · Madras

**Published by Blackie Academic & Professional, an imprint of Chapman & Hall,
2–6 Boundary Row, London SE1 8HN, UK**

Chapman & Hall, 2–6 Boundary Row, London SE1 8HN, UK

Chapman & Hall GmbH, Pappelallee 3, 69469 Weinheim, Germany

Chapman & Hall USA, Fourth Floor, 115 Fifth Avenue, New York NY 10003, USA

Chapman & Hall Japan, ITP-Japan, Kyowa Building, 3F, 2-2-1 Hirakawacho, Chiyoda-ku, Tokyo 102, Japan

DA Book (Aust.) Pty Ltd, 648 Whitehorse Road, Mitcham 3132, Victoria, Australia

Chapman & Hall India, R. Seshadri, 32 Second Main Road, CIT East, Madras 600035, India

First edition 1984

Second edition 1997

© 1984, 1997 Chapman & Hall

Typeset in 10/12pt Times by Doyle Graphics, Tullamore, Ireland
Printed in Great Britain by T.J. Press (Padstow) Ltd, Padstow, Cornwall

ISBN 0 7514 0346 6

A catalogue record for this book is available from the British Library

Library of Congress Catalog Card Number: 96-86400

∞ Printed on acid-free text paper, manufactured in accordance with
ANSI/NISO Z39.48-1992 (Permanence of Paper).

Contents

9   Proteolytic systems of dairy lactic acid bacteria   299
F. MULHOLLAND

10   Molecular genetics of dairy lactic acid bacteria   319
M.J. GASSON

Contributors

Y. Ardö The Royal Veterinary and Agricultural University, Department of Dairy and Food Science, Dairy Science, Rolighedsvej 30, DK-1958 Frederiksberg C, Denmark

J. Bakker Institute of Food Research, Reading Laboratory, Earley Gate, Reading, RG6 6BZ, UK

P.F. Fox Department of Food Chemistry, University College, Cork, Ireland

M.J. Gasson IFR, Norwich Laboratory, Norwich Research Park, Colney, Norwich, NR4 7UA, UK

J.-C. Gripon Unité de Recherche de Biochimie et Structure des Protéines, INRA, 78350 Jouy-en-Josas, France

E.B. Hansen Chr Hansen A/S, DD Division, R&D Management, Hoersholm, Denmark

B.A. Law (formerly of) Institute of Food Research, Reading Laboratory, Earley Gate, Whiteknights Road, Reading, RG6 6BZ, UK

P.L.H. McSweeney Department of Food Chemistry, University College, Cork, Ireland

V.M.E. Marshall The University of Huddersfield, School of Applied Sciences, Queensgate, Huddersfield, HD1 3DH, UK

F. Mulholland Institute of Food Research, Food Macromolecular Sciences Department, Reading Laboratory, Earley Gate, Whiteknights Road, Reading, RG6 2EF, UK

M. El Soda Laboratory of Microbial Biochemistry, Department of Dairy Technology, Faculty of Agriculture, Alexandria University, Alexandria, Egypt

A.Y. Tamine SAC-Auchincruive, Food Science and Technology Department, Ayr, KA6 5HW, UK

G. Urbach CSIRO Division of Food Science and Technology, Melbourne Laboratory, Highett, Australia 3190

Preface

The first edition of *Advances in the Microbiology and Biochemistry of Cheese and Fermented Milk* was aimed at the gap in the literature between the many excellent technical texts on the one hand, and the widely scattered scientific literature on the other. We tried to present the state of the art in pre-competitive research in a predigested, yet scientifically coherent form, and relate it to the marketable properties of fermented dairy products. In this way, researchers could use the book to mentally step back from their specializations and see how far they had progressed as a community; at the same time we hoped that R&D-based companies could use it to assess the utility (or lack of it) of the research output in setting out their research acquisition strategy for product improvement and innovation.

In a sense, the first edition could claim to have initiated Technology Foresight in its limited field before Government caught the idea, and it certainly gave the science base an opportunity to display its talents and resources as a potential source of wealth creation, well before this became an 'official' function of publicly funded science and technology.

Thus, the first edition was intended as a progressive move within the growing science and technology literature, and judged by its market success, it seems to have served precisely that purpose. This second edition strives to continue the progression, both by collecting and discussing new research findings under the first edition headings, and by introducing completely new approaches to the subject matter. For example, Chapter 1 on the subject of coagulants refects the realization over the past 12 years that these enzymes play a pivotal, rather than peripheral role in cheese maturation; the remarkable new developments in coagulant production technology, arising from advances in molecular biology, have to be viewed not only as changes in the vat stage of cheese technology, but also in the maturation stage. The initial breakdown of caseins to peptide substrates for enzymes of the starter culture and adventitious cheese flora to act on is a function of the coagulant after it has done its work in the initial formation of the curd gel. Thus, the coagulant initiates the entire cascade of protein breakdown to taste and aroma compounds, described in chemical detail in Chapter 8.

Molecular biology has also been harnessed increasingly by starter culture researchers in academe and industry to investigate and manipulate the industrially important functions of these bacteria. These developments are dealt with in several chapters, including revised coverage of the molecular genetics of lactic acid bacteria (Chapter 10) and a new Chapter (9) on culture enzymology, dealing with the enzymes of cheese bacteria which are most clearly implicated in texture and flavour development.

The increasing range and market penetration of fermented milks is reflected in increased basic research on their microbiology, especially of those milks to which health-promoting properties are assigned. Thus the original chapter on the research base underpinning these products has been extensively revised and extended to two chapters (3 and 4) covering science-driven technological developments as well as new findings in microbiology, physiology and biochemistry.

The taxonomy of bacteria has undergone such upheavals in the last 12 years, particularly arising from the development and application of nucleic acid sequences in the design of molecular probes, that there is no single authority on cheese and fermented milk bacteria, as there was when the first edition was published. For this edition, I have chosen to defer to existing reviews and formal texts for phylogenetically based classifications of the dairy lactic acid bacteria. For non-lactics, I have been fortunate to have enlisted the help of an industry-based molecular biologist involved directly in dairy culture development to select the important genera and to guide the reader through their inter-relatedness. Thus Chapter 2, dealing with taxonomy, is very brief and highly focussed; it should be used as a supplement to the original chapter (Garvie, 1984), rather than as a stand-alone work. The classification of fermented milk bacteria has been covered by the specialist authors in this area, in the context of their technology, physiology and product biochemistry (Chapter 4).

Although diagnostic and determinative microbiology has taken enormous strides since the first edition was published, this area was not specifically covered then, and will not be in this new edition. This is because the research base is generic, dealing with pathogenic and spoilage bacteria which occur in a wide range of foods, including cheese, and the reader is referred to the reviews by Grant and Kroll (1993) and Lee and Morgan (1993) to gain an appreciation of molecular probe technology applied to the detection and isolation of food pathogens. For those who need to keep abreast of the occurrence and behaviour of pathogens in cheese, the 1994 International Dairy Federation survey (Spahr and Url, 1994) is an ideal source to supplement national surveys of food poisoning outbreaks, published annually by Public Health Laboratories. Also, the Food Micromodel Database, held and managed by the Institute of Food Research, Reading, UK, is a PC-based system which predicts the growth and toxin-producing potential of a range of commonly occurring pathogens in foods, in response to selected conditions of temperature, pH, water activity, and preservative concentration (McClure et al., 1994); this has obvious application in dairy product research to determine the spoilage and poisoning potential of specific microorganisms. Specific research data on the anti-pathogen activity of cheese and fermented milk cultures appears in a number of chapters dealing with the dairy lactic acid bacteria.

Finally, I should comment on the book's coverage of flavour and texture in cheese; since the first edition was published, the range of cheese varieties under detailed and intensive investigation has increased to an extent that a single chapter is no longer appropriate. I have therefore dealt with this in the second edition by including a separate chapter (5) exclusively on soft cheeses, with their unique chemistry and microbiology, and a new chapter (6) on challenges to research from the special problems of making low-fat cheeses. All other cheese varieties feature more or less in the other chapters within the discussions of, for example, coagulants, starter culture enzymology, flavour chemistry and accelerated ripening. The only omission, if it is such, is a detailed coverage of pasta filata cheeses used as pizza toppings; most of the literature in this area is technically based, rather than research-based, and deals with process technology. I have relied on authors of the research area chapters to note developments in pasta filata, as appropriate.

Barry A. Law
March, 1996

References

Garvie, E.I. (1984) Taxonomy and identification of bacteria important in cheese and fermented dairy products, in *Advances in the Microbiology and Biochemistry of Cheese and Fermented Milk* (eds F.I. Davies and B.A. Law), Elsever Applied Science Publishers, London, New York, pp. 35–66.

Grant, K.A. and Kroll, R.G. (1993) Molecular biology techniques for the rapid detection and characterisation of food borne bacteria. *Food Sci. Technol. Today*, 7, 80–3.

Lee, H.A. and Morgan, M.R.A. (1993) Food immunoassays: application of polyclonal, monoclonal and recombinant assays. *Trends Food Sci. Technol.*, 4, 129–33.

McLure, P.J., Blackburn, C.W., Cole, M.B., Curtis, P.S., Jones, J.E. and Legan, J.D. (1994) Modelling the growth, survival and death of microorganisms in foods; the UK Micromodel approach. *Int. J. Food Microbiol.*, 23, 265–75.

Spahr, U. and Url, B. (1994) The behaviour of pathogens in cheese; summary of experimental data. *Bull. Int. Dairy Fed.*, 298, 2–16.

Preface to the first edition

The manufacture of cultured dairy products constitutes a major fermentation industry, particularly in the developed countries. For example, in 1980, 3.5 million tonnes of cheese and over 1500 million litres of yoghurt and fermented milks were produced in the EEC. Milk and milk products account for about 15% of domestic food expenditure in the UK. Although liquid milk consumption is declining in many countries, the market for cheese is increasing slowly, with a tendency towards more complex varieties whose maturation is most difficult to control. Fermented milks enjoy not only increased interest in a wider variety of products but also an overall increase in popularity. It is clear, therefore, that the fermentation of milk to nutritious, stable and interesting products is an expanding activity which requires a constant input of both new technology and fundamental research. Although the industry can call upon a comprehensive technical literature as an aid to solving some of its problems, an equivalent coverage of basic research from which to draw new ideas is not at present available. Such information is widely scattered in specialized publications. We have therefore attempted to provide one volume in which research relating to the major stages of fermented dairy product manufacture is reviewed. In this way we hope to provide some insight into possible future directions for development in the industry, based on current efforts in research laboratories.

Studies into the physico-chemical nature of cheese have been important in investigations into modified processes such as the use of concentrated milk and the increased incorporation of whey proteins. The starter cultures used in these fermentations can now be more accurately identified and their relationships determined by the use of techniques such as DNA–DNA hybridization. The physiological mechanisms by which they utilize lactose and milk proteins to produce lactic acid and other essential end products are only now being well characterized, and the genetic control of these mechanisms is a subject of increasing study. Thus modification and improvement of starter cultures for traditional fermentations and the extension of their use to new food areas are important new possibilities. The application of genetic knowledge will also be prominent among approaches being taken to eliminate or control the attack of starter cultures by bacteriophages. Elucidation of the biochemical pathways of flavour development in fermented milks has led to new possibilities for product development and in the case of cheese this knowledge has been supplemented by important new work which could substantially reduce the traditionally long ripening periods required to achieve a mature flavour. Early detection of key flavour volatiles could be important in prediction of final mature cheese quality and the use of non-sensory instrumental methods will permit more objective assessments to be made.

It is the advances in these areas with which this book is concerned. All the contributors are actively engaged in the research they describe and the individual chapters are seen as current 'state of the art' reviews directed at those working at this level in industry, government and education.

Barry A. Law
F. Lyndon Davies

1 Rennets: their role in milk coagulation and cheese ripening

P.F. FOX and P.L.H. McSWEENEY

1.1 Introduction

Cheese production is one of the oldest forms of biotechnology, dating perhaps from 6000 BC and was well established during the era of the Roman Empire. Cheese is one of the most diverse food groups: Burkhalter (1981) listed *ca.* 500 varieties produced from cows' milk, while Kalantzopoulos (1993) listed 500 more produced from sheep's and/or goats' milks. In spite of its long history, cheese still has a very vibrant image and enjoys consistent growth of about 4% p.a. Cheese enjoys epicurean status, has desirable nutritional properties, is the classical convenience food and may be consumed as the main component of a meal, as a dessert or as a food ingredient (the latter has been the major growth sector, e.g. in the US, *ca.* 30% of all cheese is now used as a food ingredient; Olson and Gould, 1995).

Cheese is one of the most scientifically interesting food groups: organic and analytical chemists, biochemists, colloid scientists, rheologists, microbiologists, molecular biologists and nutritionists find ready research challenges in cheese. Consequently, there is a very considerable scientific literature on cheese which has been compiled in a two volume text (Fox, 1993); a more concise treatment has been prepared by Fox *et al.* (1996).

This review will focus on one particular aspect of cheese chemistry, i.e. the role and significance of exogenous proteinases, commonly referred to as rennets, in cheese production.

The first step in cheese manufacture essentially involves coagulating the principal group of milk proteins, the caseins, by one of three methods:

1. Limited proteolysis by a crude proteinase (rennet); this method is employed for the vast majority of ripened and some fresh cheeses.
2. Isoelectric precipitation at $\sim$pH 4.6, used mainly for fresh cheeses, usually by *in situ* production of lactic acid by a starter culture and less frequently by direct acidification with acid, usually HCl, or acidogen, usually gluconic acid-δ-lactone.
3. Acid plus heat, i.e. acidification to $\sim$pH 5.2 with acid whey, citrus juice, vinegar or acetic acid at 80–90°C, e.g. Ricotta, Quesco Blanco.

If whole milk is used, the fat is occluded in the gel but does not participate in gel formation. When coagulated by mechanisms 1 or 2, the whey proteins are removed in the whey but are denatured in method 3 and are incorporated in the coagulum. Ultrafiltration is being used increasingly in cheese manufacture, in which case at least part of the whey proteins are recovered in the coagulum.

Cheeses produced by mechanisms 2 and 3 are consumed fresh and represent 25% of total cheese production ($\sim 3.2 \times 10^6$ tonnes p.a.; Fox *et al.*, 1996); the principal varieties are Quarg, Cottage, Queso Blanco and Fromage Frais. Rennet-coagulated cheeses include all the ripened varieties; their great diversity is facilitated by the syneretic properties of rennet-coagulated casein gels, which enables the production of cheese ranging in moisture content from ~ 30 to $\sim 55\%$ and the numerous biochemical reactions that occur during ripening, catalysed by a great diversity of enzymes. Only rennet-coagulated cheeses will be considered in this review.

A small amount of rennet-coagulated cheese is consumed fresh but the vast majority is ripened (matured) for a period ranging from 3 weeks (e.g. Mozzarella) to 2 or more years (e.g. extra-mature Cheddar or Parmesan) to develop a characteristic flavour and texture. Thus, the production of these cheeses can be divided into two phases: curd manufacture and ripening. Curd manufacture involves five principal operations: coagulation, acidification, curd syneresis/dehydration, moulding of the curd and salting.

The essential role of rennet is in milk coagulation but it also plays a major role in ripening.

Hammersten, in the 1880s, showed that the rennet coagulation of milk involves proteolysis with the formation of 'para-casein' and non-protein nitrogen, followed by Ca-induced coagulation of the former. This mechanism was confirmed by Berridge (1942) who showed that gelation of the rennet-altered para-casein micelles by Ca^{2+} is very temperature-dependent (Q_{10} *ca.* 16) and does not normally occur below *ca.* 20°C. The proteolytic stage is essentially complete before the onset of coagulation. Elucidation of the chemical and physicochemical properties of milk proteins has made it possible to explain the rennet coagulation of milk, especially the primary enzymatic phase, at the molecular level. Reviews on the rennet coagulation of milk include: Fox (1984), Fox and Mulvihill (1990), Dalgleish (1992, 1993). For a comprehensive description of various aspects of milk proteins, see Fox (1992); a very brief summary of the pertinent features is given below.

1.2   The milk protein system

Bovine and, probably, all milks, contain two distinctively different groups of proteins: the caseins, i.e. phosphoproteins insoluble at pH 4.6 and 20°C, and the whey (or non-casein) proteins. The principal whey proteins in bovine milk are β-lactoglobulin and α-lactalbumin, with lesser amounts of blood serum albumin and immunoglobulins, trace amounts of lactotransferrin and other proteins and perhaps 40 enzymes. Traditionally manufactured rennet-coagulated cheese contains very little whey proteins, which, therefore, will not be considered here.

In bovine milk, casein represents *ca.* 80% of total nitrogen and is comprised of four principal proteins, α_{s1}, α_{s2}, β and κ, in the approximate ratio, 40:10:35:12, and several minor proteins, which originate from post-translational proteolysis of the primary caseins, especially by indigenous alkaline milk proteinase, plasmin. These include γ^1-, γ^2- and γ^3-caseins and proteose peptones 5, 8-slow and 8-fast, all derived from β-casein, λ-caseins derived from α_{s1}-casein and at least 30 unidentified peptides. All the principal milk proteins are very well characterized, chemically and physically.

α_{s1}-, α_{s2}- and β-caseins contain 8–9, 10–13, 4–5 mols of phosphorus (P) per mole, respectively; consequently, they bind Ca^{2+} strongly and precipitate at $[Ca^{2+}] > 6$ mM. However, κ-casein, which usually has only one phosphate (PO_4) residue per mole, is soluble at high $[Ca^{2+}]$. It reacts hydrophobically with α_{s1}-, α_{s2}- and β-caseins and can stabilize up to 10 times its weight of these Ca-sensitive caseins against precipitation by forming coarse colloidal particles, referred to as casein micelles.

In milk, >95% of the casein exists as micelles which consist, on a dry-weight basis, of ~94% protein and 6% low-molecular weight species, mainly Ca and PO_4, with some Mg and citrate, collectively called colloidal calcium phosphate (CCP). The micelles are spherical, 50–300 nm (mean *ca.* 120 nm) in diameter, with M_r of *ca.* 10^8, i.e. a typical micelle contains *ca* 5000 monomers which have masses of 20000–24000 Da. The micelles typically bind *ca.* 2 g H_2O/g protein.

Although the structure of the micelles is not yet established, there is a widely held view that they consist of submicelles (spherical particles, M_r *ca.* 5×10^6) held together by CCP bridges, hydrophobic interactions and hydrogen bonds. The micelles dissociate on removing CCP, e.g. by Ca chelators or acidification, on adding sodium dodecyl sulphate (SDS) or urea or by increasing the pH > *ca.* 9. The precise structure of the submicelles is unknown but a widely supported view is that the Ca-sensitive α_s- and β-caseins interact hydrophobically to form the core of the submicelles, with κ-casein located predominantly on the surface. The N-terminal two-thirds of κ-casein are hydrophobic and react hydrophobically with the core proteins, leaving the hydrophilic C-terminal region projecting into the surrounding aqueous environment. The submicelles are considered to contain variable amounts of surface κ-casein and to aggregate so that the κ-casein-rich submicelles are concentrated on the surface of the micelles with the κ-casein-deficient submicelles buried within. The micelles are considered to be stabilized by a zeta potential of ~ -20 mV and by steric factors caused by the protruding C-terminal segments of κ-casein which give the micelles a 'hairy' appearance and which prevent close approach.

Milk contains at least 40 indigenous enzymes; many of these, including xanthine oxidase and alkaline phosphatase, are located mainly in the fat globule membrane, while lipoprotein lipase and the principal proteinase, plasmin, are associated with the casein micelles. Thus, some of the indigenous milk enzymes are incorporated into cheese curd and play significant roles in cheese during ripening, even in cheese made from pasteurized milk, since many of them are quite heat stable and survive, at least partially, HTST pasteurization.

1.3 Primary phase of rennet action

κ-Casein was isolated by Waugh and von Hippel (1956), who also showed that it is responsible for micelle stability and that its micelle-stabilizing properties are lost on renneting. Only κ-casein is hydrolysed to a significant extent during the primary phase of rennet action (Wake, 1959). The primary chymosin cleavage site is Phe_{105}–Met_{106} (Delfour et al., 1965), which is many times more susceptible than any other bond in milk proteins.

The Phe–Met bond of κ-casein is also preferentially hydrolysed by pepsins and the acid proteinases of *Rhizomucor miehei* and *R. pusillus*, but the acid proteinase of *Cryphonectria parasitica* preferentially cleaves Leu_{104}–Phe_{105} (Drøhse and Foltmann, 1989). However, the *Rhizomucor* and *Cryphonectria* proteinases also cleave several other bonds in κ-casein (Shammet et al., 1992a, b).

The unique sensitivity of the Phe–Met bond has aroused interest. The dipeptide, H.Phe–Met.OH, is not hydrolysed, nor are tri- or tetrapeptides containing a Phe–Met bond. However, this bond is hydrolysed in the pentapeptide, H.Ser–Leu–Phe–Met–Ala–OMe (Hill, 1968, 1969) and reversing the positions of serine and leucine, to give the correct sequence of κ-casein, increases the susceptibility of the Phe–Met bond to chymosin (Schattenkerk and Kerling, 1973). Both the length of the peptide and the sequence around the sectile bond are important determinants of enzyme–substrate interaction. Serine 104 appears to be particularly important (Hill, 1968, 1969) and its replacement by Ala in the above pentapeptide renders the Phe–Met bond very resistant to hydrolysis by chymosin (Raymond et al., 1972) but not by pepsins (Raymond and Bricas, 1979); even substituting D-Ser for L-Ser markedly reduced its sensitivity (Raymond and Bricas, 1979). Extension of the above pentapeptide from the N- and/or C-terminal to reproduce the sequence of κ-casein around the chymosin-susceptible bond increases the efficiency with which the Phe–Met bond is hydrolysed by chymosin (Visser et al., 1976, 1977) (Table 1.1). The sequence His_{98}–Glu_{129} of κ-casein includes all the residues necessary to render the Phe–Met bond

Table 1.1 Kinetic parameters for hydrolysis of κ-casein peptides by chymosin at pH 4.7

Peptide	Sequence	k_{cat} (s^{-1})	K_m (mM)	k_{cat}/K_m $(s^{-1} mM^{-1})$
S.F.M.A.I.	104–108	0.33	8.5	0.038
S.F.M.A.I.P.	104–109	1.05	9.2	0.114
S.F.M.A.I.P.P.	104–110	1.57	6.8	0.231
S.F.M.A.I.P.P.K.	104–111	0.75	3.2	0.239
L.S.F.M.A.I.	103–108	18.3	0.85	21.6
L.S.F.M.A.I.P.	103–109	38.1	0.69	55.1
L.S.F.M.A.I.P.P.	103–110	43.3	0.41	105.1
L.S.F.M.A.I.P.P.K.	103–111	33.6	0.43	78.3
L.S.F.M.A.I.P.P.K.K.	103–112	30.2	0.46	65.3
H.L.S.F.M.A.I.	102–108	16.0	0.52	30.8
P.H.L.S.F.M.A.I.	101–108	33.5	0.34	100.2
H.P.H.P.H.L.S.F.M.A.I.P.P.K.	98–111	66.2	0.026	2509.0
	98–111[a]	46.2[a]	0.029[a]	1621.0[a]
κ-Casein[b]		2–20	0.001–0.005	200–2000
L.S.F.(NO$_2$)NleA.L.OMe		12.0	0.95	12.7

[a]pH 6.6.
[b]pH 6.4.

as susceptible to hydrolysis by chymosin as it is in intact κ-casein (Hill and Hocking, 1978). Indeed, the sequence His$_{98}$–Lys$_{111}$ (or Lys$_{112}$) includes all the necessary determinants (Visser et al., 1980), and this tetradecapeptide is hydrolyzed with a k_{cat}/K_m of ca. 2 M^{-1} s^{-1}, which is very close to that for intact κ-casein. κ-Casein and κ-CN f98–111 are readily hydrolysed at pH 6.6 (and at pH 4.7) but smaller peptides are not.

The Phe and Met residues in the chymosin-sensitive bond of κ-casein are not intrinsically essential for chymosin action. There are numerous Phe and a substantial number of Met residues in all milk proteins. Neither porcine nor human κ-casein contains a Phe–Met bond [in both, the chymosin-sensitive bond is Phe–Ile (Chobert et al., 1976; Brignon et al., 1985)], yet both are readily hydrolysed by calf chymosin, although more slowly than bovine κ-casein; in contrast, porcine milk is coagulated more effectively than bovine milk by porcine chymosin (Foltmann, 1981), indicating that unidentified subtle structural features influence chymosin action. In rat and mouse κ-casein, Met is replaced by Leu (Thompson et al., 1985). Replacement of Phe in synthetic peptides by Phe (NO$_2$) or cyclohexylamine reduces k_{cat}/K_m by a factor of ca. 3- and ca. 50-fold, respectively (Visser et al., 1977); oxidation of Met$_{106}$ reduces it ca. 10-fold but substitution of Ile for Met increases it ca. 3-fold. This suggests that the sequence around the Phe–Met bond, rather than the residues in the bond itself, contains the important determinants of hydrolysis. The particularly important residues are Ser$_{104}$, the hydrophobic residues Leu$_{103}$ and Ile$_{108}$, at least one of the three histidines (residues 98, 100 or 102, as indicated by the inhibitory effect of photooxidation) and Lys$_{111}$.

Visser *et al.* (1987), using chemical and enzymatic modifications of the peptide representing the sequence His_{98}–Lys_{112} of κ-casein, attempted to identify the relative importance of residues in the sequences of 98–102 and 111–112. They suggested that the sequence Leu_{103} to Ile_{108} of κ-casein, which probably exists as an extended β-structure (Loucheux-Lefebvre *et al.*, 1978; Raap *et al.*, 1983), fits into the active site cleft of acid proteinases. The hydrophobic residues, Phe_{105}, Met_{106}, Leu_{103} and Ile_{108}, are directed to hydrophobic pockets along the active site cleft while the hydroxyl group of Ser_{104} forms part of a hydrogen bond with some counterpart in the enzyme. The sequences 98–102 and 109–111 form β-turns (Loucheux–Lefebvre *et al.*, 1978; Raap *et al.*, 1983) around the edges of the active site cleft in the enzyme–substrate complex; this conformation is stabilized by Pro residues at positions 99, 101, 109 and 110. The three His residues at positions 98, 100, 192, and Lys_{111} are probably involved in electrostatic bonding between enzyme and substrate; none appears to have a predominant role. Lys_{112} appears not to be important in enzyme–substrate binding as long as Lys_{111} is present (Visser *et al.*, 1987).

A genetically engineered mutant of κ-casein, in which Met_{106} was substituted by Phe (Oh and Richardson, 1991), i.e. the chymosin-sensitive bond was changed from Phe_{105}–Met_{106} to Phe_{105}–Phe_{106}, was hydrolysed 1.8 times faster by chymosin than the wild-type.

The significance of electrostatic interactions in chymosin–substrate complex formation is indicated by the effect of added NaCl on the rennet coagulation time (RCT) of milk: addition of NaCl up to 3 mM reduced RCT but higher concentrations had an inhibitory effect (Hamdy and Edelsten, 1970; Payens and Both, 1980; Visser *et al.*, 1980; Payens and Visser, 1981; Grufferty and Fox, 1985); it is claimed that the effect of NaCl is on the primary, enzymatic phase rather than on the aggregation of rennet-altered micelles. Increasing ionic strength (0.01–0.11) reduces the rate of hydrolysis of His_{98}–Lys_{112}, the effect becoming more marked as the reaction pH is increased but independent of ion type (Visser *et al.*, 1980; Visser and Rollema, 1986).

As well as serving to elucidate the importance of certain residues in the hydrolysis of κ-casein by chymosin, small peptides, mimicking or identical to the sequence of κ-casein around the Phe–Met bond, offer a very useful means of determining the activity of rennets in absolute units, i.e. independent of variations in the non-enzymatic phase of clotting of different milks (Raymond *et al.*, 1973; Schattenkerk and Kerling, 1973; Salesse and Garnier, 1976; de Koning *et al.* 1978; Martin *et al.*, 1981; Visser and Rollema, 1986). Since the specific activity of different rennets on these peptides varies, methods for quantifying the proportions of acid proteinases in commercial rennets have been proposed (Salesse and Garnier, 1976; de Koning *et al.*, 1978; Martin *et al.*, 1981).

1.4 Rennet and rennet substitutes

Most proteinases will coagulate milk under suitable conditions but most are too proteolytic relative to their milk-clotting activity; consequently, they hydrolyse the coagulum too quickly, causing reduced cheese yields and/or defective cheese, with a propensity to bitterness. Although plant proteinases appear to have been used as rennets since prehistoric times, gastric proteinases from calves, kids or lambs have been used traditionally as rennets, with very few exceptions.

Animal rennets are prepared by extracting the (usually) dried or salted gastric tissue with 10% NaCl and activating and standardizing the extract. Standard calf rennet contains *ca.* 50 RU/ml and is preserved by making the extract to 20% NaCl and adding sodium benzoate or sodium propionate. Chymosin (an aspartyl proteinase) represents >90% of the milk-clotting activity of good-quality calf rennet, the remaining activity being due to pepsin. As the animal ages, especially when fed solid food, the secretion of chymosin declines while that of pepsin increases.

Like many other animal proteinases, chymosin is secreted as its zymogen, prochymosin, which is autocatalytically activated on acidification to pH 2–4 by removal of a 44-residue peptide from the N-terminal of the zymogen (Foltmann, 1993).

Chymosin is well characterized at the molecular level (Foltmann, 1993). The enzyme, which was crystallized in the 1960s, is a single-chain polypeptide containing about 320 amino acid residues with a molecular mass of 35 600 Da. Its primary structure has been established and a considerable amount of information is available on its secondary structure. The molecule exists as two domains, separated by an active site cleft in which the two catalytically active aspartyl groups are located.

Calf rennet contains three chymosin isoenzymes; A and B are the principal components with lesser amounts of C (Foltmann, 1966). Chymosins A and B are produced from the corresponding zymogens, prochymosins A and B, but the origin of chymosin C is not clear. The specific activity of chymosin A, B and C is 120, 100 and 50 RU/mg, respectively (Foltmann, 1966). Chymosins A and B differ by a single amino acid substitution, Asp and Gly, respectively, at position 244 (Foltmann *et al.*, 1977) and have an optimum pH at 4.2 and 3.7, respectively (Martin *et al.*, 1980).

Owing to increasing world cheese production ($\sim$4% p.a. over the past 20 years), concomitant with a reduced supply of calf rennet (due to a decrease in calf numbers and a tendency to slaughter older calves), the supply of this enzyme has been inadequate for many years. This has led to an increased price for calf rennet and to a search for rennet substitutes. Despite the availability of numerous potentially useful milk coagulants, only six rennet substitutes have been found to be more or less acceptable for cheese production: bovine, porcine and chicken pepsins and the acid proteinases from *Rhizomucor miehei*, *R. pusillus* and *Cryphonectria parasitica*.

Chicken pepsin is the least suitable of these and is used widely only in Israel. Bovine pepsin is probably the most satisfactory and many commercial 'calf rennets' contain $\sim 50\%$ bovine pepsin; its proteolytic specificity is similar to that of calf chymosin and it gives generally satisfactory results with respect to cheese yield and quality. The activity of porcine pepsin is very sensitive to pH > 6.6 and it may be extensively denatured during cheese-making and consequently contribute little to proteolysis during cheese ripening. Although a 50 : 50 mixture of porcine pepsin and calf rennet gave generally acceptable results, porcine pepsin has been withdrawn from most markets. The proteolytic specificity of the three commonly used fungal rennets is considerably different from that of calf rennet, but the quality of most cheese varieties made using them has been fairly good. Before the introduction of genetically engineered chymosin, microbial rennets were used widely in the United States but not in most European countries or in New Zealand. The extensive literature on rennet substitutes has been reviewed (Sardinas, 1972; Ernstrom and Wong, 1974; Nelson, 1975; Sternberg, 1976; Green, 1977; Phelan, 1985).

Like chymosin, all commercially successful rennet substitutes are acid proteinases. The molecular and catalytic properties of the principal rennet substitutes are generally similar to those of chymosin (Foltmann, 1993). Acid proteinases have a relatively narrow specificity, with a preference for peptide bonds to which a bulky hydrophobic residue supplies the carboxyl group; this narrow specificity is significant for the success of these enzymes in cheese manufacture. The fact that, in cheese, these enzymes operate at a pH far removed from their optima (*ca.* 2 for porcine pepsin) is probably also significant. However, not all acid proteinases are suitable as rennets because they are too active even under the prevailing relatively unfavourable conditions. As discussed in section 1.13, the specificity of porcine and bovine pepsins on α_{s1}- and β-caseins is quite similar to that of chymosin, but the specificity of the fungal rennet substitutes are quite different.

The milk-clotting activity of the pepsins, especially porcine pepsin, is more pH-dependent than that of chymosin, while that of the fungal rennets is less sensitive (Phelan, 1985). The coagulation of milk by *C. parasitica* proteinase is also less sensitive to $[Ca^{2+}]$ than when calf rennet is used but coagulation by *Rhizomucor* proteinases is more sensitive and the rates of gel firming differ – these aspects of milk coagulation should be independent of rennet type and may indicate nonspecific proteolysis by the fungal enzymes.

The thermal stability of rennets is important when the whey is to be used for processing – the early fungal rennets were considerably more thermostable than chymosin or pepsins but the present products have been modified (oxidation of Met) and have thermal stability similar to that of chymosin. The thermal stability of all rennets increases with decreasing pH (Phelan, 1985).

Although they are relatively cheap, rennets represent the largest single industrial application of enzymes, with a world market of *ca.* 25×10^6 litres of standard rennet per annum (worth $\sim$£100 million). Therefore, it is not surprising that rennets are attractive to industrial enzymologists and bio-technologists. The gene for prochymosin has been cloned in *E. coli*, *Saccharomyces cerevisiae*, *Kluyveromyces marxianus* var. *lactis*, *Aspergillus nidulans*, *A. niger* and *Tricoderma reesei* [see Pitts *et al.*, (1992) and Foltmann (1993) for references]. The enzymatic properties of the recombinant enzymes are indistinguishable from those of calf chymosin although they may contain only one or other of the isoenzymes. The cheese-making properties of recombinant chymosins have been assessed on many cheese varieties, always with very satisfactory results (see reviews by Teuber, 1990; IDF, 1992; Fox and Stepaniak, 1993). Recombinant chymosins have been approved for commercial use in many countries; notable exceptions are Germany and Japan. Three recombinant chymosins are now marketed commercially: Maxiren, secreted by *K. marxianus* var. *lactis* and produced by Gist Brocades (the Netherlands), Chymogen (secreted by *A. niger*, produced by Hansen's, Denmark) and Chymax (secreted by *E. coli*, produced by Pfizer, USA). The genes for Maxiren and Chymogen were isolated from calf abomasum while that used for Chymax was synthesized. To date, data on the use of these enzymes are not available, but they are widely used and have taken market share from both calf rennet and especially fungal rennets.

The recombinant chymosins used to date are identical, or nearly so, to calf chymosin but there are several published studies on engineered chymosins. At present, attention is focussed on elucidating the relationship between enzyme structure and function but this work may lead to rennets with improved milk-clotting activity or modified general proteolytic activity, i.e. on α_{s1}- and/or β-casein. The natural function of chymosin is to coagulate milk in the stomach of the neonate. It was not intended for cheese-making and it is probable that the wild-type enzyme may not be the most efficient or effective proteinase to catalyse proteolysis in cheese during ripening. Therefore, it may be possible to modify chymosin to accelerate its action on specific bonds of casein during ripening and/or to reduce its activity on others, hydrolysis of which may lead to undesirable consequences, e.g. bitterness.

The pH optimum of chymosin on synthetic peptides is increased by substituting Ala for Asp_{304} or Ala for Thr_{218} (Mantafounis and Pitts, 1990; Suzuki *et al.*, 1990). Strop *et al.* (1990) expressed a point-mutated prochymosin in *E. coli*, in which a Val residue was replaced by Phe. Replacement of Val by the larger Phe led to local changes in conformation and increased the K_m for various peptide substrates 2–3-fold without affecting K_{cat}. The mutant had higher affinity than wild-type chymosin for inhibitors with small side-chains, but showed reduced binding to inhibitors with large side-chains. The results suggested that substitution of a bulky Phe for Val reduced the size of the S1 and S3 specificity pockets.

The thermal stability of *Rhizomucor* rennets can be modified by genetic approaches as an alternative to chemical modification. Yamashita *et al.* (1994) modified the gene for *R. pusillus* proteinase, expressed in *Saccharomyces cerevisiae;* variants in which Thr_{101} was replaced by Ala or Asp_{186} by Gly had reduced thermal stability and proteinase with a double substitution (101 and 186) had the lowest heat stability.

Pitts *et al.* (1994) replaced the sequence MDRNGQESML (residues 155–164) of chymosin, which forms a surface loop, by the corresponding but longer sequence of the acid proteinase of *Rhizopus chinensis,* i.e. IGKAK-NGGGGEL. The activity of the modified enzyme, $Chy_{155-165}$ Rhi, which was cloned and expressed in *Trichoderma reesei,* on a synthetic chromogenic peptide (unspecified) was similar to that of chymosin.

The gene for *R. miehei* proteinase has been cloned in and expressed by *A. oryzae* (Novo Nordisk A/S, Denmark). It is claimed that this new rennet (Marzyme GM) is free of other proteinase/peptidase activities present in fungal rennets and which may reduce cheese yield. Chen *et al.* (1996) reported excellent cheese-making results with Marzyme GM. Cloning of the gene for *R. miehei* proteinase has created the possibility for site-directed mutagenesis of the enzyme.

1.5   Immobilized rennets

Most (>90%) of the rennet activity added to cheese-milk is lost in the whey, representing an economic loss and creating problems for whey processors; both problems would be resolved if immobilized rennets could be used to coagulate milk. A further incentive for the immobilization of rennets is the possibility of producing cheese curd continuously, which should facilitate process control. As discussed in section 1.13, the small proportion of chymosin (or rennet substitute) retained in cheese curd plays a major role in cheese ripening so that if immobilized rennets were used successfully to coagulate milk, it would be necessary to add some rennet (or other proteinase) to the curd and uniform incorporation of this enzyme(s) would be problematic, as has been experienced with the use of exogenous proteinases to accelerate cheese ripening (see Fox, 1988/89).

In modern practice, most cheese-making operations are continuous or nearly so; the actual coagulation step is the only major batch operation remaining although the use of small 'batches' of milk, as is the Alpma process for Camembert, makes coagulation, in effect, a continuous process. The feasibility of continuous coagulation using cold renneting principles has been demonstrated but the technique has had very limited commercial success to date.

On the academic side, the availability of a completely immobilized, effective rennet would permit the manufacture of rennet-free curd for studies on the contribution of enzymes from different sources to cheese ripening. A number of approaches have been used to produce rennet-free curd (Fox *et al.*, 1993).

The first report on an immobilized rennet (chymotrypsin or calf rennet) is that of Green and Crutchfield (1969) who successfully immobilized the enzyme but found that leaching occurred and the free enzyme coagulated the milk. This was followed by a series of publications by workers at the University of Wisconsin, Madison, who concluded that milk could be coagulated by immobilized rennets; this and related work was reviewed by Taylor *et al.* (1976). Fox (1981) concluded that the published data did not conform with the known kinetics and characteristics of the rennet coagulation of milk and suggested that coagulation was caused by solubilized enzyme. In critical reviews on the coagulation of milk by immobilized rennets, Carlson *et al.* (1986) and Dalgleish (1987, 1993) reached similar conclusions. However, publications on the coagulation of milk by immobilized rennets continue to appear, e.g. Shah *et al.* (1995).

Even if immobilized rennets could coagulate milk, they may not be cost-competitive (rennets are cheap) and would be difficult to use in factory situations. The strategy envisaged for their use involves the passage of cold milk (e.g. 10°C) through a column of immobilized enzyme where the enzymatic phase of renneting would occur without coagulation. The rennet-altered micelles would then be coagulated by heating the milk exiting the column to ~ 30°C. Heating would have to be conducted under quiescent conditions to ensure the formation of a good gel and minimize losses of fat and protein; quiescent heating may be difficult on a very large industrial scale (e.g. many cheese factories process 10^6 L milk/day). Hygiene and phage-related problems may present serious problems, since cleaning the column by standard regimes would inactivate the enzyme. Plugging of the column and loss of activity have been problematic even on a laboratory scale, and power cuts would be disastrous as the column reactors would become plugged with cheese curd which would be difficult or impossible to remove.

In short, the prospects for the use of immobilized rennets on a commercial scale are not bright and they are not being used.

1.6   Factors affecting the hydrolysis of κ-Casein

1.6.1   pH

The pH optimum for chymosin and bovine pepsin on small synthetic peptides is *ca.* 4.7 (Hill, 1968; de Koning *et al.*, 1978; Raymond and Bricas, 1979) but is 5.3–5.5 on κ-CN f His_{98}-Lys_{111} (Visser *et al.*, 1987). Chymosin hydrolyses insulin, acid-denatured haemoglobin and Na-caseinate optimally at pH 4.0, 3.5 and *ca.* 3.5, respectively (Fish, 1957; Foltmann, 1964; Humme, 1972). The pH optimum for the first stage of rennet action in milk is ~ 6.0 at 4 or 30°C (van Hooydonk *et al.*, 1986).

1.6.2   Ionic strength

The influence of ionic strength on the primary phase of rennet coagulation was discussed in section 1.3.

1.6.3   Temperature

The optimum temperature for the coagulation of milk by calf rennet at pH 6.6 is ~ 45°C (Fox, 1969); presumably, the optimum for the hydrolysis of κ-casein is about this value. The temperature coefficient (Q_{10}) for the hydrolysis of κ-casein in solutions of Na-caseinate is *ca.* 1.8, activation energy, E_a, is $\sim 40\,000$ J/mol, and activation entropy, ΔS, is ~ -90 J/deg/mol (Nitchman and Bohren, 1955); generally similar values were reported by Garnier (1963) for the hydrolysis of isolated κ-casein by chymosin.

1.6.4   Degree of glycosylation

The efficiency of κ-casein as a substrate for chymosin decreases with the degree of glycosylation (Doi *et al.*, 1979; Vreeman *et al.*, 1986). According to Vreeman *et al.* (1986), at pH 6.6, k_{cat} decreased from ~ 43 s^{-1} for carbohydrate-free κ-casein (B-1) to ~ 25 s^{-1} for κ-casein B-7 (containing 6 mol N-acetylneuraminic acid (NANA)/mol). However, K_m was optimal for B-5 (3 mol NANA/mol). Polymerization (aggregation) markedly increased K_m with little effect on k_{cat}.

1.6.5   Heat treatment of milk

It has long been recognized that heat treatment of milk at temperatures >65°C adversely affects its rennet coagulability; if the heat treatment is very severe (>90°C for 10 min), the milk fails to coagulate on renneting. Although changes in salts equilibria are involved, the principal causative factor is intermolecular disulphide bond formation between κ-casein and β-lactoglobulin and/or α-lactalbumin (Kannan and Jenness, 1961; Morrissey, 1969; Sawyer, 1969; Shalabi and Wheelock, 1976, 1977).

Both the primary and especially the secondary phases of rennet action are inhibited in heated milk (Morrissey, 1969; Marshall, 1986; van Hooydonk *et al.*, 1987; Singh *et al.*, 1988), as is the strength of the resulting gel. The adverse effects of heating can be reversed by acidification before or after heating or by addition of CaCl$_2$ (Banks and Muir, 1985; Marshall, 1986; van Hooydonk *et al.*, 1987; Singh *et al.*, 1988; Lucey *et al.*, 1993).

1.7   Secondary (non-enzymatic) phase of coagulation and gel assembly

Hydrolysis of κ-casein during the primary phase of rennet action causes the release of highly charged, hydrophilic macropeptides representing the C-terminal segment of κ-casein, thereby reducing the zeta potential of the casein micelles from $-10/-20$ to $-5/-7\,mV$ and removing the protruding peptides from their surfaces, reducing the intermicellar repulsive forces (electrostatic and steric) and the colloidal stability of the caseinate system. When $\sim 85\%$ of the total κ-casein has been hydrolysed, the casein micelles begin to aggregate but an individual micelle cannot participate in gelation until $\sim 97\%$ of its κ-casein has been hydrolysed. Reducing the pH or increasing the temperature from the normal values (~ 6.6 and $\sim 31°C$, respectively) permits coagulation at a lower degree of κ-casein hydrolysis (Fox, 1984; Fox and Mulvihill, 1990; Dalgleish, 1992, 1993).

Coagulation of rennet-altered micelles is dependent on a critical $[Ca^{2+}]$ which may act by crosslinking micelles, possibly via serine phosphate residues, or simply by charge neutralization. Colloidal calcium phosphate is also essential for coagulation but can be replaced by increased $[Ca^{2+}]$. Partial enzymatic dephosphorylation of casein, which reduces micellar charge, reduces coagulability; interaction of casein micelles with various cationic species predisposes them to coagulation by rennet and may even coagulate unrenneted micelles. Chemical modification of histidine, lysine or arginine residues inhibits coagulation, presumably by reducing micellar positive charge. It has been suggested that coagulation occurs via electrostatic interactions between a positively charged cluster in the C-terminal region of para-κ-casein, which is exposed on removal of the macropeptide, and an unidentified, negatively charged cluster on neighbouring micelles (Fox, 1984; Fox and Mulvihill, 1990; Dalgleish, 1992, 1993).

The apparent importance of micellar charge in the coagulation of rennet-altered micelles suggests that pH should have a major influence on the secondary phase of coagulation. However, Pyne (1955) claimed that pH has essentially no effect on the coagulation process although the rate of firming of the resultant gel is significantly increased on reducing the pH (Kowalchyk and Olson, 1977).

The coagulation of renneted micelles is very temperature-dependent ($Q_{10} \sim 16$) and bovine milk does not coagulate $< \sim 18°C$ unless $[Ca^{2+}]$ is increased. The marked difference between the temperature dependence of the enzymatic and non-enzymatic phases of rennet coagulation has been exploited in studies on the effects of various factors on the rennet coagulation of milk, in attempts to develop a system for the continuous coagulation of milk for cheese or casein manufacture and in the application of immobilized rennets. The very high temperature dependence of rennet coagulation suggests that hydrophobic interactions play a major role.

The rate of firming of renneted milk gels is influenced by the type of rennet, especially under unfavourable conditions, e.g. high pH, low $[Ca^{2+}]$ (Richardson et al., 1971; Kowalchyk and Olson, 1979; Ustumol and Hicks, 1990). Perhaps such differences reflect the effect of pH on rennet activity or perhaps some general proteolysis by rennet substitutes.

The assembly of renneted micelles into a gel has been studied using various forms of viscometry, electron microscopy and light scattering. The micelles remain discrete until ca. 60% of the visual coagulation time, after which the rennet-altered micelles begin to aggregate steadily, without sudden changes, into chain-like structures which eventually link up to form a network (for reviews, see Fox, 1984; Fox and Mulvihill, 1990; Dalgleish, 1992, 1993; Green and Grandison, 1993). Aggregation of rennet altered micelles can be described by the Smoluchowski theory for diffusion-controlled aggregation of hydrophobic colloids when allowance is made for the need to produce, enzymatically, a sufficient concentration of particles capable of aggregating.

1.8 Curd tension and gel syneresis

The strength of the rennet-induced gel (curd tension, CT) is as, or perhaps more, important as the rennet coagulation time, especially from the viewpoint of cheese yield. This subject has been reviewed (Fox, 1984; Green and Grandison, 1993). Suffice it to say here that curd tension is influenced by the type of rennet; calf rennet gives a more rapid increase in CT than microbial rennets, standardized to equal rennet coagulation times (RCT), presumably reflecting differences in the extent or specificity of proteolysis.

The first and second stages of rennet coagulation and the gel assembly process are common to all rennet-coagulated cheeses, although the rate and extent of both may vary with numerous compositional and processing variables. However, the subsequent treatment of the coagulum is variety-specific. The objective of the immediate post-gelation operations is to remove sufficient whey to give cheese with the desired, characteristic composition, which in turn determines the rate, extent and pattern of ripening. Dehydration is achieved by the marked tendency of renneted milk gels to contract (syneresis) when cut or broken. The subject has been reviewed by Walstra (1993).

1.9 Industrial manipulation of rennet coagulation time

The overall RCT of milk increases markedly with increasing pH $> \sim 6.4$, to an extent characteristic of the type of rennet; porcine pepsin is the most pH-sensitive of the commonly used rennets. Addition of 1.5–2% starter reduces the pH of milk by ~ 0.15 units and cognisance must be taken of this when neutralized or concentrated starters are used. It was common practice to 'ripen' milk after starter addition, as is still practised for some varieties, e.g. Camembert. Ripening allows the starter to enter its log growth phase before renneting and the developed acid accelerates rennet action. Improved starter technology has eliminated the need for ripening, which increases the risk of phage infection. The microbiological quality of cheese-milk has improved considerably, especially with respect to mesophilic lactic acid bacteria; consequently, the pH of cheese-milk is normally higher than it was previously.

It is fairly common practice to add $CaCl_2$ ($\sim 0.04\%$) to cheese-milk. This increases $[Ca^{2+}]$ and $[CCP]$ and reduces pH, all of which reduce RCT and increase gel strength. The pH of milk may also be standardized, usually by adding gluconic acid-δ-lactone, to standardize rennet coagulation and related processes.

1.10   Proteolysis during ripening

Acid-coagulated cheeses are usually consumed fresh but the vast majority of rennet-coagulated cheeses are ripened (matured) for periods ranging from ~ 3 weeks (e.g. Mozzarella) to >2 years (e.g. extra-mature Cheddar and Parmesan); the duration of ripening is more or less inversely proportional to the moisture content of the cheese and, obviously, to the intensity of flavour desired. During ripening, a multitude of microbiological, biochemical and chemical events occur, as a result of which the principal constituents of the cheese, i.e. proteins, lipids and residual lactose, are degraded to primary, and later to secondary, products. Among the principal compounds that have been isolated from several cheese varieties are peptides, amino acids, amines, acids, thiols and thioesters (from proteins), fatty acids, methyl ketones, lactones and esters (from lipids), acids (lactic acid, acetic, propionic), carbon dioxide, esters and alcohols (from lactose). In appropriate combinations and concentrations, these compounds are responsible for the characteristic flavour of various cheeses. Literature on the biogenesis of cheese flavour has been reviewed by Fox *et al.* (1995).

While lipolysis and glycolysis are more or less important in all varieties and critical in some (e.g. blue cheeses, hard Italian varieties and Swiss-type cheeses), proteolysis is essential in all varieties, especially internal-bacterially ripened and surface-ripened cheeses in which it is probably the principal biochemical event during ripening. A good correlation exists between the intensity of Cheddar cheese flavour and the concentration of free amino acids (Aston *et al.*, 1983). Attempts have been made to develop proteolytic indices of cheese maturity (see Chapter 9 of this volume); while these correlate well with age and maturity, they fail to detect off-flavours and should therefore be regarded as complementary to organoleptic assessment of quality.

1.11   Proteolytic agents in cheese

Proteolytic enzymes from four, and in some varieties five, sources are involved in cheese ripening: (1) rennet or rennet substitute; (2) indigenous milk proteinases, particularly plasmin; (3) starter bacteria and their enzymes, which are released after the cells have lysed; (4) non-starter bacteria, i.e. organisms which either survive pasteurization of the cheesemilk or gain access to the pasteurized milk or curd during manufacture; and (5) secondary starters, e.g. propionic acid bacteria, *Br. linens*, yeasts and moulds (*Penicillium roqueforti* and *P. camemberti*), and their enzymes are of major importance in some varieties.

Techniques have been developed and used during the past 30 years which permit quantitation of the contribution of each of these five agents to the primary aspects of cheese ripening, i.e. proteolysis, lipolysis and glycolysis, and to the secondary reactions (Fox, 1989; Fox *et al.*, 1993).

1.12   Significance of secondary coagulant proteolysis

Studies on cheese with controlled microflora have shown that the coagulant is responsible for the level of proteolysis detected by gel electrophoresis and for most of the nitrogen soluble in water or at pH 4.6; however, little trichloroacetic acid (TCA)- or phosphotungstic acid (PTA)-soluble N is produced by the coagulant (Visser, 1977; Visser and de Groot-Mostert, 1977; O'Keeffe *et al.*, 1978). Proteolysis by rennet is believed (de Jong, 1976; Creamer and Olson, 1982; Creamer *et al.*, 1984) to be responsible for the softening of cheese texture early during ripening via hydrolysis of α_{s1}-casein at the bond Phe_{23}–Phe_{24}, which is sufficient to break the continuous protein matrix. Subsequent proteolysis by coagulant, plasmin and bacterial proteinases further modifies the texture. Even in surface mould-ripened cheese, e.g. Camembert, and probably in smear cheeses, coagulant is considered to be essential for the development of proper texture, although the very marked increase in pH (to 7) caused by the catabolism of lactic acids and the production of ammonia (by deamination of amino acids) is also essential (Noomen, 1983; Lenoir, 1984). The proteinases and peptidases excreted by the mould or other surface microorganisms diffuse into the cheese to only a slight extent and contribute little to proteolysis within the cheese, although peptides and amino acids produced by these enzymes in the surface layer may diffuse into the cheese.

In Swiss and hard Italian varieties, little, if any, coagulant survives the cooking process (Matheson, 1981; Boudjellab *et al.*, 1994) and plasmin plays a significant role in primary proteolysis (Steffen *et al.*, 1993).

The secondary proteolytic action of the coagulant influences flavour in three ways:

1. Some rennet-produced peptides are small enough to influence flavour. Unfortunately, some of these peptides are bitter and excessive proteolysis, e.g. due to too much or excessively proteolytic rennet or unsuitable environmental conditions, e.g. too much moisture or too little NaCl, leads to bitterness.
2. Rennet-produced peptides serve as substrates for microbial proteinases and peptidases which produce small peptides and amino acids. These contribute at least to background flavour, and perhaps, unfortunately, to bitterness if the activity of such enzymes is excessive. Catabolism of amino acids by microbial enzymes, and perhaps alterations via chemical mechanisms, leads to a range of sapid compounds – amines, acids, NH_3, thiols – which are major contributors to characteristic cheese flavours (see Chapter 8 of this volume).
3. Alterations in cheese texture appear to influence the release of flavourful and aromatic compounds, arising from proteolysis, lipolysis, glycolysis and secondary metabolic changes, from cheese during mastication which may be the most significant contribution of proteolysis to cheese flavour (McGugan *et al.*, 1979).

Since residual coagulant is important in the development of cheese texture and flavour, it is desirable to have methods available to quantify residual coagulant activity in cheese. Likewise, since the type of coagulant used in the manufacture of cheese can affect its quality and the use of certain coagulants can be subject to regulatory constraints, it is also desirable to be able to identify the coagulant used. Methodology for estimation and identification of residual coagulant activity has been reviewed by Baer and Collin (1993).

Extraction of the residual coagulant from the cheese matrix is the first step in quantification and identification of coagulants in cheese; adjustment of pH is necessary to dissociate the coagulant from the caseins. Residual activity has been quantified by determining the clotting time of milk by the cheese extract, agar diffusion assays, activity on synthetic peptide substrates, release of the glycomacropeptide from κ-casein (Zoon *et al.*, 1994) or by immunochemical methods, although these may measure enzyme concentration and not activity (for references, see Baer and Collin, 1993). Residual coagulants may be identified by characteristic patterns of casein degradation by gel electrophoresis and by various immunological methods, including ELISA (Baer and Collin, 1993; Boudjellab *et. al.*, 1994).

1.13 Specificity of chymosin and rennet substitutes in cheese

Only *ca.* 6% of the chymosin or 2–3% of microbial rennets added to cheese-milk is retained in the curd. The amount retained is influenced by the type of rennet, e.g. porcine pepsin is extensively denatured during cheese-making, pH [low pH favours retention of chymosin but not pepsins or microbial rennets (Holmes *et al.*, 1977; Creamer *et al.*, 1985) and reduces denaturation], and cooking temperature, e.g. very little coagulant survives the cooking conditions used for Swiss cheeses (Matheson, 1981; Boudjellab *et al.*, 1994). Studies on controlled microflora cheese have also shown that the coagulant is mainly responsible for primary proteolysis in most cheese varieties, e.g. as detected by polyacrylamide gel electrophoresis (PAGE) or the formation of watersoluble or pH 4.6-soluble nitrogen. The proteolytic specificity of calf chymosin on α_{s1}-, α_{s2}- and β-caseins in solution has been established (see below) and the results can, largely, be extended to cheese. The specificity of rennet substitutes on the caseins has not been established although the proteolytic patterns they produce have been compared with those produced by chymosin (Figure 1.1).

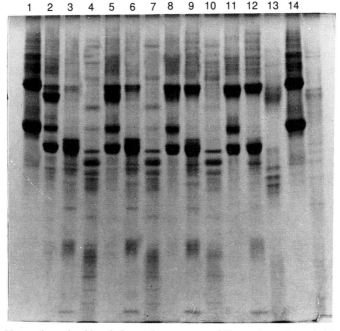

Figure 1.1 Urea polyacrylamide gel electrophoretograms of Na caseinate (lanes 1, 14) and Na caseinate hydrolysed at pH 6.5 for 6 h by 0.05 RU ml^{-1} chymosin (lanes 2, 5, 8, 11), *R. miehei* proteinase (lanes 3, 6, 9, 12) or *C. parasitica* proteinase (lanes 4, 7, 10, 13) in the presence of 0, 1, 2.5 or 5% NaCl (D.M. Rea and P.F. Fox, unpublished data).

A number of authors (Pelissier *et al.*, 1974; Creamer, 1976; Visser and Slangen, 1977; Carles and Ribadeau-Dumas, 1984) have investigated the action of chymosin on β-casein. In solution in 0.05 M Na acetate buffer, pH 5.4, chymosin cleaves β-casein at seven sites: $Leu_{192}-Tyr_{193} > Ala_{189}-Phe_{190} > Leu_{165}-Ser_{166} \geqslant Gln_{167}-Ser_{168} \geqslant Leu_{163}-Ser_{164} > Leu_{139}-Leu_{140} \geqslant Leu_{127}-Thr_{128}$ (Visser and Slangen, 1977). The Michaelis parameters, K_m and k_{cat}, for the action of chymosin on the $Leu_{192}-Tyr_{193}$ bond is 0.075 mM and $1.54 s^{-1}$, respectively, for micellar casein and 0.007 mM and $0.56 s^{-1}$ for the monomeric protein (Carles and Ribadeau-Dumas, 1984). The resulting large peptides, β-CN f1–192, f1–189, f1–163/4/5 and f1–139 are referred to as β-I^I, β-I^{II}, β-II and β-III, respectively. At low pH (2–3), the bond $Leu_{127}-Tyr_{128}$ may also be hydrolysed to yield β-IV (β-CN f1–127). The small C-terminal peptides, β-CN f193–209 and β-CN f190–209, are very hydrophobic and, consequently, bitter. They have been isolated from bitter cheese (Visser *et al.*, 1983).

Hydrolysis of the bond $Leu_{192}-Tyr_{193}$ is markedly retarded in dilute buffers (50 mM phosphate) in comparison with distilled water (Hunter, 1996). NaCl inhibits the hydrolysis of β-casein by chymosin to an extent dependent on pH; hydrolysis is strongly inhibited by 5% NaCl and completely by 10% (Mulvihill and Fox, 1978). The strong influence of ionic strength on the hydrolysability of bonds in the C-terminal region of β-casein presumably reflects the high hydrophobicity of this region. β-Casein undergoes strong temperature-dependent association (Payens and van Markwijk, 1963). Presumably, the chymosin-susceptible bonds of β-casein are inaccessible in the associated protein, e.g. β-casein becomes more susceptible to hydrolysis by chymosin relative to α_{s1}-casein as the reaction temperature is reduced (Fox, 1969). The susceptibility of isolated β-casein to hydrolysis by chymosin also decreases with increasing protein concentration (Hunter, 1996).

The primary site of chymosin action on α_{s1}-casein is $Phe_{23}-Phe_{24}$ (Hill *et al.*, 1974; Carles and Ribadeau-Dumas, 1985). Cleavage of this bond is believed to be responsible for softening of cheese texture and the small peptide product of the cleavage (α_{s1}-CN f1–23) is rapidly hydrolysed by starter proteinases. The specificity of chymosin on α_{s1}-casein in solution was studied by Pelissier *et al.*, (1974), Mulvihill and Fox (1979), Pahkala *et al.* (1989) and McSweeney et al. (1993b). The hydrolysis of α_{s1}-casein by chymosin is influenced by pH and ionic strength (Mulvihill and Fox, 1977, 1979). In 0.1 M phosphate buffer, pH 6.5, chymosin cleaves α_{s1}-casein at $Phe_{23}-Phe_{24}$, $Phe_{28}-Phe_{29}$, $Leu_{40}-Ser_{41}$, $Leu_{149}-Phe_{150}$, $Phe_{153}-Tyr_{154}$, $Leu_{156}-Asp_{157}$, $Tyr_{159}-Pro_{160}$ and $Trp_{164}-Tyr_{165}$ (McSweeney *et al.*, 1993b). These bonds are also hydrolysed at pH 5.2 in the presence of 5% NaCl (i.e. conditions similar to those in cheese), and, in addition, $Leu_{11}-Pro_{12}$, $Phe_{33}-Gly_{35}$, $Leu_{101}-Lys_{102}$, $Leu_{142}-Ala_{143}$ and $Phe_{179}-Ser_{180}$. The rate at which many of these bonds are hydrolysed depends on the ionic strength and pH, particularly $Leu_{101}-Lys_{102}$ which is cleaved far faster at the lower pH. The k_{cat} and K_m for hydrolysis of the $Phe_{23}-Phe_{24}$ bond of α_{s1}-casein by chymosin are $0.7\,s^{-1}$ and $0.37\,mM$, respectively (Carles and Ribadeau-Dumas, 1985).

In contrast to β-casein, NaCl up to 5% stimulates the hydrolysis of α_{s1}-casein and significant proteolysis occurs in the presence of 20% NaCl. Consequently, α_{s1}-casein is readily hydrolysed in cheese, initially to α_{s1}-I casein (f24–199). In Cheddar and Dutch-type cheeses, α_{s1}-casein is completely degraded to α_{s1}-I and some further products by the end of ripening. In mould-ripened cheeses, α_{s1}-casein is completely degraded to at least α_{s1}-I before the mould-ripening phase and very extensive degradation occurs thereafter due to the action of fungal proteinases and peptidases (Gripon, 1993).

α_{s2}-Casein appears to be relatively resistant to proteolysis by chymosin; cleavage sites are restricted to the hydrophobic regions of the molecule, i.e. residues 90–120 and 160–207: $Phe_{88}-Tyr_{89}$, $Tyr_{95}-Leu_{96}$, $Gln_{97}-Tyr_{98}$, $Tyr_{98}-Leu_{99}$, $Phe_{163}-Leu_{164}$, $Phe_{174}-Ala_{175}$, $Tyr_{179}-Leu_{180}$ (McSweeney et al., 1994a). Although para-κ-casein has several potential chymosin cleavage sites, it does not appear to be hydrolysed either in solution or in cheese (Green and Foster, 1974).

Mickelsen and Fish (1970) studied the proteinases of C. parasitica and R. pusillus. These enzymes were more proteolytic on various casein substrates (as measured by the liberation of non-protein N) than either pepsin or chymosin. PAGE of the enzyme-treated caseins showed extensive degradation by the C. parasitica proteinase. The authors, who also used PAGE to study proteolysis in Cheddar cheese slurries containing the above enzymes, found that the two fungal enzymes degraded caseins in slurries to a lesser extent than the corresponding casein in solution and that C. parasitica and R. pusillus proteinases produced more soluble N than pepsin or chymosin.

The activity of calf rennet, pepsin (type not specified) and proteinases from R. pusillus and B. polymyxa on casein fractions (pH 6.5 and 35% casein) was investigated by Itoh and Thomasow (1971). Rennet or pepsin produced little non-protein N (NPN; soluble in 12% TCA) from α_{s1}-casein although some proteolysis was observed using starch gel electrophoresis. R. pusillus proteinase did not increase NPN values significantly, but greater proteolysis of both caseins was observed electrophoretically. B. polymyxa proteinase hydrolysed both α_{s1}- and β-caseins very extensively. All enzymes produced considerable amounts of NPN from κ-casein.

The initial rates of hydrolysis, and rates and extents of hydrolysis after extended incubation, of whole, α-, β- and κ-caseins by crystallized chymosin, crystallized porcine pepsin and purified proteinases from R. pusillus and C. parasitica were studied by Tam and Whitaker (1972) who found that C. parasitica proteinase generally was more proteolytic than the other three enzymes on most substrates at most pH values.

The proteolytic activity of *C. parasitica*, *R. pusillus*, *R. miehei* proteinases and chymosin on various casein substrates were compared by Paquet and Alais (1978). *C. parasitica* proteinase had the highest proteolytic activity on all substrates, especially on α_s- and κ-caseins. The activity of *R. pusillus* proteinase was similar to that of chymosin except that less NPN was librated from κ-casein. *R. miehei* proteinase was slightly more active on α_s- and β-caseins than *R. pusillus* proteinase, and showed marked activity on κ-casein. Hassan *et al.* (1988) used PAGE to study the specificity of calf rennet, pepsin and the proteinases of *R. miehei* and *C. parasitica*. The microbial coagulants were more active on α_s-casein than calf rennet and buffalo casein was more susceptible to proteolysis than bovine casein. These authors found that pepsin had the highest proteolytic activity on all casein fractions, except bovine κ-casein.

1.14 Specificity of indigenous milk proteinases

1.14.1 *Plasmin*

Plasmin (fibrinolysin, EC 3.4.21.7), the principal indigenous proteinase in milk, has been the subject of much study (for reviews see Grufferty and Fox, 1988a; Bastian and Brown, 1996). The plasmin complex consists of the active enzyme, its zymogen, zymogen activators and inhibitors of the enzyme and zymogen activators, all of which are present in milk. Plasmin, plasminogen and plasminogen activators are associated with the casein micelles in milk, while the inhibitors are in the serum phase.

Plasmin is a trypsin-like serine proteinase which is optimally active at about pH 7.5 and 37°C and is highly specific for peptide bonds to which lysine, and to a lesser extent arginine, contributes to the carboxyl group. It is active on all caseins, especially on α_{s2}- and β-caseins. The primary cleavage sites in β-casein are: Lys_{28}–Lys_{29}, Lys_{105}–His_{106} and Lys_{107}–Glu_{108}, hydrolysis of which yields the γ-caseins [β-CN f29–209 (γ_1-CN), f106–209 (γ_2-CN) and f108–209 (γ_3-CN)], and proteose peptones 5 (β-CN f1–105 and f1–107), 8 slow (β-CN f29–105 and f29–107) and 8 fast (β-CN f1–28) (Eigel *et al.*, 1984); Lys_{113}–Tyr_{114} and Arg_{183}–Asp_{184} are also hydrolysed fairly rapidly (Fox *et al.*, 1994).

Plasmin cleaves α_{s2}-casein in solution at eight sites: Lys_{21}–Gln_{22}, Lys_{24}–Asn_{25}, Arg_{114}–Asn_{115}, Lys_{149}–Lys_{150}, Lys_{150}–Thr_{151}, Lys_{181}–Thr_{182}, Lys_{188}–Ala_{189} and Lys_{197}–Thr_{198} (Le Bars and Gripon, 1989; Visser *et al.*, 1989), producing about 14 peptides, three of which (f198–207, f182–207, f189–207) are potentially bitter (Le Bars and Gripon, 1989).

Although plasmin is less active on α_{s1}-casein than on α_{s2}- and β-caseins, the formation of λ-casein, a minor casein component, has been attributed to its action on α_{s1}-casein (Aimutis and Eigel, 1982). The principal plasmin cleavage sites on α_{s1}-casein in solution are $Arg_{22}-Phe_{23}$, $Arg_{90}-Tyr_{91}$, $Lys_{102}-Lys_{103}$, $Lys_{103}-Tyr_{104}$, $Lys_{105}-Val_{106}$, $Lys_{124}-Glu_{125}$, $Arg_{151}-Gln_{152}$ (Le Bars and Gripon, 1993; McSweeney et al., 1993c).

Plasmin has very low activity on κ-casein although it contains several potential sites; Eigel (1977) found no hydrolysis of κ-casein under conditions adequate for the hydrolysis of α_{s1}-casein; however, Andrews and Alichanidis (1983) reported that 4% of the peptides produced by indigenous plasmin in pasteurized milk stored at 37°C for 7 days and detectable by PAGE originated from κ-casein. The specificity of plasmin on κ-casein has not been determined.

Plasmin activity in cheese is clearly indicated by the formation of γ-caseins during ripening (Farkye and Fox, 1990, 1991). In most cheese varieties, β-casein is more resistant than α_{s1}-casein, e.g. in typical 6-month-old Cheddar or Gouda, α_{s1}-casein is usually completely hydrolysed, whereas ~ 50% of the β-casein remains (Visser and de Groot-Mostert, 1977; Thomas and Pearce, 1981). Since the primary chymosin-produced product of β-casein (β-I, β f1–189/192) is present at very low concentrations in normal Gouda or Cheddar, hydrolysis of β-casein in these cheeses is considered to be due mainly to plasmin.

Plasmin activity is much higher in high-cooked (i.e. Swiss and hard Italian) and washed-curd cheeses (e.g. Dutch types) than in Cheddar (Lawrence et al., 1987; Farkye and Fox, 1990) which was attributed by Lawrence et al. (1987) to the lower pH of Cheddar curd at whey drainage compared with Swiss, Italian or Dutch varieties. However, this could not be confirmed by Farkye and Fox (1990) who attributed the higher plasmin activity in high-cooked and washed-curd cheeses to inactivation or removal, respectively, of inhibitors of plasminogen activators. The plasmin, plasmino-gen and plasminogen activators dissociate from the micelles on acidification, which is the basis of the proposal by Lawrence et al. (1987), but dissociation does not occur until the pH is reduced to < 4.6 (Grufferty and Fox, 1988b; Farkye and Fox, 1990; Madkor and Fox, 1991). HTST pasteurization actually increases plasmin activity (Noomen, 1975), probably by inactivat-ing inhibitors of plasminogen activators such that indigenous plasminogen is activated (Farkye and Fox, 1990). The greater plasmin activity in washed-curd cheeses can be explained by the greater removal of inhibitors on washing.

Work by Visser (1977) on aseptic, rennet-free, starter-free Gouda cheese (in which only indigenous heat-stable proteinases should be active) in-dicated that plasmin or other indigenous proteinases rendered only ca. 3% of the total N soluble in 0.55% $CaCl_2$–4% NaCl and contributed very little to the formation of PTA-soluble N. These findings were essentially con-firmed for Cheddar by Farkye and Fox (1991) using 6-aminohexanoic acid to inhibit plasmin.

1.14.2   Cathepsin D

Milk contains an indigenous acid proteinase (pH optimum, 4.0), first recognized by Kaminogawa and Yamauchi (1972), who isolated and characterized the enzyme and considered it to be similar to the lysosomal acid proteinase, cathepsin D (EC 3.4.23.5). Procathepsin D in milk has been reported (Larsen *et al.*, 1993). Cathepsin D is relatively heat labile (completely inactivated by 70°C × 10 min) (Kaminogawa and Yamauchi, 1972). The specificity of cathepsin D on the caseins has not been determined, although electrophoretograms of caseins incubated with milk acid proteinase (Kaminogawa and Yamauchi, 1972) or cathepsin D (McSweeney *et al.*, 1995) indicate a specificity very similar to that of chymosin; surprisingly, cathepsin D has poor milk clotting activity (McSweeney *et al.*, 1995).

The contribution of cathepsin D to proteolysis in cheese is difficult to quantify owing to the similarity of its specificity to that of chymosin which is the most active acid proteinase in cheese. However, the formation of α_{s1}–CN f24–199 (α_{s1}-I) in aseptic rennet-free, starter-free cheese (Visser and de Groot-Mostert, 1977; Noomen, 1978; Lane and Fox, 1996) suggests that it may be active although complete inactivation of chymosin in these cheeses may not have been achieved. Igoshi and Arima (1993) isolated a proteinase from Swiss-type cheese which hydrolysed α_{s1} to α_{s1}-I casein and β to β-I which the authors considered to be the milk acid proteinase as described by Kaminogawa and Yamauchi (1972).

1.15   Proteolytic enzymes from microorganisms in cheese

All cheeses contain a range of microbial species which contribute to proteolysis, being the principal proteinases in some varieties, e.g. blue-mould varieties. The microorganisms can be grouped as:

- Starter (*Lactococcus* or *Lactococcus* plus *Leuconostoc* or *Lactobacillus* or *Lactobacillus* plus *Streptococcus*)
- Non-starter lactic acid bacteria (NSLAB; mainly mesophilic *Lactobacillus* and *Pediococcus*; also *Micrococcus*)
- Secondary cultures (*Penicillium*, *Brevibacterium*, yeasts, *Lactobacillus*).

The proteolytic system of the *Lactococcus* has been studied extensively (see Thomas and Pritchard, 1987; Monnet *et al.*, 1993; Pritchard and Coolbear, 1993; Tan *et al.*, 1993; Visser, 1993; Law and Haandrikman, 1996; Law and Mulholland, 1996, and Chapter 9 of this volume). Although *Lactococcus* are weakly proteolytic, they possess a very comprehensive proteolytic system which includes a cell wall-associated proteinase (capable of hydrolysing caseins to peptides small enough to be transported into the

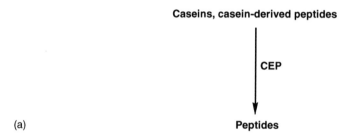

(a)

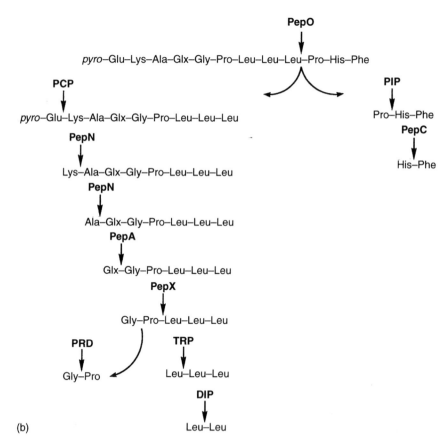

(b)

Figure 1.2 (a) Production of peptides from intact caseins and large casein-derived peptides by the cell envelope associated proteinase (CEP) of *Lactococcus*. (b) Degradation of a hypothetical peptide by the combined action of lactococcal endo- and exopeptidases. PepO, oligoendopeptidase; PIP, prolyl iminopeptidase; PCP, pyrrolidonyl carboxylyl peptidase; PepN, PepA, PepC, general, glutamyl and thiol aminopeptidase, respectively; PepX, X-proline dipeptidylaminopeptidase, PRD, prolidase; TRP, tripeptidase; DIP, dipeptidase.

cell), intracellular proteinases and endo-oligopeptidases (PepO, PepF), at least three aminopeptidases (PepN, PepA and PepC), an X-proline dipeptidylaminopeptidase (PepX), an iminopeptidase (prolidase), a tripeptidase, at least one general dipeptidase and proline-specific dipeptidases. This comprehensive proteolytic system is needed to enable *Lactococcus* to grow to high numbers in milk, which contains a low concentration of small peptides and amino acids. The proteolytic system of *Lactococcus* is summarized schematically in Figure 1.2. The specificity of the cell wall proteinase from several species on the individual caseins has been established, as are the specificities of the principal endo- and exopeptidases (Fox *et al.*, 1994, 1996).

The proteolytic system of the thermophilic *Lactobacillus* spp. is fairly well characterized and is generally similar to that of the *Lactococcus*.

The proteolytic system of the mesophilic *Lactobacillus* is not well characterized (Fox *et al.*, 1994, 1996). Several *Micrococcus* proteinases have been isolated but their specificity on the caseins is not known; *Micrococcus* peptidases have not been well characterized (Fox and McSweeney, 1996). The proteolytic systems of *Br. linens* and *Propionibacteria* have also been studied (see Fox and McSweeney, 1996, for references).

P. roqueforti and *P. camemberti* secrete potent aspartyl and metalloproteinases which have been characterized (Gripon, 1993). Some aminopeptidases from these organisms have also been characterized (Gripon, 1993).

Studies on controlled microflora cheeses have shown that the proteolytic system of starter and non-starter bacteria contribute little to primary proteolysis in cheese, e.g. as detected by PAGE, but are primarily responsible for the formation of small peptides and free amino acids (O'Keeffe *et al.*, 1976; Visser, 1977). Results from this laboratory (Lane and Fox, 1996; Lynch *et al.*, 1996) in which the relative importance of starter and NSLAB was studied suggested that the NSLAB play a similar but lesser role in proteolysis to the starter; they appear to contribute to the formation of free amino acids.

1.16 Characterization of proteolysis in cheese

The extent of proteolysis in cheese varies from very limited, e.g. Mozzarella, to very extensive, e.g. blue-mould varieties. PAGE of cheese shows marked inter-varietal differences in both the pattern and extent of proteolysis (Ledford *et al.*, 1966; Marcos *et al.*, 1979). The PAGE patterns of both the water-insoluble and water-soluble fractions of a number of varieties are shown in Figure 1.3(a, b). RP–HPLC chromatograms of the water-soluble fraction or sub-fractions thereof also show varietal characteristics (Figure 1.4). Both the PAGE and HPLC patterns become more complex as the cheese matures and are useful indices of cheese maturity and to a lesser extent of its quality and, therefore, may be potentially useful indices in the objective assessment of cheese quality (O'Shea *et al.*, 1996).

Many of the water-insoluble and water-soluble peptides in Cheddar cheese have been isolated and identified by amino acid sequencing and mass spectrometry (Fox *et al.*, 1994; McSweeney *et al.*, 1994b; Singh *et al.*, 1994, 1995, 1996).

All the principal water-insoluble peptides are produced either from α_{s1}-casein by chymosin or from β-casein by plasmin (McSweeney *et al.*, 1994b). In mature Cheddar (>6 months old), all the α_{s1}-casein is hydrolysed by chymosin at $Phe_{23}-Phe_{24}$. The peptide α_{s1}-CN f1–23 is rapidly hydrolysed at the bonds Gln_9-Gly_{10} and $Gln_{13}-Glu_{14}$ by the lactococcal cell wall proteinase (Fox and McSweeney, 1996). A significant amount of the

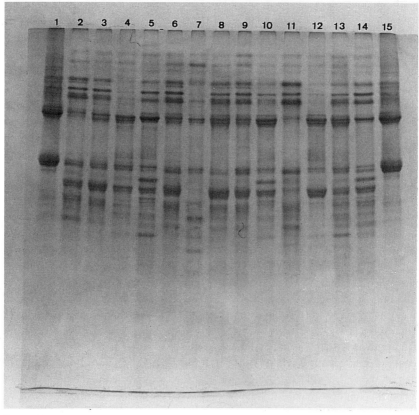

(a)

Figure 1.3 Urea polyacrylamide gel electrophoretograms of mature samples of a selection of cheeses (a) and water extracts therefrom (b). 1, Na-caseinate; 2, Appenzell; 3, Beaufort; 4, Brie; 5, Cheddar; 6, Comte; 7, Danish Blue; 8, Emmental; 9, Fontina; 10, Gouda; 11, Parmesan; 12, Port Salut; 13, Regato; 14, Svecia; 15, Na caseinate. (Reproduced from Fox, 1993.)

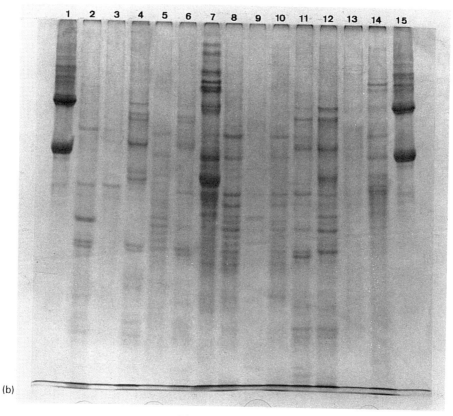

(b)

Figure 1.3 (*Continued*).

larger peptide, α_{s1}-CN f24–199, is hydrolysed at Leu_{101}–Lys_{102} to yield α_{s1}-CN f24–101 and α_{s1}-CN f102–199; a low proportion of the bonds Phe_{28}–Pro_{29} and Leu_{40}–Ser_{41} are also hydrolysed, probably in α_{s1}-CN f24–199, rather than in α_{s1}-CN because the bond Phe_{23}–Phe_{24} is hydrolysed considerably faster than any other bond in α_{s1}-CN. Surprisingly, the bond Trp_{164}–Tyr_{165}, which is hydrolysed rapidly in α_{s1}-CN in solution (McSweeney *et al.*, 1993b), does not appear to be hydrolysed in cheese; at least a peptide with Tyr_{165} as its N-terminal has not yet been identified in cheese.

In mature Cheddar, $\sim 50\%$ of the casein is hydrolysed, mainly by plasmin, to γ-caseins (β-CN f29–209, f106–209 and f108–209) and proteose peptones (β-CN f1–28, f1–105, f1–107, f29–105, f29–108). Proteose peptones are hydrolysed further, principally by lactococcal proteinase and peptidases.

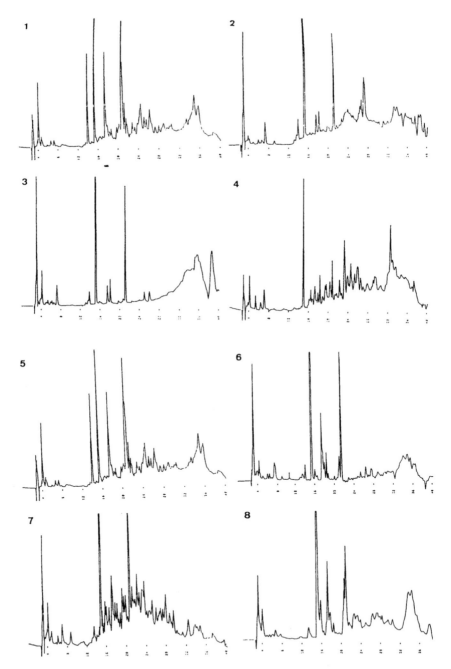

Figure 1.4 Reversephase (C_{18}) HPLC chromatograms of water-soluble extracts of 1, Cheddar; 2, Appenzell; 3, Beaufort; 4, Brie; 5, Cheddar; 6, Comte; 7, Danish Blue; 8, Emmental; 9, Cheddar; 10, Fontina; 11, Gouda; 12, Parmesan; 13, Cheddar; 14, Port Salut; 15, Regato; 16, Svecia. Acetonitrile/water gradient, detection at 230 nm. (Reproduced from Fox, 1993.)

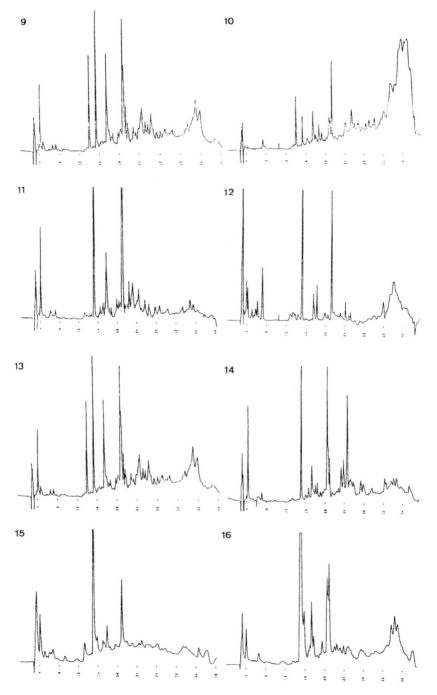

Figure 1.4 (*Continued*).

As discussed in section 1.13, β-CN in solution is hydrolysed rapidly by chymosin at $Leu_{192}-Tyr_{193}$; in fact it is the most sensitive bond in the casein system after $Phe_{105}-Met_{106}$ of κ-CN. However, this bond is not hydrolysed in cheese, possibly due to the strong inhibitory effect of ionic strength on the hydrolysis of this bond but even in salt-free cheese, little hydrolysis of this bond by chymosin occurs (Phelan *et al.*, 1973; Kelly, 1993). Presumably, β-casein in cheese interacts hydrophobically via the C-terminal region but there is no direct experimental work on this. The γ-caseins should also be quite sensitive to hydrolysis by chymosin but do not appear to be hydrolysed in cheese, presumably for the same reasons as β-casein.

It is desirable that little or no hydrolysis of the $Leu_{192}-Tyr_{193}$ bond of β-casein should occur since the peptide β-CN f193–209 is very bitter and peptides from this region are a major cause of bitterness in cheese (Visser *et al.*, 1983). One of the significant effects of an appropriate concentration of salt in cheese is the reduction of the hydrolysis of β-CN by chymosin and perhaps microbial proteinases which can also cleave bonds in this region of β-CN. An alternative approach to avoiding bitterness is to engineer β-casein in the region of the chymosin-susceptible bond, as reported by Simons *et al.* (1993), who engineered a strain of *E. coli* to produce β-CN in which $Leu_{192}-Tyr_{193}$ was altered to Pro.Pro or Leu.; the latter casein did not contain the bitter sequence, β-CN f193–209 and the $Pro_{192}-Pro_{193}$ bond was not hydrolysed by chymosin.

Although α_{s2}-casein gradually disappears from PAGE patterns (McSweeney *et al.*, 1993a), polypeptides produced from it, if any, have not been identified to date in the water-insoluble fraction of Cheddar, although four small α_{s2}-CN-derived peptides have been identified in the water-soluble fraction (Singh *et al.*, 1995). Since α_{s2}-casein is hydrolysed by both chymosin and plasmin one would expect that it would be hydrolysed quite rapidly in cheese – perhaps it is hydrolysed less selectively than is α_{s1}-casein by chymosin or β-casein by plasmin and this coupled with its low concentration (10% of total casein) may result in no α_{s2}-CN-derived peptide accumulating to a sufficient concentration to be a major water-insoluble peptide. Para-κ-casein appears to be resistant to proteolysis and no κ-casein-derived peptides have been identified.

The principal water-insoluble peptides in Dutch-type cheese have similar electrophoretic mobilities on urea–PAGE gels to those in Cheddar cheese but at different proportions (Figure 1.3(a)). Higher levels of γ-caseins are present, reflecting higher plasmin activity.

The peptide α_{s1}-CN f24–199 also appears to be formed in Swiss-type cheeses (Figure 1.3(a)) which is rather surprising since chymosin is reported (Steffen *et al.*, 1993) to be extensively inactivated by the high cook temperatures used in such varieties. Perhaps the indigenous milk acid proteinase (cathepsin D) contributes to the hydrolysis of α_{s1}-casein in Swiss cheeses. The level of γ-caseins in Swiss varieties is higher than in Cheddar, reflecting the higher plasmin activity in the former (Farkye and Fox, 1990; Steffen *et al.*, 1993).

Very extensive proteolysis occurs in Blue-mould cheeses, as shown in Figure 1.3(a). Both α_{s1}- and β-caseins are completely hydrolysed and most of the principal water-insoluble peptides have different mobilities from those in Cheddar. Initial proteolysis is due mainly to chymosin and α_{s1}-CN f24–199 is the principal large peptide formed initially, but following sporulation of the mould at about 15 days, its extracellular proteinases become dominant.

Considering the very large changes in texture that occur in Camembert and Brie during ripening, it might be expected that very extensive proteolysis occurs in these cheeses. However, the textural changes are due mainly to the increase in pH (to 6.5–7.5) resulting from the metabolism of lactic acid and the production of NH_3 and the diffusion of calcium to the surface of the cheese (in response to the increase in pH at the surface) (Karahadian and Lindsay, 1987). The extent of proteolysis is relatively limited, e.g. $\sim 20\%$ of the total N is soluble at pH 4.6 in a commercial Camembert (Khidr, 1995). Urea–PAGE shows that the proteolytic pattern in both the outer and inner zones of the cheese are generally similar to Cheddar; some very low-mobility peptides are present in the outer zone of the cheese, reflecting the action of the extracellular metalloproteinase of *P. camemberti* (Khidr, 1995).

Parmigiano-Reggiano is made from raw milk using a natural *Lactobacillus* starter and a high cook temperature; it has a low moisture content (which decreases from ~ 40 to $\sim 30\%$ during ripening) and is ripened at a relatively high temperature (18–20°C) for at least 18 months. Consequently, it undergoes extensive proteolysis which is probably due mainly to *Lactobacillus* proteinases and plasmin since chymosin is largely inactivated during cooking (Battistotti and Corradini, 1993). Typically, 35% of the total N is soluble in water and free amino acids represent $\sim 25\%$ of total N. Urea–PAGE shows (Battistotti and Corradini, 1993) that both α_{s1}- and β-caseins are degraded slowly but extensively; at the end of ripening only a few large peptides (detectable by PAGE) accumulate: γ^1-, γ^2- and γ^3-caseins, presumably produced by plasmin, and a peptide with a mobility similar to α_{s1}-CN f24–199, possibly produced by a low level of residual chymosin activity or cathepsin D.

At the other end of the proteolysis spectrum is Mozzarella (pizza) cheese, which is normally matured for about 3 weeks. Again, the principal proteolytic products detectable by PAGE are α_{s1}-CN f24–199 (reflecting residual chymosin activity in spite of the relatively high temperature during stretching) and γ-caseins (Mozzarella has high plasmin activity) (Kindstedt, 1993).

Thus, the principal water-insoluble peptides (detectable by PAGE) are common to many cheese varieties ranging from the very immature (Mozzarella) to the very mature (extra-mature Cheddar and Parmesan); the notable exception is mature blue, although young blue cheeses are similar to internal bacterially ripened varieties. These findings indicate that primary proteolysis in all varieties involves hydrolysis of α_{s1}-CN by chymosin (apparently even in high-cooked cheeses) and hydrolysis of β-CN by plasmin. Further hydrolysis of α_{s1}-CN f24–199 varies, presumably depending on the age of the cheese, its pH and in the case of blue cheeses, fungal proteinases; γ-caseins accumulate in all cheeses. Bacterial proteinases, either from starter or NSLAB, contribute little to the formation of large peptides. This may be due to the very high susceptibility of the bonds $\text{Phe}_{23}-\text{Phe}_{24}$ and $\text{Leu}_{101}-\text{Lys}_{102}$ of α_{s1}-casein to chymosin and of the bonds $\text{Lys}_{28}-\text{Lys}_{29}$, $\text{Lys}_{105}-\text{His}_{106}$ and $\text{Cys}_{107}-\text{Gln}_{108}$ of β-casein to plasmin, cleavage of which leads to the formation of the principal large, water-insoluble peptides in cheese.

During the past few years, considerable progress has been made on fractionating and characterizing the water-soluble peptides in Cheddar cheese (Fox et al., 1994; Singh et al., 1994, 1995, 1996; Breen et al., 1995). The fractionation scheme used in these studies is summarized in Figure 1.5. The water-soluble fraction (WSF) is first fractionated by diafiltration (DF) through 10-kDa polysulfone membranes. The retentate is resolved by ion-exchange chromatography on DEAE cellulose and the permeate by gel permeation chromatography on Sephadex G25 (Singh et al., 1994). All fractions are very heterogeneous on analysis by RP–HPLC. Individual peptides have been isolated from the ion exchange or Sephadex fractions by RP–HPLC using C_8 or C_{18} Nucleosil columns, and characterized by amino acid sequencing and mass spectrometry.

To date, we have isolated and characterized all the principal peptides in fractions I–IV of the DF retentate. Most of the peptides are derived from the N-terminal half of β-casein, especially from the sequence residues 53 to 91 (Figure 1.6a, b), with a smaller number from the N-terminal half of α_{s1}-CN (Figure 1.6a, b). The N-terminal of most of these peptides corresponds to a cleavage site for chymosin (α_{s1}-CN) or plasmin (β-CN) or a lactococcal cell wall proteinase. However, the N-terminal and especially the C-terminal of many peptides does not correspond precisely to known cleavage sites of chymosin, plasmin or lactococcal proteinase. This strongly suggests the action of bacterial aminopeptidases. Carboxypeptidase activity would explain why the C-terminal of many peptides does not correspond to known proteinase cleavage sites but Lactococcus spp. have not been reported to possess a carboxypeptidase and there are few reports of carboxypeptidase activity in Lactobacillus spp. (El-Soda et al., 1978). It must be presumed that other proteinases, e.g. from NSLAB, or endopeptidases (PepO, PepF) are involved, or perhaps other cleavage sites for lactococcal proteinase remain to be identified.

Fractions (from Sephadex G25) of the DF permeate contain several peptides and free amino acids. All the peptides isolated to date from the DF

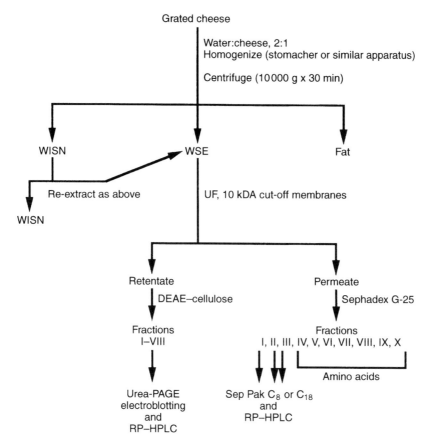

Figure 1.5 Fractionation scheme for cheese nitrogen. (Reproduced from Fox *et al.*, 1994.) WSE, water-soluble extract; WISN, water-insoluble nitrogen; UF, ultrafiltration.

permeate originate from the N-terminal half of α_{s1}-CN. The two principal peptides are α_{s1}-CN f1–9 and α_{s1}-CN f1–13, which are produced from α_{s1}-CN f1–23 (produced rapidly from α_{s1}-CN by chymosin) by lactococcal cell wall proteinase. The N-terminal sequence of these peptides is Arg.Pro.Lys.His.Pro... and should be hydrolysed by PepX. The fact that they accumulate in Cheddar [and in Gouda (Kaminogawa *et al.*, 1986; Exterkate and Alting, 1995)] suggests that either PepX is unstable or inactive in cheese or that it is unable to hydrolyse the above sequence, perhaps due to its high positive charge. The smaller peptides are difficult to isolate and especially to sequence (they do not adsorb on the membrane of the sequencer); we are now focussing attention on these small peptides.

Only four α_{s2}-CN-derived peptides have, to date, been isolated from the water-soluble fraction of Cheddar. All were internal peptides, requiring

(a)

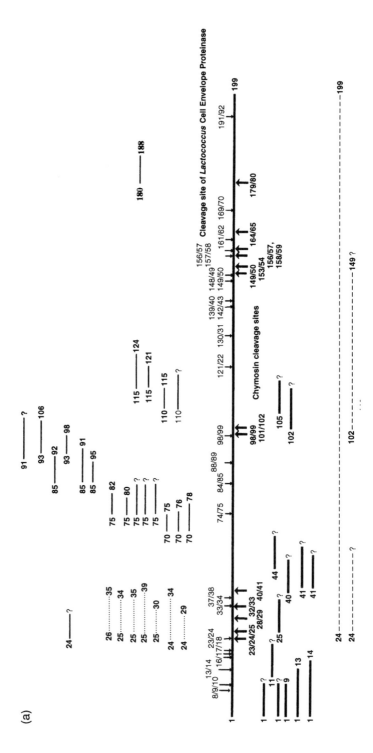

(b)

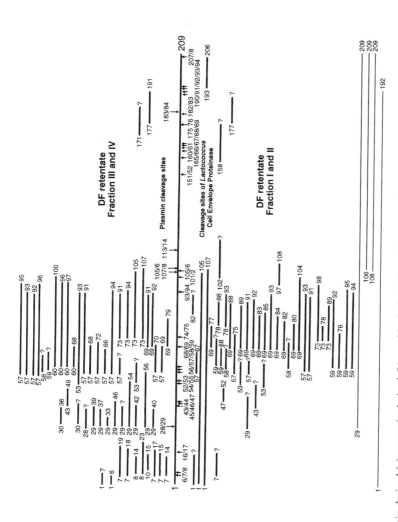

Figure 1.6 α_{s1}-Casein-derived (a) or casein-derived (b) peptides isolated from the water-soluble fraction of Cheddar cheese. (From Singh *et al*, 1994, 1995, 1996.)

cleavage of two bonds for their formation. Presumably, larger α_{s2}-CN-derived peptides are formed but are not detected on PAGE gels owing to their low concentration; α_{s2}-CN represents only 10% of total casein and if it is hydrolysed non-preferentially at a number of sites (both chymosin and plasmin can hydrolyse it in solution), the concentration of none of the products may be sufficiently high to permit their detection in comparison with the preponderance of α_{s1}- and β-CN-derived peptides since these proteins appear to be hydrolysed rather specifically by chymosin or plasmin, respectively.

It is interesting that the vast majority of the water-soluble peptides identified to date originate from the N-terminal halves of α_{s1}- and β-CNs. This is probably not surprising since the primary chymosin cleavage sites in α_{s1}-CN are Phe_{23}–Phe_{24}, Leu_{101}–Lys_{102} and to a lesser extent Phe_{28}–Pro_{29} and Leu_{40}–Ser_{41} and the primary plasmin cleavage sites in β-CN are Lys_{28}–Lys_{29}, Lys_{105}–His_{106} and Lys_{107}–Glu_{108}. It is tempting to suggest that the lactococcal and NSLAB proteinases hydrolyse the oligopeptides produced by chymosin or plasmin from α_{s1}-CN and β-CN, respectively, while the large C-terminal fragments, α_{s1}-CN f24–199, α_{s1}-CN f102–199, β-CN f29–209 (γ^1-CN), β-CN f106–209 (γ^2-CN) and β-CN f108–209 (γ^3-CN), accumulate, although these regions contain several bonds that are susceptible to lactococcal cell envelope-associated proteinase (CEP) in model systems. Few water-soluble peptides isolated to date include one of the principal chymosin or plasmin cleavage sites in α_{s1}- and β-casein, respectively, suggesting that the lactococcal and NSLAB proteinases do not extensively cleave intact α_{s1}- or β-CN, or at least preferentially cleave oligopeptides produced by chymosin or plasmin. In the case of β-CN, the resistance of bonds in the C-terminal region to hydrolysis may be explained by intermolecular hydrophobic interactions of this very hydrophobic region which render even the very chymosin-susceptible Leu_{192}–Tyr_{193} bond of β-CN resistant to chymosin in cheese. No obvious explanation can be offered for the resistance of the C-terminal half of α_{s1}-CN.

It is interesting that many of the water-soluble phosphopeptides are partially dephosphorylated. Cheese contains acid phosphatases derived from milk and starter (Fox et al., 1993) and these are clearly active in cheese. Phosphatase activity is a prerequisite for the hydrolysis of phosphopeptides which are not hydrolysed by peptidases.

It is also interesting that many of the identified peptides may be biologically active: β-f60–68 and fragments thereof are β-caseinomorphins; β-f57–70 and fragments thereof inhibit Pep O and Pep N (Stepaniak et al., 1995); β-f193–209 has immunomodulating activity (Coste et al., 1992); α_{s1}-f194–199 (Maruyama et al., 1987) and β-f193–202 (Meisel and Schlimme, 1994) inhibit angiotensin-activating enzyme.

The water-soluble fractions of other varieties are much less well characterized than that of Cheddar. RP–HPLC profiles of WSF of a selection of

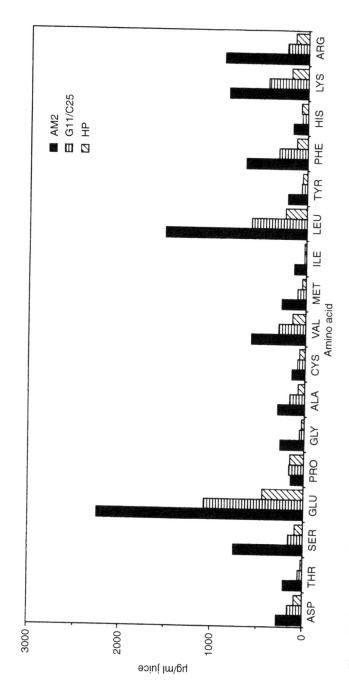

Figure 1.7 Free amino acids in Cheddar cheese made using different starters (AM2; G11/C25; HP) and ripened at 10°C for 42 days. (From Wilkinson, 1992.)

varieties are shown in Figure 1.4. Clearly, the profiles are characteristic of the variety/group, indicating that, in contrast to the water-insoluble peptides, the water-soluble peptides are unique to the variety and presumably reflect the specificity of the starter proteinases and peptidases (*Lactococcus* or *Lactobacillus*) and possibly NSLAB also.

Fractionation and identification of the small water-soluble peptides in cheese varieties is much less advanced than for Cheddar. Although not fractionated systematically, a large number of 12% TCA-soluble and insoluble peptides were identified in water extracts of Parmesan by Addeo *et al.* (1992, 1994) using fast atom bombardment mass spectrometry. Parmesan is quite an exceptional cheese; while it undergoes extensive proteolysis and has a very high concentration of free amino acids, it contains low concentrations of medium-sized peptides (Resmini *et al.*, 1988).

Although very extensive proteolysis occurs in blue cheeses and some of the larger peptides detectable by PAGE have been partially identified (Gripon, 1993), very little work has been done on the small (pH 4.6-soluble) peptides. The only study we are aware of is the partial identification of four PTA-soluble peptides from Gamonedo blue cheese by Gonzalez de Llano *et al.* (1991). Tsuda *et al.* (1993) identified the dipeptide Leu–Leu in Camembert using capillary isotacophoresis. Some of the peptides resulting from the cleavage of α_{s1}-CN f1–23 (produced by chymosin) by lactococcal cell envelope-associated proteinase have been identified in Gouda (Kaminogawa *et al.*, 1986; Exterkate and Alting, 1995). Proteolysis in Swiss-type cheeses has been studied using PAGE and RP–HPLC (Bican and Spahni, 1993; Steffen *et al.*, 1993) but as far as we are aware, small peptides have not been isolated and characterized.

The peptide profile of Cheddar and Blue are the most complex of the varieties we have studied, probably reflecting age in the case of Cheddar and fungal proteinases/peptidases in the case of Blue. The profile of Parmesan is markedly different from all other varieties studied; most compounds have very short retention times which conforms with the very high proportion of free amino acids in the water-soluble fraction of this cheese.

Significant concentrations of amino acids, the final products of proteolysis, occur in all cheeses that have been investigated. Relative to the level of water-soluble N, Cheddar contains a low concentration of free amino acids. The principal amino acids in Cheddar are Glu, Leu, Arg, Lys, Phe, Ser (Figure 1.7; Wilkinson, 1992). Parmesan contains a very high concentration of amino acids which appear to make a major contribution to its characteristic flavour (Resmini *et al.*, 1988). The presence of amino acids in cheeses clearly indicates peptidase activity; since these enzymes are intracellular, their action indicates lactococcal cell lysis. On the presumption that amino acids contribute to cheese flavour, interest is now being focused on a search for fast-lysing lactococcal strains, e.g. heat-induced (Feirtag and McKay, 1987), phage-induced (Crow *et al.*, 1995) or bacteriosin-induced (Morgan *et al.*, 1995).

1.17 Influence of chymosin on rate of cheese ripening

It is clear from the foregoing that the coagulant proteinase plays a major role in primary proteolysis in cheese. Since the proteinases and exopeptidases of the starter and NSLAB act primarily on polypeptides produced by the coagulant from α_{s1}-casein or by plasmin from β-casein, the activity of these enzymes may be rate-limiting in proteolysis, and hence flavour development, in cheese during ripening. Indeed, it has been reported that supplementation of cheese-milk with four to six times the normal level of plasmin found in milk accelerates ripening without off-flavour development (Farkye and Fox, 1992; Farkye and Landkammer, 1992). Although Exterkate and Alting (1993) suggested that chymosin activity is rate-limiting with respect to proteolysis, several authors (Stadhouders, 1960; Creamer et al., 1987; Guinee et al., 1991; Johnston et al., 1994) have reported that increasing the chymosin activity in cheese, e.g. by adding more rennet to cheese-milk, adding rennet powder to the curd at salting, or reducing the pH of whey at draining, does not accelerate flavour development and in fact causes off-flavours, especially bitterness. However, proteolysis in such cheeses has not been studied in detail and as far as we are aware, the effect of combining high starter and/or NSLAB activity with increased chymosin activity has not been investigated; such an approach to the accelerated ripening of cheese may be useful and is known to be efficacious when neutral proteinase and peptidases are combined (Law and Wigmore, 1983).

1.18 Possible future developments

The primary function of rennets in cheese-making, i.e. specific hydrolysis of κ-casein, is now well understood. The production of chymosin by engineered microorganisms has ensured an unlimited supply of good-quality rennet. Although it is likely that the enzymatic efficiency of chymosin could be improved by modifying its structure through point mutation of the gene, such developments may face regulatory restrictions. Genetically engineering κ-casein to improve its susceptibility to hydrolysis by chymosin appears to be possible but the commercial feasibility and consumer acceptability of this approach remains to be assessed. Modern *Rhizomucor* rennets appear to be acceptable alternatives to chymosin, although the specificity of the enzymes on the caseins is somewhat different.

Although only a small proportion of the rennet added to cheese-milk is retained in cheese curd, it makes a major and essential contribution to cheese ripening. The specificity of chymosin on the caseins in model systems is known and can largely be extrapolated to cheese. The specificity of microbial rennets on the caseins remains to be determined. The failure of chymosin to hydrolyse β-casein in cheese can be explained on the basis that intermolecular hydrophobic interactions renders the chymosin-susceptible bonds in the C-terminal region of this protein inaccessible to chymosin in cheese. However, a plausible explanation for the apparent failure of chymosin to hydrolyse bonds in the C-terminal half of α_{s1}-casein, which contains the second most susceptible bond, i.e. Trp_{164}–Tyr_{165}, can not be offered and warrants investigation.

The synergistic action of the three principal proteolytic systems in cheese, i.e. coagulant, plasmin and starter and perhaps NSLAB, warrants investigation in model systems and later in cheese. It may be possible to optimize the interaction between these enzyme systems, permitting the acceleration of cheese ripening.

Since the specificity of *C. parasitica* proteinase appears to be quite different from that of chymosin, they may act synergistically in cheese; e.g. it may be possible to add *C. parasitica* proteinase to curd at salting (of Cheddar). As far as we are aware, studies on this approach have not been reported.

Considering that the natural function of chymosin is to coagulate milk in the stomach, thereby improving the efficiency of digestion, it is fortuitous that it is also the most suitable proteinase for cheese ripening. It is very probable that the cheese-ripening properties of chymosin could be improved via genetic engineering, although regulatory restrictions may apply.

Now that the proteolytic activity of chymosin in both the coagulation and ripening stages of cheese manufacture are known, it appears that future research should focus on optimizing the symbiotic relationship between chymosin (or other coagulant) and the other proteolytic enzymes in cheese, especially those from the starter and NSLAB.

1.19   References

Addeo, F., Chianese, L., Salzano, F., Sacchi, R., Cappuccio, U., Ferranti, P. and Molorni, A. (1992) Characterization of the 12% trichloroacetic acid insoluble oligopeptides of Parmigiano-Reggiano cheese. *J. Dairy Res.*, **59**, 401–11.

Addeo, F., Chianese, L., Sacchi, R., Musso, S.P., Ferranti, P. and Molorni, A. (1994) Characterization of the oligopeptides of Parmigiano-Reggiano cheese soluble in 120 g trichloroacetic acid/l. *J. Dairy Res.*, **61**, 365–74.

Aimutis, W.R. and Eigel, W.N. (1982) Identification of λ-casein as plasmin-derived fragments of bovine α_{s1}-casein. *J. Dairy Sci.*, **65**, 175–81.

Andrews, A.T. and Alichanidis, E. (1983) Proteolysis of caseins and the proteose peptone fraction of bovine milk. *J. Dairy Res.*, **50**, 275–90.

Aston, J.W., Grieve, P.A., Durward, I.G. and Dulley, J.R. (1983) Proteolysis and flavour development in Cheddar cheeses subjected to accelerated ripening treatments. *Aust. J. Dairy Technol.*, **38**, 59–65.

Baer, A. and Collin, J.C. (1993) Determination of residual activity of milk-clotting enzymes in cheese. Specific identification of chymosin and its substitutes in cheese. Bulletin 284, International Dairy Federation, Brussels, pp. 18–23.

Banks, J.M. and Muir, D.D. (1985) Incorporation of the protein from starter growth medium in curd during manufacture of Cheddar cheese. *Milchwissenschaft*, **40**, 209–12.

Bastian, E. and Brown, R.J. (1996) Plasmin in dairy products. *Intern. Dairy J.*, **6**, 435–7.

Battistotti, B. and Corradini, C. (1993) Italian cheese, in *Cheese: Chemistry, Physics and Microbiology*, Vol. 2, 2nd edn (ed. P.F. Fox), Chapman & Hall, London, pp. 221–43.

Berridge, N.J. (1942). The second phase of rennet coagulation. *Nature*, 149, 194–5.

Bican, P. and Spahni, A. (1993) Proteolysis in Swiss-type cheeses: a high performance liquid chromatography study. *Intern. Dairy J.*, 3, 73–84.

Boudjellab, N., Rolet-Repecaud, O. and Collin, J.-C. (1994) Detection of residual chymosin in cheese by an enzyme-linked immunosorbent assay. *J. Dairy Res.*, 61, 101–9.

Breen, E.D., Fox, P.F. and McSweeney, P.L.H. (1995) Fractionation of peptides in a 10 kDa ultrafiltration retentate of a water-soluble extract of Cheddar cheese. *Ital. J. Food Sci.*, 11, 211–20.

Brignon, G., Chtourou, A. and Ribadeau-Dumas, R. (1985) Preparation and amino acid sequence of human κ-casein. *FEBS Lett.*, 188, 48–54.

Burkhalter, G. (1981) IDF Catalogue of Cheeses, Document 141, International Dairy Federation, Brussels.

Carles, C. and Ribadeau-Dumas, B. (1984) Kinetics of action of chymosin (rennin) on some peptide bonds of bovine β-casein. *Biochemistry*, 23, 6839–43.

Carles, C. and Ribadeau-Dumas, B. (1985) Kinetics of the action of chymosin (rennin) on a peptide bond of bovine α_{s1}-casein: comparison of the behaviour of this substrate with that of β-caseins and κ-caseins. *FEBS Lett.*, 185, 282–6.

Carlson, A., Hill, G.C. and Olson, N.F. (1986) The coagulation of milk with immobilized enzymes: a critical review. *Enzyme Microbiol. Technol.*, 8, 642–50.

Chen, C.M., Jaeggi, J.J. and Johnson, M.E. (1996) Study of Cheddar cheese making and yield using a protease from *Rhizomucor miehei* expressed in *Aspergillus oryzae*. *J. Dairy Sci.* (submitted).

Chobert, J.-M., Mercier J.C. Bahy, C. and Haze, G. (1976) Structure primaire du caseinomacropeptide des casein porcine et humane. *FEBS Lett.* 72, 173–8.

Coste, M., Rochet, V., Leonil, J., Molle, D., Bouhallab, S. and Tome, D. (1992) Identification of C-terminal peptides of bovine β-casein that enhance proliferation of rat lymphocytes. *Immunol. Lett.*, 33, 41–6.

Creamer, L.K. (1976) A further study of the action of rennin on β-casein, *N.Z. J. Dairy Sci. Technol.*, 11, 30–9.

Creamer, L.K. and Olson, N.F. (1982) Rheological evaluation of maturing Cheddar cheese. *J. Food Sci.*, 41, 631–6, 646.

Creamer, L.K., Zoerb, H.F., Olson, N.F. and Richardson, T. (1984) Hydrophobicity of α_{s1}-I, α_{s1}-casein A and B and its implications in cheese structure. *J. Dairy Sci.*, 65, 902–6.

Creamer, L.K., Lawrence, R.C. and Gilles, J. (1985) Effect of acidification of cheese milk on the resultant Cheddar cheese. *N.Z. J. Dairy Sci. Technol.*, 20, 185–203.

Creamer, L.K., Lyer, M. and Lelievre, J. (1987) Effect of various levels of rennet addition on characteristics of Cheddar cheese made from ultrafiltered milk. *N.Z. J. Dairy Sci. Technol.*, 22, 205–14.

Crow, V.L., Martley, F.G., Coolbear, T. and Roundhill, S. (1995) Influence of phage-assisted lysis of *Lactococcus lactis* subsp. *lactis* ML8 on Cheddar cheese ripening. *Intern. Dairy J.* 5, 451–72.

Dalgleish, D.G. (1987) The enzymatic coagulation of milk, in *Cheese: Chemistry, Physics and Microbiology*, Vol. 1. (ed. P.F. Fox), Elsevier Applied Science Publishers, London, pp. 63–96.

Dalgleish, D.G. (1992) The enzymatic coagulation of milk, in *Advanced Dairy Chemistry*, Vol. 1. (ed. P.F. Fox), Elsevier Applied Science, London, pp. 579–619.

Dalgleish, D.G. (1993) The enzymatic coagulation of milk, in *Cheese: Chemistry, Physics and Microbiology*, Vol. 1, 2nd edn (ed. P.F. Fox), Chapman & Hall, London, pp. 69–100.

de Jong, L. (1976) Protein breakdown in soft cheese and its relation to consistency. 1. Proteolysis and consistency of 'Noorhollandse Meshanger' cheese. *Neth. Milk Dairy J.* 30, 242–53.

de Koning, P.J., van Rooijen, P.J. and Visser, S. (1978) Application of a synthetic hexapeptide as a standard substrate for the determination of the activity of chymosin. *Neth. Milk Dairy J.*, 32, 232–44.

Delfour, A., Jolles, J., Alais, C. and Jolles, P. (1965) Caseino-glycopeptides: characterization of a methionine residue and of the N-terminal sequence. *Biochem. Biophys. Res. Commun.*, 19, 452–5.

Doi, H., Kawaguchi, V., Ibuki, F. and Kanamori, M. (1979) Susceptibility of κ-casein components to various proteases. *J. Nutr. Sci. Vitaminol.*, **25**, 33–41.

Drøhse, H.B. and Foltmann, B. (1989) Specificity of milk-clotting enzymes towards bovine κ-casein. *Biochim. Biophys. Acta*, **995**, 221–4.

Eigel, W.N. (1977) Effect of bovine plasmin on α_{s1}-B and κ-A caseins. *J. Dairy Sci.*, **60**, 1399–403.

Eigel, W.N., Butler, J.E., Ernstrom, C.A., Farrell, H.M., Jr, Harwalkar, V.R., Jenness, R. and Whitney, R.McL. (1984) Nomenclature of proteins of cow's milk: fifth revision. *J. Dairy Sci.*, **67**, 1599–631.

El-Soda M., Bergere, J.L. and Desmazeaud, M.J. (1978) Detection and localization of peptide hydrolases in *Lactobacillus casei*. *J. Dairy Res.*, **45**, 519–24.

Ernstrom, C.A. and Wong, N.P. (1974) Milk clotting enzymes and cheese chemistry, in *Fundamentals of Dairy Chemistry*, 2nd edn (eds B.H. Webb, A.H. Johnson and J.A. Alford), AVI Publishing Co. Inc., Westport, CT, pp. 662–771.

Exterkate, F.A. and Alting, A.C. (1993) The conversion of the α_{s1}-casein-(1-23)-fragment by the free and bound form of the cell-envelope proteinase of *Lactococcus lactis* subsp. *cremoris* under conditions prevailing in cheese. *System. Appl. Microbiol.*, **16**, 1–8.

Exterkate, F.A. and Alting, A.C. (1995) The role of starter peptidases in the initial proteolytic events leading to amino acids in Gouda cheese. *Intern. Dairy J.*. **5**, 15–28.

Farkye, N.Y. and Fox, P.F. (1990) Observations on plasmin activity in cheese. *J. Dairy Res.*, **57**, 413–15.

Farkye, N.Y. and Fox, P.F. (1991) A preliminary study of the contribution of plasmin to proteolysis in Cheddar cheese: cheese containing plasmin inhibitor, 6-aminohexanoic acid. *J. Agric. Food Chem.*, **39**, 786–8.

Farkye, N.Y. and Fox, P.F. (1992) Contribution of plasmin to Cheddar cheese ripening: effect of added plasmin. *J. Dairy Res.*, **59**, 209–16.

Farkye, N.Y. and Landkammer, C.F. (1992) Milk plasmin activity influence on Cheddar cheese quality during ripening. *J. Food Sci.*, **57**, 622–4, 639.

Feirtag, J.M. and McKay, L.L. (1987) Thermoinducable lysis of temperature sensitive *Streptococcus cremoris* strains. *J. Dairy Sci.*, **70**, 1779–84.

Fish, J.S. (1957) Activity and specificity of rennin. *Nature*, **180**, 345.

Foltmann, B. (1964) Studies on rennin IX: on the limited proteolysis of A-rennin and the proteolytic activity of chromatographically purified fractions of rennin. *CR. Trav. Lab Carlsberg*, **34**, 319–25.

Foltmann, B. (1966) A review on prorennin and rennin. *C.R. Trav. Lab. Carlsberg*, **35**, 143–231.

Foltmann, B. (1981) Mammalian milk-clotting proteases: structure, function, evolution and development. *Neth. Milk Dairy J.*, **35**, 223–31.

Foltmann, B. (1993) General and molecular aspects of rennets, in *Cheese: Chemistry, Physics and Microbiology*, Vol. 1, 2nd edn (ed. P.F. Fox), Chapman & Hall, London, pp. 37–68.

Foltmann, B., Pederson, V.B., Jacobsen, M., Kauffman, D. and Wybrandt, G. (1977) Complete amino acid sequence of prochymosin. *Proc. Natl Acad. Sci. USA*, **74**, 2321–4.

Fox, P.F. (1969) Milk-clotting and proteolytic activities of rennet, and bovine pepsin and porcine pepsin. *J. Dairy Res.*, **36**, 427–33.

Fox, P.F. (1981) Exogenous proteinases in dairy technology, in *Proteinases and their Inhibitors: Structure, Function and Applied Aspects* (eds V. Turk and Lj Vitale), Pergamon Press, Oxford, pp. 245–67.

Fox, P.F. (1984) Proteolysis and protein–protein interactions in cheese manufacture, in *Developments in Food Proteins–3*, (ed. B.J.F. Hudson), Elsevier Applied Science Publishers, London, pp. 691–12.

Fox, P.F. (1988/89) Acceleration of cheese ripening. *Food Biotechnol.*, **2**, 133–85.

Fox, P.F. (1989) Proteolysis during cheese manufacture and ripening. *J. Dairy Sci.*, **72**, 1379–400.

Fox, P.F. (1992) *Advanced Dairy Chemistry–1–Proteins,*. Elsevier Applied Science, London.

Fox, P.F. (1993) *Cheese: Chemistry, Physics and Microbiology*, 2nd edn, Vols. 1 and 2, Chapman & Hall, London.

Fox, P.F. and McSweeney, P.L.H. (1996) Proteolysis in cheese during ripening. *Food Rev. Intern.*, **12**(4), 3.1–3.56.

Fox, P.F. and Mulvihill, D.M. (1990) Casein, in *Food Gels* (ed. P. Harris), Elsevier Applied Science Publishers, London, pp. 121–73.

Fox, P.F. and Stepaniak, L. (1993) Enzymes in cheese technology. *Intern. Dairy J.*, **3**, 509–30.

Fox, P.F., Law, J., McSweeney, P.L.H. and Wallace, J. (1993) Biochemistry of cheese ripening, in *Cheese: Chemistry, Physics and Microbiology*, Vol. 1 (ed. P.F. Fox), Chapman & Hall, London, pp. 389–438.

Fox, P.F., Singh, T.K. and McSweeney, P.L.H. (1994) Proteolysis in cheese during ripening, in *Biochemistry of Milk Products* (eds A.T. Andrews and J. Varley), Royal Society of Chemistry, Cambridge, pp. 1–31.

Fox, P.F., Singh, T.K. and McSweeney, P.L.H. (1995) Biogenesis of flavour compounds in cheese, in *Chemistry of Structure/Function Relationships in Cheese* (eds E.L. Malin and M.H. Tunick), Plenum Press, New York, pp. 59–98.

Fox, P.F., O'Connor, T.P., McSweeney, P.L.H., Guinee, T.P. and O'Brien, N.M. (1996) Cheese: physical, biochemical and nutritional aspects. *Adv. Food Nutr. Res.*, **39**, 163–328.

Garnier, J. (1963) Etude cinetique d'une proteolyse limitee action de la presure sur la caseine κ. *Biochim. Biophys. Acta*, **66**, 366–77.

Gonzalez de Llano, D., Polo, M.C. and Ramos, M. (1991) Production, isolation of low molecular mass peptides by high performance liquid chromatography. *J. Dairy Res.*, **58**, 363–72.

Green, M.L. (1977) Review on the Progress of Dairy Science. Milk coagulants. *J. Dairy Res.*, **44**, 159–88.

Green, M.L. and Crutchfield, G. (1969) Studies on the preparation of water-insoluble derivatives of rennin and chymotrypsin and their use in the hydrolysis of casein and the clotting of milk. *Biochem J.*, **115**, 183–9.

Green, M.L. and Foster, P.D.M. (1974) Comparison of the rates of proteolysis during ripening of Cheddar cheeses made with calf rennet and swine pepsin as coagulants. *J. Dairy Res.*, **41**, 269–82.

Green, M.L. and Grandison, A.S. (1993) Secondary (non-enzymatic) phase of rennet coagulation and postcoagulation phenomena, in *Cheese, Chemistry, Physics and Microbiology*, Vol. 1, 2nd edn (ed. P.F. Fox), Chapman & Hall, London, pp. 101–40.

Gripon, J.-C. (1993) Mould-ripened cheeses, in *Cheese, Chemistry, Physics and Microbiology*, Vol. 2, 2nd edn (ed. P.F. Fox), Chapman & Hall, London, pp. 111–36.

Grufferty, M.B. and Fox, P.F. (1985) Effect of added NaCl on some physicochemical properties of milk. *Ir. J. Food Sci. Technol.*, **9**, 1–9.

Grufferty, M.B. and Fox, P.F. (1988a) Milk alkaline proteinase: a review, *J. Dairy Res.*, **55**, 609–30.

Grufferty, M.B. and Fox, P.F. (1988b) Factors affecting the release of plasmin activity from casein micelles. *N.Z. J. Dairy Sci. Technol.*, **23**, 153–63.

Guinee, T.P., Wilkinson, M.G., Mulholland, E.D. and Fox, P.F. (1991) Influence of ripening temperature, added commercial enzyme preparations and attenuated mutant (Lac⁻) *Lactococcus lactis* starter on the proteolysis and maturation of Cheddar cheese. *Ir. J. Food Sci. Technol.*, **15**, 27–52.

Hamdy, A. and Edelsten, D. (1970) Some factors affecting the coagulation strength of 3 different microbial rennets. *Milchwissenschaft*, **25**, 450–3.

Hassan, H.N., El Deeb, S.A. and Mashaly, R.I. (1988) Action of rennet and rennet substitutes on casein fractions in polyacrylamide gel electrophoresis. *Indian J. Dairy Sci.*, **41**, 485–490 [cited from *Food Sci. Technol. Abstr.*, **22**, 3P19 (1990)].

Hill, R.D. (1968) The nature of the rennin-sensitive bond in casein and its possible relation to sensitive bonds in other proteins. *Biochem. Biophys. Res. Commun.*, **33**, 659–63.

Hill, R.D. (1969) Synthetic peptide and ester substrates for rennin. *J. Dairy Res.*, **36**, 409–15.

Hill, R.D. and Hocking, V.M. (1978) A study of structural factors that affect the reactivity of the chymosin-sensitive bond of κ-casein. *N.Z. J. Dairy Sci. Technol.*, **13**, 195–204.

Hill, R.D., Lahav, E. and Givol, D. (1974). A rennin-sensitive bond in α_{s1}-casein. *J. Dairy Res.*, **41**, 147–53.

Holmes, D.G., Duersch, J.W. and Ernstrom, C.A. (1977) Distribution of milk clotting enzymes between curd and whey and their survival during Cheddar cheese making. *J. Dairy Sci.*, **60**, 862–9.

Humme, H.E. (1972) Optimum pH for limited specific proteolysis of κ-casein by rennin (primary phase of milk clotting). *Neth. Milk Dairy J.*, **26**, 180–5.

Hunter, J.N. (1996) *Hydrolysis of β-Casein by Acid Proteinases and Plasmin*, MSc Thesis, National University of Ireland, Cork.

IDF (1992) Fermentation-produced enzymes and accelerated ripening in cheesemaking. Bulletin 269. International Dairy Federation, Brussels.

Igoshi, K. and Arima, S. (1993) Acid and semi-alkaline proteinase in Swiss-type cheese. *Milchwissenschaft*, **48**, 623–5.

Itoh, T. and Thomasow, J. (1971) Action of rennet and other milk clotting enzymes on casein fractions. *Milchwissenschaft*, **26**, 671–5.

Johnston, K.A., Dunlop, F.P., Coker, C.J. and Wards, S.M. (1994) Comparisons between the electrophoretic pattern and textural assessment of aged Cheddar made using various levels of calf rennet or microbial coagulant (Rennilase 46L). *Intern. Dairy J.*, **4**, 303–27.

Kalantzopoulos, G.C. (1993) Cheeses from ewes' and goats' milk, in *Cheese: Chemistry, Physics and Microbiology*, Vol. 2, 2nd edn (ed. P.F. Fox), Chapman & Hall, London, pp. 507–53.

Kaminogawa, S. and Yamauchi, K. (1972) Acid protease of bovine milk. *Agr. Biol. Chem.*, **36**, 2351–6.

Kaminogawa, S., Yan, T.R., Azuma, N. and Yamauchi, K. (1986) Identification of low molecular weight peptides in Gouda-type cheese and evidence for the formation of these peptides from 23-N-terminal residues of α_{s1}-casein by proteinases of *Streptococcus cremoris* H61. *J. Food Sci.*, **51**, 1253–6, 1264.

Kannan, A. and Jenness, R. (1961) Relation of milk serum proteins and milk salts to the effects of heat treatment on rennet clotting. *J. Dairy Sci.*, **44**, 808–22.

Karahadian, C. and Lindsay, R.C. (1987) Integrated roles of lactate, ammonia, and calcium in texture development of mold surface-ripening cheese, *J. Dairy Sci.*, **70**, 909–18.

Kelly, M. (1993) *The Effect of Salt and Moisture on Proteolysis in Cheddar Cheese*, MSc Thesis, National University of Ireland, Cork.

Khidr, M.K.A. (1995) *Proteolysis in Camembert cheese*, MSc Thesis, National University of Ireland, Cork.

Kindstedt, P.S. (1993) Mozzarella and pizza cheese, in *Cheese: Chemistry, Physics and Microbiology*, Vol. 2, 2nd edn (ed. P.F. Fox), Chapman & Hall, London, pp. 337–62.

Kowalchyk, A.W. and Olson, N.F. (1977) Effects of pH and temperature on the secondary phase of milk clotting by rennet. *J. Dairy Sci.*, **60**, 1256–9.

Kowalchyk, A.W. and Olson, N.F. (1979) Milk clotting and curd firmness as affected by type of milk clotting enzyme, calcium chloride concentration, and season of year. *J. Dairy Sci.*, **62**, 1233–7.

Lane, C.N. and Fox, P.F. (1996) Contribution of starter and adjunct lactobacilli to proteolysis in Cheddar cheese during ripening. *Intern. Dairy J.* (in press).

Larsen, L.B., Boisen, A. and Petersen, T.E. (1993) Procathepsin D cannot autoactivate to cathepsin D at acid pH. *FEBS Lett.*, **319**, 54–8.

Law, B.A. and Mulholland, F., (1996) Enzymology of *Lactococci* in relation to flavour development from milk proteins. *Intern. Dairy J.*, **5**, 833–54.

Law, B.A. and Wigmore, A.S. (1983) Accelerated ripening of Cheddar cheese with commercial proteinase and intracellular enzymes from starter streptococci. *J. Dairy Res.*, **50**, 519–25.

Law, J. and Haandrikman, A. (1996) Proteolytic enzymes of lactic acid bacteria. *Intern. Dairy J.*, **6**, (in press).

Lawrence, R.C., Creamer, L.K. and Gilles, J. (1987) Texture development during cheese ripening. *J. Dairy Sci.*, **70**, 1748–60.

Le Bars, D. and Gripon, J.-C. (1989) Specificity of plasmin towards bovine α_{s2}-casein. *J. Dairy Res.*, **56**, 817–21.

Le Bars, D. and Gripon, J.C. (1993) Hydrolysis of α_{s1}-casein by bovine plasmin. *Le Lait*, **73**, 337–44.

Ledford, R.A., O'Sullivan, A.C. and Nath, K.R. (1966) Residual casein fractions in ripened cheese. *J. Dairy Sci.*, **49**, 1098–101.

Lenoir, J. (1984) The surface flora and its role in the ripening of cheese. Bulletin 171, International Dairy Federation, Brussels, pp. 3–20.

Loucheux-Lefebvre, M.H., Aubert, J.-P. and Jolles, P. (1978) Prediction of conformation of cow and sheep κ-caseins. *Biophys. J.*, **224**, 323–36.

Lucey, J.A., Gorry, C. and Fox, P.F. (1993) Rennet coagulation properties of heated milk. *Agric. Sci. Finl.*, **2**, 361–8.

Lynch, C.M., McSweeney, P.L.H., Fox, P.F., Cogan, T.M. and Drinan, F.D. (1996) Contribution of starter lactococci and non-starter lactobacilli to proteolysis in Cheddar cheese with a controlled microflora. *Intern. Dairy J.* (in press).

Madkor, S. and Fox, P.F. (1991) Plasmin activity in buffalo milk. *Food Chem.*, **39**, 139–56.

Mantafounis, D. and Pitts, J.E. (1990) Protein engineering of chymosin; modification of the optimum pH of enzyme catalysis. *Protein Engineering*, **3**, 605–9.

Marcos, A., Esteban, M.A., Leon, F. and Fernandez-Salguero, J. (1979) Electrophoretic patterns of European cheeses: comparison and quantitation. *J. Dairy Sci.*, **62**, 892–900.

Marshall, R.J. (1986) Increasing cheese yields by high heat treatment of milk. *J. Dairy Res.*, **53**, 313–22.

Martin, P., Raymond, M.N., Bricas, E. and Ribadeau-Dumas, B.R. (1980) Kinetic studies on the action of *Mucor pusillus*, *Mucor miehei* acid proteases and chymosins A and B on a synthetic chromophoric hexapeptide. *Biochim. Biophys. Acta*, **612**, 410–20.

Martin, P., Collin, J.-C., Garnot, P., Ribadeau-Dumas, B. and Mocquot, G. (1981) Evaluation of bovine rennets in terms of absolute concentrations of chymosin and pepsin A. *J. Dairy Res.*, **48**, 447–56.

Maruyama, S., Hitachi, H., Awaya, J., Kurono, M., Tomizuka, N. and Suzuki, H. (1987) Angiotensin I-converting enzyme inhibitory activity of the C-terminal hexapeptide of α_{s1}-casein. *Agric. Biol. Chem.*, **51**, 2557–61.

Matheson, A.R. (1981) The immunochemical determination of chymosin activity in cheese. *N.Z. J. Dairy Sci. Technol.*, **16**, 33–41.

McGugan, W.A., Emmons, D.B. and Larmond, E. (1979) Influence of volatile and non-volatile fractions on intensity of Cheddar cheese flavour. *J. Dairy Sci.*, **62**, 398–403.

McSweeney, P.L.H., Fox, P.F., Lucey, J.A., Jordan, K.N. and Cogan, T.M. (1993a) Contribution of the indigenous microflora to the maturation of Cheddar cheese. *Intern. Dairy J.*, **3**, 613–34.

McSweeney, P.L.H., Olson, N.F., Fox, P.F., Healy, A. and Højrup, P. (1993b) Proteolytic specificity of chymosin on bovine α_{s1}-casein. *J. Dairy Res.*, **60**, 401–12.

McSweeney, P.L.H., Olson, N.F., Fox, P. F., Healy, A. and Højrup, P. (1993c) Proteolytic specificity of plasmin on bovine α_{s1}-casein. *Food Biotechnol.*, **7**, 143–58.

McSweeney, P.L.H., Olson, N.F., Fox, P.F. and Healy, A. (1994a) Proteolysis of bovine α_{s2}-casein by chymosin. *Z. Lebensm. Unters. Forsch.*, **119**, 429–32.

McSweeney, P.L.H., Pochet, S., Fox, P.F. and Healy, A. (1994b) Partial identification of peptides from the water-insoluble fraction of Cheddar cheese. *J. Dairy Res.*, **61**, 587–90.

McSweeney, P.L.H., Fox, P.F. and Olson, N.F. (1995) Proteolysis of bovine caseins by cathepsin D: preliminary observations and comparison with chymosin. *Intern. Dairy J.*, **5**, 321–36.

Meisel, H. and Schlimme, E. (1994) Inhibitors of angiotensin converting enzyme derived from bovine casein (casokinins), in *β-Casomorphins and Related Peptides, Recent Developments V* (eds V. Brantl and H. Techemacher), VCH Weinheim, New York, pp. 27–33.

Mickelsen, R. and Fish, N.L. (1970) Comparing proteolytic action of milk-clotting enzymes on caseins and cheese. *J. Dairy Sci.*, **53**, 704–10.

Monnet, V., Chapot-Chartier, M.P. and Gripon, J.-C. (1993) Les peptidases des lactocoques. *Le Lait*, **73**, 97–108.

Morgan, S., O'Donovan, C., Ross, R.P., Hill, C. and Fox, P.F. (1995) Significance of autolysis and bacteriocin-induced lysis of starter cultures in Cheddar cheese ripening. Proceedings of the 4th Cheese Symposium, Moorepark, February 13–14, 1995, pp. 51–60.

Morrissey, P.A. 1969. The rennet hysteresis of heated milk. *J. Dairy Res.*, **36**, 333–341.

Mulvihill, D.M. and Fox, P.F. (1977) Proteolysis of α_{s1}-casein by chymosin: influence of pH and urea. *J. Dairy Res.*, **44**, 533–40.

Mulvihill, D.M. and Fox, P.F. (1978) Proteolysis of bovine β-casein by chymosin: influence of pH, urea and sodium chloride. *Ir. J. Food Sci. Technol.*, **2**, 135–9.

Mulvihill, D.M. and Fox, P.F. (1979) Proteolytic specificity of chymosin on bovine α_{s1}-casein. *J. Dairy Res.*, **46**, 641–51.

Nelson, J.H. (1975) Application of enzyme technology to dairy manufacturing. *J. Dairy Sci.*, **58**, 1739–50.

Nitchmann, H. and Bohren, H.U. (1955) Das Lab und seine wirkung auf das Casein der Milch. X. Eine Methode sur direction bestimmung der geschwindigkeit der primarreaktion der labgerinnung dar Milch. *Helv. Chim. Acta*, **38**, 1953–63.

Noomen, A. (1975) Proteolytic activity of milk protease in raw and pasteurized cow's milk. *Neth. Milk Dairy J.*, **29**, 153–61.

Noomen, A. (1978) Activity of proteolytic enzymes in simulated soft cheeses (Meshanger type). 1. Activity of milk protease. *Neth. Milk Dairy J.*, **32**, 26–48.

Noomen, A. (1983) The role of the surface flora in the softening of cheeses with a low initial pH. *Neth. Milk Dairy J.*, **37**, 229–32.

Oh, S. and Richardson, T. (1991) Genetic engineering of bovine casein to enhance proteolysis by chymosin, in *Interactions of Food Proteins* (eds N. Parris and R. Barford), ACS Symposium Series No. 454, American Chemical Society, Washington, DC, pp. 195–211.

O'Keeffe, A.M., Fox, P.F. and Daly, C. (1978) Proteolysis in Cheddar cheese: role of coagulant and starter bacteria. *J. Dairy Res.* **45**, 465–77.

O'Keeffe, R.B., Fox, P.F. and Daly, C. (1976) Contribution of rennet and starter proteases to proteolysis in Cheddar cheese. *J. Dairy Res.*, **43**, 97–107.

Olson, N.F. and Gould, B.W. (1995) Overview of the US cheese industry. Proceedings of the 4th Cheese Symposium (eds T.M. Cogan, P.F. Fox and R.P. Ross), Moorepark, Fermoy, Co. Cork, 13–14 February, 1995, pp. 1–11.

O'Shea, B.A., Uniacke-Lowe, T. and Fox, P.F. (1996) Objective assessment of Cheddar cheese quality. *Intern. Dairy J.* (in press).

Pahkala, E., Pihlanto-Leppälä, A., Laukkanen, M. and Antila, V. (1989) Decomposition of milk proteins during the ripening of cheese. 1. Enzymatic hydrolysis of α_s-casein. *Meijeritietleenllinen Aikakauskirja*, **47**, 39–47.

Paquet, D. and Alais, C. (1978) Action of fungal proteases on bovine casein and its constituents. *Milchwissenschaft*, **33**, 87–90.

Payens, T.A.J. and Both, P. (1980) Enzymatic clotting process. IV. The chymosin-triggered clotting of *p*-κ-casein, in *Biochemistry: Ions, Surfaces, Membranes* (ed. M. Blank), Adv. Chem. Series 188; American Chemical Society, Washington DC, pp. 129–41.

Payens, T.A.J. and van Markwijk, B.W. (1963) Some features of the association of β-casein. *Biochim. Biophys. Acta*, **71**, 517–30.

Payens, T.A.J. and Visser, S. (1981) What affects the specificity of chymosin towards κ-casein? *Neth Milk Dairy J.*, **35**, 387–9.

Pelissier, J.-P., Mercier, J.-C. and Ribadeau-Dumas, B. (1974) Etude de la proteolyse des caseines α_{s1}- et β-bovines par la presure. *Ann. Biol. Anim. Biochim. Biophys.*, **14**, 343–62.

Phelan, J.A. (1985) *Milk Coagulants – An Evaluation of Alternatives to Standard Calf Rennet*, PhD Thesis, National University of Ireland, Cork.

Phelan, J.A., Guiney, J. and Fox, P.F. (1973) Proteolysis of β-casein in Cheddar cheese. *J. Dairy Sci.*, **40**, 105–12.

Pitts, J.E., Dhanaraj, V., Dealwis, C.G., Mantafounis, D., Nugent, P., Orprayoon, P., Cooper, J.B., Newman, M. and Blundell, T.L. (1992) Multidisciplinary cycles for protein engineering: Site-directed mutagenesis and X-ray structural studies of aspartic proteinases. *Scan. J. Clin. Lab. Invest.*, **52** (suppl. 210), 39–50.

Pitts, J.E., Orprayoon, P., Nugent, P., Dhanaraj, R.V., Cooper, J.B., Blundell, T.L., Uusitalo, J. and Penttila, M. (1994) Protein engineering and preliminary x-ray analysis of Chy$_{155-165}$ Rhi loop exchange mutant, in *Biochemistry of Milk Products* (eds A.T. Andrews and J. Varley), Royal Society of Chemistry, Cambridge, pp. 72–82.

Pritchard, G. and Coolbear, T. (1993) The physiology and biochemistry of the proteolytic system in lactic acid bacteria. *FEMS Microbiol. Rev.*, **12**, 179–206.

Pyne, G.T. (1955) The chemistry of casein: a review of the literature. *Dairy Sci. Abstr.*, **17**, 531–54.

Raap, J., Kerling, K.E.T., Vreeman, H.E. and Visser, S. (1983) Peptide-substrates for chymosin (rennin)-conformational studies of κ-casein and some κ-casein-related oligopeptides by circular-dichroism and secondary structure prediction. *Arch. Biochem. Biophys.*, **221**, 117–124.

Raymond, M.N. and Bricas, E. (1979) New chromophoric peptide substrates. *J. Dairy Sci.*, **62**, 1719–25.

Raymond, M.N., Garnier, J., Bricas, E., Cilianu, S., Blasnic, M., Chaix, A. and Lefrancier, P. (1972) Studies on the specificity of chymosin (rennin). I. Kinetic parameters of the hydrolysis of synthetic oligopeptide substrates. *Biochimie*, **54**, 145–54.

Raymond, M.N., Bricas, E., Salesse, R. Garnier, J., Garnot, P. and Ribadeau-Dumas, B. (1973) A proteolytic unit for chymosin (rennin) activity based on a reference synthetic peptide. *J. Dairy Sci.*, **56**, 419–22.

Resmini, P., Pellegrino, L., Hagenboom, J. and Bertuccioli, M. (1988) Gli ammioacidi liberi nel formaggio Parmigiano-Reggiano stagionato Ricerca triennale scilla composizione e su alcune peculiari caratteristiche de formaggio Parmigiano-Reggiano. Consorzio de Formaggio Parmigiano-Reggiano, Reggio Emilia, pp. 41–57.

Richardson, G.H., Gandhi, N.R., Divatia, M.A. and Ernstrom, C.A. (1971) Continuous curd tension measurements during milk coagulation. *J. Dairy Sci.*, **54**, 182–6.

Salesse, R. and Garnier, J. (1976) Synthetic peptides for chymosin and pepsin assays: pH effect and pepsin independent determination in mixtures. *J. Dairy Sci.*, **59**, 1215–21.

Sardinas, J.L. (1972) Microbial rennets. *Adv. Appl. Microbiol.*, **15**, 39–73.

Sawyer, W.H. (1969) Complex between β-lactoglobulin and κ-casein: a review. *J. Dairy Sci.*, **52**, 1347–55.

Schattenkerk, C. and Kerling, K.E.T. (1973) Relation between structure and capacity to function as rennin substrate. *Neth. Milk Dairy J.*, **27**, 286–7.

Shah, B., Kumar, S.R. and Devi, S. (1995) Immobilized proteolytic enzymes on resinous materials and their use in milk-clotting. *Process Biochem.*, **30**(1), 63–8.

Shalabi, S.I. and Wheelock, J.V. (1976) The role of α-lactalbumin in the primary phase of rennet action on heated casein micelles. *J. Dairy Res.*, **43**, 331–5.

Shalabi, S.I. and Wheelock, J.V. (1977) Effect of sulphydryl-blocking agents on the primary phase of chymosin action on heated casein micelles and heated milk. *J. Dairy Res.*, **44**, 351–5.

Shammet, K.M., Brown, R.J. and McMahon, D.J. (1992a) Proteolytic activity of some milk-clotting enzymes on κ-casein. *J. Dairy Sci.*, **75**, 1373–9.

Shammet, K.M., Brown, R.J. and McMahon, D.J. (1992b) Proteolytic activity of proteinases on macropeptide isolated from κ-casein. *J. Dairy Sci.*, **75**, 1380–8.

Simons, G., van den Heuval, W. Raynan, T., Frijters, A., Rutten, G., Slangen, C.J., Groenen, M., de Vos, W. and Siezen, R.J. (1993) Overproduction of bovine β-casein in *Escherichia coli* and engineering of its main chymosin cleavage site. *Protein Engineering*, **6**, 763–70.

Singh, H., Shalabi, S.I., Fox, P.F., Flynn, A. and Barry, A. (1988) Rennet coagulation of heated milk: influence of pH adjustment before or after heating. *J. Dairy Res.*, **55**, 205–15.

Singh, T.K., Fox, P.F., Højrup, P. and Healy, A. (1994) A scheme for the fractionation of cheese nitrogen and identification of principal peptides. *Intern. Dairy J.*, **4**, 111–122.

Singh, T.K., Fox, P.F. and Healy, A. (1995) Water-soluble peptides in Cheddar cheese: isolation and identification of peptides in diafiltration retentate. *J. Dairy Res.*, **62**, 629–40.

Singh, T.K., Fox, P.F. and Healy, A. (1996) Isolation and identification of further peptides in the diafiltration retentate of the water-soluble of Cheddar cheese. *J. Dairy Res.*, **63**, (in press).

Stadhouders, J. (1960) The hydrolysis of protein during the ripening of Dutch cheese. Enzymes and bacteria involved. *Neth. Milk Dairy J.*, **14**, 83–110.

Steffen, C., Eberhard, P., Bosset, J.O., Ruegg, M. (1993) Swiss-type varieties, in *Cheese: Chemistry, Physics and Microbiology, Vol. 2 (Major Cheese Groups)*, 2nd edn (ed. P.F. Fox), Chapman & Hall, London, pp. 83–110.

Stepaniak, L., Fox, P.F., Sørhaug, T. and Grabska, J. (1995) Effect of peptides from the sequence 58-72 of β-casein on the activity of endopeptidase, aminopeptidase and x-prolyl-dipeptidylaminopeptidase of *Lactococcus lactis* ssp. *lactis* MG 1363. *J. Agric. Food Chem.*, **43**, 849–53.

Sternberg, M. (1976) Microbial rennets. *Adv. Appl. Microbiol.*, **20**, 135–57.

Strop, P., Sedlacek, J., Stys, J., Kaderabkova, Z., Blaha, I., Pavlickova, L., Pohl, J., Fabry, M., Kostka, V. Newman, M., Frazao, C., Shearer, A., Tickle, I.J. and Blundell, T.L. (1990) Engineering enzyme subsite specificity: preparation, kinetic characterization and X-ray analysis at 2.0-Angstrom resolution of Val_{111}Phe site-mutated calf chymosin. *Biochemistry*, **29**, 9863–71.

Suzuki, J., Hamu, A., Nishiyama, M., Horinouchi, S. and Beppu, T. (1990) Site-directed mutagenesis reveals functional contribution of Thr_{218}, Lys_{220} and Asp_{304} in chymosin. *Protein Engineering*, **4**, 69–71.

Tam, J.J. and Whitaker, J.R. (1972) Rates and extents of hydrolysis of several caseins by pepsin, rennin, *Endothia parasitica* protease and *Mucor pusillus* protease. *J. Dairy Sci.*, **55**, 1523–31.

Tan, P.S.T., Poolman, B. and Konings, W.N. (1993) Proteolytic enzymes of *Lactococcus lactis*. *J. Dairy Res.*, **60**, 269–86.

Taylor, M.J., Richardson, T. and Olson, N.F. (1976) Coagulation of milk with immobilized proteases: a review. *J. Milk Food Technol.*, **39**, 864–71.

Teuber, M. (1990) Production of chymosin (EC 3.4.23.4) by microorganisms and its use for cheesemaking. Bulletin 251, International Dairy Federation, Brussels, pp. 3–15.

Thomas T.D. and Pearce, K. (1981) Influence of salt on lactose fermentation and proteolysis in Cheddar cheese. *N.Z. J. Dairy Sci. Technol.*, **16**, 253–9.

Thomas, T.D. and Pritchard, G. (1987) Proteolytic enzymes and dairy starter cultures. *FEMS Microbiol. Rev.*, **46**, 245–68.

Thompson, M.D., Dave, J.R. and Nakkasi, H.L. (1985) Molecular cloning of mouse mammary gland κ-casein – comparison with rat κ-casein and rat and human γ-fibrinogen. *DNA*, **4**, 263–71.

Tsuda, T., Yamada, M. and Nakazawa, Y. (1993) Measurement of lower molecular weight peptides in Camembert cheese using a computer simulation system of capillary isotachophoresis. *Milchwissenschaft*, **48**, 74–8.

Ustumol, Z. and Hicks, C.C. (1990) Effect of milk-clotting enzymes on cheese yield. *J. Dairy Sci.*, **73**, 8–16.

van Hooydonk, A.C.M., Boerrigter, I.J. and Hagendoorn, H.G. (1986) pH-induced physicochemical changes of casein micelles in milk and their effect on renneting. 2. Effect of pH on renneting of milk. *Neth. Milk Dairy J.*, **40**, 297–313.

van Hooydonk, A.C.M., de Koster, P.G. and Boerrigter, I.J. (1987) The renneting properties of heated milk. *Neth. Milk Dairy J.*, **41**, 3–18.

Visser, F.M.W. (1977) Contribution of enzymes from rennet, starter bacteria and milk to proteolysis and flavour development in Gouda cheese. 3. Protein breakdown: analysis of the soluble nitrogen and amino acid fractions. *Neth. Milk Dairy J.*, **31**, 210–39.

Visser, F.M.W. and de Groot-Mostert, A.E.A. (1977) Contribution of enzymes from rennet, starter bacteria and milk to proteolysis and flavour development in Gouda cheese. 4. Protein breakdown: a gel electrophoretical study. *Neth. Milk Dairy J.*, **31**, 247–64.

Visser, S. (1993) Proteolytic enzymes and their relation to cheese ripening and flavor: an overview. *J. Dairy Sci.*, **76**, 329–50.

Visser, S. and Rollema, H.S. (1986) Quantification of chymosin action of nonlabeled κ-casein-related peptide substrates by ultraviolet spectrophotometry: description of kinetics by the analysis of progress curves. *Anal. Biochem.*, **153**, 235–41.

Visser, S. and Slangen, K.J. (1977) On the specificity of chymosin (rennin) in its action on bovine β-casein. *Neth. Milk Dairy J.*, **31**, 16–30.

Visser, S., van Roijen, P.J., Schattenkerk, C. and Kerling, K.E.T. (1976) Peptide substrates for chymosin (rennin). Kinetic studies with peptides of different chain length including parts of the sequence 101–112 of bovine κ-casein. *Biochim. Biophys. Acta*, **438**, 265–72.

Visser, S., van Roijen, P.J., Schattenkerk, C. and Kerling, K.E.T. (1977) Peptide substrates for chymosin (rennin). Kinetic studies with bovine κ-casein-(103–108) hexapeptide analogues. *Biochim. Biophys. Acta*, **481**, 171–6.

Visser, S., van Rooijen, P.J. and Slangen, C.J. (1980) Peptide substrates for chymosin (rennin). Isolation and substrate behaviour of two tryptic fragments of bovine κ-casein. *Eur. J. Biochem.*, **108**, 415–21.

Visser, S., Hup, G., Exterkate, F.A. and Stadhouders, J. (1983) Bitter flavour in cheese. 2. Model studies on the formation and degradation of bitter peptides by proteolytic enzymes from calf rennet, starter cells and starter cell fractions. *Neth. Milk Dairy J.*, **37**, 169–80.

Visser, S., Slangen, C.J. and van Rooijen, P.J. (1987) Peptide substrates for chymosin (rennin). Interaction sites in κ-casein-related sequences located outside the (103–108) hexapeptide region that fits into the enzyme's active-site cleft. *Biochem. J.*, **244**, 553–8.

Visser, S., Slangen, K.J., Alting, A.C. and Vreeman, H.J. (1989) Specificity of bovine plasmin in its action on bovine α_{s2}-casein, *Milchwissenschaft*, **44**, 335–9.

Vreeman, W.J., Visser, S., Slangen, C.J. and van Reel, J.A.M. (1986) Characterization of bovine κ-casein fractions and the kinetics of chymosin-induced macropeptide release from carbohydrate-free and carbohydrate-containing fractions determined by high performance gel permeation chromatography. *Biochem. J.*, **240**, 87–97.

Wake, R.G. (1959) Studies on casein. V. The action of rennin on casein. *Aust. J. Biol. Sci.*, **12**, 479–89.

Walstra, P. (1993) The syneresis of curd, in *Cheese: Chemistry, Physics and Microbiology*, Vol. 1, 2nd edn (ed. P.F. Fox), Chapman & Hall, London, pp. 141–91.

Waugh, D.F. and von Hippel, P.H. (1956) κ-Casein and the stabilization of casein micelles. *J. Am. Chem. Soc.*, **78**, 4576–82.

Wilkinson, M.G. (1992) *Studies on the Acceleration of Cheese Ripening*, PhD Thesis, National University of Ireland, Cork.

Yamashita, T., Higashi, S., Higashi, T., Machida, H., Iwasaki, S., Nishiyama, M. and Beppu, T. (1994) Mutation of a fungal aspartic proteinase, *Mucor pusillus* rennin, to decrease thermostability for use as a milk coagulant. *J. Biotechnol.*, **32**, 17–28.

Zoon, P., Ansems, C. and Faber, E.J. (1994) Measurement procedure for concentration of active rennet in cheese. *Neth. Milk Dairy J.*, **48**, 141–50.

2 Classification and identification of bacteria important in the manufacture of cheese

B.A. LAW and E.B. HANSEN

2.1 Introduction

Referring to the Preface to this, the Second Edition of *Advances in the Microbiology and Biochemistry of Cheese and Fermented Milk*, we present this chapter as a supplement to, rather than a replacement of the chapter on the subject written by Dr Ellen Garvie for the First Edition.

Since the First Edition of this book was published there has been increased interest in the secondary bacteria in cheese as a vehicle for increasing the diversity of flavour notes in order to break away from the limitations imposed by defined strain starter cultures. This has put even greater importance on the correct and consistent identification of the lactococci, streptococci, lactobacilli, pediococci, brevibacteria, coagulase-negative staphylococci and 'micrococci'. In revising this chapter we have tried to take this into account and included additional information on the molecular phylogeny of these groups in relation to their likely efficacy as flavour adjuncts, added as controlled supplements to the acid-producing starter bacteria.

Anyone involved in the microbiology, biochemistry, or molecular genetics of bacteria used in, or functionally present in fermented dairy products when the First Edition was published would not be able to recognize many of the currently-used generic groupings. The biggest single change has of course been in the genus *Streptococcus*, and for an authoritative review of its present status, there is little to be gained here by trying to add to, or paraphrase the work of Hardie and Whiley (1995). However, for those requiring a fingertip reference, this chapter will attempt to summarize the changes which have taken place in the classification of the range of dairy bacteria, including non-lactics. (For a summary of fermented milk microbial classification, see Chapter 4.)

2.2 Molecular phylogeny as a basis for bacterial taxonomy

Bacterial taxonomy is currently being revised in order to bring it into agreement with phylogenetic relatedness. The analysis of bacterial phylogeny can now be based on sequence comparison of proteins and nucleic acids. Ribosomal RNA (rRNA) has been particularly useful due to the ubiquitous presence of ribosomes in living organisms. This has allowed the establishment of the phylogenetic relationships among all bacteria (for a review, see Olsen *et al.*, 1994). Oligonucleotide probes whose design and syntheses are based on complementarity to rRNA sequences allow specific detection of bacteria of a species or a genus. This has been exploited successfully for lactic acid bacteria (LAB) by Schleifer *et al.* (1995). Due to the high copy number of ribosomes, these methods can be used without enrichment to identify starter cultures in fermented products, and to monitor the population balance and purity of commercial mixed and multiple-strain starter cultures.

In the field of phylogeny, sequence probe methods have taken away many of the uncertainties associated with phenotypic identification, and revealed many enlightening and technologically-significant relationships (and non-relationships) which have helped, and will in the future help culture technologists to formulate consistently generic, species and strain combinations to meet changing demands for flavour, texture and health-promoting properties in fermented dairy products.

2.3 Current classification of cheese bacteria

The former genus *Streptococcus* has been divided up into four genera, *Enterococcus*, *Lactococcus*, *Streptococcus* and *Vagococcus*. Phylogenetically, these bacteria group together with the other typical LAB such as lactobacilli, leuconostocs and pediococci in the low mol % G + C phylum, whereas other important dairy bacteria such as bifidobacteria, propionibacteria and brevibacteria belong to a distinctly different phylum (Actinomycetes) having high mol % G + C (Schleifer *et al.*, 1995).

The genus *Lactococcus* contains the traditional mesophilic cheese starter cultures widely used in soft, semi-hard and hard (cheddar-type) cheeses (*Lc. lactis* subsp. *lactis*, subsp. *cremoris* and subsp. *lactis* biovar. *diacetylactis*). They are grouped as the non-motile mesophilic streptococci carrying group N antigen. The motile group N carriers, including some notable fish pathogens, make up *Vagococcus*. *Enterococcus* contains the 'faecal streptococci' (*En. faecalis*) and 17 other species, as well as some atypical streptococci formerly identified in Greek yoghurt as *S. thermophilus*. However, the commercial thermophilic starter cultures containing strains of *S. thermophilus* use true streptococci belonging to the oral group of *Streptococcus sensu stricto*, related to *S. salivarius*.

Interestingly, the traditionally distinct lactobacilli, leuconostocs and pediococci are intermixed phylogenetically, and this reflects their metabolic and technological overlaps when they are used to (for example) promote aroma-forming (hetero-)fermentations alone or in combination to generate

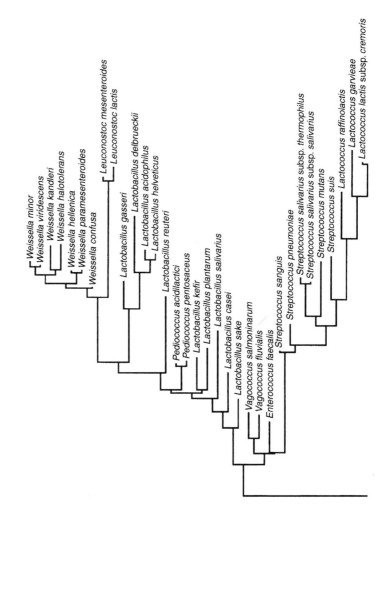

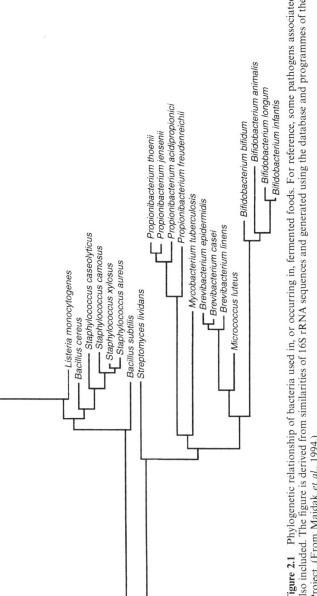

Figure 2.1 Phylogenetic relationship of bacteria used in, or occurring in, fermented foods. For reference, some pathogens associated with dairy foods are also included. The figure is derived from similarities of 16S rRNA sequences and generated using the database and programmes of the Ribosomal Database Project. (From Maidak *et al.*, 1994.)

dicarbonyls and organic acids in dairy products. A classification which would immediately distinguish homofermentative strains and species from heterofermentatives for the benefit of culture technologists is not possible. The consensus appears to be to put obligate homofermenters into a *Lactobacillus* group containing *Lb. delbruckii* and *Lb. acidophilus*, and place facultatives with known heterofermentative capacity in a *Lb. casei–Pediococcus* group (the six pediococcus species plus *Lb. casei, plantarum, curvatus, sake* and *salivarius*). This leaves the obligately heterofermentative lactobacilli grouped together with the genus *Leuconostoc*. This group is now divided into the genera *Leuconostoc* (*sensu stricto*) and *Weissella*, the latter taking in *Leuconostoc paramesenteroides* and the heterofermentative lactobacilli, *Lb. confusus, minor, kandleri, halotolerans* and *viridescens* (Collins *et al.*, 1993).

Despite these remaining 'imperfections' in the classification of LAB, the dairy strains in current use are generally well defined, but the science base may be challenged to improve genus, species and strain definition as technologists and flavour scientists search for new biological diversity in cultures to lead them to product innovation. For example, there is a significant potential for new product development using the more biochemically versatile dairy bacteria such as heterofermentative lactobacilli, leuconostocs, pediococci, propionibacteria, bifidobacteria, 'micrococci' and brevibacteria in non-conventional applications involving pre-grown biomass, rather than surface smear or adventitious surface growth, or as sources of exogenous flavour-producing enzymes. The phylogeny of these dairy bacteria has benefited from the application of molecular probes, and a summary of current inter-relatedness is made with the LAB included for completeness (Figure 2.1).

2.4   Taxonomy in relation to safety and technology of dairy bacteria

As far as the usefulness of these bacteria is concerned, some conclusions can be drawn from the following chapters of this book, and by considering the available evidence/data on their ability to grow and metabolize or otherwise transform substrates in cheese. For example, it is beginning to emerge that the early tentative findings of Law and Sharpe (1978) that LAB may produce small quantities of volatile sulphur compounds in cheese were correct following the isolation of a cystathionine lyase from a strain of lactococcus by Alting *et al.* (1995). Also, 'micrococci' are suspected of mediating in the formation of cheesy thioesters, lactobacilli can produce aromatic pyrazenes, and brevibacteria have wide-ranging amino acid-catabolizing metabolic pathways as well as the appropriate substrate-generating enzymes. However, culture technologists who are responsible for producing cultures of these bacteria to reliable commercial specifications must know precisely which genus and/or species they are working with in order to use their skill and experience to develop fermentation and QA/QC procedures. Also, bacteria loosely classified as micrococci or coryneform/ brevibacteria group may have close relatives which are potentially pathogenic, and must be well-identified before use in cheese, however efficacious they may be as flavour-producers.

The safety of LAB is based on the long history of safe use for the production of feremented foods. A similar record of safe use is also present for some commonly used non-lactic starters. Bacteria from the natural intestinal flora of humans and other mammals have recently been introduced as starter cultures. These bacteria do not have a history of safe use, but rather a history of safe co-existence. It would be very desirable if the assessment of safety could be based on other methods than history of safe use, as the conservative nature of this criterion excludes the introduction of new bacterial species as starter cultures.

Taxonomy is very important in this context, as the discussion is meaningful only if bacterial species are well-defined and natural units. Taxonomy can also have a predictive value for properties and phenotypes of bacteria, as is the case if the trait relates to a particular set of genes. However, the ability to cause disease in humans is not related to one gene or set of genes, and consequently we find pathogenic bacteria scattered throughout the entire kingdom of Bacteria. Some bacteria are pathogenic due to production of toxins, some because they are able to invade the tissues, and others because they have mechanisms to avoid the various natural defence systems. The definition of safe bacterial taxa is difficult, as pathogenicity is not confined to, nor universally exhibited by, all of the members of a particular taxon. For each new strain introduced into food products it is necessary to evaluate the phylogenetic distance to strains with a history of safe use and the distance to known pathogens. If a genus contains numerous species with a safe history of use and no pathogenic strains, it will be reasonable to assume that the genus is safe. If the genus contains known pathogens it is necessary to evaluate the case more carefully. If pathogenic strains of the same species are known we would advise against the use in food, even if the particular strain appears safe.

In evaluating the safety of bacterial strains it is also necessary to keep in mind that among the consumers an increasing number will be immunocompromised due to medication or infection.

2.5 Conclusions

In summary, the traditional classification methods for dairy bacteria which are discussed at length in the First Edition remain useful to dairy technologists as guides to gross genetic relatedness, morphology, surface properties and sugar fermentations; molecular phylogeny is proving valuable in confirming many of the more detailed biochemical, enzymological and metabolic differences between the genera and species of dairy bacteria, and facilitating the selection of the appropriate bacteria for particular technologies.

2.6 References

Alting, A.C., Engels, W.J.M., van Schalwijk, S. and Exterkate, F.A. (1995) Purification and characterisation of cystathione lyase from *L. lactis* subsp. *cremoris* B78 and its possible role in flavour development in cheese. *Appl. Environm. Biol.*, **61**, 4037–42.

Collins, M.D., Semelis, J., Metaxopoulus, J. and Wallbanks, S. (1993) A taxonomic study of leuconostoc-like organisms from fermented sausage: description of a new genus *Weissella*. *J. Appl. Bacteriol.*, **75**, 595–603.

Hardie, J.M. and Whiley, R.A. (1995) The genus Streptococcus, in *The Genera of Lactic Acid Bacteria*, Vol. 2 (eds B.J.B. Wood and W.H. Holzapfel), Blackie Academic & Professional, London, pp. 55–124.

Law, B.A. and Sharpe, M.E. (1978) The formation of methanethiol by bacteria isolated from raw milk and Cheddar cheese. *J. Dairy Res.*, **45**, 267–75.

Maidak, B.L., Larsen, N., McCaughey, M.J., Overbeek, R., Olsen, G.J., Fogel, K., Blandy, J. and Woese C.R. (1994) The Ribosomal Database Project. *Nucleic Acids Res.*, **22**, 3485–587.

Olsen, G. J., Woese, C.R. and Overbeek, R. (1994) Minireview: The winds of (evolutionary) change: Breathing new life into microbiology. *J. Bacteriol.*, **176**, 1–6.

Schleifer, K,-H., Ehrmann, M., Beimfohr, C., Brockman, E., Ludwig, W. and Amann, R. (1995) Application of molecular methods for the classification and identification of lactic acid bacteria. *Intern. Dairy J.*, **5**, 1081–94.

3 Microbiology and technology of fermented milks

A.Y. TAMIME and V.M.E. MARSHALL

3.1 Introduction

There are numerous fermented milk products which are manufactured in many countries of the world (Campbell-Platt, 1987; Kurmann *et al.*, 1992), but few are of commercial significance. Cheese-making and fermented milks production are one of the oldest methods practised by man for the preservation of a highly perishable and nutritional foodstuff (i.e. milk) into products with an extended shelf-life. The exact origin(s) of fermented milks making is difficult to establish, but according to Pederson (1979), fermented milks were produced some 10–15000 years ago as man's way of life changed from being a 'food gatherer' to a 'food producer'. This included the domestication of animals such as the cow, sheep, goat, buffalo and camel. It is likely that this transition may have occurred at different times in different parts of the world. However, archaeological evidence shows some civilizations (e.g. the Sumarians and Babylonians in Mesopotamia, the Pharoes in north-east Africa and the Indians in Asia) were well advanced in agricultural and husbandry methods, and in the production of fermented milks such as yoghurt.

It is possible that modern fermented milks production evolved as follows:

1. The methods of production manufacturing involved the constant use of the same vessels or the addition of fresh milk to an ongoing fermentation relying mainly on the indigenous microflora to sour the milk.
2. The heating of milk over an open fire to concentrate the milk slightly followed by seeding the cool milk (blood or ambient temperature) with previous day sour milk.
3. The preparation of these products with the use of defined microorganisms since the early 1900s.

Thus, the nature of these products differed from one area to another depending on the microflora used and climatic conditions of the region. For example, in the sub-tropical conditions in the Middle East, the thermophilic lactic acid microflora (optimum growth at 40–45°C) might be expected to be dominant over mesophilic species (optimum growth at 30°C) in northern Europe. In some regions, such as the Balkans and the Caucasian mountains, a range of products like kefir and koumiss were derived from lactic acid and yeast fermentation.

The manufacturing stages of fermented milks are still a complex process combining: microbiology and enzymology, chemistry and biochemistry, and physics and engineering. There is a considerable degree of similarity in technological aspects, and only some will be discussed. The fermentation processes of milk only involving specific microflora will be discussed in detail, and the physiology and the metabolic routes to flavour production will be covered in Chapter 5.

3.1.1 The diversity of fermented milks

Throughout the world around 400 names are applied to traditional and industrialized fermented milk products (Kurmann *et al.*, 1992). These products have different local names but are practically the same. A more accurate list would include only a few varieties, especially taking into account the type of milk used and the microbial species which dominate(s) the flora. According to Robinson and Tamime (1990) fermented milks can be divided into three broad categories based on the metabolic products, which are lactic fermentations, yeast–lactic fermentations and mould–lactic fermentations (Table 3.1). Closely related products are manufactured from fermented milks by: (i) de-wheying to concentrate the product which could resemble soft cheese (e.g. labneh, ymer or skyr); (ii) drying of cereal/

Table 3.1 Scheme of classification for fermented milk products. (Adapted from Robinson and Tamime, 1990.)

Category	Typical examples
I. Lactic fermentations	
A. Mesophilic	Buttermilk
	Cultured buttermilk
	Långofil
	Täetmjolk
	Ymer[a]
B. Thermophilic	Yoghurt, Laban, Zabadi
	Labneh[a]
	Skyr[a]
	Bulgarian buttermilk
C. Therapeutic	Biogarde®
	Bifighurt®
	Acidophilus milk
	Cultura-AB®
	Yakult
II. Yeast–lactic fermentations	Kefir
	Koumiss
	Acidophilus–yeast milk
III. Mould–lactic fermentations	Villi

[a]Concentrated fermented milk products.

fermented milk mixture (e.g. kishk or trahana); and (iii) freezing fermented milk to resemble ice cream.

As early as 1900 Conn (Sandine *et al.*, 1972) attached importance to bacterial growth in the development of aroma in cultured products. He concluded that cream ripening involved more than souring when he found that addition of acid did not accomplish the same results as bacterial growth. He suggested that acid and flavour were the results of different fermentations and that aroma was separate from flavour. The truth of these observations is still evident today when we make a careful choice in blending mixed starter strains to bring out the best flavour and aroma in our fermented dairy products. Fermentation of milk is carried out by microorganisms (bacteria, yeasts, moulds or combinations of these) which are capable of fermenting lactose predominantly to lactic acid. This is responsible for the sharp, refreshing taste of all fermented milks and although relatively non-volatile it serves as an excellent background for the more distinctive flavours and aromas characteristic of each fermented milk (Desmazeaud, 1990). Although the predominant metabolite is lactic acid, the minor metabolites are also vital for product quality and identity. Also important is a cultured milk of good consistency, and organisms producing polymers (ropy strains) may be added to improve texture, or may be necessary for starter integrity, e.g. kefiran and kefir grain formation. Consistency may be destroyed by gas production if fermentation is heterofermentative, but the presence of carbon dioxide is a requirement for the alcoholic milk beverages, kefir and koumiss, as it imparts a 'sparkling' character. All these organoleptic qualities are consequences of multiple fermentations and it is accepted that selection of the multi-strain or more often multi-species inoculum is necessary to obtain a good fermented product.

3.1.2 Annual production figures

There are no data available in the dairy industry to indicate the true world production for each type of fermented milk produced every year in each country. Table 3.2 shows total fermented milks production in some selected countries, and the pattern of production over the past decade is steadily increasing. The popularity of fermented milks in many countries may be due, in part, to tradition, but their characteristic properties (flavour, aroma, appearance, texture and possibly health aspects) are more attractive to consumers than fresh milk.

Production data for fermented milks for countries in the Middle East, North Africa, Balkans and the rest of the world are not available. However, yoghurt and strained yoghurt (e.g. labneh) are very popular in these countries, and in 1992 the world production figures for fermented milks could be estimated as ~ 20 M tonnes. Yoghurt and yoghurt-related

Table 3.2 Total production ($\times 1000$ tonnes) figures of different types of fermented milks in some selected countries. (Source IDF 1983, 1984a, 1993, 1994.)

Country	Buttermilk[a]		Yoghurt		Others	
	1982	1992	1982	1992	1982	1992
Austria	15.4	16.7	47.4	67.6	17.5	20.1
Australia	b	–	30.0	76.4	–	–
Belgium	27.9	22.4	47.8[c]	57.4	d	33.2
Canada	14.6	13.3	40.9	92.8	–	–
Switzerland	6.6	12.0	94.1	118.5	–	–
Chile	–	–	22.6	54.3	–	–
Czech and Slovak	72.8	24.5	31.5	52.9	41.2	79.1
Germany[e]	155.0	228.0	432.0	916.0	81.0	55.0
Denmark	52.0	33.3	48.0	44.9	34.0	37.2
Spain	–	–	199.0[c]	313.6[c]		
Finland	13.0	5.9	39.0	60.4	148.5	115.7
France	–	–	602.9[c]	989.2[c]		
UK	–	–	131.5	269.0	–	9.0
Greece	–	2.0	53.5[f]	60.0[g]	2.5	8.0
Hungary	0.3	0.6	11.0	16.2	75.3	14.0
Ireland	–	11.0	8.0	13.0	–	–
Israel	–	–	29.1	55.9	37.0	54.0
India	NR	a	2574.0[f]	3950.0[g]	–	–
Iceland	1.8	4.7	1.3	2.4	3.2	4.2
Italy	–	–	80.0[c]	270.0[c]		
Japan	–	–	283.3	533.0	148.2	476.0
USSR	–	–	–	–	1544.9	1411.8[g, h]
Luxembourg	NR	0.7[g]	1.9	2.8[g]	–	–
Netherlands	143.0	155.4	247.0	329.5	–	–
Norway	NR	113.7	12.9	27.8	51.6	43.2
Sweden	0.8	–	35.9	61.6	175.9	185.6
USA	423.7	368.0	283.0	491.0	–	51.0
Total	926.9	1012.2	5387.6	8926.2	2360.8	2597.1

[a]Figures are for natural and cultured buttermilk (refer to text).
[b]Indicate no figures or product is not manufactured.
[c]Yoghurt figures include other types of fermented milks.
[d]Empty spaces indicate combined production figures with yoghurt.
[e]Figures for 1982 represent Federal Republic only.
[f]Production figures for 1981.
[g]Production figures for 1991.
[h]Indicates production figures for Russia and Lithuania only.
NR, not reported.

products are the most widely manufactured fermented milks in the world (Table 3.2).

Production data for buttermilk in most countries is not well classified because (i) traditional or natural buttermilk is the by-product during the manufacture of ripened cream butter; (ii) cultured buttermilk is manufactured from skimmed milk with the addition of butter flakes; and (iii) buttermilk may not be fermented (i.e. sweet). The latter aspect may be true for the production data in India. In 1991 it was reported that butter milk production was 19.5 M tonnes (IDF, 1993), with no indication of whether it was sweet or traditional. Such production data is not included in Table 3.2 because figures for 1981 were not reported.

3.2   Microbiology of fermented milks

3.2.1   *Starter organisms and types*

A hundred years ago starter cultures were unknown, but throughout this century dairy microbiologists have unravelled the mixes of organisms responsible for much of the flavour and aroma. Table 3.3 lists the many organisms which have been isolated from various fermented milks, together with their metabolic products. This gives a clue to their role in flavour production. Some of the starters are mesophilic, others are thermophilic and some may contain desirable yeasts and mould (Marshall, 1993). Recently, a standard of identity of lactic acid starter cultures, which are used for the manufacture of fermented milks and cheese, has been drafted as a **Standard** by the International Dairy Federation (IDF, 1991). For further characteriz-ation and for more information about distinguishing features the reader is referred to Lodder (1970) and *Bergey's Manual of Systematic Bacteriology* (Sneath *et al.*, 1986).

Organisms are selected and combined depending on their use. For example fast acid-producing, thermophilic organisms are used for yoghurt manufacture; buttermilk requires a slower rate of acid production, and for kefir a gentle release of carbon dioxide at low temperatures is necessary. Starters of mixed species are now frequently available commercially, but analysis of growth, balance and metabolic activity is essential if the products are to be manufactured with consistently good quality. Standards for propagation by today's starter manufacturers need to be high. Some starters are more sensitive to poor sub-culturing than others. For example, lactic acid production by mixed buttermilk cultures is poor if incubation is too long before harvesting and making the bulk starter. Lactic acid accumulates most rapidly during the exponential phase of growth and further incubation results in loss of the lactococci relative to the leuconostoc. Sensitivity to lactic acid or nisin of some species of lactococci is well established, and if mixed starters are to contain these, then attention must be given to the time of sub-culturing and harvesting when preparing starters to ensure that the correct balance is maintained. Yoghurt starters would appear to be more stable to changes of balance. Quite different ratios of rods:cocci can be tolerated, the unbalanced culture soon making up the deficit, and there is a

Table 3.3 Starter cultures for milk fermentations and their principal metabolic products

Starter organism	Important metabolic products
I. Lactic acid bacteria	
A. Mesophilic	
Lactococcus lactis subsp. *lactis*	L(+) lactate
biovar. *diacetylactis*	L(+) lactate, diacetyl
subsp. *cremoris*	L(+) lactate
Leuconostoc mesenteroides subsp. *mesenteroides*	D(−) lactate, diacetyl
subsp. *cremoris*	D(−) lactate, diacetyl
subsp. *dextranicum*	D(−) lactate, diacetyl
Pediococcus acidilactici	DL lactate
B. Thermophilic	
Streptococcus thermophilus	L(+) lactate, acetaldehyde, diacetyl
Lactobacillus delbrueckii subsp. *delbrueckii*	D(−) lactate
subsp. *bulgaricus*	D(−) lactate, acetaldehyde
subsp. *lactis*	D(−) lactate
Lactobacillus fermentum	DL lactate
Lactobacillus helveticus	DL lactate
Lactobacillus kefir	DL lactate
Lactobacillus kefiranofaciens	DL lactate
C. Therapeutic	
Lactobacillus acidophilus	DL lactate
Lactobacillus paracasei subsp. *paracasei*	L(+) lactate
biovar. *shirota*	L(+) lactate
Lactobacillus rhamnosus	L(+) lactate
Lactobacillus reuteri	DL lactate, CO_2
Bifidobacterium adolescentis	L(+) lactate, acetate
Bifidobacterium bifidum	lactate, acetate
Bifidobacterium breve	L(+) lactate, acetate
Bifidobacterium infantis	lactate, acetate
Bifidobacterium longum	L(+) lactate, acetate
II. Miscellaneous bacteria	
Acetobacter aceti	
III. Yeasts	
Candida kefyr	Ethanol, CO_2
Saccharomyces unisporus	
Saccharomyces cerevisiae	Ethanol, CO_2
Saccharomyces exiguus	
Kluyveromyces marxianus	
VI. Moulds	
Geotrichum candidum	

self-regulatory capacity of a mixed lactococcal starter when grown in continuous culture (Marshall, 1986; Juillard *et al.*, 1987). Methods of propagation are therefore important (Tamime, 1990; Anon., 1995).

To keep and maintain a good starter for a particular fermented product it is essential to know what is expected in terms of flavour and aroma. This knowledge is not always available and many products are poor because of this. Knowledge of the biochemical pathways leading to flavour production can help in making the right choice of starter. This is the area in which the dairy microbiologist can make a great contribution. Chapter 4 attempts to cite some examples of what can be expected of a given organism in terms of its ability to produce flavour when grown in milk.

3.2.2 Technology of manufacture

Over the past few decades there have been many publications on the technology of fermented milks. Tamime and Robinson (1988) reviewed the various technological aspects of fermented milks under one heading because of the universal popularity of yoghurt. It is appropriate, therefore, to consider a similar approach in this chapter, and only the technological development over the past decade will be discussed in detail. The following are recommended for further reading (Rasic and Kurmann, 1978; De, 1980, Kosikowski, 1985; Tamime and Robinson, 1985; Chandan, 1989; Robinson and Tamime, 1990, 1993; Anon., 1995). Periodically, the International Dairy Federation publishes monographs updating the technological and scientific aspects of fermented milks (IDF, 1984b, 1986, 1988, 1992). The principal stages of production of yoghurt and fermented milks are shown in Figure 3.1. Some major problems in large-scale production include de-wheying, texture and consistency or firmness, and the manufacturing stages described in the subsequent sections help to overcome these problems.

(a) Milk reception. Most dairy factories use liquid milk as the main ingredient, mainly delivered in bulk road tankers. The reception tests applied to milk (i.e. chemical and bacteriological) ensure compliance with existing legal standards. Harmonization of milk and dairy products regulations within the European Union (EU) has been enforced since the 15th of June 1995. These regulations are similar in principle and intent worldwide in that the reception tests will relate to the compositional quality of milk, the microbial count, somatic cell count, freedom from antibiotic residues and the reception temperature.

(b) Standardization of the fat content. The major market share of natural and fruit yoghurts in most countries is taken by low-fat products; however, there is also a clear market for full-fat yoghurt. In the UK, low-fat and skimmed yoghurts contain ~ 1.5 and $< 0.5\,\mathrm{g}\,100\,\mathrm{g}^{-1}$ fat, respectively (Tamime et al., 1987), and full-fat yoghurt ranging between 3.5 to $\sim 5.0\,\mathrm{g}\,100\,\mathrm{g}^{-1}$. The Greek-style or strained yoghurt may contain $\sim 10\,\mathrm{g}\,100\,\mathrm{g}^{-1}$. The next step in manufacture involves passage of milk through a centrifugal separator to give skimmed milk and cream. The latter component is used for an automatic in-line standardization of the fat content to the desired level in order to comply with existing statutory instrument in any one country. Comparative study of the compositional

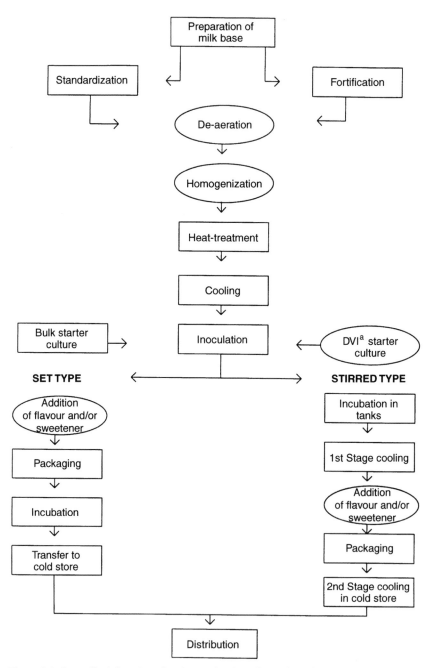

Figure 3.1 Generalized flowchart for the production of set/stirred fermented milks. Steps in ovoids are optional (refer to text). [a]Direct-to-vat inoculation.

standards for yoghurt in the EU member states has been reported by Pappas (1988).

(c) Fortification of milk solids not fat (MSNF). The main objective of fortification of the MSNF content of the milk base is to produce a viscous product, and in particular to increase the level of protein. The average composition of commercial yoghurts is around 14% MSNF, of which ~5% is protein and ≤1.5% is fat. The methods available to increase the MSNF are:

- boiling of the milk (i.e. not used commercially)
- addition of dairy powders
- concentration by evaporation under vacuum (EV)
- concentration by using membrane filtration, for example, ultrafiltration (UF) or reverse osmosis (RO)

Dairy powders. The incorporation of skimmed milk powder (SMP) with a high shear mixer is widely used method for increasing the MSNF in the milk base at ~40°C (Tamime and Robinson, 1985; Bøjgaard, 1987; Robin-

Table 3.4 Comparison of compositional quality (g 100 g^{-1}) of different powders used for the manufacture of fermented milks[a]

Powder	Protein	Lactose	Fat
Whole milk			
Commercial	26.3	39.4	26.3
Retentate[b]	41.7	9.3	41.7
Skimmed milk			
Commercial	36.1	52.9	0.6
Retentate[b]	62.8	23.9	0.9
Retentate[b]	74.3	11.8	1.4
Retentate[b,c]	80.5	5.5	1.5
Whey			
Commercial	12.2	78.0	1.3
Demineralized	14.5	80.5	1.0
Protein concentrate[b]	35.0	55.0	d
Protein concentrate[b]	73.2	12.0	0.2
Buttermilk	34.1	51.0	5.0
Na-caseinate	87.3	–	0.2

[a]Data compiled from Tamime and Robinson (1985), Holland *et al.* (1989), Tamime and Kirkegaard (1991) and Guinee *et al.* (1994).
[b]Ultrafiltration and diafiltration processing were used to reduce the lactose level in the retentate before drying.
[c]Commercial product by DMV bv, Veghel in The Netherlands.
[d]Constituent not present.

son and Tamime, 1993; Anon., 1995). However, other types of dairy powder could be used; Table 3.4 shows the differences in the chemical composition of these dry raw ingredients. The fortification target is to increase the level of protein in the milk to $\sim 5\,g\,100\,g^{-1}$. The use of UF retentate powder (whole milk, skimmed or whey protein concentrates) raises only the protein and/or fat levels (El-Samragy et al., 1993a, b), but not the lactose (Table 3.4). The final product will have lower lactose content when compared with milk fortified with whole milk powder, SMP or buttermilk powder. These powders should be free from antibiotics as all dairy starter cultures are sensitive to their presence (Cogan, 1972; Soback, 1981; Park et al., 1984; Orberg and Sandine, 1985; Ramakrishna et al., 1985; Juillard et al., 1987; Larsen and Anón, 1989; Yondem et al., 1989).

The fortification with whey protein powders should be limited to 1–2% because it may cause flavour defects (Tamime and Robinson, 1985), due to cysteine residues of β-lactoglobulin which increase the sulphydryl content in the milk after heating to give an oxidized off-flavour (Marshall, 1986). Also, high heating coagulates whey protein concentrates which may cause problems, and in some instances the milk/whey mixture is only heated to $\sim 80°C$ for 30 min for the production of yoghurt (Abd Rabo et al., 1988). Firmer gel was obtained after fermentation by either mesophilic or thermophilic lactic acid bacteria when the basic milk was fortified with 4% whey protein concentrates (Jelen, 1993).

The specifications of the SMP for recombination are important, and can influence the quality of the fermented milk product. The specifications first proposed by the American Dry Milk Institute (ADMI) [currently known as American Dairy Products Institute (ADPI)] have been accepted internationally, and the latest specifications have been published by ADPI (1990). Critical reviews of dairy powder specifications including an update of standards have been reported by Sjollema (1988) and Kjaergaard-Jensen (1990). According to Wilcek (1990), some specific requirements of SMP used for recombination are: (i) whey protein nitrogen index, 4.5–5.9; (ii) cystein number, 38–48; (iii) thiol number, 7.5–9.4; and (iv) heat number 80–83. Thus, these specifications classify the powders as 'medium' heat which is recommended for the production of fermented milks.

Other dried ingredients that may be added at the recombination stage include sugar and stabilizers. These additives are governed by statutory regulations (Pappas, 1988).

Concentration of the milk base.   This is performed by EV, RO or UF. In some countries, for example Denmark, the addition of powder(s) is not permitted, and the basic milk is concentrated under vacuum (EV) or by using membrane technology (RO or UF). In EV or RO water is removed from the milk, and the milk is concentrated to $14–15\,g\,100\,g^{-1}$ total solids (TS), and for UF to $\sim 12\,g\,100\,g^{-1}$ TS maintaining the protein level at

$\sim 5\,g\,100\,g^{-1}$ (Tamime and Robinson, 1985). The EV equipment (i.e. single effect) is part of the milk-processing installation of a fermented milk production line, while RO and UF equipment are installed separately (Anon., 1995). The application of membrane technology in the dairy industry has been reviewed extensively by Renner and Abd El-Salam (1991).

Irrespective of the method used, the increase in TS has little detrimental effect on starter culture's metabolism, but it increases the buffering capacity of milk, and it may promote growth of some microorganisms such as *Lactobacillus acidophilus* and *S. thermophilus* (Marshall, 1986).

Cheese-making from UF milk (i.e. 3- to 5-fold concentration) can affect the starter cultures activity due to the high buffering capacity of the retentate. Some of the recommended precautionary steps include: (i) the use of highly active starter culture; (ii) increase the inoculation rate; and (iii) addition of nutrients to UF retentate (Tamime and Kirkegaard, 1991). This does not arise during the manufacture of UF-fermented milks except when producing yoghurt 'cheese' from UF retentate, known in the Lebanon as labneh anbaris, containing $> 35\,g\,100\,g^{-1}$ TS.

(d) Filtration and de-aeration. Filtration of the recombined dairy powders is a recommended procedure for the removal of undissolved and/or scorched particles from the milk base. This can be achieved by using: (i) an in-line conical filter fitted inside the stainless pipe-line; this method of filtration is only suitable for small-scale production lines; or (ii) centrifugal clarifiers or duplex stainless steel or nylon filters are used if large volumes of milk are involved. Illustrations of recombination equipment and filtering method have been reported by Robinson and Tamime (1993). The reasons for removing undissolved particles from the milk are to reduce the wear and tear of the homogenizer orifice and minimize the deposition of milk solids in the plate heat-exchanger.

De-aeration of the milk base is also recommended in order to remove air which is incorporated during the recombination stage, to provide the appropriate growth conditions for the starter cultures (e.g. *Lb. acidophilus* and *Bifidobacterium* spp.), and to reduce heat-exchanger fouling.

(e) Homogenization. The basic process involves passing the milk at 60–70°C under high pressure (15–18 MPa) through a small orifice, and the shearing effect reduces the average diameter of fat globule to $< 2\,\mu m$. These small globules are less inclined to coalesce into larger units and rise to the surface. Homogenization of the basic milk is optional for the manufacture of viili and 'crusty' full-fat yoghurt. As a result, the cream rises to the surface during fermentation and, on cooling, sets to give these products their well-defined physical characteristics.

Before applying any processing to any type of mammalian full-fat milk, there are no interactions between the major milk components, i.e. the proteins (β-lactoglobulin (B-Lg), α-lactalbumin (α-La) and the caseins), the fat and lactose (Walstra and Jenness, 1984). The fat constituent in raw milk is encapsulated within a membrane made of protein, lipids and phospholipids (Mulder and Walstra, 1974). Heat- and pressure-induced processing cause chemical and physical changes in the milk fat globules. The former change is due to the fatty acids residue; however, the effect of homogenization and heating results in complex interactions between the milk components. Mulder and Walstra (1974), Dalgleish and Sharma (1993), Sharma and Dalgleish (1994), and van Boekel and Walstra (1995) have reviewed the effect of these physical changes on the quality of many dairy products. The possible changes applicable to homogenization at $\leqslant 70°C$ before heat treatment are:

- Fat surface area is increased, the size of the globule is decreased, and the composition of the membrane is different.
- Adsorption of surface-active material (mainly proteins) occurs, in part, on the fat surface.
- The turbulent effect of homogenization favours mainly casein micelles adsorption over serum protein ($\sim 5\%$) covering 25% of surface area of the fat globule.
- In recombination (i.e. the milk fat is homogenized into the skimmed milk) the resulting fat globule membrane consists only of serum protein.
- The homogenized fat globules act as large casein micelles (i.e. because the membrane consists mainly of caseins) which increases the effective casein concentration, and hence, participates in caseins reaction such as acid precipitation.
- The increased number of small fat globules enhances the ability of the milk to reflect light, and as a result the fermented milk appears 'whiter'.
- The risk of syneresis (i.e. separation of free whey onto the surface of set fermented milk) is reduced, and the firmness of the end-product is increased.

(f) Heat treatment. Heating of milk is the most widely used unit operation in the manufacture of dairy products. The temperatures applied range from $\sim 50°C$ (thermization) to $150°C$ for ultra-high temperature (UHT) sterilization processes. This subject has been extensively studied in relation to the functional properties of dairy products, for example, the heat-stability of UHT milk, evaporated milk or milk powders. The heat treatment of the milk base for the manufacture of fermented milk can vary depending on time and temperature combinations. Some examples are: (i) 30 min at 85°C; (ii) 5 min at 90–95°C; and (iii) 3 s at 110°C (Tamime and Robinson, 1985). Incidentally, these heating temperatures (i.e. 85–120°C) are similar to the preheat treatment of milks for the manufacture of evaporated milk and heat-stable powders. Only the heat-induced changes in milk, which are relevant to the functional properties of fermented milks, will be discussed.

At temperatures above 70°C, physical and chemical changes that can occur in the milk base are complex and multifunctional. In many factories the formulation of the milk base differs and the processing conditions are dictated by the available plant specifications. The impact of processing are summarized below.

Microorganisms and indigenous milk enzymes. The heat treatment of the milk base is sufficient to kill most if not all, the vegetative cells of microorganisms associated with raw milk, but spore-formers will remain (Gilmour and Rowe, 1990). The destruction and/or elimination of pathogens will occur at 85–95°C. The bacterial load is reduced, and this ensures that the heated milk provides less competition and a good growth medium for the starter cultures.

Over 60 indigenous enzymes have been identified in raw milk. Some of these enzymes are heat-labile, while others can survive high heat treatment of the milk. Activities of milk enzymes have been used as useful indicators of diseases or physiological changes in the udder of the mammalian, of processing conditions applied to milk and of factors influencing the flavour and quality of dairy products. The survival of these enzymes has not been identified as a significant problem in the fermented milk industry, although some enzymes may be present. The role of these enzymes in dairying have been reviewed critically elsewhere (Fox, 1991; Farkye and Imafidon, 1995).

Heating of homogenized milk. Caseins are heat stable while the whey proteins (β-Lg and α-La) are denatured at the temperatures employed for the processing the milk base (Dannenberg and Kessler, 1988a, b; de Wit, 1990; Pearce, 1994; Law, 1995). α-Lg reacts with other milk components when denatured, while β-La undergoes heat-induced interactions only after severe heat treatment (Dalgleish and Sharma, 1993; Sharma and Dalgleish, 1994). Possible interactions are:

- Self-association of denatured β-Lg molecules to form larger aggregates (Xiong *et al.*, 1993).
- Associations between β-Lg and κ-casein involving hydrophobic interactions of exposed –SH groups as a result of heating the milk (Haque and Kinsella, 1988; Noh and Richardson, 1989; Dalgleish, 1990).
- Interactions between whey proteins and fat globule membrane proteins may not be explained solely by disulphide linkages (Kim and Jimenez-Flores, 1995) as one fat globule membrane protein (49 kDa) appeared extensively modified after heat treatment.
- Interactions between the κ-casein and the fat globule membrane proteins as a consequence of adsorption of the former to the fat globule surface; this may also result losses of triacylglycerols and a change in lipid content

upon heating milk at 80°C for 20 min (Houlihan *et al.*, 1992a, b; Singh, 1993; van Boekel and Walstra, 1995).

- Interaction of β-Lg with homogenized milk fat globule surface may displace the adsorbed micellar caseins (Dalgleish and Sharma, 1993; Sharma and Dalgleish, 1994).
- A binding of colloidal calcium phosphates and other ions by the caseins, although this shift in the ionic constituents is not critical in acid gel formation (Schmidt and Poll, 1986; Aoki *et al.*, 1990; Wahlgren *et al.*, 1990; Holt, 1995; Zhang and Aoki, 1995), but can cause a delay of the enzymic coagulation of milk during cheese-making.
- Aggregation of casein micelles into larger particles and also dissociation of casein micelles forming soluble caseins at 100°C or above (Singh, 1993).

The protein/fat interactions in recombined milk are dependent on many factors such as: (i) larger size milk fat globules lower the protein load which consists mainly of whey proteins; (ii) increasing the protein content in skimmed milk–the protein load increases to reach a maximum ~ 6 mg per m^2 of fat surface; (iii) alteration of the ratio of whey protein to casein in the skimmed milk decreases the adsorbed protein load slightly, but markedly influences its composition; (iv) the disintegration of casein micelles after the removal of colloidal calcium phosphate before recombination results in a decrease of the protein load on the fat globule surface and alters the composition of adsorbed protein; and (v) by increasing the SNF from 10 to 20% TS in recombined skimmed milk at pH 6.5 to 7.1 before heating, the extent of κ-casein dissociation increases, and a similar effect occurs when the milk is heated at 120°C for 2–11 min at pH 6.5 only (Singh and Creamer, 1991; Singh *et al.*, 1993).

Quantitative study and relative rates of irreversible denaturation of whey proteins (immunoglobulins, serum albumin/lactoferrin, β-Lg and α-La) of cows', goats' and sheep's milks on heating at 70 to 90°C has been recently studied by Law (1995). Differences were found in the relative amounts and concentrations of the individual whey proteins in the milk of the three species. Compared with cows' milk, sheep's milk contained considerably more whey protein (i.e. immunoglobulin and β-Lg contents were 0.22 and 0.66 g 100 g^{-1}, respectively). On heating, the order of denaturation of all species were immunoglobulin > serum albumin/lactoferrin $> \beta$-Lg > α-La. However, at 90°C the order of ease of denaturation of whey proteins was sheep > goat > bovine (Law, 1995). Figure 3.2 illustrates the pattern of whey protein denaturation of different mammalian milks. For further information on protein characterization and effect of heating refer to Law and Tziboula (1992, 1993), Law *et al.* (1993, 1994), Law and Brown (1994) and Brown *et al.* (1995).

Miscellaneous changes. Heating of recombined milk at different temperatures can either stimulate or inhibit the activity of starter cultures due

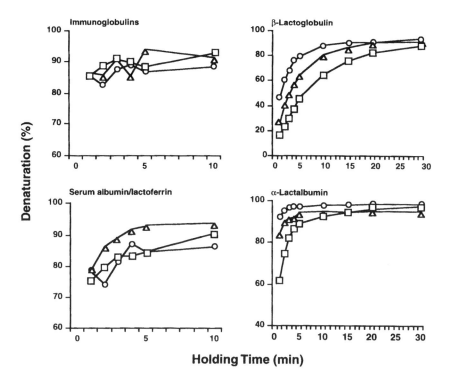

Figure 3.2 The effect of heating at 90°C on the extent of denaturation of different fractions of whey protein. ■, cows' milk; △, goats' milk; ●, sheep's milk. Results are means of three different milks. (Adapted from Law, 1995.)

to release of denatured serum protein nitrogen $(0.15-0.20 \, \text{mg ml}^{-1})$ or cysteine $(10-20 \, \mu\text{g ml}^{-1})$ (Greene and Jezeski, 1957a, b, c). The cycle of stimulation and/or inhibition is as follows:

- Stimulation between 62°C for 30 min and 72°C for 40 min.
- Inhibition between 72°C for 45 min, 82°C for 10–120 min and 90°C for 1–45 min.
- Stimulation between 90°C for 60–180 min and 120°C for 15–30 min.
- Inhibition at 120°C for > 30 min.

The same authors have recommended the use of high-heated powders; however, these observations may not be applicable at the present time in view of development in powder manufacture technology and selection of starter cultures strains.

Heating of milk may reduce the amount of oxygen present, lowering the redox potential, which encourages starter culture growth. Furthermore, improvement in the smoothness of the coagulum and reduction in syneresis is an important desirable consequence of heating milk at < 100°C (Robinson and Tamime, 1990). Preheating of the milk base may also destroy some heat-labile vitamins, but may improve its digestibility in the intestinal tract for some individuals when compared with unheated milk (Puhan, 1988).

Such heat-induced changes in the milk constituents during the heat treatment of the milk base can cause fouling of the surfaces of processing equipment. This in turn can influence the operational time of the heat-exchanger before cleaning is required. The denatured β-Lg plays a major role in the adsorption of the milk constituents onto the heating surfaces, and relatively few studies in the field have been reported (Kessler, 1981; Dannenberg and Kessler, 1988c; de Jong et al., 1992; Gotham et al., 1992; de Jong and van der Linden, 1993; Hinrichs and Kessler, 1995). The operational time of a plate heat-exchanger processing fresh liquid milk is longer when compared with heating recombined milks.

(g) Starter culture addition.   The heated milk base is cooled to the desired incubation temperature required in relation to the type of starter culture used. Some examples are:

- Yoghurt: at 40–45°C for $2\frac{1}{2}$–3 h, i.e. the short incubation method or at 30°C for 18 h, i.e. the long incubation method when using liquid culture; direct-to-vat inoculation (DVI) culture may require $\sim$5 h at 42°C.
- Probiotic: at 37°C for few hours or up to 2 days depending on the organism(s) used.
- Buttermilk: at 20–30°C for up to 10–20 h.

The starter inoculation rate may vary from 2–3%, 10% to 30% for the production of yoghurt, probiotic fermented milk products or koumiss, respectively (Berlin, 1962; Marshall, 1987; Tamime and Robinson, 1988; Tamime et al., 1995a).

(h) Gel formation.   Desirable gel formation in milk, mediated by coagulant, acid and/or a combination of salt and heat is widely used in the dairy industry for the manufacture of cheese and fermented milks. Comparatively, the coagulant-induced gelation in milk has been heavily researched, and this topic is reviewed in Chapter 1. It could be argued, however, that although the production of fermented milks does not involve addition of coagulant enzymes, proteinases originating from starter cultures may have a role. Hence, it should be understood that these products are not simply acid gels. Furthermore, the enzymes themselves may contribute to the denatured protein matrix which is also relevant to the gel properties of fermented milk products.

The main difference between acid- and coagulant-induced milk gels is that the permeability of the former gel does not change during the first 24 h after gelation while in coagulant-induced gels it increases continuously during the same period (van Vliet et al., 1989). Furthermore, the milk gel formed using coagulant enzymes is more robust when compared with acid-induced gel which is fragile and shatters very easily (Walstra and van Vliet, 1986).

Muir and Hunter (1992) have provided sensory attributes for the evaluation of fermented milks. Texture criteria, which include sensory attributes such as firmness, viscosity and extent of syneresis, have been identified as important for consumer acceptability. Although the microstructure of fermented milk products has been well studied, little published data are available on the mechanism(s) of acid-induced gelation in milk by starter culture at 20–45°C.

Casein micelles are composed of different protein fractions (α_{s1}-, α_{s2}-, β-, κ- and γ-) in combination with calcium phosphate (i.e. micellar or colloidal). It is established that during the fermentation of milk the calcium (and to a lesser extent magnesium and citrate) content in the serum increases as the pH is lowered (Pouliot $et\ al.$, 1989; Le Graet and Brulé, 1993) due to solubilization of the micellar Ca-phosphate. Alteration in the physical nature of the casein micelles will play a major role in acid-induced gelation of milk. Direct acidification of milk using HCl or glucono-δ-lactone (GDL) and the addition of calcium chelating agents are different techniques used to study the gelation of milk under controlled conditions without the metabolic interference of the starter cultures (Roefs $et\ al.$, 1985; Holt $et\ al.$, 1986; van Hooydonk $et\ al.$, 1986; Visser $et\ al.$, 1986; Roefs, 1987; Bremer $et\ al.$, 1990; Bringe and Kinsella, 1990, 1991; Banon and Hardy, 1991; Horne and Davidson, 1993).

Studies of casein micelle dissociation and aggregation during acid-induced gelation of milk suggest that the mechanisms involved are pH-, ion concentration- and temperature-dependent (Aoki $et\ al.$, 1986, 1987a, b, 1988; Holt $et\ al.$, 1986). Predominantly, β-casein dissociates from the casein micelles at low pH (van Hooydonk $et\ al.$, 1986); however, dissociation of other casein fractions (κ-, α_{s1}-, α_{s2}) from the micelles have been reported by Roefs $et\ al.$ (1985), Roefs (1987) and Dalgleish and Law (1988). The same authors also observed that the dissociated β-casein was more extensive than other caseins, and the amounts and proportions of the dissociated caseins to the serum were pH- and temperature-dependent. At pH 5.6 all major caseins were prone to dissociation, and the dissociation occurred at the outer rather than inner layers of the sub-micelles (van Hooydonk $et\ al.$, 1986). Solubilization of the micellar Ca-phosphate occurs at pH $\leqslant$ 5.3, and there is a linear relationship between $Ca^{2+} + Mg^{2+}$ and inorganic phosphate (P_i) + citrate (cit). The binding of ionic calcium and magnesium to casein appears independent of pH between 5.6 to 6.7. Calcium binding may involve carboxyl groups; however, a decrease in pH also affects spatial properties because of electrostatic interactions between positively and negatively charged groups (van Hooydonk $et\ al.$, 1986). Dalgleish and Law (1989) have observed a similar pattern of mineral solubilization due to pH-

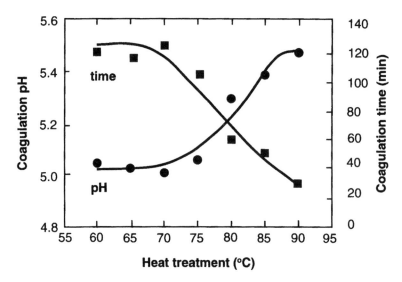

Figure 3.3 Effect of pre-heating temperature on the coagulation time (■) and coagulation pH (△) for acid-induced gelation of skimmed milk. Gelation was measured using diffusing wave spectroscopy (DWS). Glucono-δ-lactone (GDL) concentration was 1.5% (w/v), and incubation temperature was 30°C. All milks were heated for 10 min at the designated temperature. (Adapted from Horne and Davidson, 1993.)

and temperature-induced conditions, but could not suggest a universal relationship to describe the dissociation of salt ions and of caseins from the micelles. Lowering the pH reduces the repulsive forces, and allows for hydrophobic interactions causing the casein micelles to coagulate. However, pre-heating the skimmed milk to 90°C followed by acidification at 30°C using slow hydrolysis of GDL shifted the coagulation pH to a value higher than 5.5, and shortened the coagulation time (Horne and Davidson, 1993). Figure 3.3 illustrates the effect of heat treatment on coagulation time and pH.

Quiescent heating of casein solutions made with reconstituted SMP or Na-caseinate, and acidified at 0–2°C produced physically stable suspensions of casein particles (Roefs *et al.*, 1990a, b). Gelation occurs above 10°C, and lowering the temperature to 4°C after gel formation had the following characteristics: (i) the casein particles formed a complex irreversible structure; (ii) the acid-induced gel was formed subject to an activation Gibbs energy which decreased on increasing the temperature; (iii) if the gel was good at > 10°C, the dynamic moduli, G' and G″, linearly increased with the logarithm of time over at least a week; and (iv) the gel network consisted of large and small agglomerates of casein particle aggregates in the form of strands and nodes, with void spaces around 1–10 μm; this suggests that the strands and nodes are made of concentrated protein (~ 25%) with a modulus of about $10^5 \, \text{N}^{-1}$ (Roefs *et al.*, 1990 a, b).

Dissociation of casein micelles in milk has been induced by other means such as salt solutions ($CaCl_2$, $MgCl_2$ or NaCl) or calcium-chelating agents like EDTA, hexametaphosphate, oxalate, citrate or othophosphate (Holt *et al.*, 1986; Aoki *et al.*, 1988; Rollema and Brinkhuis, 1989; Bringe and Kinsella, 1991; Johnston and Murphy, 1992). Dialysis against phosphate-free and Ca-phosphate buffers decreased both the colloidal Ca phosphate and P_i depending on the type of buffer used before casein dissociation occurred. Holt *et al.* (1986) reported dissociation through break-up of linkages between the caseins and the inorganic components. However, the dissociation of the casein micelles in simulated milk ultrafiltrate dialysed against imidazole buffer is dependent on the ester phosphate content (Aoki *et al.*, 1988). The addition of Ca^{2+}, Mg^{2+} and Na^+ ions, which are associated with casein phosphates and carboxyl groups, tends to increase the hydrogen ion concentration due to reduced repulsive hydration forces between micelles. Hence, the attractive hydration forces cause coagulation because hydrogen ions displace bound Ca^{2+}, Mg^{2+} and Na^+ ions in the casein micelles. Ions (Cl, NO_3, Br and SCN) binding to lysine, arginine and histidine groups also decrease the repulsive hydration forces between colloidal ions of the casein micelles (Bringe and Kinsella, 1991).

The current published research on the mechanism(s) of direct acid addition to induce gelation of milk provide some limited knowledge; however, since the milk base for the manufacture of fermented milks is prepared in different ways and subjected to homogenization and high heat treatment, the properties of the fermented-induced gel may differ. Roefs *et al.* (1985) concluded that, because low-heat SMP was used in their study:

'because of the dependency of results on the history of samples, both in terms of pH and temperature, it will take painstaking studies to determine precisely what changes occur'.

(i) Cooling and miscellaneous handling.   Cooling is one of the most popular methods used to control the metabolic activity of the starter culture and its enzymes. Fermented milks may be of set-type where the fermentation stage takes place in the retail container or stirred-type where the milk is incubated in bulk (see Figure 3.1). The only difference between set- and stirred-type products is the rheological properties of the coagulum. The gel is either in the form of semi-solid continuous mass, while the stirred-type has the gel broken at the end of fermentation before cooling and further processing.

The process of cooling fermented milks may be carried out using one of the following approaches:

- One-phase cooling to $\leqslant 10°C$.
- Two-phase cooling to $\sim 20°C$, the final cooling taking place in a refrigerated cold store.

However, according to White (1995) the cooling of yoghurt entails four basic phases: (i) shock cooling from 42°C to 30°C; (ii) dysgentical stage to 20°C; (iii) lact-less-phase to 14.5°C; and (iv) holding phase at 2–4°C. This approach of cooling may be difficult to apply under industrial situations, unless stages (ii) and (iii) are combined where cooling of the packaged yoghurt takes place in a 'chill' tunnel (i.e. to ≤20°C) before transferring it to the cold store (Anon., 1995).

Little technological development has taken place in the past decade. The practical considerations for handling the coagulum including addition of fruit and/or flavours until it reaches the cold store have been well documented; for further reading refer to Rasic and Kurmann (1978), Tamime and Robinson (1985) and Robinson and Tamime (1990, 1993).

It is evident that all these sections have much in common, and thus, the technical aspects which will vary are: (i) the organisms that constitute the starter culture; (ii) temperature and period of fermentation; (iii) the inoculation rate; and (iv) production of bulk starter. These aspects will be considered in detail in subsequent sections. The technology of fermented milk products have been recently reviewed by Driessen and Loones (1990, 1992), Nilsson and Hallström (1990) and Puhan et al. (1994). A comprehensive scheme summarizing the new developments in technology including products with special microorganisms is as follows:

- Membrane techniques such as UF and RO provide the possibility of utilizing the required properties and of avoiding the unwanted properties of microbial metabolites.
- Separate cultivation of starter cultures provides the possibility of combining microorganisms needing different conditions for their proliferation, for example, mesophilic and thermophilic strains.
- By applying automatic pH control systems in the incubation tanks to end the fermentation process, more consistent products can be achieved.
- By applying a cooler on top of the filler, high viscosity can be achieved in stirred-type fermented milks.
- The manufacture of set-type fermented milks becomes more flexible by applying in-line inoculation.
- An overhead pressure of sterile air in the fermentation tank has been proved to be very effective in protecting starters against contamination with other microorganisms and bacteriophages.

3.3 Fermented milks with lactic acid

This type of fermented milk is the most popular product in the dairy industry (see Table 3.1). The starter cultures employed are mainly lactic acid bacteria, and some examples are shown in Table 3.3. In general, most of the fermented milks (e.g. yoghurt, ymer, dahi, buttermilk or therapeutic products) are made using mixed starter cultures, but sometimes a single strain organism is employed, for example, during the manufacture of Bulgarian buttermilk and acidophilus milk.

3.3.1 Use of mesophilic strains

Microflora of the genera *Lactococcus*, *Leuconostoc* and *Pediococcus* are used for the manufacture of products denoted as 'mesophilic' lactic acid fermentation. The term mesophilic is used for strains whose growth optima is ~20–30°C.

The starter cultures of the genus *Lactococcus* consist of *Lac. lactis* subsp. *lactis*, subsp. *cremoris* and *Lac. lactis* subsp *lactis* biovar *diacetylactis*. These cultures were previously known as 'mesophilic lactic streptococci' belonging to Lancefield group N (Lancefield, 1933). The microorganisms are characterized by spheres of ovoid cells occurring singly, in pairs or in chains, and are often elongated in the direction of the chain. The lactococci are Gram-positive, microaerophilic bacteria, produce L(+) lactate from lactose, and only *Lac. lactis* biovar *diacetylactis* produces diacetyl. The *Lactococcus* spp. do not have flagella, nor do they form endospores, and some strains (i.e. ropy) are able to excrete extracellular polysaccharide material (Teuber, 1995).

The *Leuconostoc* spp. (*Leu. mesenteroides* subsp. *mesenteriodes*, subsp. *cremoris* and subsp. *dextranicum*) are related phenotypically to the genera *Lactobacillus* and *Pediococcus*, and share many features with heterofermentative lactobacilli. They have complex nutritional requirements, some strains produce exopolysaccharide, and in spite of their wide practical applications they have not been studied extensively when compared with *Lactobacillus* spp. and *Lactococcus* spp. Some strains (*Leu. mesenteriodes* subsp. *cremoris*) have a long generation time (48 h at 22–30°C) when compared with *Leu. mesenteriodes* subsp. *mesenteriodes*, which have a short generation time with 24 h of incubation at 30°C. The cell morphology varies with growth medium, and in glucose medium they are elongated resembling lactobacilli. The cells are coccoid in shape, Gram-positive, asporogenus, non-motile, and occur singularly or in pairs which may form short to medium-length chains. The *Leuconostoc* spp. used as dairy starter culture mainly produce D(−) lactate and diacetyl when grown in milk (Dellaglio *et al.*, 1995).

Pediococcus acidilactici is the only strain of this genus to be used in dairy starter cultures. The pediococci divide alternately in two perpendicular directions to form tetrades which differentiate them morphologically from other lactic acid bacteria. The cells are invariably Gram-positive cocci and of uniform size, produce DL-lactate, are non-motile, do not form spores, and are not capsulated. This organism had other synonyms such as *Pediococcus lindneri*, *Pediococcus cerevisiae* and *Streptococcus lindneri* (Simpson and Taguchi, 1995).

(a) Traditional or natural buttermilk. Many countries favour ripened cream butter, and traditional or natural buttermilk is the by-product of butter-making after churning ripened cream. The starter culture is composed of a mixture of *Lactococcus* spp. and *Leuconostos* spp. This type of buttermilk is popular in northern European countries, and is consumed in place of fresh milk either as a beverage or with cereal (Marshall, 1987; Tamime and Robinson, 1988).

The heat treatment of cream for ripening is normally carried out at 90–95°C for 15 s or flash with no holding at 105–110°C. Such high temperatures denature β-Lg, exposing the –SH groups which act as an antioxidant and can enhance starter culture growth. Vacuum treatment of the cream during cooling is optional, possibly for the removal of cooked flavours; however, the production of lactic acid and aroma development by the lactococci and *Leuconostoc* spp. will mask such undesirable effects (Wilbey, 1994). The main sensory attributes of this type of buttermilk are: (i) taste (lactic acid or sour); (ii) aroma (buttery flavour due to diacetyl) and (iii) presence of some small butter granules.

It is possible to suggest that the worldwide production of this type of fermented milk is decreasing due to the fact that butter production is seasonal, and hence, such product will not be always available to the consumers. Also, use of the 'NIZO method' where starter culture distillate is injected into the butter in the working section of the buttermaker, means that there is no need to ripen the cream.

(b) Cultured buttermilk. This is a fermented skimmed milk ($\sim 9.5\%$ SNF) which is preheated, homogenized (17–20 MPa), heated at 85°C for 30 min or 95°C up to 5 min, cooled to 22°C and inoculated with mesophilic starter cultures (Anon., 1967; Lundstedt, 1975; Vedamuthu, 1978, 1985; White, 1978; Lundstedt and Corbin, 1983a, b; Kosikowski, 1984, 1985; Meriläinen, 1987; Rash, 1990). The SNF and fat contents in Californian buttermilk may range between 7.4 to 11.4 and 0.25 to 1.9%, respectively (Green *et al.*, 1992). The fresh skimmed milk is normally fortified with SMP, and if milk fat is added, two-stage homogenization (1st stage at 13.8 MPa, 2nd stage at 3.5 MPa) is recommended (Kosikowski, 1984). The starters used in the manufacture of buttermilk consist of combinations of two types: *Lac. lactis* subsp. *lactis* and subsp. *cremoris* are the main lactic acid producers and *Lac. lactis* biovar *diacetylactis* and *Leu. mesenteroides* subsp. *cremoris* are responsible for the aroma or flavour producton (Hammond, 1969; Kosikowski, 1984; Vedamuthu, 1985; Marshall, 1986). The importance of mixed cultures for buttermilk production was reported by Babel (1967), and recently a commercial starter culture supplier is providing a Direct Vat Inoculation (DVI) frozen concentrated culture comprising selected strains of *Lactococcus* spp. and *Leuconostoc* spp. for acid-, flavour- and viscosity-producing microflora (Anon., 1992). Such cultures have to be hydrated for 10–15 min

in the processed milks, followed by incubation at 22.8°C for 12–14 h in order to reach a pH of 4.65.

Flavour defects in buttermilk are mainly attributed to factors influencing the activity of the microflora, including the oxygen and citrate levels. Slow and very fast acid development by the starters may not provide for diacetyl flavour to develop (Frank, 1984). *Lac. lactis* biovar *diacetylactis* and *Leu. mesenteroides* subsp. *cremoris* are able to metabolize citrate in milk to synthesize diacetyl with CO_2 as a by-product, and when the citrate level is depleted, the diacetyl is reduced to acetoin. The addition of 0.05–0.1% citric acid to the milk has been recommended by Hammond (1969), Vedamuthu (1978, 1985) and Frank (1984). Excess accumulation of acetaldehyde by some lactococci causes an undesirable characteristic known as 'green' flavour; the inclusion of *Leu. mesenteriodes* subsp. *cremoris* helps to catabolize it (cited by Marshall, 1984, 1986). Some *Lactococcus* spp. (psychrotrophes and mesophiles) are able to reduce diacetyl in buttermilk during storage at refrigerated temperatures, and strain selection of starters is important (Hogarty and Frank, 1982). Plasmid profiling of commercial mixed starter cultures (*Lac. lactis* subsp. *lactis*, biovar *diacetylactis* and subsp. *cremoris*) before and after fermenting the milk were studied by Thompson and Collins (1989), and suggest that the strains predominant in the fermented milks differed from the original culture. Strains which appeared to have lost plasmids during storage were less competitive than those with full plasmid complement. Such results may indicate a possible competitive advantage of certain cryptic plasmids (Thompson and Collins, 1989). It may be possible to suggest that the loss of plasmids in certain lactococci strains during the storage of fermented milk may be advantageous to minimize the catabolism of diacetyl.

Increasing the oxygen content in milk allows maximum production of diacetyl (Frank, 1984). Incorporation of air during the cooling stage leads to reduced viscosity and whey syneresis; this method is not recommended and citrate addition is preferred. Rapid cooling is not desirable since it arrests acid accumulation and loss of diacetyl flavour (Vedamuthu, 1985).

To mimic the traditional buttermilk, incorporation of butter granules can be achieved by using one of the following methods: (i) the addition of freeze-dried butter flakes or granules; (ii) the use of churning methods; and (iii) dripping melted cream or anhydrous milk fat into cold buttermilk (Anon., 1967; Hammond, 1969; Kosikowski, 1984, 1985). The addition of salt at a rate of 0.1% is optional, and it is normally added during the cooling stage. Alternatively, potassium chloride could be used to replace sodium chloride without significantly affecting the taste or acceptability of the product (Demott *et al.*, 1986). Off-flavour development due to light irradiation is not detectable in buttermilk (Hoskin, 1989).

Advances in cultured buttermilk technology may include progress in the following scientific fields.

1. The application of UF technology which can increase the calcium content and improve the viscosity of buttermilk (Mann, 1986).
2. Development in methods of production by using direct acidification and citrate fermentation using only *Leu. mesenteriodes* subsp. *cremoris* (Walker and Gilliland, 1987) or combinations of fermentation and direct acidification of the milk (Gettys and Davidson, 1985); however, the direct acidification process may not be acceptable in view of the therapeutic properties of fermented milk and possible loss of diacetyl during storage.
3. Developments in Holland for the production of stable diacetyl content have been achieved either using interactive fermentation system by means of a membrane dialysis or genetic engineering to increase production of α-acetolactate from which diacetyl is derived in *Lactococcus* spp. (Klaver *et al.*, 1992b; de Vos, 1993).
4. Reduction of buttermilk consumption in the USA over the past two decades is compensated by increase in yoghurt consumption, and by using different blends of mammalian milks and flavouring the buttermilk (Bachmann and Karmas, 1988) or the use of probiotic organisms in the starter culture may revive the popularity of buttermilk among consumers.

Products closely related to buttermilk are produced in different countries. For example, Saya® is a carbonated ($\sim 0.4\%$ CO_2) German fermented milk product made from a mixed starter of *Lac. lactis* subsp. *lactis* and two strains of *Leuc. mesenteroides* subsp. *dextranicum*. Protease enzymes are added with the starter culture, and the product is enriched with vitamins A, B and C (Kurmann *et al.*, 1992). Junket or Ylette® are produced in Denmark using the following blend of mesophilic lactic acid bacteria: *Lac. lactis* subsp. *cremoris* (75%), *Lac. lactis* biovar *diacetylactis* (20%) and *Leu. mesenteriodes* subsp. *cremoris* (5%) (Larsen, 1982).

(c) Nordic or Scandinavian sour milks.   The traditional buttermilk products in Norway, Sweden and other neighbouring countries are slimy or ropy; herbs are also used to produce thick fermented milks. These products are known as långfil (the latter translated means long milk), tättfil, långmjölk, filmjölk, täetmjölk, tätmjölk and filbunk (which is made from full-fat milk to be consumed immediately). These products were made locally on farms where tradition has it that in addition to the indigenous undefined mesophilic lactic acid bacteria present in milk, the inoculum was enriched by rubbing the interior of milk pails with leaves of locally grown butterwort (*Pinguicula vulgaris*). In some instances the leaves of *Dorsera* spp. are used, and both plants are known locally as täätegräs or tätgräs (meaning thickening grass); thus, täetmjölk or tätmjölk means ropy or thick milk. The butterworth is known as hleypisgras or lyfjargas (in Iceland), Unduløvugréás (in Faroes) and yirni' girse (in Shetland) (Fenton, 1976; Alm and Larrson, 1983; Bertelsen, 1983). The use of such grasses may introduce

a slime-producing microorganism (*Alcaligenes viscosum*) whose natural habitat is the leaf, but according to Nilsson and Nilsson (1958) it is not used today to produce långfil. Other closely related products known as viili (in Finland), ymer (in Denmark) and skyr (in Iceland) will be discussed under separate sections because viili contains a mould *Geotrichum candidum*, and ymer and skyr are concentrated products.

The role of the microflora is important in any fermented dairy product, and the nature of the slime-producing organism(s) has been debated for years in the Scandinavian countries. *Bacterium lacticus longi* was isolated in 1899 from långfil (cited by Marshall, 1986), whereas other researchers have identified the cultures as variants of *Lac. lactis* subsp. *lactis* and subsp. *cremoris* capable of producing slime (Macy, 1923; Sundman, 1953; Macura and Townsley, 1984), and have suggested the name *Lac. lactis* biovar *longi*. A major problem for the commercial manufacture of this milk is that the ropiness is easily lost. Many producers feel that if they could rely on a consistent slime-producing starter, their product would achieve the quality needed for increased sales. The slime production and stability has been shown to depend on temperature and time of transfer when propagating cultures. The ropy variants excrete slime up for to 24 h during growth in milk at 18°C; viscosity stays constant for the next 24 h and then declines (Macura and Townsley, 1984). Recent studies on the isolation and characterization of capsule and/or slime production by mesophilic lactococci have been reported by Kontusaari *et al.* (1980), Forsèn *et al.* (1985, 1989), Neve *et al.* (1988), Toba *et al.* (1991a) and Nakajima *et al.* (1990); for further discussion refer to Chapter 4.

Studies carried out by Linné in 1980 (Alm and Larsson, 1983) on ropy milk made by the traditional process resulted in a flavourful product with an apple-like aroma, a pH of 4.4 with a good consistency, and no sign of syneresis. The shelf-life was up to 3 months at 8°C and milk sugar was reduced by one-third, with the folic acid content much higher than for other fermented milk products. The total viable count was 3.4×10^8 cfu g^{-1}, and that of the aroma-producing bacteria was about 1.1×10^7 cfu g^{-1}.

The Nordic sour milk products, for example filmjölk, are manufactured from fat-standardized milk ranging from 0.5–3.0% (Anon., 1995). The SNF content of the milk is not normally fortified, and the full-fat milk base is pre-heated to 78°C and deaerated in a vacuum vessel where the temperature drops to ~ 70°C. The milk is transferred to a separator for direct standardization of the fat level desired in the end-product, followed by homogenization at a pressure of 17.5–20.0 MPa at 70°C. The milk is heated to 90–95°C and held at this temperature for 3–6 min, cooled to 20°C, inoculated with starter culture (1–2% bulk starter) consisting of a mixture of *Lac. lactis* subsp. *lactis*, subsp. *cremoris*, biovar *diacetylactis* and *Leu. mesenteriodes* subsp. *cremoris* including slime-forming strains, (Kahala *et al.*, 1993), agitated for 10 min and fermented for 20 h at 20°C. At 0.8–0.9% lactic acid the coagulum is cooled, packaged and moved to the cold store. Breaking the filmjölk at lower acidity results in whey separation, and if the milk base is not deaerated the typical faults are granulation, lumpiness and reduced consistency (Anon., 1995). By contrast, täetmjölk has a shelf-life of up to 10 months, and contains $\sim 1.0\%$ lactic acid, 0.3–0.5% alcohol, traces of acetic acid, and is saturated with CO_2 (Oberman, 1985).

The increase of folic acid content in the Nordic ropy milk is mainly attributed to the starter culture activity rather than the presence of tätgräs in the milk. Alm (1982a, b) has studied the effect of fermentation on vitamin B content of milk and fermented milk products in Sweden, including ropy milk without the addition of tätgräs. Such a product was manufactured by Alm (1982b) from cows' milk (3% fat, w/v) which was heated to 90°C for 3 min, cooled, inoculated with a mixed culture of *Lac. lactis* biovar *longi* and *Leu. mesenteriodes* subsp. *cremoris*, and incubated at 17–18°C for 20–22 h (pH range 4.5–4.6). The reported increase in the folic acid content of this type of ropy milk was more than 2-fold after 1 day's fermentation. Similar increases in the level of folic acid content of other fermented milks has been reported by different researchers in many parts of the world.

(d) Cultured cream.   Cultured cream, sometimes known as sour cream, is highly viscous product with a flavour and aroma of buttermilk. The fat content in the cream varies from 10–12% or 20–30%, and the method of consumption of cultured cream is more akin to that of normal cream. The product should have a pleasant acidic taste, 'buttery' aroma, and a nutty meat-like flavour.

A typical method for production would involve standardization of whole milk with cream to the desired fat content or fortification of the cream with 2–3% milk solids. Optionally, citric acid or sodium citrate is added to enhance the metabolic activity of the starter culture (Anon., 1987). The cream base is warmed to 60–70°C, and for low-fat cream (10–12%) it is homogenized at 15–20 MPa, while lower pressure (10–12 MPa) is used for cream with fat content of 20–30% (Anon., 1995). In general, increases in homogenization temperature improve the consistency of the retail product. The homogenized cream is heated to 80°C for 30 min or 90°C for 5 min. After cooling to 18–21°C, the cream is inoculated with 1–2% bulk starter culture, and fermented for 18–20 h to an acidity of 0.8%. The starter culture consists of *Lac. lactis* subsp. *lactis*, subsp. *cremoris*, biovar *diacetylactis* and *Leu. mesenteriodes* subsp. *cremoris*.

After fermentation, the viscosity of the cultured cream is very high, but becomes thinner during mechanical treatment, cooling, pumping and packaging. For this reason some manufacturers produce a set-type product where the cream is incubated in the retail container in order to produce a markedly thicker product. Chymosin may be added in small quantities with the starter cultures to produce a thicker body sour cream gel. Excess amount of coagulant leads to a rough or grainy texture and increases syneresis due to hydrolysis of κ-casein and/or κ-casein/β-Lg interaction (Kosikowski, 1985; Lee and White, 1993). Alternatively, improvement in the body and texture of cultured cream can be achieved by the addition of stabilizers (Hylmar, 1976; Lee and White, 1992, 1993) or by UF of buffalo's milk (Hofi, 1990). In the latter case, the UF cultured cream was fermented with a mixed starter culture (*Lac. lactis* biovar *diacetylactis* and subsp. *cremoris*) at a ratio of 2:1. However, Yugoslavian cultured creams from eight different manufacturers contained yoghurt microflora beside the *Lactococcus* spp., and the majority of the samples have been contaminated with *Bacillus* spp., *Micrococcus* spp., *Escherichia coli*, mould and *Odium lactis*, suggesting a poor manufacturing process (Terzid-Vidojevid, 1991).

Developments in cultured cream technology may include (i) high-heat treatment after fermentation followed by homogenization and hot-packaging; a process similar to the manufacturing of Cream cheese (Kosikowski, 1985); and (ii) the use of *Lb. acidophilus* at a rate of 5% to ferment cream $(40–45\,g\,100\,g^{-1})$ and 5% of added sugar in the form of syrup (Kurmann *et al.*, 1992).

(e) Miscellaneous products.   A wide range of indigenous fermented milks are produced in rural areas of different countries (De, 1980; Campbell-Platt, 1987; Kurmann *et al.*, 1992; Driar, 1993; Mutukumira *et al.*, 1995). In the majority of these products spontaneous fermentation of milk is the normal practice, and gourds made from dried fruit of plants (*Lagenaria peucantha*) are widely used. The inside of the gourd is smoothed with glowing splints of the *Olea africana* tree (Kimonye and Robinson, 1991). In some countries these traditional products have been scientifically studied and commercially developed, and some examples follow:

Kenya.   Maziwa lala is a traditional fermented milk; the commercial product is called mala. The milk, which is unfortified, is heated to 85°C for 30 min or 90°C for 15 min, cooled to incubation temperature and inoculated with a mixed starter culture (*Lac. lactis* subsp. *cremoris*, biovar *diacetylactis* and *Leu. mesenteriodes* subsp. *cremoris*). Optionally, the fermented milk is flavoured with fruit juices, sweetened with the addition of sugar and stabilized with pectin, gelatin or Na-caseinate (Shalo and Hansen, 1973; Kurmann *et al.*, 1992; Wanghof *et al.*, 1992).

Iria ri matii is another traditional Kenyan fermented milk product, and laboratory development of such product was made successfully using 2% of *S. thermophilus* (Kimonye and Robinson, 1991). Susa is made from camels' milk and by using a heterofermentative mesophilic culture (B-CH: 40 from Chr. Hansen's Laboratorium in Denmark), the product was highly rated by consumers when compared with homofermentative type starter culture (Farah *et al.*, 1990).

Nigeria.   Nono is a traditional product made with undefined microflora in northern Nigeria. Fresh nono samples subjected to heat treatment indicated that the surviving lactic acid bacteria were *Lb. delbrueckii* subsp. *bulgaricus*, *Lb. helveticus*, *Lb. plantarum* and *Lac. lactis* subsp. *cremoris* (Atanda and Ikenebomeh, 1989a), and methods for flavour improvement of nono have been reported by Atanda and Ikenebomeh (1988,1989b). Also, the addition of 0.25 g benzoate per 100 g to nono extended its shelf-life at 4°C rather than 25°C (Olasupo and Azeez, 1992). Organoleptically acceptable nono has been produced by using mixed starter culture (*Lac. lactis* biovar *diacetylactis*, subsp. *cremoris* and *Lactobacillus brevis*), and the incorporation of yeast adversely affected acid development and production of diacetyl (Okagbue and Bankole, 1992).

Zimbabwe.   Fermented milk products in some sub-Saharan countries including Zimbabwe have been reviewed by Mutukumira *et al.* (1995). Amasi is an example of fermented milk product which is made in the rural region of Zimbabwe. After natural fermentation of the milk, some of the whey is removed and to improve the consistency of the product, naturally fermented cream is added. A typical chemical composition (g 100 g^{-1}) of amasi is total solids 16.5, fat 5.9 and protein 4.6; the pH is ~ 3.9 (Mutukumira, 1995). The product contained $8-10 \log_{10}$ cfu g^{-1} and was heavily contaminated with coliforms, yeasts and moulds. Mesophilic starter culture similar to the one used to produce filmjölk in Nordic countries is used to produce lacto from pasteurized milk (Feresu and Muzondo, 1990). However, traditional fermented milks were rated as more acceptable when compared with lacto (Feresu and Muzonda, 1989), and the lactic acid bacterial isolates belonged to the genera of *Lactobacillus* (e.g. *Lb. helveticus*, *Lb. paracasei* subsp. *paracasei*, *Lb. paracasei* subsp. *pseudoplantarum*, *Lb. delbrueckii* subsp. *lactis*) while only four lactococci isolates were found in lato (Feresu and Muzondo, 1990; Feresu, 1992). It is safe to conclude that the Zimbabwean naturally fermented milk should be classified as a thermophilic rather than a mesophilic product.

Ethiopia and Somalia.   Ergo or irgo and ititu are traditionally fermented milks made in Ethiopia (Kurmann *et al.*, 1992; Mutukumira *et al.*, 1995). Little information is available on ergo, but recently Ashenafi (1992, 1993, 1994) has studied the fate of different pathogenic microorganisms in this product, and the fermentation of milk was initiated with lactic acid bacteria belonging to the genera *Lactobacillus* and *Streptococcus*. He recommended that pasteurization of the milk, smoking the containers with olive wood splinters, and inoculation with 3-day-old ergo as starter culture produces a better product. *Lb. plantarum* has been identified to be the predominant lactic acid bacterial species *in situ* (Kassaye *et al.*, 1991).

Suusaac is a fermented product made from camel's milk in Somalia, and the use of defined starter culture results in a better quality (Mutukumira *et al.*, 1995).

Morocco. Lben is a traditional Moroccan fermented milk which is processed from different mammalian milks. The milk is fermented spontaneously at 18–24°C (i.e. depending on the season) for 24–48 h, churned to remove the butter granules for the manufacture of smen (i.e. local name for butter) and followed by the addition of water (∼10%). A typical average composition (g 100 g^{-1}) of Iben is total solids 9.1, fat 0.9, and protein 2.6 (Tantaoui-Elaraki and El Marrakchi, 1987). The main lactic acid bacteria that have been identified in Iben are *Lac. lactis* subsp. *lactis*, biovar *diacetylactis*, *Leuconostoc lactis*, *Leu. mesenteroides* subsp. *cremoris*, subsp. *dextranicum*; lactobacilli are present in low numbers, and yeasts, moulds and coliforms are also present (Tantaoui-Elaraki *et al.*, 1983).

3.3.2 Use of thermophilic strains

Some of the microflora of the genera *Streptococcus*, *Lactobacillus*, *Bifidobacterium* and *Enterococcus* are used for the production of fermented milks denoted, in part, as 'thermophilic' lactic acid fermentation. The term thermophilic is used for the products employing microorganisms whose growth optima range from 37 to 45°C.

Certain starter cultures of the genus *Lactobacillus* (*Lb. acidophilus*, *Lb. paracasei* subsp. *paracasei*, *Lb. reuteri*, *Lb. rhamnosus*, *Lb. kefir* and *Lb. kefiranofaciens*), *Bifidobacterium* spp. and *Enterococcus* spp., which are classified as probiotic organisms and/or found in kefir grains, will be discussed separately in sections 3.3.3 and 3.4.1. Thus, only two starter cultures belonging to the genera *Lactobacillus* (*Lb. delbrueckii* subsp. *bulgaricus*) and *Streptococcus* (*S. thermophilus*) are widely used for the manufacture of thermophilic-types of fermented milks.

The taxonomic status of *S. thermophilus* has fluctuated since the early 1980s due to the close relationship with *Streptococcus salivarius*, and as a consequence it was denoted as a subspecies. In 1991 a separate species status was re-proposed, based on genetic and phenetic criteria (for further detail, see the review by Hardie and Whiley, 1995). The cells are spherical or ovoid, ⩽1 µm in diameter and form chains. Such streptococci are Gram-positive, anaerobic homofermentative bacteria, produce L(+) lactate, acetaldehyde and diacetyl from lactose in milk, and some strains produce exopolysaccharide. No growth occurs at 15°C, but most strains are able to grow at 50°C, and require B-vitamins and some amino acids for enhanced growth rate.

Lb. delbrueckii subsp. *bulgaricus* is represented in Group I or Aa – the obligately homofermentative lactobacilli; the letter (a) indicates the affiliation to *Lb. delbrueckii* group. This organism ferments fewer sugars, produces D(+) lactate and acetaldehyde from lactose in milk, and some strains produce exopolysaccharide. The cells are rods with rounded ends, of 0.5–0.8 × 2–9 μm, and occur singly or in short chains. Slight growth occurs at < 10°C, and most strains are able to grow at 50–55°C (Hammes and Vogel, 1995).

The associative growth rather than the symbiotic relationship between *S. thermophilus* and *Lb. delbrueckii* subsp. *bulgaricus* (see Chapter 4) means that if a mixed starter culture is used the milk coagulates more quickly when compared with single culture. On some occasions these microorganisms are used separately to manufacture different fermented milk products.

(a) Bulgarian buttermilk.   This is a Bulgarian fermented milk product produced using *Lb. delbrueckii* subsp. *bulgaricus* alone (Marshall, 1984, 1986; van den Berg, 1988). Other synonyms for such a product are Bulgaricus milk, Bulgaricus cultured buttermilk or Bulgarian milk, and different mammalian milks are used to make set-type fermented milk. On some occasions cows' skimmed milk has been used, and the starter culture may also contain *S. thermophilus* or a cream culture (Robinson and Tamime, 1990; Kurmann *et al.*, 1992). This fermented milk product has a 'clean' flavour and sharp acidic taste reminiscent of yoghurt (Marshall, 1984), suggesting that the lactobacilli metabolize some of the milk components to acetaldehyde.

The manufacturing techniques are similar to yoghurt, but the inoculation rate of the starter culture ranges between 2–5%, and the titratable acidity up to 1.4% lactic acid. Bulgarian buttermilk containing sucrose is freeze-dried and made into pharmaceutical tablets containing 2.5×10^9 cfu g^{-1} of viable cells. When fed to babies, their faeces contain 1×10^9 cfu g^{-1} viable cells (Kurmann *et al.*, 1992).

(b) Use of S. thermophilus.   Kehran, karan or heran (in Siberia) and lapte-akru (in Rumania) are fermented milks made with a single starter culture of *S. thermophilus*. In the latter product, the fat content in the milk is standardized to 10 g 100 g^{-1}, homogenized, heated to 95–98°C for 3–5 h and cooled to ∼40°C. The culture is added at a rate of 5%, incubated, and followed by stirring and cooling. The coagulum can be either sweetened by the addition of sugar or fruit syrup (Kurmann *et al.*, 1992).

(c) Use of different combinations of starter cultures.   Mixed starter cultures such as *S. thermophilus*, *Lac. lactis* biovar *diacetylactis* and *Lb. helveticus* or *Lb. delbrueckii* subsp. *bulgaricus* in a ratio 1 : 1 : 1 is used for the production of katyk (in Kazakhstan). For the manufacture of lyubitelskii (in the former USSR), the processed skimmed milk is inoculated first with *S. thermophilus*

culture followed with *Lac. lactis* biovar *diacetylactis* in a ratio of 1:4, respectively (Kurmann *et al.*, 1992).

(d) Yoghurt. There are many different types of yoghurt that are produced worldwide, and they can be divided into various categories. The sub-divisions are usually created on the basis of:

- existing or proposed legal standards (full, medium or low fat);
- methods of gel production (set, stirred or fluid/drinking);
- flavours (natural, fruit or flavoured); and
- post-incubation processing (heat treatment, freezing, drying or concentration).

It can be seen from this review that it is difficult to cover all the aspects in detail. However, drinking yoghurt could be considered as stirred yoghurt having low viscosity where the milk (g $100\,g^{-1}$) contains fat 1.5 and MSNF 8. After incubation and cooling, the coagulum is mixed with 30–35% water and 1% salt (Akin and Rice, 1994). Another example is the production of silivri or crusty layer yoghurt; the unhomogenized milk (sheep, cow or buffalo) is heated to 95°C for 30 min, and subsequently cooled slowly in a stainless steel vessel. The containers are placed in an air-ventilated cabinet at 40–45°C for 1 h for the cream layer to form. Inoculation with starter culture is with a syringe without breaking the cream layer, and at pH 4.5 the containers are removed to the cold store (Akin and Rice, 1994). Furthermore, frozen yoghurt resembles ice-cream while dried yoghurt is mainly utilized in a wide range of food preparations, but not to be reconstituted and consumed as yoghurt. Thus, Tamime and Deeth (1980) have proposed a scheme of classification that covers all types of yoghurt based on the physical state of the product (Table 3.5). The manufacturing stages of yoghurt are similar (Figure 3.1), but the post-incubation processing differs, and in this section only, the product(s) of group I (Table 3.5) will be reviewed while the 'bio' yoghurts and concentrated fermented milks are discussed elsewhere in this Chapter (Sections 3.3 and 3.6). Developments in

Table 3.5 Proposed scheme of classification of yoghurt and related products[a]

Physical state	Product	Grouping
Viscous/liquid	Yoghurt	I
Semi-solid	Strained yoghurt	II
Solid	Soft/hard frozen yoghurt	III
Powder	Dried yoghurt	IV

[a]Adapted from Tamime and Deeth (1980) and Robinson and Tamime (1990).

yoghurt technology have been reported by Driessen and Loones (1990) and Nilsson and Hallström (1990).

Processing aspects.   The use of milk containing high counts of psychro-trophic bacteria (Riber, 1989) and somatic cell (Rogers and Mitchell, 1994) can influence the quality of yoghurt. Milk containing bacterial proteases from psychrotrophic bacteria and cellular-derived plasmin for the manufac-ture of yoghurt resulted in a product with substantially different physical properties (i.e. firmness, syneresis, apparent viscosity, water-holding capacity and protein hydration) when compared with the untreated milk (Gassem and Frank, 1991). However, the organoleptic quality of yoghurt made from milk containing somatic cell count $<250 \times 10^3$ cells ml^{-1} was superior to a similar product manufactured from milk of a higher cell count (Rogers and Mitchell, 1994). Thermization of milk before storage at $\leqslant 2°C$ and clarifica-tion before heat treatment of the milk base tends to minimize the above-mentioned defects in yoghurt.

Fortification of MSNF of the milk base is highly recommended since the rheological and organoleptic properties of yoghurt are improved. A wide range of dried dairy ingredients could be used (Table 3.4) and/or different techniques employed to concentrate the milk. Some aspects of the functional properties of whey protein concentrates have been reported by Wilmsen (1991), Dybing and Smith (1991), Barbut (1995) and Pearce (1995). The physical properties of the coagulum are primarily influenced by the type of powder used [SMP, buttermilk powder (BMP), whey protein concentrate (WPC), Na-caseinate or UF retentate] (Tamime *et al.*, 1984; Tamime and Robinson, 1988; Rohm, 1993a; Rohm and Schmid, 1993; Guinee *et al.*, 1994, 1995). Differences in syneresis and firmness of yoghurt have been reported by these authors despite the fact that the protein in the milk base had been adjusted to $\sim 5 g 100 g^{-1}$, and Morris *et al.* (1995) have reported no significant differences in firmness of yoghurts fortified with SMP and WPC. Furthermore, whole-milk powders that have been made from pre-heated milk at 85°C before drying and post-heat-treatment of the reconstituted milk base have produced a firmer yoghurt gel with increased water-holding capacity (i.e. decreased syneresis values) when compared with parallel milk pre-heated at 100°C for powder production (McKenna and Anema, 1993).

Since the microstructure of the yoghurt consists of a protein matrix made up of micellar chain (medium to short) and clusters with the fat globules embedded in the matrix, the ratio of casein:non-casein protein ratio in the milk is important. Ratios of 2.9:1 to 4.6:1 have been reported (Modler and Kalab, 1983; Modler *et al.*, 1983; Tamime *et al.*, 1984). The latter authors have recommended a ratio of 3.3:1 because at higher ratios fusion of casein micelles occur, and this influences the mouthfeel of the yoghurt. Figure 3.4 illustrates the detailed microstructure of yoghurt fortified with SMP and Na-caseinate. Furthermore, by using UF retentate powders (whole or

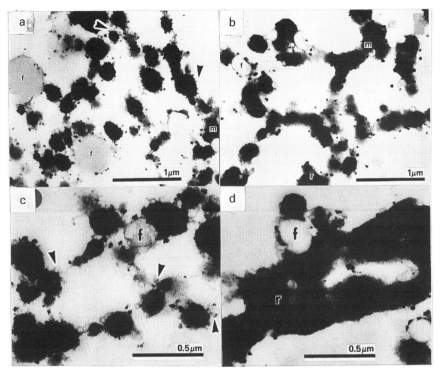

Figure 3.4 Microstructural detail using transmission electron microscopy (TEM) of casein micelles chains and clusters in yoghurts prepared from skimmed milk fortified with SMP (a,c) and Na-caseinate (b,d). Arrows point to 'spikes' on casein micelle surfaces. f, fat globules; m and r, simple and complex casein micelle chains, respectively. (After Tamime *et al.*, 1984. Reproduced by courtesy of *Scanning Microscopy International.*)

skimmed) the 'original' ratio of casein to non-casein protein is maintained, and good-quality yoghurt can be made (i.e. firm body and minimal syneresis) (Mistry and Hassan, 1992). Reduction in heat-stability of the milk base has been associated with high addition of whey protein concentrates (i.e. casein:whey protein ratios of 20:80 or 40:60) where the milk coagulated during heating at 93°C (Jelen *et al.*, 1987), and the yoghurt had clumpy appearance.

Yoghurts made from EV, UF and RO milks have been reported elsewhere (Tamime *et al.*, 1984; Guirguis *et al.*, 1987a; Mehanna *et al.*, 1988; Becker and Puhan, 1989; Puhan, 1990; Savello and Dargan, 1995), and the overall conclusion is that the UF technique has been highly recommended for the production of yoghurt.

Homogenization and high-heat treatment of the milk base increases the hydrophilic properties of the coagulum and stability of the gel due to

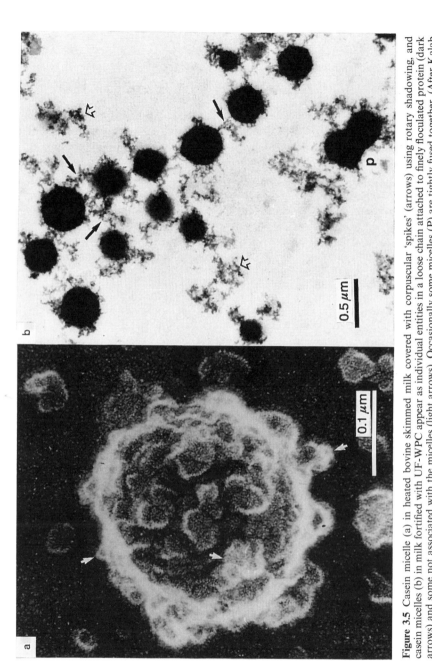

Figure 3.5 Casein micelle (a) in heated bovine skimmed milk covered with corpuscular 'spikes' (arrows) using rotary shadowing, and casein micelles (b) in milk fortified with UF–WPC appear as individual entities in a loose chain attached to finely floculated protein (dark arrows) and some not associated with the micelles (light arrows). Occasionally some micelles (P) are tightly fused together. (After Kalab *et al.*, 1982 and Modler and Kalab, 1983; reproduced by courtesy of *Milchwissenschaft* and *Journal of Dairy Science*.)

denaturation of β-Lg and α-La, and their association with κ-casein (Parnell-Clunies et al., 1988; Dannenberg and Kessler, 1988d, e; Mottar et al., 1989; Kessler et al., 1990). However, the microstructure of heated casein micelles contain large numbers of small particles of irregular shapes attached to the micelle surface (Figure 3.5(a)), and finely flocculated protein surrounding casein micelles (or not associated as separate entities) (Figure 3.5(b)) in yoghurt milk fortified with UF-WPC (Kalab et al., 1982; Modler and Kalab, 1983; Kalab and Caric, 1990; Kalab, 1992). In a recent study (Hollar et al., 1995), a WPC mixture ($16\,g\,TS\,100\,g^{-1}$) was dialysed against simulated UF milk containing calcium and heated; the denaturation of the protein was influenced by: (i) the calcium content, which as it decreased resulted in more soluble aggregates and less soluble precipitate being formed; (ii) the pH, which as it increased (5.8 to 7.0), caused more protein denaturation and the reverse aspects mentioned in (i) occurred; (iii) α-La denatured more extensively than β-Lg at 66 and 71°C; and (iv) the addition of low-heat SMP limited whey protein denaturation in WPC.

Starter organisms.   At present, the commercial process of yoghurt making uses defined mixed starter cultures (*Lb. delbrueckii* subsp. *bulgaricus* and *S. thermophilus*). Fermentation of the milk takes place at 40–45°C (short-set) and at 30°C (long-set), and the latter growth temperature may not be favourable for the lactobacilli. Recently, zabadi, which is the Egyptian equivalent of yoghurt, has been produced at 30 or 35°C with improved firmness and smoothness, minimum syneresis and with pleasant flavour (Mehanna, 1991).

The principal metabolic products of these organisms are lactate, aroma compounds (e.g. acetaldehyde and diacetyl) and sometimes exopolysaccharide. Thus, careful selection of different strains of *Lb. delbrueckii* subsp. *bulgaricus* and *S. thermophilus* may provide the manufacturer with the following broad options of yoghurt starter cultures:

1. Flavour producer (e.g. high, medium or low).
2. Polymer producer (e.g. high, medium or low).

Recently, texture studies and organoleptic properties of a wide range of commercially available yoghurt starter cultures have been reported by Rohm (1993b), Rohm and Kovac (1994, 1995) and Rohm et al. (1994). They concluded the following: (i) significant differences between products have been found in each sensory category except gel firmness; (ii) multiple regression analysis of hedonic scores have been mainly influenced by 'flavour' and 'ropiness' attributes showing positive and negative weightings, respectively; and (iii) interactions with long relaxation times have contributed less to viscoelasticity when ropy cultures were used, and attachment of ropy bacterial strains to the protein matrix has decreased the firmness of

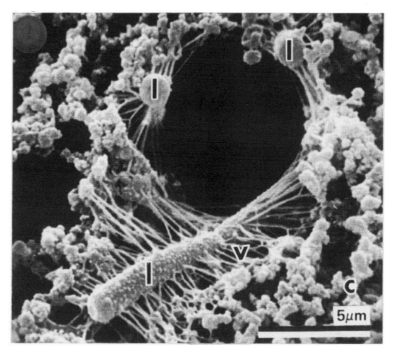

Figure 3.6 Exopolysaccharide production by yoghurt starter cultures; filaments of slime (v) attaching lactobacilli (I) to the protein matrix (c). (After Tamime *et al.*, 1984; reproduced by courtesy *of Scanning Microscopy International.*)

yoghurt. Figure 3.6 illustrates an example of the microstructure of yoghurt made with ropy starter culture.

In some countries the yoghurt-equivalent products may contain other organisms beside *Lb. delbrueckii* subsp. *bulgaricus* and *S. thermophilus*. For example, in India (dahi) other bacteria are used (*Lb. plantarum* and *Lac. lactis* subsp. *lactis*; Marshall, 1984), while Chander *et al.* (1992) have used only mixed cultures comprising *Lac. lactis* biovar *diacetylactis* and *cremoris*. *Lb. plantarum* is a heterofermentative lactobacillus, producing acetate in addition to lactate, and the lactococci produce only diacetyl and no acetaldehyde. However, in Australia, yoghurt has been produced by using yoghurt starter plus *Lb. helveticus* LB1 or *S. thermophilus* TS2 and *Lb. helveticus* LB1 (Giurguis *et al.*, 1987b; Rogers and Mitchell, 1994), while in Switzerland, ACO-yoghurt is made by using a mixed starter culture consisting of *Lb. delbrueckii* subsp. *bulgaricus*, *S. thermophilus* and *Lb. acidophilus* (Tamime and Robinson, 1988).

Advances in yoghurt technology may include progress in many scientific fields. Yoghurt has been manufactured by dialysis fermentation to obtain a product with a smooth structure and mildly acidic character (Klaver et al., 1992a). The possibility of continuous yoghurt processing has been studied by Prevost and Divies (1988a, b) and Ho and Mittal (1995). Reduction of the sodium content in yoghurt has been achieved by processing the milk in an ion-exchanger without affecting the starter culture activity or quality of the product (Nakazawa et al., 1990). The lactose content in yoghurt can be reduced by either UF treatment of the milk followed by enzyme hydrolysis (Rasic et al., 1992) or by using hydrolysing enzymes only (Whalen et al., 1988; Shah et al., 1993). The observed enhanced starter culture activity in lactose-hydrolysed milk was attributed to the carry-over of residual proteolytic enzymes in the β-galactosidase preparation. However, β-galactosidase activity in yoghurt can be increased by 5- to 6-fold when the pH of the product has been increased from 4.4 to 7.0 by the addition of 5 M calcium hydroxide for 6 h, and then allowing the pH to drop to 4.4 (Kotz et al., 1994). Yoghurt has been made using a monoculture of Lb. delbrueckii subsp. bulgaricus to ferment treated milk with β-galactosidase, glucose oxidase $(2.5\,\mu\,\text{ml}^{-1})$ and hydrogen peroxide $(0.15\,\text{g}\,100\,\text{g}^{-1})$ in order to replace the slow growing of S. thermophilus, and the product has been highly rated by the sensory panellists (Tahajod and Rand, 1993). Finally, good-quality, low-calorie yoghurt has been produced using starch-based fat substitutes and microparticulated whey proteins (Tamime et al., 1994, 1995b; Barrantes et al. 1994a, b, c, d). The starch-based fat substitutes could not be found in the microstructure of yoghurt because they were highly soluble except the P-fibre preparations and microparticulated whey proteins (Tamime et al., 1995b, 1996).

Yoghurt made from different milks. Sheep's, goats' and buffalo's milks have been used for the manufacture of yoghurt, and as mentioned elsewhere, the milk is processed in a similar manner to cows' milk. However, the casein components in these milks differ (Table 3.6), and as shown in Figure 3.2 the extent of whey protein denaturation is different which, as a consequence, can affect the rheological properties of yoghurt. Reviews on the fermentation of goats' and sheep's milks have been reported elsewhere (IDF, 1984b, 1986; Kehagias, 1987; Anifantakis, 1990; Abrahamsen and Rysstad, 1991).

Improved firmness and reduction in syneresis of sheep's yoghurt have been achieved by homogenization of the milk. A trend between soft fat in the milk and the viscosity of set-type natural yoghurt during the lactation season have been reported by Muir and Tamime (1993). Heat treatment of sheep's milk at 91°C for 30 s has reduced the fermentation period when compared with cows' milk, and the use of mixed cultures consisting of S. thermophilus and Lb. acidophilus to replace the yoghurt starter culture has resulted in a superior product (Kisza et al., 1993).

The total solids content in goats' milk can vary between 11.7–15.9 g 100 g^{-1} (Robinson and Vlahopoulou, 1988), and improvement in the

Table 3.6 The composition of the casein fractions in milks expressed as (%) of total casein

Casein	Cow[a]	Goat[a]	Sheep[a]	Buffalo[b]
Minor	6.2	2.5	4.8	_[c]
κ-casein	9.4	13.2	8.8	–
β-casein	37.5	50.1	43.5	50.5
α_{s1}-casein	33.0	18.4	}42.9	}26.1
α_{s2}-casein	13.9	15.8		

[a]Data compiled from Law (1995).
[b]Data compiled from Laxminarayana and Dastur (1968); γ-casein content reported was 23.4.
[c]Data not reported.

consistency of the yoghurt can be achieved by the addition of goats' powder or by concentrating the milk (EV, UF and RO). EV of goats' milk may reduce the goaty flavour in the milk, and the manufactured yoghurt may have a wider consumer acceptability. However, UF of goats' milk has improved the consistency of the yoghurt (Marshall and El-Bagoury, 1986). It has been observed that some inhibition of the yoghurt starter culture in goats' milk could be associated with milk containing strong 'goaty' flavour or higher concentration of free fatty acids (Abrahamsen and Rysstad, 1991). Inoculation rates ($\leqslant 1.5\%$) of the yoghurt starter culture have been recommended by Vlahopoulou et al. (1994) to produce a firmer gel, but other researchers have used $2-3\%$ (Marshall and El-Bagoury, 1986; El-Samragy, 1988). Low levels of acetaldehyde in goats' yoghurt have been attributed to relatively high concentration of glycine in the milk, which inhibits the enzyme involved to convert threonine to acetaldehyde and glycine (Abrahamsen and Rysstad, 1991); however, the addition of threonine to goats' milk stimulated acetaldehyde production (Marshall and El-Bagoury, 1986). *Lb. delbrueckii* subsp. *bulgaricus* and *S. thermophilus* produce some carbon dioxide in cows' milk, but higher levels have been reported in goats' milk. Furthermore, the citrate content in goats' milk is very low when compared with cows' milk, and as a consequence such milk may not be suitable for diacetyl production by mesophilic lactococci (Abrahamsen and Rysstad, 1991).

Buffalo's milk utilization for the manufacture of yoghurt has been extensively studied by Indian and Egyptian researchers. Some recently reported work includes: (i) UF of buffalo's milk followed by diafiltration (water or skimmed milk) tend to reduce the lactose content, and hydrolysis of lactose with β-galactosidase virtually resulted in a yoghurt free from lactose (Haggag and Fayed, 1988; Khorshid et al., 1993); and (ii) treatment of salted whey ($7-8\,\text{g}\,100\,\text{g}^{-1}$ CaCl) by UF, diafiltration and diluting the WPC with sweet whey have been used to fortify buffalo's milk for the manufacture of yoghurt (Abd El-Salam et al., 1991).

3.3.3   Use of probiotic strains

It is evident from the existing scientific publications (see the review by Renner, 1991) that a broad range of human health claims are associated with metabolites produced by lactic acid bacteria. When considering these claims, a degree of caution should be used for the following reasons: (i) results of animal studies may not be applicable to humans; (ii) *in vitro* studies may not be true *in vivo*; (iii) lack of coordinated efforts between technologists, microbiologists and clinicians; (iv) lack of proper modelling of the research work and possibly the interpretation of the data; and (v) the definition of probiotic is somewhat presumptuous. Nevertheless, fermented milks were the first probiotic products manufactured for human consumption (Fuller, 1994). The definition of probiotics in humans and animals by Fuller (1989) states:

'A live microbial feed supplement which beneficially affects the host animal by improving its intestinal microbial balance',

and such a definition emphasizes the role of probiotics in human health.

The intestinal microflora of healthy human beings consists of a diverse and complex ecosystem of bacterial populations. A vast literature exists on this topic, and some comprehensive data have been recently reviewed by Tamime *et al.* (1995a). Around 400 types of bacteria have been detected in the faeces of humans (Mitsuoka, 1990a, b, 1992; Tannock, 1992; Lichtenstein and Goldin, 1993; Mikelsaar and Mänder, 1993), and *Bifidobacterium* spp. and *Lactobacillus* spp. are present in the intestinal tract. The available clinical studies suggest that a high level of some desirable microorganisms whose natural habitat is the human gut, and the presence of bifidobacteria in the colon, is a desirable feature (see the review by Tamime *et al.*, 1995a). Thus, the idea of dietary supplementation with dairy products containing such organisms has gained credibility in many countries.

(a) Products containing bifidobacteria.   Taxonomic information including nomenclature and classification of the genera *Bifidobacterium* were extensively reported (Rasic and Kurmann, 1983; Scardovi, 1986; Bezkorovainy and Miller-Catchpole, 1989; Modler *et al.*, 1990; Romond and Romond, 1990; Biavati *et al.*, 1992; Ballongue, 1993; Modler, 1994; Sgorbati *et al.*, 1995).

Currently, 24 different species of bifidobacteria have been identified, and only five species (*Bifidobacterium adolescentis, breve, bifidum, infantis* and *longum*) have attracted attention in the dairy industry for the manufacture of therapeutic/probiotic fermented milk products (Driessen and de Boer, 1989; Kurmann and Rasic, 1991; Rambaud *et al.*, 1994; Tamime *et al.*, 1995a). Figure 3.7 shows some examples of the cellular morphology of five different species of bifidobacteria.

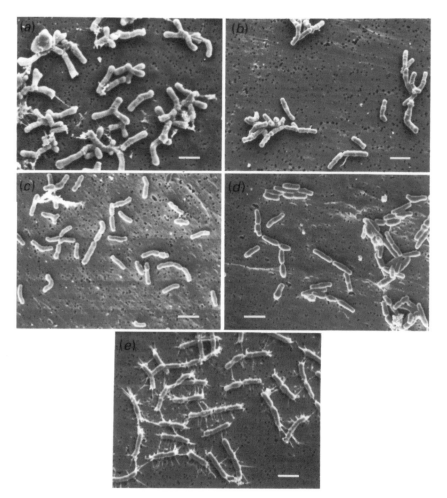

Figure 3.7 Cellular morphology of five different species of *Bifidobacterium*. (After Tamime *et al.*, 1995a) (D. Roy and D. Montpetit, personal communication; reproduced by courtesy of Food Research and Development Centre, Canada; reproduced by courtesy of *Journal of Dairy Research*.)

Over the past decade there has been a tremendous increase in the markets of Europe, North America and many other countries for dairy products (fermented or sweet) containing *Bifidobacterium* spp. originating from humans. A list of commercially available fermented milk products in different markets containing bifidobacteria is shown in Table 3.7. It is evident that *Bifidobacterium* spp. are not widely used as single strain starter cultures, possibly due to: (i) the slow rate of acid development when grown in milk; and (ii) the undesirable taste of acetic acid (see Chapter 4). The

Table 3.7 Commercial and developed fermented milk products containing bifidobacteria

Product	Microflora[a]											
	1	2	3	4	5	6	7	8	9	10	11	12
Cultura-AB®, Biomild®, Diphilus®, and Lünebest®	✓											
Acidophilus bifidus yoghurt	↔[b]	✓						✓				
Bifidus Active (BA®)		✓						✓				
Bifidus milk	↔	✓										
Bifidus milk with yoghurt flavour	↔	✓	↔									
Bifidus yoghurt[c], Mil-Mil E® or Biobest[®,d]	↔	✓				✓						
Bifighurt®		✓					✓					
Biogarde® or ABT	✓					✓		✓				
Biokys®	✓				✓			✓				
Biomild®	✓			✓				✓			✓	
Mil-Mil®	✓					✓		✓				
Ofilus® 'nature'	✓							✓				
'double douceur'									✓	✓		
Progurt®		✓							✓	✓	✓	
BRA		✓									✓	✓

[a] 1 → 4, *B. bifidum, longum, infantis, breve*, respectively; 5, *Bifidobacterium* spp. (not identified – see text); 6, *S. thermophilus*; 7, *Lb. delbreuckii* subsp. *bulgaricus*; 8, *Lb. acidophilus*; 9, *Lac. lactis* biovar *diacetylactis*; 10, *Lac. lactis* subsp. *cremosis*; 11, *P. acidilactici*; 12, *Lactobacillus reuteri*.

[b] ↔ Indicates that either species have been used.

[c] With or without *Lb. acidophilus*.

[d] The product contains 'biogerm' grains and fruits.

Data compiled from Tamime and Robinson (1988), Rogelj (1994), Rothchild (1995) and Tamime *et al.* (1995a).

strains that are widely used are *B. bifidum* and *B. longum*, in combination with other lactic acid bacteria, and these therapeutic organisms should ideally be of human origin and have a viable cell count at the time of consumption of $> 10^6$ cfu ml^{-1} (Tamime *et al.*, 1995a).

The manufacturing stages are similar to yoghurt in terms of fortification of the milk base, homogenization and high-heat treatment. The viable counts in the product are influenced by the rate of inoculation and incubation temperature. The common practice is to use DVI starter cultures in milk supplemented by nutrients, based on the recommendations of the

Table 3.8 Selection of different cryogenic compounds employed during the production of freeze-dried *Bifidobacterium* spp.

Microorganism	Processing conditions	Survival rate
B. longum M-101, ML3 *B. breve* M-161 *Lb. acidophilus* BR	Grow organism in nutrient culture broth; centrifuge and resuspend in cryoprotecting medium containing sucrose, sodium glutamate and gelatin; freeze-dry at $-40°C$ and 30°C.	$10^{10}–10^{12}$ cfu g^{-1}
B. bifidum ATCC[a] 11146 15696 *B. longum* *N. infantis*	Skimmed milk (SNF, 10–20 g 100 ml^{-1}) is fortified with: (i) modified MRS (0.2 g 100 ml^{-1}) or (ii) cysteine (0.5 g 100 ml^{-1}) plus pyruvic acid (0.5 g 100 ml^{-1}) or ascorbic acid (0.2 g 100 ml^{-1}); grow culture, centrifuge, resuspend in cryoprotecting medium (g 100 ml^{-1}: SNF 5, sucrose 8, gelatin 1.5) and freeze-dry ($-30°C$ and 20°C).	6.0×10^8 cfu g^{-1}
B. bifidum (three strains) *B. bifidum* NCFB[b] 1452 1453 *B. bifidum* NDRI[c]	Several growth media were studied; centrifuge cultured milk, wash cells with 0.1 M-phosphate butter and resuspend in sterilized skim milk (SNF, 10 g 100 ml^{-1}); plus dimethyl sulphoxide (1 g 100 ml^{-1}); freeze-dry at $-30°C$ and 20°C.	$10^{10}–10^{11}$ cfu g^{-1}
B. infantis ATCC 27920	The culture was grown in commercial MRS medium containing lactose (2.5 g 100 ml^{-1}) at constant pH 6.0 using 10 M-NH$_4$OH and temperature maintained at 37°C; after ~ 12 h fermentation, the cells were harvested by centrifugation, microfiltration or ultrafiltration and suspended in three cryoprotective media containing skimmed milk, sucrose and gelatin or combination of these ingredients plus ascorbic acid and freeze-dried at $-40°C$ and 30°C for 12 h followed by 4 h at 40°C.	

[a]American Type Culture Collection, USA.
[b]National Collection of Food Bacteria, UK (formerly National Collection of Dairy Organisms, NCDO).
[c]National Dairy Research Institute, India.
After Tamime *et al.* (1995a). Reproduced by courtesy of *Journal of Dairy Research*.

starter culture suppliers. However, Klaver *et al.* (1993) have provided a general bulk starter medium (i.e. skimmed milk, yeast extract, pepsin hydrolysed milk, corn extract and whey protein) which is suitable for growing bifidobacteria. The techniques used for the concentration of cell biomass of *Bifidobacterium* spp. and other microflora have been recently reviewed by Tamime *et al.* (1995a); some examples are illustrated in Table 3.8.

The main processing features of the products listed in Table 3.7 are as follows (Tamime *et al.*, 1995a):

Acidophilus bifidus yoghurt. This product is similar to bifidus yoghurt or Lünebest®, and the processed milk at 40–42°C is inoculated with separate cultures (yoghurt, *Lb. acidophilus*, *B. bifidum* or *longum*). Different starter tanks are required for the production of bulk starter cultures (Driessen and de Boer, 1989; Driessen and Loones, 1990, 1992) or alternatively some of the starter cultures (e.g. *Lb. acidophilus* and *Bifidobacterium* spp.) could be used as DVI.

Other closely related products are Biobest® and Mil-Mil E® (Tamime and Robinson, 1988; Tamime *et al.*, 1995a). The former product contains 'biogerm' grains, and the yoghurt culture is added at a rate of $\sim 0.1\,\text{g}\,100\,\text{ml}^{-1}$. Biobest® is manufactured in Germany while Mil-Mil E® is a Japanese product.

Bifidus milk. Cows' milk $(15–20\,\text{g}\,\text{TS}\,100\,\text{g}^{-1})$ is inoculated with a single or pure culture of *B. bifidum* or *longum* at a rate of 10% followed by incubation at 37°C until the pH reaches ~ 4.5. The product characteristics are:

- it has a mild, acid and slightly spicy taste;
- the viable count of bifidobacteria is $10^8–10^9$ cfu ml^{-1} with a decline of 2 log cycles during storage.

A similar product was developed in the UK by Marshall *et al.* (1982a). A blend of UF skimmed milk and UF Cheddar cheese whey is used $(\sim 15\,\text{g}\,\text{TS}\,100\,\text{g}^{-1})$ heated to 80°C for 30 min, cooled to 37°C, fortified with threonine $(0.1\,\text{g}\,100\,\text{g}^{-1})$, inoculated with 2% single strains starter of *Bifidobacterium* spp. and incubated for 24 h. The cooled products could be described as:

- the products made with *B. bifidum*, *infantis* and *longum* were similar and resemble yoghurt;
- the fermented milks have a 'walnut' flavour, pleasant acidity and acetaldehyde content $\sim 28\,\mu\text{g}\,\text{g}^{-1}$;
- the initial viable counts was 10^9 cfu ml^{-1}, but dropped after storage (to $<10^5$ *B. infantis*, 10^6 *B. longum* and 10^7 *B. bifidum*);

- the product made with *B. adolescentis* was unacceptable as the pH was 5.1.

In India, Misra and Kuila (1992) have produced a bifidus fermented milk containing 10^8 cfu g^{-1} of *B. bifidum*. The single strain starter culture was used at a rate of 10%, and incubated at 37°C.

Bifighurt®.   This fermented milk product is similar to bifidus milk or bifidus yoghurt, but the starter culture consists of *B. longum* and *S. thermophilus*. The bifidobacterial count in the product is 10^7 cfu ml^{-1}, and only L(+) lactic acid is produced.

Biogarde® and ABT®.   These products are somewhat similar and the starter culture consists of *B. bifidum* (human origin), *Lb. acidophilus* (human origin) and *S. thermophilus*. Originally when Biogarde® was developed in Germany, the inoculation rate of the mixed bulk starter culture was 10–20 g 100 ml^{-1}. Currently a culture nutrient medium (1.5 g 100 ml^{-1}) is added to the starter milk, and only 6 g 100 ml^{-1} inoculation rate is required. The final product contains 10^7–10^8 cfu ml^{-1} of *Lb. acidophilus*, 10^6–10^7 cfu ml^{-1} of *B. bifidum* and an abundance of streptococci. DVI cultures of Biogarde® and ABT® are widely used in order to ensure high numbers of viable cells.

The French product Ofilus® 'nature' (3.6 g fat 100 g^{-1}) is made by using a starter culture similar to Biogarde® or ABT®. Ofilus® 'double douceur' (10 g fat 100 g^{-1}) however, has a milder taste because *S. thermophilus* is replaced with *Lac. lactis* subsp. *cremoris*. In clinical studies, Marteau *et al.* (1990) suggest that consumption of Ofilus® containing viable cells of *B. bifidum* B1 (10^8 cfu g^{-1}) increases the activity of β-glucosidase, which is implicated in the colonic fermentation of cellulose. The same authors recommended further studies to determine the mechanics of these modifications and the fate of bifidobacteria in the intestine.

Cultura-AB®, Biomild® and Diphilus®.   These are generic names of products made using mixed cultures of *B. bifidum* (in some instances other species of bifidobacteria may be added) and *Lb. acidophilus*. It is recommended that protein-enriched milk should be used for the manufacture of Cultura AB®, and that the milks be homogenized and heat-treated, cooled to 37°C, inoculated with DVI starter culture (bulk starter optional), and incubated for 16 h followed by cooling (Tamime *et al.*, 1995a). The viable cell counts of freshly made Cultura AB® is $\geqslant 10^8$ cfu ml^{-1} for both starter organisms, and the product shelf-life is ~ 20 days.

Biokys®.   This is a fermented cows' milk (15 g TS 100 g^{-1}) developed in Czechoslovakia and has similar therapeutic properties to the pharmaceutical product Femilact®. The milk is homogenized, heat-treated, cooled to 30°C and inoculated with mixed starter culture (2–5 g 100 ml⁻¹, *B. bifidum*, *Lb. acidophilus* and *P. acidilactici* at a ratio of 1 : 0.1 : 1). At the desired acidity, the coagulum is stirred, cooled and packaged. At present, Biokys® is manufactured using a different mixed combination of starter cultures consisting of cream DL culture, *Lac. lactis* subsp. *cremoris* (ropy-producing strain), *B. longum*, *B. bifidum* and/or *Lb. acidophilus* (Tamime *et al.*, 1995a).

Progurt®. This is a Chilean high solids product made from partially concentrated fermented skimmed milk (a process similar to the traditional method for the manufacture of ymer in Denmark – see section 3.6.1), and mixed with buttermilk and cream (Schacht and Syrazynski, 1975). The processed milk base is inoculated with 1–3 g 100 g⁻¹ mesophilic starter culture (*Lac. lactis* biovar *diacetylactis* and subsp. *cremoris* at a ratio of 1 : 1), and incubated for 12–18 h. The gel is heated to <45°C to separate the whey, standardized with cream, mixed with buttermilk, and *Lb. acidophilus* and *B. bifidum* (0.5–1 g 100 ml⁻¹) are added before homogenization, cooling and packaging.

Miscellaneous products.   Bifilakt® or Bifilact® products are made with mixed flora of unknown species of bifidobacteria and lactobacilli. Bifilakt® had been developed for the treatment of gastrointestinal diseases in children. The processed milk is inoculated at a rate of 5 g 100 g⁻¹, and fermented for 18–20 h at 37°C. The packaged product is sweet in taste (i.e. pH ∼5.9), stable for a week at <10°C, and the viable cell count is 10^8–10^9 cfu ml⁻¹.

BRA is a Swedish fermented milk product which has been recently launched on the market (Rothchild, 1995). A mixed starter culture is used to ferment the milk, and consists of *B. bifidum*, *Lb. reuterii* (isolated from the gastrointestinal tract of humans), and *Lb. acidophilus*.

Bifidobacteria have been used in the development of a wide range of dairy products; some examples are listed in Table 3.9. However, the same organisms have been used alone or in combinations with other probiotic microorganisms for the preparation of pharmaceutical products (Table 3.10), and dried baby food formulae (Kurmann *et al.*, 1992).

Isolation and characterization of *Bifidobacterium* spp. in commercial yoghurts sold in Europe revealed that many products contain *Bifidobacterium animalis* (this organism is not of human origin) beside other bifidobacteria, and in some instances the viable counts of bifidobacteria in some products were $<10^2$ cfu ml⁻¹ (Reuter, 1990; Dibb *et al.*, 1991; Anon., 1993; Iwana *et al.*, 1993; Roy *et al.*, 1994). Consequently, the probiotic and/or 'bio' efficacy of such products is questionable.

(b) Products containing lactobacilli.   The potential 'health-promoting' species of lactobacilli include *Lactobacillus paracasei* subsp. *paracasei*, subsp.

Table 3.9 The use of bifidobacteria in some dairy products

Product	Comments
1. Frozen cultured ice-cream	
Biogarde®	The starter culture viable cells are similar to those reported for the fermented milk product (see text).
Hard type	The survival rates of *Lb. acidophilus* and *B. bifidum* after 17 weeks storage were 4×10^6 and 1×10^7 cfu ml^{-1}, respectively; this corresponds to a decrease of 2 and 1 log cycles respectively. A freeze-dried preparation of *B. longum* (1.7×10^9–2.5×10^{11} cfu ml^{-1}) have been added to the ice-cream mix before freezing; after 11 weeks of storage, the count had declined by 1 log cycle.
Soft-serve type	The numbers of *B. bifidum* have remained the same before and after freezing (i.e. 6 h holding at $-5°C$), but the viable cell count of *Lb. acidophilus* has increased under the same processing conditions.
2. Strained yoghurt	UF of warm fermented milk (cows', ewes' and goats') with *B. bifidum* Bb-12 have been used to produce strained yoghurt; the bifidobacterial counts ranged between 4.6×10^5 to 4.1×10^7 cfu g^{-1} depending on the type of milk used.
3. Fresh cheese	The Biobest® culture has been used during the manufacture of Thermo Quarg; *B. bifidum* has been used: (i) in Cottage cheese where it was added at the creaming stage before packaging; and (ii) during the manufacture of Tvarog in combination with a cream starter culture.
4. Gouda cheese	A mixed starter culture consisting of *Lb. acidophilus* Ki and *Bifidobacterium* spp. Bo has been used successfully to manufacture Gouda cheese; during 9 weeks' storage both cultures averaged 0.2–5×10^7 and 6–18×10^8 cfu g^{-1}, respectively, and their survival was dependent on the salt concentration in the cheese; body and texture, appearance and firmness of the experimental cheeses varied when compared with a reference cheese; the overall acceptability by the panellists was less preferred, but highly rated as a 'new' type of cheese.
5. Cultured Cottage cheese dressing	Milk base medium containing 12% UF retentate powder has been used to grow *B. infantis* ATCC 27920G to acidify the cream dressing (14% fat) to pH 4.5 in ~ 10 h; the viable count of bifidobacteria was $8.86 \log_{10}$ cfu g^{-1} and much lower ($6.47 \log_{10}$ cfu g^{-1}) if salt was added.
6. Growth in UF milk	*B. bifidum* ATCC 15696 and *B. longum* ATCC 15708 growth in UF milk (between 1- to 5-fold) have counts of 10^8–10^9 cfu ml^{-1} due to buffering capacity of the milk and the pH remained ~ 5.4; such behaviour of bifidobacteria suggest their use for bulk starter production, fermented milk or some cheese varieties.
7. Milk powder	*Bifidobacterium* spp. alone or in combination of *Lb. acidophilus* and *P. acidlactici* have been used as formula feed for infants.

Data compiled from Rasic and Kurmann (1983), Kurmann *et al.* (1992), Tamime *et al.* (1995a), Blanchette *et al.* (1995) and Gomes *et al.* (1995).

Table 3.10 Some examples of pharmaceutical preparations[a] containing bifidobacteria

Country	Product	Microorganism
Japan	Bifider®	*B. bifidum*
France	Bifidogène®	*Bifidobacterium* spp.
	Synerlac®	*Lb. acidophilus*
		B. bifidum
		Lb. delbrueckii subsp. *bulgaricus*
	Lyobifidus®	*B. bifidum*
Yugoslavia	Liobif®	*B. bifidum*
Switzerland	Infloran Berna®	*Lb. acidophilus*
		B. bifidum
USA	Life Start Two®	*B. bifidum*
	Life Start Original®	*B. infantis*
Germany	Eugalan® ⎫	
	Euga-Lein® ⎬	*Bifidobacterium* spp.
	Lactopriv ⎭	
	Omniflora®	*B. longum*
		Lb. acidophilus
		E. coli

[a]Products as freeze-dried tablets.
After Kurmann *et al.* (1992) and O'Sullivan *et al.* (1992).

paracasei biovar *shirota*, subsp. *rhamnosus* and *Lb. acidophilus*. *Lactobacillus* strain GG is similar to *Lb. paracasei* subsp. *rhamnosus*. Recently, the genus *Lactobacillus* has been reviewed by Pot *et al.* (1994) and Hammes and Vogel (1995), and the taxonomy of this genus is till under investigation. Nevertheless, the latest classification of the species mentioned above could be briefly described as follows:

- *Lb. acidophilus* is represented in Group I or Aa–the obligately homofermentative lactobacilli affiliated to *Lb. delbrueckii* group.
- *Lb. paracasei* subsp. *rhamnosus* is elevated to the species level as *Lb. rhamnosus*; this strain and *Lb. paracasei* subsp. *paracasei* are represented in Group II or Bb–the facultative heterofermentative lactobacilli affiliated phylogenetically to the *Lb. paracasei*–Pediococcus group.
- *Lb. paracasei* subsp. *paracasei* biovar *shirota* is the strain which was isolated from the human intestine by Dr Minora Shirota, the founder of Yakult company in Japan, and should be possibly designated as *Lb. paracasei* subsp. *paracasei* and/or *Lb. rhamnosus*.

Examples of some fermented milk products containing lactobacilli other than the yoghurt starter culture may include the following.

Yakult. This is a therapeutic fermented milk product originating from Japan where the milk TS is rather low ($\sim 3.7\%$), and has 14% added sugars (S. Mutsuhashi, personal communication). It is a beverage made with a single strain culture containing high numbers of *Lb. paracasei* subsp.

paracasei. The precise manufacturing details of the manufacture of yakult are not readily available outside Japan. It is most likely that the fermentation time is around 4 days under extremely sanitary conditions. The product contains $>10^8$ cfu ml^{-1} of viable cells of *Lb. paracasei* subsp. *paracasei* (Kurmann *et al.*, 1992).

The milk base is heated to a very high temperature (140°C for 3–4 s) and in the presence of added sugars the colour becomes slightly brown, indicating a Maillard reaction. Sometimes the product is flavoured with vegetable juices. Yakult is manufactured under license in the Far East, California, Mexico, Brazil and Australia.

A closely related product is called Yakult Miru-Miru which is made with a mixed starter culture consisting of *B. bifidum*, *B. breve* and *Lb. acidophilus*. Alternatively, a different starter culture blend is used where *B. bifidum* is replaced by *Lb. paracasei* subsp. *paracasei*. The chemical composition is (g 100 g^{-1}) fat 3.1, protein 3.1, lactose 4.5, added sugars 6.1 and ash 0.7 (Tamime and Robinson, 1988).

Yoke.   SMP is rehydrated to $<8\%$ SNF and blended with sugar, calcium, pantothenic acid, vitamin B$_6$ and linolenic acid. The mixture is heated to 50°C, homogenized, heated to 90°C, and followed by sterilization at 100°C for 65 min. After cooling to 37°C, the mixture is inoculated with mixed starter culture (*Lb. paracasei* subsp. *paracasei*, *Lb. acidophilus* and *S. thermophilus*), incubated for 48 h, cooled, homogenized, and before filling, vitamin C is added aseptically. The lactobacillus viable count is 18×10^8 cfu g^{-1}.

Calpis.   This is a Japanese sour milk product prepared from skimmed milk using a starter culture containing *Lb. helveticus* and *Saccharomyces cerevisiae* (Nakamura *et al.*, 1995a,b). Only the laboratory method of manufacture is available. The current interest of Calpis is the possible role of two peptides naturally present in it (Val-Pro-Pro and Ile-Pro-Pro) as inhibitors of angiotensin I-converting enzyme (ACE) and their antihypertensive activity in rats.

Acidophilus milk.   An unfermented milk product known as 'sweet acidophilus milk' was promoted in the 1970s in the USA. The product has the same flavour as normal milk because little acid develops under refrigerated storage. Fermented acidophilus milk can be made using a single starter culture of *Lb. acidophilus*. Strains isolated from the intestinal tract, however, grow slowly in milk, only increasing 5-fold in 18–24 h with a developed acidity of $\leqslant 0.8\%$ (Marshall *et al.*, 1982b; Marshall, 1986). Interest in these strains comes from the possible benefits for alleviation of intestinal and other disorders. Publications by Nahaisi (1986), Welch (1987), Gilliland (1989), Gilliland and Walker (1990), Sellars (1991) and Salji (1992) review the 'health-promoting' properties of *Lb. acidophilus*. The role of *Lactobacillus* spp. (most likely a *Lb. acidophilus* with ability to colonize the vagina) and lactate-gel (a pharmaceutical preparation containing growth substrates of lactobacilli) in the treatment of bacterial vaginosis, on survival of HIV in the female genital tract, and on diarrhoea and immunological complaints have been reported by many researchers (Andreshch *et al.*, 1986; Martins *et al.*, 1988; Holst and Brandenberg, 1990; Marteau *et al.*, 1990; Perdigón *et al.*, 1990, 1992; Klebanoff and Coombs, 1991; Klebanoff *et al.*, 1991; Coconnier *et al.*, 1993).

When acidophilus milk was first launched in different Western markets, the product was not well received by the consumers (even by the health-conscious). The reasons could be associated with lack of belief in a correlation of 'acidophilus' products with 'good health' but the poor quality control during production with respect to microbiological quality and organoleptic properties may also have contributed. Improvements in the processing conditions and the blending of *Lb. acidophilus* with other starter organisms have resulted in an increase in market share.

A general method for production of acidophilus milk is as follows: milk (whole or skimmed) is heated to 95°C for 60 min, cooled to 37°C and held for 3–4 h, re-heated to 95°C for 10–15 min, cooled to 37°C and inoculated with 2–5% bulk starter culture. The processed milk is incubated for up to 24 h or to 1% lactic acid, cooled to 5°C, packaged and finally transferred to the cold store (Chandan, 1982). However, modern methods of streamlining this process are: (i) the milk is homogenized at 14.5 MPa of pressure, heated to 95°C for 60 min, cooled to 37°C and inoculated with DVI culture; the incubation time is 12–16 h or to $\sim$0.65% lactic acid; and (ii) the UHT process of 140–145°C for 2–3 s would provide a milk where the proliferation of undesirable contaminants is avoided (Chandan, 1982; Alm, 1983). Mixed ropy and non-ropy strains of *Lb. acidophilus* (i.e. ratio 1 : 4) may also be used, for example, in the former USSR for the production of moskowski (Kurmann *et al.*, 1992).

The retail product should contain 5×10^8 cfu ml^{-1} *Lb. acidophilus* at the time of consumption (i.e. 14–21 days after manufacture). To ensure such quality, it is critical to cool the product at $\sim$0.65% lactic acid and maintain it at $\leqslant$5°C during distribution and retailing (Tamime and Robinson, 1988). The stability of the microflora in acidophilus milk may also be achieved by: (i) the addition of yeast extract or 'V medium' to milk to improve the growth of *Lb. acidophilus* (Alm, 1981, 1982a, b, c, d, e, f); (ii) fortification of the milk proteins with concentrated skimmed milk (Alm, 1983); and (iii) the use of mixed inocula of *Lb. acidophilus* and yoghurt starter culture to stabilize cell numbers and improve the flavour. Such approaches rely on careful selection of *Lb. acidophilus* strains, the acidity should be monitored so that survival of *Lb. acidophilus* is assured for at least 14 days (Gilliland and Speck, 1977; Hull *et al.*, 1984; Roberts *et al.*, 1984; Johnston *et al.*, 1987; Robinson, 1987).

Lb. acidophilus added to yoghurt before packaging may survive poorly due to hydrogen peroxide produced by *Lb. delbrueckii* subsp. *bulgaricus* (Gilliland and Speck, 1977), but addition of catalase (240 U ml^{-1}) can prevent such inhibition (Hull *et al.*, 1984). Alternatively, the addition of high cell numbers of *Lb. acidophilus* and/or *Bifidobacterium* spp. to pasteurized milk before packaging will provide viable cell counts $>2 \times 10^6$ cfu ml^{-1} at the end of 2 weeks shelf-life (Tamime *et al.*, 1995a).

Other milks may also be used, and good-quality acidophilus milk has been produced from buffalo's milk with viable cell counts of 6.4–8.1 $\times$ 10^8 cfu ml^{-1} at pH ~4.0 (Rao and Gandhi, 1988). Mixed cultures of *Lb. acidophilus* and *Lb. paracasei* subsp. *paracasei* grown in soy milk substrate had high numbers of *Lb. acidophilus*, but the addition of *S. thermophilus* reduced the final count (de Valdez and de Giori, 1993). An alternative means of providing the consumer with viable *Lb. acidophilus* is as a spray-dried blend preparation. The powder is made from acidophilus milk, tomato juice and sugar, and the product has a viable cell count of 22.9 $\times$ 10^7 cfu g^{-1}, representing a survival rate of 14.8% (Prajapati *et al.*, 1986, 1987).

A probiotic sour milk product resembling cultured buttermilk is called 'A-38' fermented milk or 'A-Fil' milk (Kurmann *et al.*, 1992). The culture has been developed by Chr. Hansen's Laboratories (Denmark) and consists of *Lactococcus* spp. and *Lb. acidophilus*. The milk base is enriched to give a protein content 3.8–3.9 g 100 g^{-1}, and the method of production is similar to Cultura-AB®.

A few pharmaceutical freeze-dried preparations of *Lb. acidophilus* are available in different markets; some examples are listed in Table 3.11. A dietetic product, produced in the former USSR, is made from skimmed milk fortified with 2 g corn oil 100 g^{-1}. The oil is heated to 50°C, mixed with a small portion of skimmed milk at 35°C, homogenized and then mixed with the rest. Sucrose is added, followed by heating to 90–95°C for 10–15 min, cooled to 37°C, inoculated with 5% *Lb. acidophilus* starter culture and incubated for a few hours. The acidity of final product is ~1.0% lactic acid (Kurmann *et al.*, 1992).

Lactobacillus rhamnosus. This culture is used to manufacture the fermented milk products Gefilac™ or Gefilus® which are available in Finland (Salminen *et al.*, 1991a,b; Salminen, 1993, 1994). The strain is of human origin, and was named *Lactobacillus* GG (after the two researchers S. Gorbach and B. Goldin). Previously, the organism was known as *Lb. paracasei* subsp. *paracasei* (Dong *et al.*, 1987), but it has recently been designated *Lb. rhamnosus* ATCC 53103 (Meurman *et al.*, 1994, 1995).

Lactose-hydrolysed milk is used for the manufacture of Gefilus®, and the product is unflavoured, sweetened with fructose and the fat content is ~1.5% g 100 g^{-1}. It has a firm consistency, refreshing and slightly sour

Table 3.11 Freeze-dried pharmaceutical tablets containing lactobacilli species

Product	Country	Microorganism	Viable cell count (cfu g^{-1})
Enpac®	UK	*Lb. acidophilus*[a]	Large numbers
Laccilla®	UK	*Lb. acidophilus*	$\geqslant 10^8$
Lactinex®	USA	*Lb. acidophilus + Lb. delbrueckii* subsp. *bulgaricus*	NR[b]
Megadophilus®	USA	*Lb. acidophilus*	5×10^9
Ribolac®	Switzerland	*Lb. acidophilus*[c]	$\geqslant 10 \times 10^7$

[a]The strain used is resistant to 10 different antibiotics.
[b]NR: not reported.
[c]The lactobacillus is resistant to antibiotics and sulfa drugs; the tablet contains also different vitamins.
Data compiled from Kurmann *et al.* (1992).

taste. Kahala *et al.* (1993) reported that the starter culture of Gefilus® consisted of the three strains of *Lactococcus* spp., *Leu. mesenteriodes* subsp. *cremoris* and *Lactobacillus* GG.

This *Lactobacillus* strain fulfils the requirements of a probiotic microorganism as, according to Salminen *et al.* (1991a,b) it:

- is stable in acid and bile;
- is able to attach avidly to human intestinal mucosal cells;
- survives passage through the stomach and upper bowel, and colonizes the human intestinal tract;
- produces antimicrobial substance(s) that inhibits the growth of harmful bacteria;
- colonizes the intestine during penicillin, ampicillin or erythromycin treatment;
- enhances intestinal immunity;
- reduces levels of intestinal bacterial enzymes that are implicated in large bowel cancer;
- prolongs survival in lethally irradiated rodents by suppressing bacteraemia.

The process of isolation and characterization of such *Lactobacillus* strains from humans has been patented in many countries (Gorbach and Goldin, 1989). A review concerned with *Lactobacillus* GG has been published by Salminen *et al.* (1993). Recent clinical studies report: (i) isolation of *Lb. rhamnosus* from the saliva of patients 1 week after a 7-day regime where 250 g of fermented milk was consumed twice daily (Meurman *et al.*, 1995); (ii) an inhibitory substance from *Lb. rhamnosus* capable of inhibiting *Streptococcus sobrinus* (Meurman *et al.*, 1994), and a compound resembling microcin (low-molecular weight <1 kDa) with an activity broader than that of bacteriocin (Silva *et al.*, 1987); (iii) colonization of human intestinal tract using tablets of lyophilized *Lactobacillus* GG (Saxelin *et al.*, 1991, 1993); (iv) colonization of premature babies with *Lactobacillus* GG where no adverse nutritional effects were observed (Stansbridge *et al.*, 1993); (v) survival of *Lactobacillus* GG in the gastrointestinal tract of some patients receiving ampicillin (Goldin *et al.*, 1992); (vi) effective treatment of colitis diarrhoea caused by *Clostridium difficile* (Gorbach, 1990), antibiotic-associated diarrhoea (Siitonen *et al.*, 1990), travellers diarrhoea (Oksanen *et al.*, 1990) and children's acute diarrhoea (Isolauri *et al.*, 1991); (vii) possible stabilization of gut mucosal barrier and stimulation of antigen-specific immune response in rats after long-term challenge with bovine milk proteins (Isolauri *et al.*, 1993a); and (viii) decrease in faecal β-glucuronidase, nitroreductase and glycocholic acid hydrolase activities in the colon of healthy female adults after consumption of yoghurt made with *Lactobacillus* GG (Ling *et al.*, 1994). Finally, Isolauri *et al.* (1993b) have reported that implantation of *Lactobacillus* GG in suckling rats may counteract rotavirus infection and associated intestinal dysfunction.

   (c) Products containing enterococci.   Organisms such as *Enterococcus faecium* and *Enterococcus faecalis* are among the microflora that colonize the intestinal tract of humans (Drasar and Barrow, 1985). Since 1929, Paraghurt® tablets containing *Ent. faecium* and yoghurt starter cultures have been used in Denmark for the treatment of irritable colon, steatorrhoea and as prophylatic treatment for diarrhoea (Fris-Møller and Hey, 1983). Each tablet of Paraghurt® contains (cfu g^{-1}) 10^7–10^8 *Ent. faecium*, $\sim 10^3$ *S. thermophilus* and $\sim 10^2$ *Lb. delbrueckii* subsp. *bulgaricus*. A similar product is Bioflorine® (Kurmann *et al.*, 1992). However, at the latest report on the workshop on 'The Safety of Lactic Acid Bacteria' held in Germany in 1994 under the auspices of the Lactic Acid Bacteria Industrial Platform (LABIP), the risk factors for pathogenicity due to lactic acid bacteria were discussed; the safety status of certain bacterial strains used as robiotic starter cultures (*Lb. rhamnosus*, *Ent. faecium* and *Ent. faecalis*) warrants further careful surveillance (Adams and Marteau, 1995).

   A yoghurt-like product is produced in Egypt using single or mixed strains of *Ent. faecalis* 19 and 22 and/or *Lb. delbrueckii* subsp. *bulgaricus* (Fayed *et al.*, 1989), and Enterococci strains have been isolated from laban rayeb (Egyptian fermented milk product; see El-Gendy, 1983, 1986). An experimental product made with 1.5% starter of each of the *Ent. faecalis* strains was efficient in the producton of acid, carbonyl compounds, exhibited higher proteolytic activity and was highly acceptable by the taste panellists.

   Recently, a product called Gaio® has been marketed in Denmark and some other European countries. The bacterial culture is called Causido®, and consists of *Ent. faecium* (of human origin) and two strains of *S. thermophilus*. This culture originated from Kiev, and the viable counts in a fresh product are: $\sim 2 \times 10^8$ cfu ml^{-1} *Ent. faecium* and $\sim 7 \times 10^8$ cfu ml^{-1}

S. thermophilus (Hølund, 1993; Hougaard, 1993, 1994). Clinical studies on the hypocholesterolaemic effect of Gaio® in healthy middle-aged men in Denmark have been reported by Agerbaek *et al.* (1995).

3.4   Fermented milks with alcohol and lactic acid

These alcoholic milk beverages may be described as lactic/sour in taste, with ethanol content as high as 2%. They have a foaming and effervescent characteristic as a result of CO_2 production. The microflora of the starter cultures is less well-defined when compared with fermented milks described above, although yeasts and lactic acid bacteria are always present. Typical examples of such fermented milks are kefir, koumiss and acidophilus-yeast milk. They originated in the Caucasian mountains and in the Steppes to the north and north east of the Caucasus to Mongolia. These fermented milks are widely produced in the former USSR, and to a very limited volume in some western European countries.

3.4.1   Kefir

The starter culture is in the form of kefir grains characterized by irregular, folded and uneven surfaces. The grains may be white or yellowish in colour, and have an elastic consistency (Koroleva, 1983, 1988a, 1991). The diameter of the kefir grain may range between 1–6 mm or more depending on the extent of agitation during growth in milk (Koroleva, 1991); however, when the grains are recovered from milk and washed with water, they are of variable sizes ranging from 0.5–3.5 cm in diameter, resembling cauliflower florets in shape and colour (Kosikowski, 1985; Marshall, 1986). The exact origin of the kefir grains is unknown, and according to legend the kefir grains were given to the Orthodox people living in the Caucasian mountains by Mahomet (possibly the Prophet Mohammed) who told them how to use it; he strictly forbade them to give away the secrets of kefir preparation or pass the grains to anybody. Thus, the kefir grains were known as 'Mahomet grains' (Koroleva, 1991), and sometimes called 'the gift of the Gods' (Kosikowski, 1985).

(a) Microflora of kefir grains.   The microflora of the kefir grains is complex and not always constant, consisting of undefined species of bacteria and yeasts (IDF, 1991). Numerous species have been found associated with the kefir grains; Table 3.12 shows the organisms that have been identified. It is evident that the type of microorganisms present in kefir grains is dependent on: (i) source and country of origin; and (ii) the culturing techniques used to identify the various species. For example, *Acetobacter* spp. in Spanish kefir grains has been considered a contaminant (Angulo *et al.*, 1993) while

Table 3.12 Microflora of kefir grains

Microflora	Viable count (cfu g^{-1})[a]
I. Yeasts	
Saccharomyces cerevisiae	
delbreuckii	10^6
florentinus	
exiguus	
fragilis	
carlbergensis	
globus	NR[c]
dairensis	
unisprous	
kefyr[b]	
Kluyveromyces marxianus	
lactis	
Candida kefyr	10^8
pseudotropicalis	
tenuis	
holmii	
valida	
friedrichii	
Mycotorula kefyr	
lactis	NR
lactosa	
Torulopsis kefyr	
holmii	
Cryptococcus kefyr	
Torulaspora delbreuckii	
Pichia fermentans	
II. Lactic acid bacteria	
Lactobacillus kefir	10^9
brevis	10^6
acidophilus	10^8
cellobiosus	10^6
helveticus	$10^4 - 10^5$
delbreuckii subsp. *bulgaricus*	$10^4 - 10^5$
paracasei subsp. *paracasei*	
subsp. *tolerans*	
subsp. *pseudoplantarium*	
subsp. *alactosus*	
rhamnosus	
kefiranofaciens	
kefirgranum	NR
parakefir	
viridescens	
fermentum	
gasseri	
plantarum	
Lac. lactis subsp. *lactis*	NR
subsp. *cremoris*	10^6
biovar. *diacetylactis*	
Leu. mesenteriodes subsp. *dextranicum*	10^6
S. thermophilus	NR

Table 3.12 (*Continued*)

Microflora	Viable count (cfu g^{-1})[a]
III. Acetic acid bacteria	
Acetobacter aceti	10^8
rasens	10^8
IV. Mould	
G. candidum[d]	NR
V. Contaminants	
Pediococcus spp.	
Micrococcus spp.	
Bacillus spp.	NR
E. coli	
Ent. durans	

[a]Colony-forming-unit per g.
[b]Previously known as kefir.
[c]Not reported (NR).
[d]Delft kefir grains only or contaminant.
Data compiled from Ueda *et al.* (1982); Molska *et al.* (1983); Kandler and Kunath (1983); Meriläinen (1984); Oberman (1985); Koroleva (1988a, 1991); Marshall (1986, 1987); Fujisawa *et al.* (1988); Hosono *et al.* (1990); Rohm *et al.* (1992); Mukai *et al.* (1992); Angulo *et al.* (1993); Takizawa *et al.* (1994).

Koroleva (1991) reported that *Acetobacter aceti* and *rasens* improve the flavour and consistency of kefir. Toba *et al.* (1991b) have isolated a capsular polysaccharide-producing *Lactobacillus kefiranofaciens* K$_1$ from kefir, and a fermented milk was prepared from the isolated strain which had a ropy consistency and was resistant to syneresis. The presence of white mould (*G. candidum*) in kefir grains may be desirable for certain markets (Kosikowski, 1985; Marshall, 1986, 1987), but a contaminant in other countries. The role of *Enterococcus durans* in kefir production is not well established, and should be considered a contaminant.

Some of the yeast genera shown in Table 3.12 are the old nomenclature, and the current generic names of these yeast have been reviewed by Kreger-van Rij (1984). The latest nomenclature are:

Old	New
Saccharomyces delbrueckii	*Torulaspora delbrueckii*
Saccharomyces florentinus	*Zygosaccharomyces florentinus*
Saccharomyces fragilis	*Kluyveromyces marxianus* var. *fragilis*
Saccharomyces carlbergenisis	*Saccharomyces cerevisiae*
Saccharomyces globus	*Saccharomyces cerevisiae*
Saccharomyces kefyr	*Kluyveromyces marxianus* var. *marxianus*
Kluyveromyces marxianus	*Kluyveromyces marxianus* var. *marxianus*

Kluyveromyces lactis	*Kluyveromyces marxianus* var. *lactis*
Mycotorula kefyr	*Kluyveromyces marxianus* var. *marxianus*
Cryptococcus kefyr	*Kluyveromyces marxianus* var. *marxianus*
Torulopsis holmii	*Candida holmii*

Furthermore, the same yeast species may be known by two generic names depending on spore formation (i.e. imperfect or perfect stage). For example, *Cand. kefyr* and *Cand. valida* (imperfect) are synonymous with *K. marxianus* var. *marxianus* and *Pichia membranaefaciens* (perfect), respectively (Kreger-van Rij, 1984).

As the indigenous microflora of kefir grains is variable (Table 3.12), different sources of kefir grains will have different proportions of lactic acid bacteria and yeasts (Figure 3.8). Vayssier (1978a, b) has proposed preparing kefir with proportions of: *Lac. lactis* subsp. *lactis*, biovar *diacetylactis*, *Leu. mesenteriodes* (possibly) subsp. *mesenteroides* or *dextranicum*, *Lactobacillus paracasei* subsp. *alactosus*, *Lb. brevis/Lactobacillus cellobiosus* (each 10^8 cfu g^{-1}), *Saccharomyces florentinus* (10^6 cfu g^{-1}) (acetic acid bacteria were excluded). Recently, Koroleva (1991) has provided a different proportion of these microorganisms in the kefir bulk starter culture based on years of experience in the former USSR to manufacture kefir with the typical taste, flavour, and aroma, and consistency. The types of microflora and counts are: lactococci (10^8-10^9 cfu ml^{-1}), *Leuconostoc* spp. referred to as heterofermentative lactic acid streptococci (10^7-10^8 cfu ml^{-1}), thermophilic lactobacilli (10^5 cfu ml^{-1}), acetic acid bacteria (10^5-10^6 cfu ml^{-1}) and yeasts (10^5-10^6 cfu ml^{-1}); the starter culture acidity must be in the range of 0.95–1.0% lactic acid. The morphology of kefir grains (Figure 3.9) shows the varying types which ranged from thin flat sheet (Figure 3.9(a)) to cauliflower floret forms (Figure 3.9(b–d)) (Marshall *et al.*, 1984).

A similar culture to kefir grains has been used domestically in rural parts of Northern Ireland, and known locally as a 'buttermilk plant' (Thompson *et al.*, 1988, 1990). Traditionally, milk is diluted with an equal volume of water, seeded with 'buttermilk plant', incubated overnight at ambient temperature, and the product is used in baking to replace buttermilk, rather than consumed as a fermented milk. The microbial flora consisted of predominantly rod-shaped bacteria (*Lactobacillus* spp.), some coccoid bacteria (mainly *Lac. lactis* subsp. *lactis* and subsp. *cremoris*), and yeasts [*Sacc. cerevisiae* (lactose-negative) and *Candida kefyr* (lactose-utilizer)]. The diluted milk, after fermentations, contained 1.5% lactic acid (of which the concentration of D(−) isomer was 4-fold more than the L(+) isomer), and 0.3% ethanol (Thompson *et al.*, 1990). It has been suggested by the same authors that the origin of the 'buttermilk plant' was kefir grain imported to Northern Ireland. The modification of the microbial flora of 'buttermilk plant' when compared with a kefir grain could be due to the lower

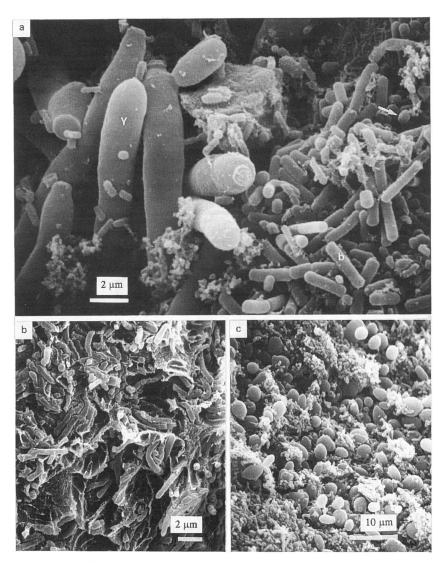

Figure 3.8 Scanning electron micrographs of kefir grains from different sources. (a) Russian kefir grains showing a variety of microorganisms; (b), rod-shaped bacteria; y, yeast cells; arrows, cocci (M. Kalab, personal communication). (b) Russian kefir grain predominantly colonized by lactobacilli (M. Kalab, personal communication). (c) Chr. Hansen's Laboratorium kefir grains showing a wide range of indigenous microflora (Toba *et al.*, 1990a; reproduced by courtesy of *International Journal of Food Microbiology.*)

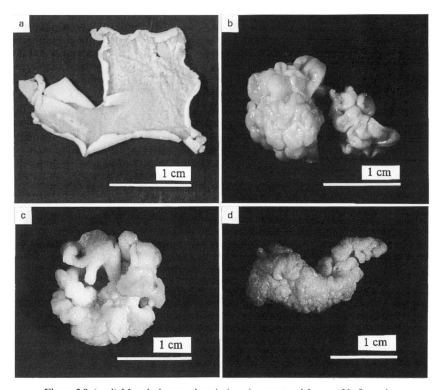

Figure 3.9 (a–d) Morphology and variations in structural forms of kefir grains.

incubation or ambient temperature in Northern Ireland, leading to a fermentation with different characteristics.

(b) Production of kefir bulk starter culture. For the manufacture of kefir in the former USSR, either starter I or II could be used; Figure 3.10 illustrates schematically the different manufacturing stages (Koroleva, 1988a). It is highly recommended that starter I is used for kefir production, but starter II is only used in factories if the equipment for separating and washing the kefir grains is not available.

In view of the need to maintain the appropriate ratio of the various microflora in the kefir grains, and to produce a quality product (Koroleva, 1988a, 1991) the starter culture handling requirements are:

- Use good-quality, low count skimmed milk, and heat-treat to 95°C for 10–15 min.

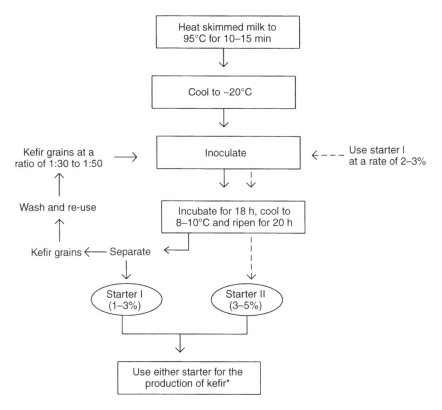

Figure 3.10 Preparation of different starters for the production of kefir. *Refer to text for further information. (Adapted from Koroleva, 1988a, 1991.)

- Use kefir grains to inoculate the milk at a ratio of 1:30 to 1:50; at ratios <1:30 the microflora balance will change; daily transfer is recommended.
- Agitate twice during the incubation period of starter I after 16 and 22 h, respectively; ferment the milk at 20°C since at temperatures >25°C the balance of lactic acid bacteria, yeasts and *Acetobacter* spp. will change, favouring an increase in the *Lactobacillus* counts and a reduction in the counts of the other microorganisms, mainly the yeasts and *A. aceti*, which can affect the flavour and aroma development in kefir.
- Slow cooling of the coagulum (10–12 h) is desirable for allowing the microorganisms to develop the right intensity of flavour.
- Excessive washing of the grains will upset the microbial balance and decrease their activity; the starter activity can be restored after further cultivation of the grains for 3–5 days (Rasic, 1987; Koroleva, 1988a, 1991).

As the production of kefir involves recovery of the grains and their re-use after washing, large-scale production is somewhat laborious. The estimated production time is ~ 24 h. In Europe there have been developments of freeze-dried starter for commercial kefir production which removes the need to recover grains. Pettersson (1984) has selected strains of homolactic lactococci, lactobacilli and *Cand. kefyr*, and each was cultivated separately at constant pH, concentrated and freeze-dried. The final blend consisted of *Lac. lactis* subsp. *lactis* (75%), *Lac. lactis* biovar *diacetylactis* (24%), lactobacilli (0.5%) and yeast (0.1%). Using this starter, a consistent kefir was obtained which was judged to have better flavour and aroma (i.e. more diacetyl and less yeast flavour), but the product had less ethanol and CO_2, due to low number of yeast in the starter culture.

Klupsch (1984) has patented a single-use kefir starter consisting of lactococci, lactobacilli and yeast. The milk is inoculated with 2% mixed starter culture containing *Lac. lactis* subsp. *lactis*, biovar *diacetylactis*, subsp. *cremoris*, and also *Lb. acidophilus*, *Lb. brevis* and *Lb. delbrueckii* subsp. *lactis*, and then incubated at 24–27°C until the pH falls to 4.4. After cooling the fermented milk a culture of *Cand. kefyr* is added, but no further incubation takes place. The so-called kefir, therefore, contains yeasts but has not been fermented by them and, similar to Pettersson's starter mentioned above, has little ethanol content or gassiness. The products have been developed to avoid CO_2 production which ultimately causes blowing of laminated paper board cartons used for packaging fermented milks. Furthermore, the use of such cultures removes the authentic effervescent and aroma qualities of kefir, and it is arguable whether these products could be called 'kefir'.

Freeze-drying of kefir grains causes cellular injury to the yeasts and reduces their viable counts in the dried culture. Addition of a mixture consisting of 20% sucrose solution and sterilized starch to the kefir grains (wet) has been shown to protect the yeast during freeze-drying (Kramkowska *et al.*, 1986). Such kefir starter culture produces a product analogous to kefir made with grains, and the preparation of such kefir starter is less laborious and less complicated when compared with kefir grains. Since the presence of CO_2 in kefir is essential, Duitschaver *et al.* (1987) have developed a blend of pure starter cultures consisting of *S. thermophilus* + *Lb. delbrueckii* subsp. *bulgaricus* (3%), *Lb. acidophilus* (3–5%) and a mixed culture of *Lac. lactis* subsp. *lactis* + *Leuconostoc* spp. (1%) to ferment the milk. Sucrose ($0.6 \, g \, 100 \, g^{-1}$) and yeast culture (1%) are added to the cold fermented milk before filling in glass bottles and crown-capping. The sensory properties indicated that this fermented milk had similar characteristics to traditional kefir prepared with grains, the process is suitable for commercial-scale production, and the product had good storage life, i.e. no deterioration in quality after 42 days at 5°C.

Recent developments in kefir starter technology involve the use of immobilized yeast cells (Clementi *et al.*, 1989) or multistarter cells (Gobetti and Rossi, 1994) for kefir production. The results are encouraging for the release of appropriate cells and production of desirable metabolites in whey and milk, respectively. These techniques are suitable for large-scale production of kefir containing the desired level of lactic acid, CO_2, ethanol, diacetyl and acetoin.

(c) Methods of manufacture. Traditionally, the Caucasian inhabitants made a refreshing drink called kefir from raw cows' or goats' milk using kefir grains as the starter culture. The milk is placed in leather sacks, seeded with grains, placed in the sun during the day and taken into the house at night. The sack is normally hung near the door, and anyone coming in or going out has to push the sack in order to mix the contents. As some of the kefir is consumed, more fresh milk was added (Roginski, 1988; Koroleva, 1991). This method of production is referred to as on-going fermentation where the level of acid, ethanol and CO_2 content depended on the holding time. The leather sacks are washed and new grains are used if the kefir develops unacceptable flavours and taste.

Cows' milk for kefir manufacture is not fortified and the fat content is standardized to range between <0.1 and 3.2%. The milk is warmed to 70°C, homogenized at 12.5–20 MPa pressure, heated to 85–87°C for 10 min or to 90–95°C for 2–3 min, cooled to ~22°C and inoculated with starter I or II (see Figure 3.10). The fermentation period is for 8–12 h until the acidity reaches ~1% lactic acid followed by agitation of the coagulum and slow cooling for 10–12 h. Before packaging, the product is agitated again, filled in retail containers and ripened in the cold store (Eller, 1971; Koroleva, 1988a,b, 1991; Kurmann *et al.*, 1992; Kroger, 1993; Bottazzi *et al.*, 1994). However, quick cooling to 4–6°C in a plate heat-exchanger at ~pH 4.5 has been reported by Anon. (1995).

According to Koroleva (1991), by using starter I the microflora in kefir is similar to that reported above. The product should have an homogenous consistency, refreshing taste and the alcohol (~0.1%) and CO_2 contents are relatively low when compared with a traditional kefir (e.g. ~2% ethanol). A similar method of kefir production is widely used in Poland (Libudzisz and Piatkiewicz, 1990). Kefir high in alcohol and CO_2 contents is obtained by fermenting the milk at temperatures ranging from 4 to 15°C which favours the yeast microflora. However, incubation at 25°C is recommended for optimum production of ethanol and volatile acids (Liu and Moon, 1983).

Other methods for the manufacture of kefir have been reviewed by Tamime and Robinson (1988) and include: (i) multiple stage heat treatment of milk (i.e. heating to 87°C cooling to 77°C, re-heating to 87°C, cooling to 77°C and holding for 30 min); this helps to improve the rheological properties of the kefir and reduce syneresis; (ii) fermentation of the milk at 25°C; this improves the viscosity and flavour of kefir; (iii) the addition of Na-caseinate to milk for the production of a dietetic kefir called osobyi; and

(iv) UHT milk; this is used successfully for the manufacture of kefir which has similar characteristics to a product made with HTST milk (Merin and Rosenthal, 1986).

A Canadian group of researchers (Duitschaver *et al.*, 1987) have developed a starter culture for the production of kefir and an alternative procedure of manufacture consisting of blending yoghurt, acidophilus milk and cultured buttermilk at a ratio of 30:30:40, respectively, followed by secondary yeast fermentation after bottling. Consequently, the same researchers (Duitschaver *et al.*, 1988) have evaluated five different procedures for making kefir as follows: (i) using the pure culture developed and sequential yeast fermentation as reported by Duitschaver *et al.* (1987); (ii) a direct-set method; (iii) simultaneous batch lactic acid and yeast fermentation; (iv) similar to (iii) but followed by additional incubation in the bottle; and (v) using a kefir grain-free starter. The kefir produced using procedure (i) was superior in quality to other kefirs, and was highly acceptable by the tasters (Duitschaver *et al.*, 1988). Furthermore, sweetening the kefir with xylitol (3%) and fruit-flavouring with peach was highly rated with tasters when compared with plain-flavoured kefir (Duitschaver *et al.*, 1991). This approach is compatible with the popularity of sweetened and flavoured yoghurt in Europe and North America.

The vitamin B-complex of kefir has recently been studied by Kniefel and Mayer (1991) using 10 different samples of kefir grains and different mammalian milks (cow, goat, sheep and mare). The kefir starter cultures have either synthesized or catabolized certain vitamins during growth, and the pattern was influenced by the origin of the kefir grain and the type of milk used. On average the results could be summarized as follows: (i) thiamin concentration has increased by 24% only in sheep's milk kefir; (ii) folic acid has increased by 22, 71 and 44% in cows', ewes' and goats' milk kefirs, respectively; and (iii) pyrodoxine content has increased by 89, 20 and 40% in ewes', goats' and mares' milk. Similar results were also reported by Laukkanen *et al.* (1988).

3.4.2 *Koumiss, kumys or coomys*

Koumiss was made originally from mares' milk whose chemical composition is ~90% moisture, 2.1% protein (of which 1.2% casein and 0.9% whey proteins), 6.4% lactose, 1.8% fat and 0.3% ash (Doreau and Boulot, 1989; Doreau *et al.*, 1990; Pagliarini *et al.*, 1993; Lozovich, 1995). The name is derived from a tribe called Kumanes who inhabited the area along the Kumane or Kuma river in the Asiatic Steppes, or it is of a Tartar origin. In some instances ass's or camels' milks are used for koumiss production. The product prepared from mares' milk is a beverage because the milk does not coagulate. It is milky grey in colour, light, fizzy, and has a sharp alcohol and acidic taste. The main metabolites after fermentation are 0.7–1.8% lactic acid, 0.6–2.5% ethanol and 0.5–0.9% CO_2 (Eller, 1971; Oberman, 1985).

Mother culture

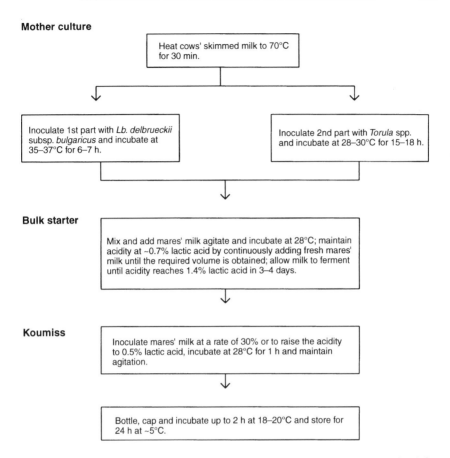

Figure 3.11 A schematic illustration showing the principles of manufacture of koumiss. (After Berlin, 1962.)

(a) Traditional methods.   Originally, koumiss was prepared from unheated mares' milk which was allowed to ferment in smoked horses hide called tursuks or burduks. The freshly-drawn mares' milk is filled into the containers, agitated with a special apparatus, and if seeded with koumiss the product is ready for consumption within 3–8 hours. As the kumiss is consumed, the tursuk is filled with mares' milk to provide an on-going fermentation. The microflora is not well defined, but consists mainly of thermophilic lactobacilli *Lb. delbrueckii* subsp. *bulgaricus, Lb. acidophilus,* lactose-fermenting yeasts (*Saccharomyces lactis, Torula koumiss*) non-lactose-fermenting yeast (*Saccharomyces cartilaginosus*) and non-carbohy-drate-fermenting yeast (*Mycoderma* spp.) (Eller, 1971; Oberman, 1985; Koroleva, 1988a, 1991). Although lactococci are present in Mongolian koumiss, their presence in starter cultures in other countries may not be

desirable because the quick development of lactic acid inhibits the growth of yeasts (Koroleva, 1991).

To preserve the koumiss starter culture in rural areas of the Asiatic Steppes and Mongolia, the tursuks are stored in a cold place from the previous season containing goats' milk koumiss. At the beginning of the following lactation period of the mares, the tursuks are filled gradually with mares' milk, and in 5 days the starter culture is reactivated. Later koumiss of good quality is used as a starter culture (Koroleva, 1991).

(b) Commercial production systems. The traditional process of koumiss production is difficult to control, and Berlin (1962) has developed a process (Figure 3.11) when the starter culture is prepared, and the product manu-factured at the same time. The inoculation rate of the starter culture is ~30%, which is possibly the highest inoculation rate used in any fermented milk or cheese production.

Heat treatment of mares' milk for koumiss making is not mandatory, and the desirable effect of whey protein denaturation on gel formation in fermented milks is not required as koumiss is a liquid. Recently, Bonomi *et al.* (1994) have studied the thermal sensitivity of mares' milk proteins, and reported that α-La and β-Lg had much lower thermal sensitivities when compared with bovine counterparts. These whey proteins did not denature or solubilize significantly at temperatures <100°C. Thus, the heat treatment of mares' milk for koumiss production maintains public health safety and provides a suitable growth medium for the starter culture without any competition from undesirable microorganisms.

Koumiss is classified into three categories depending on the extent of fermentation (Berlin, 1962):

Flavour category	Acidity (%)	Alcohol (%)
Mild	0.6–0.8	0.7–1.0
Medium	0.8–1.0	1.1–1.8
Strong	1.0–1.2	1.8–2.5

The normal-type koumiss may have viable cell counts of 4.97×10^7 cfu ml^{-1} and 1.43×10^7 cfu ml^{-1} of bacteria and yeast, respectively.

Developments in starter for koumiss have been reported by many researchers which included: (i) the use of *Lac. lactis* subsp. *lactis*, *Lb. delbrueckii* subsp. *bulgaricus*, *Sacc. lactis* and *A. aceti* (0.02%) for flavour development, and at the same time the technology has been improved to extend the shelf-life of the product to 14 days (Oberman, 1985); (ii) the therapeutic property of koumiss against *Mycobacterium tuberculosis* is said to be enhanced by using cows' milk and a starter culture consisting of *Lb. delbrueckii* subsp. *bulgaricus*, *Lb. acidophilus* and *Sacc. lactis* (Koroleva,

1988b, 1991); and (iii) the presence of *Lactobacillus leichmanii* (currently known as *Lb. delbrueckii* subsp. *lactis* because of high phenotypic and genotypic similarities) in koumiss has been reported by Aguirre and Collins (1993), but no significance was given for the use of such microorganism.

Due to shortage of mares' milk, the industrial scale of koumiss can be made in one of the following ways (Eller, 1971; Koroleva, 1988b, 1991):

1. Skimmed milk is fortified with 2.5% sucrose, heated to 90°C for 2–3 min and cooled to ∼28°C. Inoculate the starter culture at a rate of 10%, stir for 15–20 min, and incubate at ∼26°C for 5–6 h until acidity reaches 0.8–0.9% lactic acid. The coagulum is agitated for 10–15 min, aerated, cooled and followed with further agitation for 10–15 min without aeration. After ∼2 h the temperature is lowered to 16–18°C, the koumiss is homogenous, slightly viscous and foamy, and filled in bottles. The final ripening takes place at 4°C and the acidity ranges from 1.0–1.5% lactic acid.

2. A dried powder, made from whole milk, skimmed milk and cheese whey, is recombined at 50°C, heated to 85–87°C with 5–10-min holding, homogenized at 10–12 MPa and cooled to incubation temperature. Starter culture (10%) and ascorbic acid (0.02%) are added to the milk, fermented for 3–4 h and constantly stirred until acidity reaches 0.8% lactic acid. The coagulum is cooled to ∼17°C, stirred constantly for 1–2 h, bottled and capped with aluminium foil and finally ripened at 6–8°C. Mild-, medium- or strong-flavour koumiss is produced depending on acidity (1.0–1.3% lactic acid) and ethanol (0.6–1.6%).

3. Blend 5 parts whole cows' milk with 8 parts UF rennet whey (i.e. 2-fold concentration of the protein content) to produce milk similar in composition to mares' milk [1.5% fat, 2.0% protein, 5.0% lactose and 0.7% ash (Puhan and Gallman, 1980)]. The milk is hydrolysed with β-D-galactosidase, heated to 95°C for 15 min, cooled, inoculated with mixed starter culture and fermented in two steps (lactic acid production for 15 h followed by alcoholic fermentation at 15°C for ∼20 h). The product is ripened at 10°C for 40 h and stored later at <5°C. The product was milder when compared with Mongolian koumiss, and upon opening the crown-cork bottles, the CO_2 pressure caused excessive foaming. The traditional product in Mongolia is called airag which contains 2.5–3.0 g alcohol $100\,g^{-1}$, and current estimated *per capita* consumption is 5 kg per year (Kurmann *et al.*, 1992).

4. When cows' milk is blended at a ratio of 1:1 with clarified whey and sweetened with 2.5% sucrose, the chemical composition of the mixture (i.e. 1.8% protein, 1.7% fat and 7.0% sugar) resembles mares' milk (Guan and Brunner, 1987). The modified milk is heated to 80°C for 20 min, cooled to 28°C and inoculated with starter culture (5–10%) consisting of *Lac. lactis* subsp. *lactis*, *Lb. delbrueckii* subsp. *bulgaricus* and *Kluyveromy-*

ces lactis or *Kluyveromyces fragilis*. The mixture is agitated vigorously for 10 min, incubated for 12–15 h at 26°C until the acidity reaches 1.0% lactic acid, passed through the homogenizer without a pressure and packed in glass containers. The koumiss is ripened for 2 h at 20–25°C and stored at 4°C. Organoleptic assessment of this type of koumiss drink was not reported.

3.4.3   Acidophilus–yeast milk

Little is known regarding the science and technology of this fermented milk beverage, which is only produced in the former USSR. Whole or skimmed milk is heated to 90–95°C for 10–15 min, cooled to 35°C, and inoculated with 3–5% mixed starter culture (*Lb. acidophilus* and *Sacc. lactis*). The milk is bottled and first stage of fermentation takes place at 35°C until acidity reaches 0.8% lactic acid, followed by the 2nd stage fermentation at 10–17°C for 6–12 h (Skorodumova, 1959; Eller, 1971; Koroleva, 1991). As antici-pated, *Lb. acidophilus* produces the acid while the ethanol and CO_2 are produced by the yeasts. The final product is stored at <8°C until consumed. Alternatively, acidophilus–yeast milk can be produced in bulk in a large fermentation tank (Eller, 1971), but the CO_2 content may be reduced due to pumping before packaging. The product is described as viscous, slightly acidic and sharp with a yeasty taste.

The product has been developed for the treatment of certain intestinal and other diseases where the *Lactobacillus* strain has a high antimicrobial activity against undesirable microflora in the intestinal tract. *Sacc. lactis* also possesses an antibiotic activity against *Myc. tuberculosis*. A 3-day-old product has the highest antimicrobial activity (Koroleva, 1991).

Subramanian and Shankar (1985) achieved high numbers of viable cells of *Lb. acidophilus* (6.2–9.8×10^8 cfu ml^{-1}) in the presence of lactose-fer-menting *Sacc. fragilis* or *Candida pseudotropicalis*. The milk is heated to 90°C for 20 min, and coagulation achieved in <20 h at 33 or 37°C. The consistency of the coagulum is improved by fortification of the milk with 1.5% skimmed milk powder and 0.5% agar. The latter compound prevents the break-up of the curd caused by the production of CO_2.

The packaging of yeast–lactic fermented products in hermetically sealed, laminated paper board cartons or plastic containers may pose the problem of consumer rejection of swollen packages due to a build-up of pressure as a result of CO_2 formation inside the container; recently a breathing membrane which allows CO_2 to escape has been developed in Switzerland (Tamime and Robinson, 1988).

3.4.4   Acidophiline or acidophilin

Acidophiline could be considered similar to acidophilus–yeast milk, but the starter culture consists of *Lb. acidophilus*, *Lac. lactis* subsp. *lactis* and kefir yeast (Eller, 1971) or kefir starter (Koroleva, 1991). This milk beverage can be made from whole or skimmed milk and is sweetened. Acidophiline has been used to treat colitis, enterocolitis, dysentery and other intestinal diseases (Koroleva, 1991).

Characterization of acidophiline using different strains of *Lb. acidophilus* R, I, Ch-2 and H in combination with *Lac. lactis* subsp. *lactic* C-10 and kefir grains have been reported by Sharma and Ghandi (1981, 1983). They concluded that: (i) the *Lactobacillus* strain H plus the other starter cultures had good desirable acidity and antibacterial activity against *Micrococcus flavus*, *E. coli*, *Bacillus subtilus* and *Staphylococcus aureus*; and (ii) fortification of the milk with SMP, sucrose, and/or cream provided improved the sensory properties.

3.5   Fermented milks with mould and lactic acid

3.5.1   Background

Mould contamination of fermented milk products is generally undesirable, but in Finland the mould *G. candidum* is deliberately added with the starter culture to ferment one particular milk product. The product is called viili, and the presence of the mould results in growth on the entire surface of the coagulated milk to give a unique characteristic when compared with other types of fermented milks.

3.5.2   Viili

This Finnish cultured milk product is similar to the Nordic fermented milks, but as the starter culture is enriched with a mould the product has a different taste, aroma and appearance. The different types of viili are: (i) low-fat ($\sim 2.5\%$) known as kevytviili; (ii) whole fat (3.9%) called viili; and (iii) cream viili containing 12% fat (Laukkanen *et al.*, 1988). Normally, the product is natural, but in the early 1980s, fruit-flavoured and sweetened products were introduced and marketed under the name of marjaviili (Tamime and Robinson, 1988).

Standardized milk is heated to 83°C for 20–25 min without homogenization, cooled to 20°C and inoculated with 3–4% starter culture comprising *Lac. lactis* subsp. *lactis*, subsp. *cremoris*, biovar *diacetylactis*, *Leu. mesenteroides* subsp. *cremoris* and the mould *G. candidum* (Meriläinen, 1984; Marshall, 1986). It is then packed into the retail cups, incubated at 18–20°C for ~ 24 h until the acidity is 0.9% lactic acid and finally cooled. During the incubation period, the fat rises to the surface where the mould grows to give a velvet-like appearance to the product. Incidentally, the marjaviili consists of three layers, i.e. fruit-flavour(s) at the bottom, coagulated milk and cream layer plus mould growth on the surface. Viili is mildly sour, aromatic in taste, stretchy, but can be cut easily with a spoon, and the flavour is similar to other types of buttermilk with a slight musty aroma attributed to *G. candidum*. The velvet-like layer of the mould may have the advantage of preventing the growth of spoilage organisms (Marshall, 1986).

The ropy *Lactococcus* strains isolated from långfil and viili have different plasmid profiles (ropyness is carried on a 17-Md plasmid in the Swedish strain and on 30-Md plasmid in the Finnish strain). Comparative restriction endonuclease analysis and DNA/DNA-hybridization suggest that the plasmids share homologous DNA regions. However, the Swedish ropy strains harboured a conjugative 45-Md lactose plasmid (Neve *et al.*, 1988). The cell surface proteins of slime-forming lactococci consisted of polypeptides of molecular weights 26 and 42 kDa, present in both ropy strains, which could be associated with slime excretion (Kontusaari and Forsèn, 1988), and lipotechoic acid from the membrane structure was also detectable on the surface of intact cells (Forsèn *et al.*, 1985).

The microstructure of viili (Figure 3.12) made with slime-forming starter culture showed that the ropy material forms a network attaching the bacterial cell to the protein matrix (Toba *et al.*, 1990b). The slime filaments in viili are thicker when compared with the microstructure of yoghurt made with ropy cultures of *Lb. delbrueckii* subsp. *bulgaricus* and *S. thermophilus* which may be due to the difference in the chemical constituents of the slime. The viili cultures produce slime made of polysaccharide and protein while

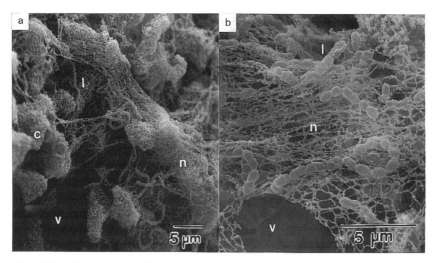

Figure 3.12 Microstructure of fermented milk made with ropy lactococci strain isolated from viili. (a) Low magnification; (b), high magnification. c, aggregates of casein micelles; l, lactococci; n, slime network; v, void space. (After Toba *et al.*, 1990b; reproduced by courtesy of *International Journal of Food Microbiology*.)

the yoghurt cultures produce a phospho-polysaccharide slime (Toba *et al.*, 1990b, and for further detail refer to Chapter 4).

3.6 Concentrated fermented milks

Concentrated or strained fermented milk products are produced in many countries. The rural method of manufacture involves draining the fermented milk (skimmed or full-fat) using a cloth bag, animal skin, or earthenware vessel. The product(s) is known as labneh or lebneh (in the Lebanon and most Arab countries), tan or than (in Armenia), torba or süzme (in Turkey), leben zeer (in Egypt), chakka or shrikhand (in India), stragisto or sakoulas (in Greece), mastou or mast (in Iraq and Iran), basa, zimme or kisela mleko-slano (Bulgaria and Yugoslavia), Greek-style (in the UK), ymer (in Denmark) and skyr (in Iceland) (El-Gendy, 1983, 1986; De, 1980; Tamime and Robinson, 1988; Kurmann *et al.*, 1992; Akin and Rice, 1994).

At present the industrial methods for the manufacture of concentrated fermented milk products could be summarized as follows:

- Traditional method using the cloth bag.
- Mechanized techniques include:
  Nozzle separators
  Membrane filtration (i.e. ultrafiltration)
- Recombination or product formulation.

Most of the concentrated fermented milk products mentioned above are made using mixed starter culture of lactic acid bacteria. In some instances yeast is employed, and examples follow.

3.6.1 Ymer

A Danish concentrated cultured milk product containing (g $100 \, g^{-1}$) fat 3.5, and SNF 11, including protein 5–6. The st)arter culture consists of *Lac. lactis* biovar *diacetylactis* and *Leu. mesenteriodes* subsp. *cremoris* (Anon., 1969). A similar product, lactofil, is made in Sweden.

Traditionally, skimmed milk is heated to 90–95°C for 3 min, cooled to 19–23°C, inoculated with a mesophilic starter culture (e.g. BD type from Chr. Hansen's Laboratorium in Denmark), and incubated for 16–18 h or until the pH reaches 4.5. The coagulum is cut and warmed indirectly in the fermentation tank to $\sim 45°C$ in around 2 h. This allows syneresis to occur, and 50% of the original volume is removed as whey. Cream (36 g fat $100 \, g^{-1}$) is then blended with the product, homogenized (4.9–9.8 MPa) at 35–45°C, partially cooled in a plate heat-exchanger and packaged (Tamime and Robinson, 1988; Robinson and Tamime, 1993).

At present, nozzle separators (quarg separator) and the UF process are used to produce ymer commercially. In the former method the fermented skimmed milk is thermized to 56–60°C for 3 min, cooled to 37°C, and separated. The ymer leaves the machine through nozzles of the quarg separator at the periphery of the bowl and discharged into a cyclone from which it is forwarded by a positive displacement pump to be mixed with cream, cooled and packaged (Anon., 1995). Figure 3.13 illustrates schematically the production line of quarg.

The manufacturing stages of ymer using the UF process are:

- **First method:** Skimmed milk is heated to 92°C for 15 s, cooled to 55°C and ultrafiltered to the desired concentration. Standardize the retentate with cream to 3.5 g fat 100 g^{-1}, homogenize (19.6 MPa) at 65°C, heat to 85°C for 5 min, cool to 20–22°C, add starter culture and incubate for 20 h. Stir coagulum, cool to 5°C and store for 24 h, package and remove to cold store (Samuelsson and Ulrich, 1982). The difference in the consistency of ymer made by the traditional and UF methods have been reported by Mogensen (1980). The brittle characteristic of UF ymer is due to retention of calcium and processing conditions, and some recommendations have been proposed.
- **Second method:** Concentrate skimmed milk to a protein content of 6 g 100 g^{-1} by UF at 50°C. Standardize the fat content in the retentate to 3.5 g 100 g^{-1} (Danish standard), and then homogenize the milk (13.7 MPa) at 74°C, de-aerate and heat to 95–100°C for 1 min. The processed milk is cooled to 22°C inoculated with a mesophilic starter culture (i.e. consisting of flavour- and aroma-producing strains, and incubated for 20–22 h. Next day the coagulum is mixed gently for 1 h, homogenized at 4.9 MPa to impart a smooth texture, and then cooled to 12°C for packaging (J. Kirkegaard, personal communication).
- **Third method:** At present it is possible to UF warm fermented milk for the manufacture of ymer; a process similar to production of strained yoghurt or labneh.

3.6.2 Skyr

This is an Icelandic concentrated fermented skimmed milk product, with a chemical composition (g 100 g^{-1}) of TS 17.5, fat 0.2, protein 12.7, lactose 3.9 and ash 0.8 (Tamime and Robinson, 1988). The traditional product, which employs the cloth bag method, has higher TS content (20.8 g 100 g^{-1}).

Little is known of the microflora of skyr, but according to Pétursson (1949) it consists of thermophilic lactic acid bacteria (*S. thermophilus, Lb. delbrueckii* subsp. *bulgaricus, Lb. helveticus*) and lactose-fermenting yeast. The commercial method of production uses nozzle separators and can be summarized as follows. Skimmed milk is heated to 90°C for 30 min, cooled

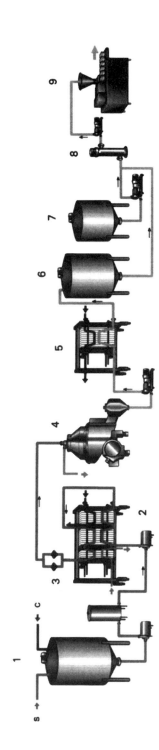

Figure 3.13 Flowchart for mechanized production of quarg. 1, Ripening tank (s, skimmed milk; c, starter culture); 2, Plate heat exchanger for thermization; 3, Filter system; 4, Quarg separator; 5, Plate heat exchanger; 6, Intermediate tank; 7, Cream tank; 8, Dynamic mixer; 9, Filling machine. After Anon., 1995 (reproduced by courtesy of Tetra Pak Processing Systems AB.)

to 40°C, inoculated with mixed starter culture and fermented for 4–6 h until the pH reaches 4.7. Sometimes chymosin is added with starter culture (i.e. 2 drops $10 l^{-1}$) to improve the quality of the product. Cool to 18°C and continue secondary fermentation for 18 h (pH 4.1), heat to 67°C for 15 s, cool to 35–40°C, concentrate using a quarg separator, partially cool to 10°C (optional add cream and or fruit flavours), package and store. In order to improve the yield, the permeate is ultrafiltered (17.5 g TS 100 g^{-1}), heated to 80°C, cooled to 40°C, homogenized, cooled to 10°C and finally blend with the concentrated product before packaging (Gudmundsson, 1987). The chemical composition of such skyr is similar to that reported earlier, but it also contains ethanol (0.3–0.5%), CO_2, flavour compounds such as acetaldehyde and diacetyl, and acetic acid is also present. The anticipated yield of skyr is ∼2 kg from every 10 l of skimmed milk used.

3.6.3   Chakka and shrikhand

Chakka is an Indian-type fermented milk which is made by straining dahi (Indian yoghurt) using a cloth bag. If the curd is sweetened with sugar, the product is known as shrikhand or shrikhand wadi (De, 1980; Patel and Chakraborty, 1988; Punjrath, 1991). Mixed strains of mesophilic lactic acid bacteria are used to ferment the milk, and for the production of an 'acidic'-type dahi, yoghurt culture is used. A mixed culture LF-40 (*Lac. lactis* subsp. *lactis* and biovar *diacetylactis*) has been accepted as the most suitable by many shrikhand manufacturers in India (Patel and Abd El-Salam, 1986). Basket centrifuge at 900 g for 90 min has been used for the production of chakka (Patel and Chakraborty, 1985).

In a separate study (Khanna *et al.*, 1982) the mutant strain PM of *Lac. lactis* biovar *diacetylactis* has been used successfully to produce chakka from milk (cows' or buffalo) and from recombination, and these products have been more highly rated by the panel of judges. The shelf-life of skirkhand can be improved by post-fermentation heat treatment at ∼70°C where the product has remained acceptable after 15 days' storage at 35–37°C or >70 days at 8–10°C (Prajapati *et al.*, 1992, 1993).

3.6.4   Strained yoghurt or labneh

The methods available for the manufacture of strained yoghurt have been reviewed by Tamime and Robinson (1988), Salji (1991), Robinson and Tamime (1993) and Tamime (1993).

The chemical composition (g 100 g^{-1}) of strained yoghurt varies depending on country of origin or existing legal standards, and for example, TS and fat content range between 20 to 28 and 7 to 10, respectively (Tamime and Robinson, 1988; Tamime, 1993). Whole milk is normally used to prepare the yoghurt followed by partial removal of the whey using a cloth bag. Factors that can affect whey drainage and yield are: (i) fortification of the milk solids helps to increase the yield; (ii) straining the fermented milk (pH ~4.8) at 5°C rather than 25°C, also contributes to the high yield due to greater retention of moisture; (iii) polymer-producing starter cultures produce very viscous strained yoghurt, but longer time is required for the whey extraction; (iv) higher losses of milk solids occur in the whey as the pressure increases, and the permeability rate of whey is not constant during the concentration period; and (v) the yield of goats' milk strained yoghurt is higher when compared with similar product made from cows' milk (Tamime and Robinson, 1978; Hamad and Al-Sheikh, 1989; Kehagias *et al.*, 1992). A study on the vacuum filtration of yoghurt confirms the reported results mentioned above (Akin *et al.*, 1995).

UF milk retentate ($\sim 22 \, \text{g TS} \, 100 \, \text{g}^{-1}$) has been used to manufacture strained yoghurt in a way similar to the production of set-type yoghurt (see Figure 3.1). Using such a method, no whey drainage takes place after the fermentation period. El-Samragy and Zall (1988) reported that an acceptable product was manufactured using UF retentate, but such type of strained yoghurt had a tendency to crack, was considerably less elastic, and excessive whey syneresis occured when the product was broken with a spoon (Tamime *et al.*, 1989a). However, UF of warm yoghurt has resulted in a product similar to traditional strained yoghurt (Tamime *et al.*, 1989b), or the addition of whey protein concentrate to UF retentate has prevented whey syneresis in strained yoghurt (Mahfouz *et al.*, 1992).

Yoghurt starter cultures are normally employed for the production of strained yoghurt; however, mesophilic lactic acid bacteria, *Lb. acidophilus*, *Bifidobacterium*, spp. and *Ent. faecalis* have been used in different combinations to ferment the milk (El-Samragy *et al.*, 1988; Mahdi *et al.*, 1990; Abou-Donia *et al.*, 1992a, b). Vitamin B_{12} and folic acid contents in strained yoghurt have been increased by 21% and 28%, respectively, when using *Propionibacterium freudenreichii* subsp. *shermanii* in combination with mesophilic starters (*Lac. lactis* subsp. *lactis* 75%, subsp. *cremoris* 15% and *Leu. mesenteriodes* subsp. *cremoris* 10%) (Khattab, 1991).

Quarg separators (Figure 3.13) have been used successfully for industrial-scale production using skimmed milk as the basic raw material and cream is added to the concentrated milk before packaging. Recent developments in the design of separators have made it feasible to use fermented, whole milk for the production of strained yoghurt (Lehmann *et al.*, 1991). A typical chemical composition ($\text{g} \, 100 \, \text{g}^{-1}$) for strained yoghurt is total solids 24% and fat 9.6% (about 40% fat-in-dry-matter); the composition of the whey is $6.1 \, \text{g} \, 100 \, \text{g}^{-1}$ total solids, consisting mainly of lactose and minerals, but also about $0.5 \, \text{g} \, 100 \, \text{g}^{-1}$ fat. Capacities of such separators are up to 6.5 tonnes h^{-1}, depending on the composition of the milk used and the acidity of the fermented milk before concentration.

An alternative system for the manufacture of strained yoghurt is by UF of warm yoghurt after fermentation (Tamime *et al.*, 1989b). Figure 3.14 shows the production line for the manufacture of strained yoghurt by UF

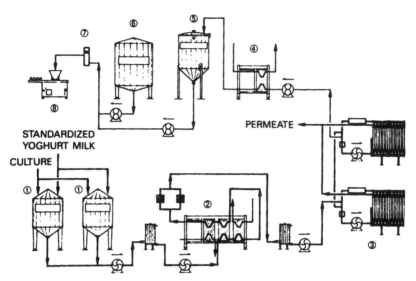

Figure 3.14 Simplied flow chart for the manufacture of strained yoghurt by ultrafiltration (UF). 1, Fermentation tanks; 2, Plate heat-exchanger; 3, Two- to four-stage UF plant; 4, Plate cooler; 5, Buffer tank; 6, Fruit tank (optional); 7, In-line mixer; 8, Packaging machine. (After Tamime, 1993; reproduced by courtesy of Tetra Pak Processing Systems AB and Academic Press Ltd.)

(Tamime, 1993). The effect of processing temperatures on the quality of UF strained yoghurt was reported by Tamime *et al.* (1991b), and UF at 40–50°C have been recommended. With UF at higher temperature, the yoghurt microorganism counts have been reduced by 2 $\log_{10}$ cycles. However, the firmness and microstructure of UF strained yoghurt was influenced by the processing conditions (Tamime *et al.*, 1991a); processing at $\geqslant 50°C$ the product becomes firmer due to more complex micellar chain compared with simple aggregates of protein particles. The structures of goats' and sheep's strained yoghurt were similar to each other (Figure 3.15(b–d)) and less uniform than a similar product from cows' milk (Figure 3.15(a)) (Tamime *et al.*, 1991c).

Ultrafiltration of warm yoghurt at temperatures >45°C increases the fouling rate of UF membranes (Attia *et al.*, 1991a, b) which can reduce the flux rate and may affect the processing conditions in large-scale operations where the equipment requires to be washed more frequently. In a recent report, Sachdeva *et al.* (1992a, b, c, d) have studied the performance (i.e. flux rate, energy consumption and optimal operational temperatures) of plate and frame, spiral wound, hollow fibre and mineral membrane modules for the manufacture of quarg from: (i) partially acidified milk (pH 6) followed by ultrafiltration and fermentation of the retentate, and (ii) concentration of coagulated skimmed milk (pH 4.6). Better quality quarg was obtained by

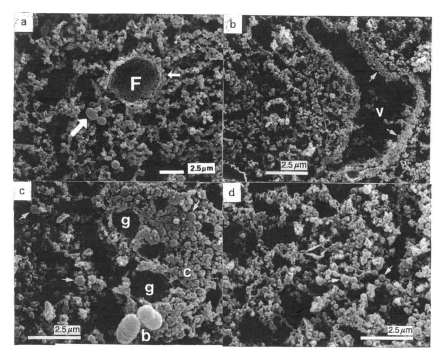

Figure 3.15 Microstructure (SEM) of strained yoghurt made from different milks. (a) UF cows' strained yoghurt: F, medium-sized fat globule; residue of fat globule membrane (small arrow) and streptococci (large arrow). (b) UF sheep's strained yoghurt: residue of fat globule membrane (asterisks); V, void space; compact protein particles (arrows). (c) Goats' traditional strained yoghurt: c, compact casein particle clusters; g, fat globules; b, bacteria; large casein micelles having smooth surface (arrows). (d) UF goats' strained yoghurt showing large casein micelles (arrows). After Tamime *et al.*, 1989a, c; reproduced by courtesy of *Scanning Microscopy International.*)

UF of coagulated milk at 50°C because in the former method the quarg had a bitter taste due to retention of calcium.

Strained yoghurt can be manufactured from recombined dairy ingredients. The process involves reconstitution of powders (whole, skimmed, high protein and/or caseinate) in water, and blending it with AMF, stabilizer(s), and salt (optional). The milk base is processed in a similar way to the production of set- or stirred-type yoghurt. In the latter type, after the fermentation stage, the product is partially cooled to 20°C, packaged and final cooling to 5°C takes place in the cold store (Kjaergaard-Jensen *et al.*, 1987; Al-Kanhal, 1993; Tamime, 1993).

A closely related product is yoghurt-cheese (35–50 g TS 100 g^{-1}) where the curd is shaped into balls and suspended in vegetable oil. Heat treatment of the product at 65°C had effectively reduced the total viable count after

1-year storage at 20°C (Tamime and Crawford, 1984), and a reduction of 2 to $4\log_{10}$ cycles were observed for total counts, lactic acid bacteria and yeasts and moulds after 6 months' storage period (Rao et al., 1987). In the latter study the storages of goats' yoghurt-cheese had significantly improved the sensory qualities.

3.7 Conclusions

It is evident from this review that different types of fermented milks are produced in many countries and share similar properties, especially the products made in the same geographical region. Primarily, the artisan or traditional methods of manufacture have provided technologists and microbiologists with the basic knowledge of processing. Over the years, selective screening and identification of starter culture microflora have provided the industry with a wide range of options for the manufacture of different products that are acceptable to the consumers. The future developments in fermented milks beyond the second millenium may include the following:

- Production of dairy powders and/or other ingredients that have specific functional characteristics to improve the quality of fermented milks.
- Improvement in selection and blending of starter organisms for flavour production; the role of biotechnology in this area should not be overlooked in terms of bacteriocin production and resistance against bacteriophages.
- Greater use of probiotic microorganisms of human origin in the manufacture of fermented milks, but developments may be influenced by regulatory bodies in some countries, or demonstrable efficacy in the human gastrointestinal tract.
- Improvement in automatic process control during the different manufacturing stages of fermented milks, and the wider use of UF to modify or avoid certain unwanted properties of milk.
- Possible re-vitalization of the continuous process for the manufacture of yoghurt which was developed in the 1970s by NIZO in Holland.
- Greater enhancement of the nutritional value of fermented milk products by vitamin supplementation or the use of starter organisms capable of vitamin synthesis, and fortification of milk with a wide range of ingredients with potential health benefits (e.g. therapeutic peptides).
- Greater development in design improvements of protein fractions in milk coupled through the ability to express them in cattle similar to that achieved with β-Lg (Batt et al., 1994).
- Diversification in product development of fermented milks in terms of low-calorie products and the use of fat-substitutes.

Acknowledgements

SAC , Research and Consultancy Services (A.Y.T.) receives financial support from the Scottish Office of Agriculture, Environment and Fisheries Department (SOAEFD).

3.8 References

Abd El-Salam, M.H., El-Shibiny, S., Mahfouz, M.B., El-Dien, H.F., El-Atriby, H.M. and Antila, V. (1991) Preparation of whey protein concentrate from salted whey and its use in yoghurt. *Journal of Dairy Research*, **58**, 503–10.

Abd Rabo, F.H.R., Partridge, J.A. and Furtado, H.M. (1988) Production of yoghurt utilizing ultrafiltration retentate of salted whey as a partial substitution of milk. *Egyptian Journal of Dairy Science*, **16**, 319–29.

Abou-Donia, S.A., Attia, I.A., Khattab, A.A. and El-Khadragy, S.M. (1992a) Characteristics of labneh manufactured using different lactic starter cultures. *Egyptian Journal of Food Science*, **20**, 1–12.

Abou-Donia, S.A., Khattab, A.A., Attia, I.A. and El-Khadragy, S.M. (1992b) Effect of modified manufacturing process of labneh on its chemical composition and microbiological quality. *Egyptian Journal of Food Science*, **20**, 13–23.

Abrahamsen, R.K. and Rysstad, G. (1991) Fermentation of goat's milk with yoghurt starter culture bacteria: a review. *Cultured Dairy Products Journal*, **26**(3) 20–6.

Adams, M.R. and Marteau, P. (1995) On the safety of lactic acid bacteria from food. *International Journal of Food Microbiology*, **27**, 263–4.

ADPI (1990) *Standards for Grades of Dry Milks Including Methods of Analysis*, Bulletin No. 916 (revised), American Dairy Products Institute, Chicago.

Agerbaek, M., Gerdes, L.U. and Richelsen, B. (1995) Hypocholesterolaeimic effect of a new fermented milk product in healthy middle-aged men. *European Journal of Clinical Nutrition*, **49**, 346–52.

Aguirre, M. and Collins, M.D. (1993) Lactic acid bacteria and human clinical infection. *Journal of Applied Bacteriology*, **75**, 95–107.

Akin, N. and Rice, P. (1994) Main yoghurt and related products in Turkey. *Cultured Dairy Products Journal*, **29**(3) 23–9.

Akin, N., Rice, P. and Holdich, R. (1995) The vacuum filtration of yoghurt. *Cultured Dairy Products Journal*, **30**(2) 2–4.

Al-Kanhal, H.A. (1993) Manufacturing methods and the quality of labneh. *Egyptian Journal of Dairy Science*, **21**, 123–31.

Alm, L. (1981) The effect of fermentation on the biological value of milk proteins using rats: a study of Swedish fermented milk products. *Journal of Science Food and Agriculture*, **32**, 1247–53.

Alm, L. (1982a) Effect of fermentation on lactose, glucose, galactose content in milk and suitability of fermented milk products for lactose intolerant individuals. *Journal of Dairy Science*, **65**, 346–52.

Alm, L. (1982b) The effect of fermentation of B-vitamin content of milk in Sweden. *Journal of Dairy Science*, **65**, 353–9.

Alm, L. (1982c) Effects of fermentation on curd size and digestibility of milk proteins *in-vitro* of Swedish fermented milk products. *Journal of Dairy Science*, **65**, 509–14.

Alm, L. (1982d) Effect of fermentation on L(+) and D(−) lactic acid in milk. *Journal of Dairy Science*, **65**, 515–20.

Alm, L. (1982e) Effect of fermentation on milk fat of Swedish fermented milk products. *Journal of Dairy Science*, **65**, 521–30.

Alm, L. (1982f) Effect of fermentations on proteins of Swedish fermented milk products. *Journal of Dairy Science*, **65**, 1696–704.

Alm, L. (1983) Arla acidofilus – an updated product with a promising future. *Nordisk Mejeriindustri*, **10**, 395–7.

Alm, L. and Larsson, I. (1983) From antiquity into the future: the Nordic ropy milk, a product with a long history. *Nordisk Mejeriindustri*, **10**, 398–400.

Andersch, B., Forssman, L., Lincoln, K. and Torstensson, P. (1986) Treatment of bacterial vaginosis with an acid cream: a comparison of lactate-gel and metronidazole. *Gynecologic and Obstetric Investigation*, **21**, 19–25.

Angulo, L., Lopez, E. and Lema, C. (1993) Microflora present in kefir grains in the Galician region (North-West of Spain). *Journal of Dairy Research*, **60**, 263–7.

Anifantakis, E.M. (1990) Manufacture of sheep's milk products, in *Proceedings of the XXIII International Dairy Congress*, Vol. 1., Mutual Press, Ottawa, pp. 420–32.

Anon. (1967) *Manual for Milk Plant Operators*, Milk Industry Foundation, Washington DC, pp. 628–33.

Anon. (1969) How to make ymer–Danish dairy delicacy. *Food Engineering*, **41**(5) 48–50.

Anon. (1987) *Cultured Sour Cream*, Technical Bulletin No. SC 9724, Microfile Technics, Sarasota.

Anon. (1992) New cultures benefit buttermilk, sour cream. *Dairy Foods*, **93**(7) 50.

Anon. (1993) Yoghurt: how healthy is it? *Which*, April, 38–41.

Anon. (1995) *Dairy Processing Handbook*, revised edition, Tetra Pak Processing Systems AB, LP Grafisca, Lund.

Aoki, T., Kako, Y. and Imamura, T. (1986) Separation of casein aggregates cross-linked by colloidal calcium phosphate from bovine casein micelles by high performance gel chromatography in the presence of urea. *Journal of Dairy Research*, **53**, 53–9.

Aoki, T., Kawahara, A., Kako, Y. and Imamura, T. (1987a) Role of individual milk salt constituents in cross-linking by colloidal calcium phosphate in artificial casein micelles. *Agricultural and Biological Chemistry*, **51**, 817–21.

Aoki, T., Yamada, N., Tomita, I., Kako, Y. and Imamura, T. (1987b) Caseins are cross-linked through their ester phosphate groups by colloidal calcium phosphate. *Biochimica et Biophysica Acta*, **911**, 238–43.

Aoki, T., Yamada, N., Kako, Y. and Imamura, T. (1988) Dissociation during dialysis of casein aggregates cross-linked by colloidal calcium phosphate in bovine casein micelles. *Journal of Dairy Research*, **55**, 189–95.

Aoki, T., Umeda, T. and Kako, Y. (1990) Cleavage of the linkage between colloidal calcium phosphate and casein on heating milk at high temperature. *Journal of Dairy Research*, **57**, 349–54.

Ashenafi, M. (1992) Growth potential and inhibition of *Bacillus cereus* and *Staphylococcus aureus* during the souring of ergo, a traditional fermented milk. *Ethopian Journal of Health Development*, **6**(2) 23–9.

Ashenafi, M. (1993) Fate of *Salmonella enteritidis* and *Salmonella typhimurium* during the fermentation of ergo, a traditional Ethiopian sour milk. *Ethopian Medical Journal*, **31**(2) 91–8.

Ashenafi, M. (1994) Fate of *Listeria monocytogenes* during the sour of ergo, a traditional Ethiopian fermented milk. *Journal of Dairy Science*, **77**, 696–702.

Atanda, O.O. and Ikenebomeh, M.J. (1988) Changes in the acidity and lactic acid content of 'nono' a Nigerian cultured milk product. *Letters in Applied Microbiology*, **6**, 137–8.

Atanda, O.O. and Ikenebomeh, M.J. (1989a) Effect of heat treatments on the microbial load of 'nono'. *Letters in Applied Microbiology*, **9**, 233–5.

Atanda, O.O. and Ikenebomeh, M.J. (1989b) An attempt to improve the flavour of 'nono' by the use of starter cultures. *Letters in Applied Microbiology*, **9**, 17–19.

Attia, H., Bennasar, M. and Tarods de la Fuente, B. (1991a) Study of the fouling of inorganic membranes by acidified milks using scanning electron microscopy and electrophoresis: I. membrane with pore diameter 0.2 μm. *Journal of Dairy Research*, **58**, 39–50.

Attia, H., Bennasar, M. and Tarods de la Fuente, B. (1991b) Study of the fouling of inorganic membranes by acidified milks using scanning electron microscopy and electrophoresis: II. membrane with pore diameter 0.8 μm. *Journal of Dairy Research*, **58**, 51–65.

Babel, F.J. (1967) Techniques for cultured products. *Journal of Dairy Science*, **50**, 431–3.

Bachmann, M.R. and Karmas, E. (1988) Novel cultured buttermilk compositions and method of preparation. *United States Patent*, US 4 748 025.

Ballongue, J. (1993) Bidifobacteria and probiotic action, in *Lactic Acid Bacteria* (eds S. Salminen and A. von Wright), Marcel Dekker, New York, pp. 357–428.

Banon, S. and Hardy, J. (1991) Study of acid milk coagulation by an optical method using light reflection. *Journal of Dairy Research*, **58**, 75–84.

Barbut, S. (1995) Cold gelation of whey proteins. *Scandinavian Dairy Information*, **9**(2) 20–2.

Barrantes, E., Tamine, A.Y. and Sword, A.M. (1994a) Production of low-calorie yoghurt using skim milk powder and fat-substitutes: 3. microbiological and organoleptic qualities. *Milchwissenschaft*, **49**, 205–8.

Barrantes, E., Tamime, A.Y. and Sword, A.M. (1994b) Production of low-calorie yoghurt using skim milk powder and fat-substitutes: 4. rheological properties. *Milchwissenschaft*, **49**, 263–6.

Barrantes, E., Tamime, A.Y., Davies, G. and Barclay, M.N.I. (1994c) Production of low-calorie yogurt using skim milk powder and fat-substitutes: 2. compositional quality. *Milchwissenschaft*, **49**, 135–9.

Barrantes, E., Tamime, A.Y., Muir, D.D. and Sword, A.M. (1994d) The effect of substition of fat by microparticulate whey protein on the quality of set-type, natural yogurt. *Journal of the Society of Dairy Technology*, **47**, 61–8.

Batt, C.A., Brady, J. and Sawyer, L. (1994) Design improvements of β-lactoglobulin. *Trends in Food Science & Technology*, **5**, 261–5.

Becker, T. and Puhan, Z. (1989) Effect of different processes to increase the milk solids non-fat content on the rheological properties of yoghurt. *Milchwissenschaft*, **44**, 626–9.

van den Berg, J.C.T. (1988) *Dairy Technology in the Tropics and Subtropics*, Pudoc, Wageningen, pp. 157–8.

Berlin, P.J. (1962) Kumiss, in *Annual Bulletin*, Part IV, Section A, International Dairy Federation, Brussels, pp. 4–16.

Bertelsen, E. (1983) Fermented milk products in the Scandinavian countries. *Nordisk Mejeriindustri*, **10**, 393–5.

Bezkorovainy, A. and Miller-Catchpole, R. (1989) *Biochemistry and Physiology of Bifidobacteria*, CRC Press, Boca Raton.

Biavati, B., Sgorbati, B. and Scardovi, V. (1992) The genus *Bifidobacterium*, in *The Prokaryotes – A Handbook on the Biology of Bacteria: Ecophysiology, Isolation, Identification, Applications*, Vol. I, 2nd edn (eds A. Barlows, H.G. Trüper, M. Dworkin, W. Harder and K.-H. Schleifer), Springer-Verlag, New York, pp. 816–33.

Blanchette, L., Roy, D. and Gauthier, S.F. (1995) Production of cultured Cottage cheese dressing by bifidobacteria. *Journal of Dairy Science*, **78**, 1421–9.

Bøjgaard, S.E. (1987) Recombination of dairy ingredients into fermented products including cheese, butter and ice cream, in *Milk – The Vital Force*, Proceedings of the XXII International Dairy Congress, D. Reidel Publishing Company, Dordrecht, pp. 259–68.

Bonomi, F., Iametti, S., Pagliarini, E. and Solaroli, G. (1994) Thermal sensitity of mare's milk proteins. *Journal of Dairy Research*, **61**, 419–22.

Bottazzi, V., Zacconi, C., Garra, P.G., Dalavalle, P. and Parisi, M.G. (1994) Kefir: microbiology, chemistry and technology. *Industria del Latte*, **30**, 41–62.

Bremer, L.G.B., Bijsterbosch, B.H., Schrijvers, R., van Vliet, T. and Walstra, P. (1990) On the fractal nature of the structure of acid casein gels. *Colloids and Surfaces*, **51**, 159–70.

Bringe, N.A. and Kinsella, J.E. (1990) Acidic coagulation of casein micelles: mechanisms inferred from spectrometric studies. *Journal of Dairy Research*, **57**, 365–75.

Bringe, N.A. and Kinsella, J.E. (1991) Effects of cations and anions on the rate of the acidic coagulation of casein micelles: the possible roles of different forces. *Journal of Dairy Research*, **58**, 195–209.

Brown, J.R., Law, A.J.R. and Knight, C.H. (1995) Changes in casein composition of goats' milk during the course of lactation: physiological inferences and technological implications. *Journal of Dairy Research*, **62**, 431–9.

Campbell-Platt, G. (1987) *Fermented Foods of the World*, Butterworth, London.

Chandan, R.C. (1982) Other fermented dairy products, in *Prescott and Dunn's Industrial Microbiology*, 4th edn (ed G. Reed), AVI, Wesport, pp. 113–84.

Chandan, R.C. (ed.) (1989) *Yogurt – Nutritonal and Health Properties*, National Yogurt Association, Virginia.

Chander, H., Batish, V.K., Mohan, M., Chand, R. and Singh, R.S. (1992) Effect of heat processing on bacterial quality of dahi – an Indian fermented dairy product. *Cultured Dairy Products Journal*, **27**(2) 8–9.

Clementi, F., Gobbetti, M. and Rossi, J. (1989) Carbon dioxide synthesis by immobilized yeast cells in kefir production. *Milchwissenschaft*, **44**, 70–4.

Coconnier, M.-H., Bernet, M.-F., Chauvière, G. and Servin, A.L. (1993) Adhering of heat-killed human *Lactobacillus acidophilus*, strain LB, inhibits the process of pathogenicity of diarrhoeagenic bacteria in cultured human intestinal cells. *Journal of Diarrhoeal Diseases Research*, **11**, 235–42.

Cogan, T.M. (1972) Susceptibility of cheese and yoghurt starter bacteria to antibiotics. *Applied Microbiology*, **23**, 960–5.

Conn, H.W. (1900) *Annual Report of the Storrs School*, cited by Sandine *et al.* (1972).

Dalgleish, D.G. (1990) Denaturation and aggregation of serum proteins and caseins in heated milk. *Journal of Agricultural and Food Chemistry*, **38**, 1995–9.

Dalgleish, D.G. and Sharma, S.K. (1993) Interactions between milk fat and milk proteins – the effect of heat on the nature of the complexes formed, in *Protein and Fat Globule Modifications by Heat Treatment, Homogenization and Other Technological Means for High Quality Dairy Products*, IDF Special Issue No. 9303, International Dairy Federation, Brussels, pp. 7–17.

Dalgleish, D.G. and Law, A.J.R. (1988) pH-induced dissociation of bovine casein micelles: I. analysis of liberated caseins. *Journal of Dairy Research*, **55**, 529–38.

Dalgleish, D.J. and Law, A.J.R. (1989) pH-induced dissociation: II. mineral solubilization and its relation to casein release. *Journal of Dairy Research*, **56**, 727–35.

Dannenberg, F. and Kessler, H.-G. (1988a) Application of reaction kinetics to the denaturation of whey proteins in milk. *Milchwissenschaft*, **43**, 3–7.

Dannenberg, F. and Kessler, H.-G. (1988b) Thermodynamic approach to kinetics of β-lactoglobulin denaturation in heated skim milk and whey. *Milchwissenschaft*, **43**, 140–2.

Dannenberg, F. and Kessler, H.-G. (1988c) Reaction kinetics of the denatured whey proteins in milk. *Journal of Food Science*, **53**, 258–63.

Dannenberg, F. and Kessler, H.-G. (1988d) Effect of denaturation of β-lactoglobulin on texture studies of set-style nonfat yoghurt: 1. syneresis. *Milchwissenschaft*, **43**, 632–5.

Dannenberg, F. and Kessler, H.-G. (1988e) Effect of denaturation of β-lactoglobulin on texture studies of set-style nonfat yoghurt: 2. firmness and flow properties. *Milchwissenschaft*, **43**, 700–4.

De, S. (1980) *Outlines of Dairy Technology*, Oxford University Press, Delhi.

de Jong, P. and van der Linden, H.J.L.J. (1993) Process design on heat-induced transformations of milk components, in *Protein & Fat Globule Modification by Heat Treatment, Homogenization & Other Technological Means for High Quality Dairy Products*, IDF Special Issue No. 9303, International Dairy Federation, Brussels, pp. 277–84.

Dellaglio, F., Dicks, L.M.T. and Torriani, S. (1995) The genus *Leuconostoc*, in *The Genera of Lactic Acid Bacteria*, vol. 2 (eds B.J.B. Wood and W.H. Holzapfel), Blackie Academic & Professional, London, pp. 235–78.

Demott, B.J., Hitchcock, J.P. and Davidson, P.M. (1986) Use of sodium substitute in cottage cheese and buttermilk. *Journal of Food Protection*, **49**, 117–20.

Desmazeaud, M.J. (1990) Role des culture de microorganismes dans la flaveur et la texture des produits laitièrs fermentes, in *Proceedings of the XXIII International Dairy Congress*, vol. 2, Mutual Press, Ottawa, pp. 1555–77.

de Valdez, G.F. and de Giori, G.S. (1993) Effective of soy milk as food carrier for *Lactobacillus acidophilus*. *Journal of Food Protection*, **56**, 320–2.

de Vos, W.M. (1993) Engineering lactic acid bacteria for improved food fermentations, in *Developing Agricultural Biotechnology in The Netherlands* (eds D.H. Vuijk, J.J. Dekkers and H.C. van der Plas), Pudoc Scientific Publishers, Wageningen, pp. 231–5.

de Wit, J.N. (1990) Thermal stability and functonality of whey proteins. *Journal of Dairy Science*, **73**, 3602–12.

Dibb, S., Hunter, R. and Mann, K. (1991) Consumer checkout: alive or dead? bio-yogurt exposed. *The Food Magazine*, Oct./Dec., 9–10.

Dong, M.-Y., Chang, T.-W. and Gorbach, S.L. (1987) Effects of feeding *Lactobacillus* GG on lethal irradiation in mice. *Diagnostic Microbiology and Infectious Disease*, **7**, 1–7.

Doreau, M. and Boulot, S. (1989) Recent knowledge on mare milk production: a review. *Livestock Production Science*, **22**, 213–35.

Doreau, M., Boulot, S., Barlet, J.P. and Patureau-Mirand, P. (1990) Yield and composition of milk from lactating mares: effect of lactation stage and individual differences. *Journal of Dairy Research*, **57**, 449–54.

Drasar, B.S. and Barrow, P.A. (1985) *Intestinal Microbiology*, Van Nostrand Reinhold (UK), Wokingham.

Driar, H.A. (1993) *The Indigenous Fermented Foods of the Sudan*, CAB International, Wallingford.

Driessen, F.M. and de Boer, R. (1989) Fermented milks with selected intestinal bacteria: a healthy trend in new products. *Netherlands Milk and Dairy Journal*, **43**, 367–82.

Driessen, F.M. and Loones, A. (1990) Developments in the fermentation process: liquid, stirred and set fermented milks, in *Proceedings of the XXII International Dairy Congress*, vol. 3, Mutual Press, Ottawa, pp. 1937–53.

Driessen, F.M., and Loones, A. (1992) Developments in the fermentation process (liquid, stirred and set fermented milks), in *New Technologies for Fermented Milks*, IDF Buttletin No. 277, International Dairy Federation, Brussels, pp. 28–40.

Duitschaver, C.L., Kemp, N. and Emmons, D. (1987) Pure culture formulation and procedure for the producton of kefir. *Milchwissenschaft*, **42**, 80–2.

Duitschaver, C.L., Kemp, N. and Emmons, D. (1988) Comparative evaluation of five procedures for making kefir. *Milchwissenschaft*, **43**, 343–5.

Duitschaver, C.L., Toop, D.H. and Buteau, C. (1991) Consumer acceptance of sweetened and flavoured kefir. *Milchwissenschaft*, **46**, 227–9.

Dybing, S.T. and Smith, D.E. (1991) Relation of chemistry and processing procedures to whey protein functionality: a review. *Cultured Dairy Products Journal*, **26**(1) 4–12.

El-Gendy, S.M. (1983) Fermented foods of Egypt and the Middle East. *Journal of Food Protection*, **46**, 358–67.

El-Gendy, S.M. (1986) Fermented Foods of Egypt and the Middle East, in *Indigenous Fermented Foods of Non-Western Origin* (ed C.W. Hesseltine and H.L. Wang), J. Creamer, Berlin, pp. 169–92.

Eller, H. (1971) *The Technology of Sour Milk Products*, the Ministry of Meat and Milk Industry of the Estonian S.S.R., Tallinn, pp. 3–39.

El-Samragy, Y.A. (1988) The manufacture of zabadi from goat milk. *Milchwissenschaft*, **43**, 92–4.

El-Samragy, Y.A. and Zall, R.R. (1988) Organoleptic properties of the yoghurt-cheese labneh manufactured using ultrafiltration. *Dairy Industries International*, **53**(3) 27–8.

El-Samragy, Y.A., Fayed, E.O., Aly, A.A. and Hagrass, A.E.A. (1988) Properties of labneh-like product manufactured using *Enterococcus* starter cultures as novel dairy fermentation bacteria. *Journal of Food Protection*, **51**, 386–90.

El-Samragy, Y.A., Hansen, C.L. and McMahon, D.J. (1993a) Production of ultrafiltered skim retentate powder: 1. composition and physical properties. *Journal of Dairy Science*, **76**, 388–92.

El-Samragy, Y.A., Hansen, C.L. and McMahon, D.J. (1993b) Production of ultrafiltered skim retentate powder: 2. functional properties. *Journal of Dairy Science*, **76**, 2886–90.

Farah, Z., Streiff, T. and Bachmann, M.R. (1990) Preparation and consumer acceptability tests of fermented camel milk in Kenya. *Journal of Dairy Research*, **57**, 281–3.

Farkye, N.Y. and Imafidon, G.I. (1995) Thermal denaturation of indigenous milk enzymes, in *Heat-Induced Changes in Milk*, 2nd edn (ed. P.F. Fox), IDF Special Issue No. 9501, International Dairy Federation, Brussels, pp. 331–48.

Fayed, E.O., Hagrass, A.E.A., Aly, A.A. and El-Samragy, Y.A. (1989) Use of enterococci starter culture in the manufacture of a yogurt-like product. *Cultured Dairy Products Journal*, **24**(1) 16–23.

Fenton, A. (1976) *Scottish Country Life*, John Donald Publishers, Edinburgh.

Feresu, S. (1992) Fermented milk products in Zimbabwe, in *Applications of Biotechnology to Traditional Fermented Foods*, National Academy Press, Washington, DC, pp. 80–5.

Feresu, S. and Muzondo, M.I. (1989) Factors affecting the development of two fermented milk products in Zimbabwe. *MIRCEN Journal*, **5**, 349–55.

Feresu, S. and Muzondo, M.I. (1990) Identification of some lactic acid bacteria from two Zimbabwean fermented milk products. *World Journal of Microbiology and Biotechnology*, **6**, 178–86.

Forsèn, R., Niskasaari, K. and Niemitola, S. (1985) Immunochemical demonstration of lipoteichoic acid as a surface-exposed plasma membrane of slime-forming, encapsulated *Streptococcus cremoris* from fermented milk product 'viili'. *FEMS Microbiology Letters*, **26**, 249–53.

Forsèn, R., Niskasaari, K., Tasanen, L. and Numiaho-Lassila, E.-L. (1989) Studies on slimy lactic acid fermentation: detection of lipoteichoic acid containing membrane antigens of *Lactococcus lactis* ssp. *cremoris* strains by crossed immunoelectrophoresis. *Netherlands Milk and Dairy Journal*, **43**, 383–93.

Fox, P.F. (ed.) (1991) *Food Enzymology*, Elsevier Applied Science Publishers, London.

Frank, J.F. (1984) Improving the flavor of cultured buttermilk. *Cultured Dairy Products Journal*, **19**(3) 6–9.

Fris-Møller, A. and Hey, H. (1983) Colonization of the intestinal canal with *Streptococcus faecium* preparation (Paraghurt®). *Current Therapeutic Research*, **33**, 807–15.

Fujisawa, T., Adachi, S., Toba, T., Arihara, K. and Mitsuoka, T. (1988) *Lactobacillus kefiranofaciens* sp. nov. isolated from kefir grains. *International Journal of Systematic Bacteriology*, **38**, 12–14.

Fuller, R. (1989) A review – probiotics in man and animals. *Journal of Applied Bacteriology*, **66**, 365–78.

Fuller, R. (1994) Probiotics: an overview, in *Human Health: the Contribution of Microorganisms* (ed. S.A.W. Gibson), Springer-Verlag, London, pp. 63–73.

Gassem, M.A. and Frank, J.F. (1991) Physical properties of yogurt made from milk treated with proteolytic enzymes. *Journal of Dairy Science*, **74**, 1503–11.

Gettys, S.C. and Davidson, P.M. (1985) A comparison of buttermilks made using fermentation, direct acidification and a combination of both. *Journal of Dairy Science*, **68**, 620–7.

Gilliland, S.E. (1989) Acidophilus milk products: a review of potential benefits to the consumers. *Journal of Dairy Science*, **72**, 2483–94.

Gilliland, S.E. and Speck, M.L. (1977) Instability of *Lactobacillus acidophilus* in yogurt. *Journal of Dairy Science*, **60**, 1394–8.

Gilliland, S.E. and Walker, D.K. (1990) Factors to consider when selecting a culture of *Lactobacillus acidophilus* as a dietary adjunct to produce a hypo-cholesterolemic effect in humans. *Journal of Dairy Science*, **73**, 905–11.

Gilmour, A. and Rowe, M.T. (1990) Micro-organisms associated with milk, in *Dairy Microbiology – The Microbiology of Milk*, vol. 1, 2nd edn (ed. R.K. Robinson), Elsevier Applied Science Publishers, London, pp. 37–75.

Gobetti, M. and Rossi, J. (1994) Batchwise fermentation with Ca-alginate immobilized multistarter cells for kefir production. *International Dairy Journal*, **4**, 237–49.

Goldin, B.R., Gorbach, S.L., Saxelin, M., Barakat, S., Gulatieri, L. and Salminen, S. (1992) Survival of *Lactobacillus* species (strain GG) in human gastrointestinal tract. *Digestive Diseases and Sciences*, **37**, 121–8.

Gomes, A.M.P., Malcata, F.X., Klaver, F.A.M. and Grande, H.J. (1995) Incorporation and survival of *Bifodobacterium* sp. strain BO and *Lactobacillus acidophilus* strain Ki in a cheese product. *Netherlands Milk and Dairy Journal*, **49**, 71–95.

Gorbach, S.L. (1990) Lactic acid bacteria and human health. *Annals of Medicine*, **22**, 37–41.

Gorbach, S.L. and Goldin, B.R. (1989) *Lactobacillus* strains and methods for selection. *United States Patent*, US 4 839 281.

Gotham, S.M., Fryer, P.J. and Pritchard, A.M. (1992) β-lactoglobulin denaturation and aggregation reactions and fouling deposit formation: a DSC study. *International Journal of Food Science and Technology*, **27**, 313–27.

Green, B., Jensen, L. and Park, K. (1992) Nutrient composition of eight Californian milk products based on analysis conducted in 1990–91. *Dairy Food and Environmental Sanitation*, **12**, 669–73.

Greene, V.W. and Jezeski, J.J. (1957a) Studies on starter metabolism: I. the relationship between starter activity and predrying heat history of reconstituted nonfat dry milk solids. *Journal of Dairy Science*, **40**, 1046–52.

Greene, V.W. and Jezeski, J.J. (1957b) Studies on starter metabolism: II. the influence of heating milk on the subsequent response of starter cultures. *Journal of Dairy Science*, **40**, 1053–61.

Greene, V.W. and Jezeski, J.J. (1957c) Studies on starter metabolism: III. Studies on cysteine-induced stimulation and inhibition of starter cultures in milk. *Journal of Dairy Science*, **40**, 1062–71.

Guan, J. and Brunner, J.R. (1987) Koumiss produced from a skim milk-sweet whey blend. *Cultured Dairy Products Journal*, **22**(1), 23.

Gudmundsson, B. (1987) Skyr. *Scandinavian Dairy Industry*, **1**(4) 240–2.

Guinee, T.P., Mullins, C.G. and Reville, W.J. (1994) Rheology and syneretic properties of yoghurt stabilized with different dairy ingredients, in *Second Food Ingredients Symposium* (ed. M.K. Keogh), National Dairy Products Research Centre, Fremoy, pp. 73–86.

Guinee, T.P., Mullins, C.G., Reville, W.J. and Cotter, M. (1995) Physical properties of stirred-curd unsweetened yoghurts stabilized with different dairy ingredients. *Milchwissenschaft*, **50**, 196–200.

Guirguis, N., Verteeg, K. and Hickey, M.W. (1987a) The manufacture of yoghurt using reverse osmosis concentrated skim milk. *The Australian Journal of Dairy Technology*, **42**, 7–10.

Guirguis, N., Hickey, M.W. and Freeman, R. (1987b) Some factors affecting nodulation in yoghurt. *The Australian Journal of Dairy Technology*, **42**, 45–7.

Haggag, H.F. and Fayed, A.E. (1988) Production of zabadi from ultrafiltered buffalo's milk. *Food Chemistry*, **30**, 29–36.

Hamad, A.M. and Al-Sheikh, S.S. (1989) Effect of milk solids concentration and draining temperature on the yield and quality of labneh (concentrated yoghurt). *Cultured Dairy Products Journal*, **24**(1) 25–8.

Hammes, W.P. and Vogel, R.F. (1995) The genus *Lactoacillus*, in *The Genera of Lactic Acid Bacteria*, vol. 2 (eds B.J.B. Wood and W.H. Holzapfel), Blackie Academic & Professional, London, pp. 19–54.

Hammond, L.A. (1969) Cultured buttermilk, in *Dairy Fermentation Technology* (ed. P.M. Linklater), The University of New South Wales, Kensington, pp. 131–40.

Haque, Z. and Kinsella, J.E. (1988) Interaction between heated κ-casein and β-lactoglobulin: predominance of hydrophobic interactions in the initial stages of complex formation. *Journal of Dairy Research*, **55**, 67–80.

Hardie, J.M. and Whiley, R.A. (1995) The genus *Streptococcus*, in *The Genera of Lactic Acid Bacteria*, vol. 2 (eds B.J.B. Wood and W.H. Holzapfel), Blackie Academic & Professional, London, pp. 55–124.

Hinrichs, J. and Kessler, H.-G. (1995) Thermal processing of milk-processes and equipment, in *Heat-Induced Changes in Milk*, 2nd edn (ed. P.F. Fox), IDF Special Issue No. 9501, International Dairy Federation, Brussels, pp. 9–21.

Ho, S. and Mittal, G.S. (1995) Feasibility of continuous yogurt processing. *Milchwissenschaft*, **50**, 146–50.

Hofi, M. (1990) Sour cream from buffalo's milk by ultrafiltration [Abstract], in *Brief Communications of XXIII International*, vol. 2, Mutual Press, Ottawa, p. 510.

Hogarty, S.L. and Frank, J.F. (1982) Low-temperature activity of lactic streptococci isolated from buttermilk. *Journal of Food Protection*, **45**, 1208–11.

Holland, B., Unwin, I.D. and Buss, D.H. (1989) *Milk Products and Eggs*, 4th Supplement to McCance and Widdowson's, *The Composition of Foods*, 4th edn, The Royal Society of Chemistry, Cambridge.

Hollar, C.M., Parris, N., Hsieh, A. and Cockley, K.D. (1995) Factors affecting the denaturation and aggregation of whey proteins in heated whey protein concentrate mixtures. *Journal of Dairy Science*, **78**, 260–7.

Holst, E. and Brandberg, Å. (1990) Treatment of bacterial vaginosis in pregnancy with lactate-gel. *Scandinavian Journal of Infectious Diseases*, **22**, 625–6.

Holt, C. (1995) Effect of heating and cooling on the milk salts and their interaction with casein, in *Heat-Induced Changes in Milk*, 2nd edn (ed. P.F. Fox), IDF Special Issue No. 9501, International Dairy Federation, Brussels, pp. 105–33.

Holt, C., Davies, D.T. and Law, A.J.R. (1986). Effects of calcium phosphate content and free calcium ion concentration in the milk serum on the dissociation of bovine casein micelles. *Journal of Dairy Research*, **53**, 557–72.

Hølund, U. (1993) Cholesterol-lowering effect of a new fermented milk product. *Scandinavian Dairy Information*, **7**(4) 10–11.

Horne, D.S. and Davidson, C.M. (1993) Influence of heat treatment on gel formation in acidic milk, in *Protein & Fat Globule Modifications by Heat Treatment, Homogenization & Other Technological Means for High Quality Dairy Products*, IDF Special Issue No. 9303, International Dairy Federation, Brussels, pp. 267–76.

Hoskin, J.C. (1989) Susceptibility of cultured buttermilk to light irradiation. *Cultured Dairy Products Journal*, **24**(1) 14–15.

(Hosono, A., Tanabe, T. and Otani, H. (1990) Binding properties of lactic acid bacteria isolated from kefir milk with mutagenic amino acid pyrolysates. *Milchwissenschaft*, **45**, 647–51.

Hougaard, E. (1993) Gaio – new revolutionary product from MD Foods. *Scandinavian Dairy Information*, **7**(4), 8–9.

Hougaard, E. (1994) Gaio has created an entirely new market. *Scandinavian Dairy Information*, **8**(2), 10.

Houlihan, A.V., Goddard, P.A., Kitchen, B.J. and Masters, C.J. (1992a) Changes in the structure of the bovine milk globule membrane on heating whole milk. *Journal of Dairy Research*, **59**, 321–9.

Houlihan, A.V., Goddard, P.A., Nottingham, S.M., Kitchen, B.J. and Masters, C.J. (1992b) Interactions between the bovine milk fat globule membrane and skim milk components on heating. *Journal of Dairy Research*, **59**, 187–95.

Hull, R.R., Roberts, A.V. and Mayes, J.J. (1984) Survival of *Lactobacillus acidophilus* in yoghurt. *The Australian Journal of Dairy Technology*, **39**, 164–6.

Hylmar, B. (1976) [Manufacture of culture cream]. *Prümysl Potravin*, **27**(4), 224 (in Czech). Cited in *Dairy Science Abstracts* (1976), **38**, 651 [6143].

IDF (1983) *Consumption Statistics for Milk and Milk Products (1981)*, Bulletin No. 160, International Dairy Federation, Brussels, pp. 1–3.

IDF (1984a) *Consumption Statistics for Milk and Milk Products (1982)*, Bulletin No. 173, International Dairy Federation, Brussels, pp. 1–3.

IDF (1984b) *Fermented Milks*, Bulletin No. 179, International Dairy Federation, Brussels.

IDF (1986) *Production and Utilization of Ewe's and Goat's Milk*, Bulletin No. 202, International Dairy Federation, Brussels.

IDF (1988) *Fermented Milks – Science and Technology*, Bulletin No. 227, International Dairy Federation, Brussels.

IDF (1991) *Lactic Acid Starters – Standard of Identity*, Standard (provisional) No. 149, International Dairy Federation, Brussels.

IDF (1992) *New Technologies for Fermented Milks*, Bulletin No. 277, International Dairy Federation, Brussels.

IDF (1993) *Consumption Statistics for Milk and Milk Products (1991)*, Bulletin No. 282, International Dairy Federation, Brussels, pp. 1–3.

IDF (1994) *Consumption Statistics for Milk and Milk Products (1992)*, Bulletin No. 295, International Dairy Federation, Brussels, pp. 1–3.

Isolauri, E., Juntunen, M., Rantanen, T., Sillanaukee, P. and Koivula, T. (1991) A human *Lactobacillus* strain (*Lactobacillus casei* sp. strain GG) promotes recovery from acute diarrahea in children. *Paediatrics*, **88**, 90–7.

Isolauri, E., Majamaa, H., Arvola, T., Rantala, I., Virtanen, E. and Arvilommi, H. (1993a) *Lactobacillus casei* strain GG reverses increased intestinal permeability induced by cow milk in suckling rats. *Gastroenterology*, **105**, 1643–50.

Isolauri, E., Kaila, M., Arvola, T., Majamaa, H., Rantala, I., Virtanen, E. and Arvilommi, H. (1993b) Diet during rotavirus enteritis affecting jejunal permeability to macromolecules in suckling rats. *Pediatric Research*, **33**, 548–53.

Iwana, H., Masuda, H., Fujisawa, T., Suzuki, H. and Mitsuoka, T. (1993) Isolation and identification of *Bifidobacterium* spp. in commercial yogurts sold in Europe. *Bifidobacteria Microflora*, **12**, 39–45.

Jelen, P. (1993) Heat stability of dairy systems with modified casein-whey protein content, in *Protein & Fat Globule Modifications by Heat Treatment, Homogenization & Other Technological Means for High Quality Dairy Products*, IDF Special Issue No. 9303, International Dairy Federation, Brussels, pp. 259–66.

Jelen, P., Buchheim, W. and Peters, K.-H. (1987) Heat stability and use of milk with modified casein: whey protein content in yoghurt and cultured milk products. *Milchwissenschaft*, **42**, 418–21.

Johnston, D. E. and Murphy, R.J. (1992) Effects of some calcium-chelating agents on the physical properties of acid-set milk gels. *Journal of Dairy Research*, **59**, 197–208.

Johnston, M.C., Ray, B. and Bhowmik, T. (1987) Selection of *Lactobacillus acidophilus* strains for use in 'acidophilus products'. *Antoine van Leeuwenhock*, **53**, 215–31.

Juillard, V., Spinnler, H.E., Desmazeaud, M.J. and Boquien, C.Y. (1987) Phénomènes de cooperation et d'inhibition entre les bactéries lactique utilisées en industrie latière. *Le Lait*, **67**, 149–72.

Kahala, M., Pahkala, E. and Philauto-Lepälä, E. (1993) Peptides in fermented Finnish milk products. *Agriculture Science of Finland*, **2**, 379–86.

Kalab, M. (1992) Food structure and milk products, in *Encyclopedia of Food Science and Technology*, Vol. 2 (ed. Y.H. Hui), John Wiley & Sons, New York, pp. 1170–96.

Kalab, M. and Caric, M. (1990) Food microstructure – evaluation of interactions of milk components in food systems, in *Proceedings of the XXIII International Dairy Congress*, vol. 2, Mutual Press, Ottawa, pp. 1457–80.

Kalab, M., Phipps-Todd, B.E. and Allan-Wojtas, P. (1982) Milk gel structure: XIII. rotary shadowing of casein micelles for electron microscopy. *Milchwissenschaft*, **37**, 513–18.

Kandler, O. and Kunath, P. (1983) *Lactobacillus kefir* sp. nov., a component of the microflora of kefir. *Systemic Applied Microbiology*, **4**, 286–94.

Kassaye, T., Simpson, B.K., Smith, J.P. and O'Connor, C.B. (1991) Chemical and microbio- logical characteristics of ititu. *Milchwissenschaft*, **46**, 649–53.

Kehagias, C.H. (1987) Fermented milk products in developing countries with emphasis on those produced from ewe's and goat's milk, in *Milk – The Vital Force*, Proceedings of XXII International Dairy Congress, D. Reidel Publishing Company, Dordrecht, pp. 683–90.

Kehagias, C., Kalavritinos, L. and Triadopoulou, C. (1992) Effect of pH on the yield and solids recovery of strained yoghurt from goat and cow milk. *Cultured Dairy Products Journal*, **27**(3), 10–14.

Kessler, H.-G. (1981) *Food Engineering and Dairy Technology*, (Translator M. Wotzilka), Verlag A. Kessler, Freising.

Kessler, H.-G., Aguilera, J.M., Dannenberg, F. and Kulkarni, S. (1990) Milk foods via new technologies, in *Proceedings of the XXIII International Dairy Congress*, vol. 3, Mutual Press, Ottawa, pp. 2005–20.

Khanna, S., Kaul, A., Kuila, R.K. and Singh, J. (1982) Studies on the utilization of mutants of *Streptococcus lactis* subspecies *diacetylactis* for preparation of fermented milk products 'Chakka'. *Cultured Dairy Products Journal*, **17**(4), 24–5.

Khattab, A.A. (1991) Studies in some β-complex vitamins in labneh. *Egyptial Journal of Dairy Science*, **19**, 231–42.

Khorshid, M.A., Saada, M.Y., Khalil, S.A., Mahran, G.A. and Hofi, M.A. (1993) Low lactose zabadi, in *Protein & Fat Globule Modifications by Heat Treatment, Homogenization & Other Technological Means for High Quality Dairy Products*, IDF Special Issue No. 9303, Interna- tional Dairy Federation, Brussels, pp. 430–9.

Kim, H.-H.Y. and Jimenez-Flores, R. (1995) Heat-induced interactions between the proteins of milk fat globule membrane and skim milk. *Journal of Dairy Science*, **78**, 24–35.

Kimonye, J.M. and Robinson, R.K. (1991) Iria ri Matii – a traditional fermented milk from Kenya. *Dairy Industries International*, **56**(2) 34–5.

Kisza, J., Domagala, J., Wszolek, M. and Kolczak, T. (1993) Yoghurts from sheep milk. *Acta Academiae Agriculturae Technical Olstenensis*, **25**, 75–87.

Kjaergaard-Jensen, G. (1990) Milk powders specifications in relation to the products to be manufactured, in *Recombination of Milk and Milk Products*, IDF Special Issue No. 9001, International Dairy Federation, Brussels, pp. 104–25.

Kjaergaard-Jensen, G., Ipsen, R.H. and Ilsøe, C. (1987) Functionality and applications of dairy ingredients in dairy products. *Food Technology*, **41**(10) 66–72.

Klaver, F.A.M., Kingma, F. and Weerkamp, A.H. (1992a) Interactive fermentation of milk by means of a membrane dialysis fermenter: yoghurt. *Netherlands Milk and Dairy Journal*, **467**, 31–44.

Klaver, F.A.M., Kingma, F., Martin, J., Timmer, K. and Weerkamp, A.H. (1992b) Interactive fermentation of milk by means of a membrane dialysis fermenter: buttermilk. *Netherlands Milk and Dairy Journal*, **46**, 19–30.

Klaver, F.A.M., Kingma, F. and Weerkamp, A.H. (1993) Growth and survival of bifidobacteria in milk. *Netherlands Milk and Dairy Journal*, **47**, 151–64.

Klebanoff, S.J. and Coombs, R.W. (1991) Viricidal effect of *Lactobacillus acidophilus* on human immunodeficiency virus type 1: possible role in heterosexual transmission. *Journal of Experimental Medicine*, **174**, 289–92.

Klebanoff, S.J., Hillier, S.L., Eschenbach, D.A. and Waltersdorph, A.M. (1991) Control of the microflora of the vagina by H_2O_2-generating lactobacilli. *The Journal of Infectious Diseases,* **164**, 94–100.

Klupsch, H.J. (1984) Verfahren zum herstellen von kefir. *German Federal Republic Patent,* DE 33 00 122 Al.

Kniefel, W. and Mayer, H.K. (1991) Vitamin profiles of kefirs made from milk of different species. *International Journal of Food Science and Technology,* **26**, 423–8.

Kontusaari, S. and Forsèn, R. (1988) Finnish ferment milk 'viili': involvement of two cell surface proteins in production of slime by *Steptococcus lactis* ssp. *cremoris. Journal of Dairy Science,* **71**, 3197–202.

Kontusaari, S. Neimitalo, S. and Forsèn, R. (1980) The surface components of lactic streptococci isolated from 'viili'. *Kemia-Kemi,* **7**, 771.

Koroleva, N.S. (1983) Special products (kefir, koumyss, etc), in *Proceedings of XXI International Dairy Congress,* vol. 2, Mir Publishers, Moscow, pp. 146–52.

Koroleva, N.S. (1988a) Starter for fermented milks: Kefir and kumys starters, in *Fermented Milks–Science and Technology,* IDF Bulletin No. 227, International Dairy Federation, Brussels, pp. 35–40.

Koroleva, N.S. (1988b) Technology of kefir and kymys, in *Fermented Milks–Science and Technology,* IDF Bulletin No. 227, International Dairy Federation, Brussels, pp. 96–100.

Koroleva, N.S. (1991) Products prepared with lactic acid bacteria and yeasts, in Therapeutic Properties of Fermented Milks, (ed. R.K. Robinson), Elsevier Applied Science, London, pp. 159–79.

Kosikowski, F.V. (1984) Buttermilk and related fermented milks, in *Fermented Milks,* IDF Bulletin No. 179, International Dairy Federation, Brussels, pp. 116–19.

Kosikowski, F.V. (1985) *Cheese and Fermented Milk Foods,* 2nd edn, Kosikowski and Associates, New York.

Kotz, C.M., Furne, J.K., Savaiano, D.A. and Levitt, M.D. (1994) Factors affecting the ability of a high β-galactosidase yogurt to enhance lactose absorption. *Journal of Dairy Science,* **77**, 3538–44.

Kramkowska, A., Fesnak, D., Kornacki, K. and Bauman, B. (1986) Production, characterization and use of expendable kefir starter culture in the kefir production process. *Acta Biotechnologica,* **6**, 167–74.

Kreger-van Rij, N.J.W. (1984) *The Yeasts–A Taxonomic Study,* 3rd edn, Elsevier Science Publishers, Amsterdam.

Kroger, M. (1993) Kefir. *Cultured Dairy Products Journal,* **28**(2), 26–9.

Kurmann, J.A. and Rasic, J.L. (1991) The health potential products containing bifidobacteria, in *Therapeutic Properties of Fermented Milks* (ed. R.K. Robinson), Elsevier Applied Science Publishers, London, pp. 117–57.

Kurmann, J.A., Rasic, J.L. and Kroger, M. (1992) *Encyclopedia of Fermented Fresh Milk Products,* Van Nostrand Reinhold, New York.

Lancefield, R.C. (1933) A serological differentiation of human and other groups of hemolytic streptococci. *Journal of Experimental Medicine,* **57**, 571–95.

Larsen, J.B. (1982) The content of carbohydrates and lactic acid in cultured liquid milk products, in *Brief Communications of XXI International Dairy Congress,* vol. 1, Book 1, Mir Publishers, Moscow, pp. 299–300.

Larsen, R.F. and Anón, M.C. (1989) Interaction of antibiotics and water activity on *Streptococcus thermophilus* and *Lactobacillus bulgaricus. Journal of Food Science,* **54**, 922–4, 939.

Laukkanen, M., Antila, P., Antila, V. and Salminen, K. (1988) The water-soluble vitamin contents of Finnish liquid milk products. *Meijeritieteellinen Aikakauskirja,* **XLVI**(1) 7–24.

Law, A.J.R. (1995) Heat denaturation of bovine, caprine and ovine whey proteins. *Milchwissenschaft,* **50**, 384–8.

Law, A.J.R. and Brown, J.R. (1994) Compositional changes in caprine whey proteins. *Milchwissenschaft,* **45**, 674–8.

Law, A.J.R. and Tziboula, A. (1992) Quantitative fraction of caprine casein by cation-exchange FPLC. *Milchwissenschaft,* **47**, 558–62.

Law, A.J.R. and Tziboula, A. (1993) Fractionation of caprine κ-casein and examination of polymorphism by FPLC. *Milchwissenschaft,* **48**, 68–71.

Law, A.J.R., Leaver, J., Banks, J.M. and Hornes, D.S. (1993) Quantitative fractionation of whey proteins by gel permeation FPLC. *Milchwissenschaft,* **48**, 663–6.

Law, A.J.R., Hornes, D.S., Banks, J.M. and Leaver, J. (1994) Heat-induced changes in the whey proteins and caseins. *Milchwissenschaft*, **49**, 125–9.

Laxminarayana, H. and Dastur, N.N. (1968) Buffaloe's milk and milk products: part 1. *Dairy Science Abstracts*, **30**, 177–86.

Lee, F.Y. and White, C.H. (1992) Effect of ultrafiltered milk retentates on physiochemical and sensory properties of stabilized lowfat sour cream [abstract]. *Journal of Dairy Science*, **75** (suppl. 1), 306.

Lee, F.Y. and White, C.H. (1993) Effect of rennin on stabilized lowfat sour cream. *Cultured Dairy Products Journal*, **28**(3) 4–13.

LeGraet, Y. and Brulé, G. (1993) The mineral equilibria in milk: effect of pH on ionic strength. *Le Lait*, **73**, 51–60.

Lehmann, H.R., Dolle, E. and Büker, H. (1991) *Processing Lines for the Production of Soft Cheese*, Technical Bulletin No. 8, Westfalia Separator AG, Olde.

Libudzisz, Z. and Piatkiewicz, A. (1990) Kefir production in Poland. *Dairy Industries International*, **55**(7) 31–3.

Lichtenstein, A.H. and Goldin, B.R. (1993) Lactic acid bacteria and intestinal drug and cholesterol metabolism, in *Lactic Acid Bacteria* (eds S. Salminen and A. von Wright), Marcel Dekker, New York, pp. 227–35.

Ling, W.H., Korpela, R., Mykkänen, H., Salminen, S. and Hänninen, P. (1994) *Lactobacillus* strain GG supplementation decreases clonic hydrolytic and reductive enzyme activities in healthy female adults. *Journal of Nutrition*, **124**, 18–23.

Liu, J.A.P. and Moon, N.J. (1983) Kefir–a new fermented milk product. *Cultured Dairy Products Journal*, **18**(3) 11–12.

Lodder, J. (ed.) (1970) *The Yeasts–A Taxonomic Study*, 2nd edn, North-Holland Publishing Company, Amsterdam.

Lozovich, A. (1995) Medical uses of whole and fermented mare milk in Russia. *Cultured Dairy Products Journal*, **30**(1) 18–21.

Lundstedt, E. (1975) All you want to know about buttermilk. *Cultured Dairy Products Journal*, **10**(4) 18–22.

Lundstedt, E. and Corbin, E.A. (1983a) Controlled fermentation of buttermilk. *Cultured Dairy Products Journal*, **18**(3) 6–8.

Lundstedt, E. and Corbin, E.A. (1983b) Controlled fermentation of buttermilk. *North European Dairy Journal*, **49**, 135–9.

Macura, D. and Townsley, P.M. (1984) Scandinavian ropy milk: identification and characterization of endogenous ropy lactic streptococci and their extracellular excretion. *Journal of Dairy Science*, **67**, 735–44.

Macy, H. (1923) A ropy milk organism isolated from the Finnish 'piima' or 'fiili'. *Journal of Dairy Science*, **6**, 127–30.

Mahdi, H.A., Tamime, A.Y. and Davies, G. (1990) Some aspects of the production of 'labneh' by ultrafiltration using cow's, sheep's and goat's milk. *Egyptian Journal of Dairy Science*, **18**, 345–67.

Mahfouz, M.B., El-Dein, H.F., El-Atriby, H.M. and Al-Khamy, A.F. (1992) The use of whey protein concentrate in the manufacture of concentrated yoghurt (labneh). *Egyptian Journal of Dairy Science*, **20**, 9–20.

Mann, E.J. (1986) Buttermilk. *Dairy Industries International*, **51**(10) 9–10.

Marshall, V.M.E. (1984) Flavour development in fermented milks, in *Advances in the Microbiology and Biochemistry of Cheese and Fermented Milk*, 1st edn (eds F.L. Davies and B.A. Law), Elsevier Applied Science Publishers, London, pp. 153–86.

Marshall, V.M.E. (1986) The microflora and production of fermented milks, in *Progress in Industrial Microbiology–Micro-organisms in the Production of Food*, vol. 23, (ed. M.R. Adams), Elsevier Science, Amsterdam, pp. 1–44.

Marshall, V.M.E. (1987) Fermented milks and their future trends: I. microbiological aspects. *Journal of Dairy Research*, **54**, 559–74.

Marshall, V.M.E. (1993) Starter cultures for milk fermentation and their characteristics. *Journal of the Society of Dairy Technology*, **46**, 49–56.

Marshall, V. and El-Bagoury, E. (1986) Use of ultrafiltration and reverse osmosis to improve goat's milk yogurt. *Journal of the Society of Dairy Technology*, **39**, 65–6.

Marshall, V.M., Cole, W.M. and Mabbit, L.A. (1982a) Fermentation of specially formulated milk with single strains of bifidobacteria. *Journal of the Society of Dairy Technology*, **35**, 143–4.

Marshall, V.M., Cole, W.M. and Vega, J.R. (1982b) A yoghurt-like product made by fermenting ultrafiltered milk containing elevated whey proteins with *Lactobacillus acidophilus*. *Journal of Dairy Research*, **49**, 665–70.

Marshall, V.M., Cole, W.M. and Brooker, B.E. (1984) Observations on the structure of kefir grains and distribution of the microflora *Journal of Applied Bacteriology*, **57**, 591–7.

Marteau, P., Pochard, P., Flourié, B., Pellier, P., Santos, L., Desjeux, J.-F. and Rambaud, J.-C. (1990) Effect of chronic ingestion of fermented dairy product containing *Lactobacillus acidophilus* and *Bifiodbacterium bifidum* on metabolic activities of the colonic flora in humans. *American Journal of Clinical Nutrition*, **52**, 685–8.

Martins, J., Krohn, M.A., Hillier, S.L., Stamm, W.E., Holmes, K.K. and Eschenbach, D.A. (1988) Relationships of vaginal *Lactobacillus* species, cervical *Chlamydia trachomatis*, and bacterial vaginosis to preterm birth. *Journal of Obstetrics and Gynaecology*, **71**, 89–95.

McKenna, A.B. and Anema, S.G. (1993) Effect of thermal processing during whole milk powder manufacture and after its reconstitution on set-yoghurt properties, in *Protein & Fat Globule Modifications by Heat Treatment, Homogenization & Other Technological Means for High Quality Dairy Products*, IDF Special Issue No. 9303, International Dairy Federation, Brussels, pp. 307–16.

Mehanna, A.S. (1991) An attempt to improve some properties of zabadi by applying low temperature long incubation period in the manufacturing process. *Egyptian Journal of Dairy Science*, **19**, 221–9.

Mehanna, N.M., El-Dien, H.F and Mahfouz, M.B. (1988) Composition and properties of yoghurt from ultrafiltered milk. *Egyptian Journal of Dairy Science*, **16**, 223–32.

Meriläinen, V.T. (1984) Microorganisms in fermented milks: other microorganisms, in *Fermented Milks*, IDF Bulletin No. 179, International Dairy Federation, Brussels, pp. 89–93.

Meriläinen, V.T. (1987) Yoghurt and cultured buttermilk, in *Milk – The Vital Force*, XXII International Dairy Congress, D. Reidel Publishing Company, Dordrecht, pp. 661–72.

Merin, U. and Rosenthal, I. (1986) Production of kefir from UHT milk. *Milchwissenschaft*, **41**, 395–6.

Meurman, J.H., Antila, H. and Salminen, S. (1994) Recovery of *Lactobacillus* strain GG (ATCC 53103) from saliva of healthy volunteers after consumption of yoghurt prepared with the bacterium. *Microbial Ecology in Health and Disease*, **7**, 295–8.

Meurman, J.H., Antila, H., Korhonen, A. and Salminen, S. (1995) Effect of *Lactobacillus rhamnosus* strain GG (ATCC 53103) on the growth of *Streptococcus sobrinus* in-vitro. *European Journal of Oral Sciences*, **103**, 253–8.

Mikelsaar, M. and Mändar, R. (1993) Development of individual lactic acid microflora in the human microbial ecosystem, in *Lactic Acid Bacteria* (eds S. Salminen and A. von Wright), Marcel Dekker, New York, pp. 237–93.

Misra, A.K. and Kuila, R.K. (1992) Use of *Bifidobacterium bifidum* in the manufacture of bifidus milk and its antibacterial activity. *Le Lait*, **72**, 213–20.

Mistry, V.V. and Hassan, H.N. (1992) Manufacture of nonfat yogurt from a high milk protein powder. *Journal of Dairy Science*, **75**, 947–57.

Mitsuoka, T. (1990a) Role of intestinal flora in health with special reference to dietary control of intestinal flora, in *Microbiology, Applications in Food Biotechnology*, (eds B.H. Nga and Y.K. Lee), Elsevier Applied Science, London, pp. 135–48.

Mitsuoka, T. (1990b) Beneficial microbial aspects (probiotics), in *Proceedings of the XXIII International Dairy Congress*, vol. 2, Mutual Press, Ottawa, pp. 1226–37.

Mitsuoka, T. (1992) The human gastrointestinal tract, in *The Lactic Acid Bacteria in Health and Disease*, vol. 1 (ed. B.J.B. Wood), Elsevier Applied Science Publisher, London, pp. 69–114.

Modler, H.W. (1994) Bifidogenic factors – sources, metabolism and applications. *International Dairy Journal*, **4**, 383–407.

Modler, H.W. and Kalab, M. (1983) Microstructure of yogurt stabilized with milk proteins. *Journal of Dairy Science*, **66**, 430–7.

Modler, H.W., Larmond, M.E., Lin, C.S., Froehlich, D. and Emmons, D.B. (1983) Physical and sensory properties of yogurt stabilized with milk proteins. *Journal of Dairy Science*, **66**, 422–9.

Modler, H.W., McKellar, R.C. and Yaguchi, M. (1990) Bifidobacteria and bifidogenic factors. *Canadian Institute of Food Science and Technology Journal*, **23**(1) 29–41.

Mogensen, G. (1980) Production and properties of yoghurt and ymer made from ultrafiltered milk. *Desalination*, **35**, 213–22.

Molska, I., Moniuszko, I., Komorowska, M. and Meriläinen, V. (1983) Characteristics of bacilli of *Lactobacilli casei* species appearing in kefir grains. *Acta Alimentaria Polonica*, **IX**(XXXIII), 79–88.

Morris, H.A., Ghaleb, H.M., Smith, D.E. and Bastian, E.D. (1995). A comparison of yogurts fortified with nonfat dry milk and whey protein concentrates. *Cultured Dairy Products Journal*, **30**(1), 2–4, 31.

Mottar, J., Bassier, A., Joniau, M. and Baert, J. (1989) Effect of heat-induced association of whey proteins and casein micellses on yogurt texture. *Journal of Dairy Science*, **72**, 2247–56.

Muir, D.D. and Hunter, E.A. (1992) Sensory evaluation of fermented milks: vocabulary development and the relations between sensory properties and composition and between acceptability and sensory properties. *Journal of the Society of Dairy Technology*, **45**, 73–80.

Muir, D.D. and Tamime, A.Y. (1993) Ovine milk: 3. effect of seasonal variations on properties of set and stirred yogurts. *Milchwissenschaft*, **48**, 509–13.

Mukai, T., Onose, Y., Toba, T. and Itoh, T. (1992) Presence of glycerol teichoic acid in the cell wall of *Lactobacillus kefiranofaciens*. *Letters in Applied Microbiology*, **15** 29–31.

Mulder, H. and Walstra, P. (1974) *The Milk Fat Globule: Emulsion Science as Applied to Milk Products and Comparable Foods*, Commonwealth Agricultural Bureau, Farnham Royal.

Mutukumira, A.N. (1995) properties of amasi, a natural fermented milk produced by small-holder milk producers in Zimbabwe. *Milchwissenschaft*, **50**, 201–5.

Mutukumira, A.N., Narvhus, J.A. and Abrahamsen, R.K. (1995) Review of traditionally-fermented milk in some sub-Saharan countries: focusing on Zimbabwe. *Cultured Dairy Products Journal*, **30**(1) 6–11.

Nahaisi, M.H. (1986) *Lactobacillus acidophilus:* therapeutic properties, products and enumeration, in *Developments in Food Microbiology*, vol. 2 (ed. R.K. Robinson), Elsevier Applied Science Publishers, London, pp. 153–78.

Nakajima, H., Toyoda, S., Toba, T., Itoh, T., Mukai, T., Kitazawa, H. and Adachi, S. (1990) A novel phosphopolysaccharide from slime-forming *Lactococcus lactis* subspecies *cremoris* SBT 0495. *Journal of Dairy Science*, **73**, 1472–7.

Nakamura, Y., Yamamoto, M., Sakai, K. and Takano, T. (1995a) Antihypersensitive effect of sour milk and peptides isolated from it that are inhibitors to angiotensin I-converting enzyme. *Journal of Dairy Science*, **78**, 1253–7.

Nakamura, Y., Yamamato, N., Sakai, K., Takano, T., Okubo, A. and Yamazaki, A. (1995b) Purification and characterization of angiotensin I-converting enzyme inhibitors from sour milk. *Journal of Dairy Science*, **78**, 777–83.

Nakazawa, Y., Asano, J. and Tokimura, A. (1990) Manufacture and chemical properties of low sodium yoghurt. *Milchwissenschaft*, **45**, 88–91.

Neve, H., Geis, A. and Teuber, M. (1988) Plasmid-encoded functions of ropy lactic acid streptococcal strains from Scandinavian fermented milk. *Biochimie*, **70**, 437–42.

Nilsson, L.-E. and Halström, B. (1990) Fermented milks engineering and process automation aspects, in *Proceedings of the XXIII International Dairy Congress*, vol. 3, Mutual Press, Ottawa, pp. 1953–9.

Nilsson, R. and Nilsson, G. (1958) Studies concerning Swedish ropy milk, the antibiotic qualities of ropy milk. *Archive für Mikrobiologie*, **31**, 191–7.

Noh, B. and Richardson, T. (1989) Incorporation of radiolabelled whey proteins into casein micelles by heat processing. *Journal of Dairy Science*, **72**, 1724–31.

Oberman, H. (1985) Fermented milks, in *Microbiology of Fermented Foods*, vol. 1 (ed. B.J.B. Wood), Elsevier Applied Science Publishers, London, pp. 167–95.

Okagbue, R.N. and Bankole, M.O. (1992) Use of starter cultures containing *Streptococcus diacetilactis*, *Lactobacillus brevis* and *Saccharomyces cerevisiae* for fermenting milk for production of Nigerian nono. *World Journal of Microbiology and Biotechnology*, **8**, 251–3.

Oksanen, P.J., Salminen, S., Saxelin, M., Hämäläinen, P., Ihantola-Vormisto, A., Muuras-niemi-Isoviita, L., Nikkari, S., Oksanen, T., Pörsti, I., Salminen, E., Sittonen, S., Stuckey, H., Toppila, A. and Vapaatalo, H. (1990). Prevention of traveller's diarrhoea by *Lactobacillus* GG. *Annals of Medicine*, **22**, 53–6.

Olasupo, N.A. and Azeez, M.K. (1992) Nono – a Nigerian study. *Dairy Industries International*, **57**(4), 37.

Orberg, P.K. and Sandine, W.E. (1985) Survey of antimicrobial resistance in lactic streptococci. *Applied and Environmental Microbiology*, **49**, 538–42.

O'Sullivan, N.G., Thornton, G., O'Sullivan, G.C. and Collins, J.K. (1992) Probiotic bacteria: myth or reality? *Trends in Food Science & Technology*, **3**, 309–14.

Pagliarini, E., Solaroli, G. and Peri, C. (1993) Chemical and physical characteristics of mare's milk. *Italian Journal of Food Science*, **5**, 323–32.

Pappas, C.P. (1988) A comparative study of laws and regulations on compositonal require- ments for yogurt in EC member states. *British Food Journal*, **90**, 195–8.

Park, S.Y., Kim, J.H., Kwon, I.K. and Kim, H.U. (1984) A study on antibiotic susceptibility of lactic acid bacteria isolated in Korea. *Korean Journal of Dairy Science*, **6**, 78–84.

Parnell-Clunies, E., Kakuda, Y., Irvine, D. and Mullen, K. (1988) Heat-induced protein changes in milk processed by vat and continuous heating systems. *Journal of Dairy Science*, **71**, 1472–83.

Patel, R.S. and Abd El-Salam, M.H. (1986) Shrikhand – an Indian analogue of western quarg. *Cultured Dairy Products Journal*, **21**(1), 6–7.

Patel, R.S. and Chakraborty, B.K. (1985) Reduction of curd-forming period in shrikhand manufacturing process. *Lait*, **65**, 55–64.

Patel, R.S. and Chakraborty, B.K. (1988) Shirkhand – aa review. *Indian Journal of Dairy Science*, **41**, 109–15.

Pearce, R.J. (1994) Food functionality: success or failure for dairy based ingredients. Paper presented at 24th International Dairy Congress, in Australia (available on CD ROM).

Pearce, R.J. (1995) Food functionality: success or failure for dairy based ingredients. *The Australian Journal of Dairy Technology*, **50**, 15–23.

Pederson, C.S. (1979) *Microbiology of Food Fermentation*, 2nd edn, AVI, Connecticut, pp. 1–29.

Perdigón, G., Alvarez, A. and Medici, M. (1992) Systematic and local augmentation of immune response in mice by feeding with milk fermented with *Lactobacillus acidophilus* and/or *Lactobacillus casei*, in *Foods, Nutrition and Immunity* (eds M. Paubert-Braquet, Ch. Dupont and R. Paoletti), Karger, Basel, pp. 66–76.

Perdigón, G., Nader de Macias, M.E., Alvarez, A., Oliver, G. and Pesce de Ruiz Holgado, A.A. (1990) Prevention of gastrointestinal infection using immunobiological methods with milk fermented with *Lactobacillus casei* and *Lactobacillus acidophilus*. *Journal of Dairy Research*, **57**, 255–64.

Pettersson, H.-E. (1984) Freeze-dried concentrated starter for kefir, in *Fermented Milks*, IDF Bulletin No. 179, International Dairy Federation, Brussels, pp. XVI–XVII.

Pétursson, S. (1949) The Icelandic skyr: microflora, manufacture and economy, in *Proceedings of XII International Dairy Congress*, Ivar Hoeggströms, Stockholm, pp. 419–22.

Pouliot, Y., Boulet, M. and Paquin, P. (1989) Observations on heat-induced salt balance changes in milk: I. effect of heating time between 4 and 90°C. *Journal of Dairy Research*, **56**, 185–92.

Pot, B., Ludwig, W., Kersters, K. and Schleifer, K.-H. (1994) Taxonomy of lactic acid bacteria, in *Bacteriocins of Lactic Acid Bacteria* (eds L. De Vuyst and E.J. Vandamme), Blackie Adacemic & Professional, London, pp. 13–90.

Prajapati, J.B., Shah, R.K. and Dave, J.M. (1986) Nutritional and therapeutical benefits of a blended-spray dried acidophilus. *Cultured Dairy Products Journal*, **21**(2), 16–21.

Prajapati, J.B., Shah, R.K. and Dave, J.M. (1987) Survival of *Lactobacillus acidophilus* in blend–spray dried acidophilus preprations. *The Australian Journal of Dairy Technology*, **42**, 17–21.

Prajapati, J.P., Upadhyay, K.G. and Desai, H.K. (1992) Comparative quality appraisal of heated shirkhand stored at ambient temperature. *The Australian Journal of Dairy Technology*, **47**, 18–22.

Prajapati, J.P., Upadhyay, K.G. and Desai, H.K. (1993) Quality appraisal of heated shirkhand stored at refrigerated temperature. *Cultured Dairy Products Journal*, **28**(2), 14–17.

Prevost, H. and Divies, C. (1988a) Continuous pre-fermentation of milk by entrapped yoghurt bacteria: 1. development of process. *Milchwissenschaft*, **43**, 621–5.

Prevost, H. and Divies, C. (1988b) Continuous pre-fermentation of milk by entrapped yoghurt bacteria: 2. data for optimization of the process. *Milchwissenschaft*, **43**, 716–19.

Puhan, Z. (1988) Treatment of milk prior to fermentation, in *Fermented Milks – Science and Technology*, IDF Bulletin No. 227, International Dairy Federation, Brussels, pp. 66–74.

Puhan, Z. (1990) Developments in the technology of fermented milk products. *Cutured Dairy Products Journal*, **25**(2), 4–9.

Puhan, Z. and Gallmann, P. (1980) Ultrafiltration in the manufacture of kumys and quark. *Cultured Dairy Products Journal*, **15**(1), 12–16.

Puhan, Z., Driessen, F.M., Jelen, P. and Tamime, A.Y. (1994) Fresh products: yoghurt, fermented milks, Quarg and fresh cheese. *Mljekarstvo*, **44**, 285–98.

Punjrath, J.S. (1991) Indigenous milk products of India – the related research and technology requirements in process equipment. *Indian Dairyman*, **43**, 75–87.

Ramakrishna, Y., Singh, R.S. and Anand, S.K. (1985) Effect of streptomycin on lactic cultures. *Cultured Dairy Products Journal*, **20**(3), 12–13.

Rambaud, J.C., Bouhnick, Y. and Marteau, P. (1994) Dairy products and intestinal flora, in *Dairy Products in Human Health and Nutriton* (eds. M. Serrano-Rios, A. Sastre, M.A. Prerez Jeuz, A. Estrala and C. de Sebastian), A.A. Balkema, Rotterdam, pp. 389–99.

Rao, S.M. and Ghandi, D.N. (1988) Studies on various quality characteristics of acidophilus sour milk from buffalo milk. *Cultured Dairy Products Journal*, **23**(2), 21–6.

Rao, D.R., Alhajali, A. and Chawan, C.B. (1987) Nutritional sensory and microbiological qualities of labneh made from goat milk and cow milk. *Journal of Food Science*, **52**, 1228–30.

Rash, K. (1990) Compositional elements affecting flavour of cultured dairy foods. *Journal of Dairy Science*, **73**, 3651–6.

Rasic, J.L. (1987) Other products, in *Milk – The Vital Force*, proceedings of the XXII International Dairy Congress, D. Reidel Publishing Company, Dordrecht, pp. 673–82.

Rasic, J.L. and Kurmann, J.A. (1978) *Yoghurt – Scientific Grounds, Technology, Manufacture and Preparations*, Technical Dairy Publishing House, Copenhagen.

Rasic, J.L. and Kurmann, J.A. (1983) *Bifidobacteria and their Role*, Birkhäuser Verlag, Basel.

Rasic, J.L., Kosikowski, F.V. and Bozic, Z. (1992) Nutrient yoghurt from low lactose milk using a combined lactase-UF retentate procedure. *Milchwissenschaft*, **47**, 32–5.

Renner, E. (1991) Cultured dairy products in human nutrition, in *Cultured Dairy Products in Human Nutrition – Dietary Calcium and Health*, IDF Bulletin No. 255, International Dairy Federation, Brussels, pp. 2–24.

Renner, E. and Abd El-Salam, M.H. (1991) *Application of Ultrafiltration in the Dairy Industry*, Elsevier Applied Science Publishers, London.

Reuter, G. (1990) Bifidobacteria cultures as components of yoghurt-like products. *Bifidobacteria Microflora*, **9**, 107–18.

Riber, R.F. (1989) Three major areas that cause defects in cultured dairy products. *Cultured Dairy Products Journal*, **24**(4), 6–9.

Roberts, A.V., Mayes, J.J. and Hull, R.R. (1984) Acidophilus yoghurt manufacture, in *Proceedings of a Dairy Culture Review Conference*, Technical Publication No. 27, Australian Society of Dairy Technology, Glenelg North, pp. 51–3.

Robinson, R.K. (1987) Survival of *Lactobacillus acidophilus* in fermented products. *South African Journal of Dairy Science*, **19**, 25–7.

Robinson, R.K. and Tamime, A.Y. (1990) Microbiology of fermented milks, in *Dairy Microbiology – The Microbiology of Milk Products*, vol. 2, 2nd edn (ed. R.K. Robinson), Elsevier Applied Science Publishers, London, pp. 291–343.

Robinson, R.K. and Tamime, A.Y. (1993) Manufacture of yoghurt and other fermented milks, in *Modern Dairy Technology – Advances in Milk Products*, vol. 2, 2nd edn (ed. R.K. Robinson), Elsevier Applied Science Publishers, London, pp. 1–48.

Robinson, R.K. and Vlahopoulou, I. (1988) Goat's milk utilization for fermented milk products. *Dairy Industries International*, **53**(12), 33–5.

Roefs, S.P.F.M. (1987) Structure of acid casein gels: a study of gel formed after acidification in the cold. *Netherlands Milk and Dairy Journal*, **41**, 99–101.

Roefs, S.P.F.M., Walstra, P., Dalgleish, D.G. and Horne, D.S. (1985) Preliminary note on the change in casein micelles caused by acidification. *Netherlands Milk and Dairy Journal*, **39**, 119–22.

Roefs, S.P.M., de Groot-Mostert, A.E.A. and van Vliet, T. (1990a) Structure of acid casein gels: 1. formation and model of gel network, *Colloids and Surfaces*, **50**, 141–59.

Roefs, S.P.F.M., van Vliet, T., van den Bijaart, H.J.C.M., de Groot-Mostert, A.E.A. and

Walstra, P. (1990b) Structure of casein gels made by combined acidification and rennet action. *Netherlands Milk and Dairy Journal*, **44**, 159–88.

Rogers, S.A. and Mitchell, G.E. (1994) The relationship between somatic cell count, composition and manufacturing properties of bulk milk: 6. Cheddar cheese and skim milk yoghurt. *The Australian Journal of Dairy Technology*, **49**, 70–4.

Roginski, H. (1988) Fermented milks. *The Australian Journal of Dairy Technology*, **43**, 37–46.

Rogelj, I. (1994) Lactic acid bacteria as probiotics. *Mljekarstro*, **44**, 277–84.

Rohm, H. (1993a) Influence of dry matter fortification on flow properties of yogurt: 2. time-dependent behaviour. *Milchwissenschaft*, **48**, 614–17.

Rohm, H. (1993b) Viscoelastic properties of set-style yogurt. *Rheology*, **3**, 173–82.

Rohm, H. and Kovac, A. (1994) Effects of starter cultures on linear viscoelastic and physical properties of yogurt gels. *Journal of Texture Studies*, **25**, 311–29.

Rohm, H. and Kovac, A. (1995) Effects of starter cultures on small deformation rheology of stirred yogurt. *Lebensmittel-Wissenschaft und–Technologie*, **28**, 319–22.

Rohm, H. and Schmid, W. (1993) Influence of dry matter fortification on flow properties of yoghurt: 1. evaluation of flow curves. *Milchwissenschaft*, **48**, 556–60.

Rohm, H., Elishkases-Lechner, F. and Bräuer, M.ⁿ (1992) Diversity of yeasts in selected dairy products. *Journal of Applied Bacteriology*, **72**, 370–6.

Rohm, H., Kovac, A. and Kneifel, W. (1994) Effects of starter cultures on sensory properties of set-style yoghurt determined by quantitative descriptive analysis. *Journal of Sensory Studies*, **9**, 171–86.

Rollema, H.A. and Brinkhuis, J.A. (1989) A H-NMR study of bovine casein micelles; influence of pH, temperature and calcium ions on micellar structure. *Journal of Dairy Research*, **56**, 417–25.

Romond, C. and Romond, M.B. (1990) Produits fermentes par *Bifidobacterium*, in *Proceedings of XXIII International Dairy Congress*, vol. 2, Mutual Press, Ottawa, pp. 1255–64.

Rothchild, P. (1995) Internal defences. *Dairy Industries International*, **60**(2), 24–5.

Roy, D., Berger, J.L. and Reuter, G. (1994) Characterization of dairy-related *Bifidobacterium* spp. based on their *β*-galactosidase electrophoretic patterns. *International Journal of Food Microbiology*, **23**, 55–70.

Sachdeva, S., Reuter, H., Prokopek, D. and Klobes, H. (1992a) Ultrafiltration of heated acidic and coagulated skim milk with different modules: Part 1. plate and frame module. *Keiler Milchwirtschaftliche Forschungsberichte*, **44**, 17–26.

Sachdeva, S., Reuter, H., Prokopek, D. and Klobes, H. (1992b) Ultrafiltration of heated acidic and coagulated skim milk with different modules: Part 2. spiral wound module. *Keiler Milchwirtschaftliche Forschungsberichte*, **44**, 27–34.

Sachdeva, S., Reuter, H., Prokopek, D. and Klobes, H. (1992c) Ultrafiltration of heated acidic and coagulated skim milk with different modules: Part 3. mineral membrane module. *Keiler Milchwirtschaftliche Forschungsberichte*, **44**, 35–46.

Sachdeva, S., Reuter, H., Prokopek, D. and Klobes, H. (1992d) Ultrafiltration of heated acidic and coagulated skim milk with different modules: Part 4. hollow fibres module. *Keiler Milchwirtschaftliche Forschungsberichte*, **44**, 47–54.

Salji, J. (1991) Concentrated yoghurt: a challenge to our industry. *Food Science and Technology Today*, **5**(1), 18–19.

Saliji, J. (1992) Acidophilus milk products: food with a third dimension. *Food Science and Technology Today*, **6**, 142–7.

Salminen, S. (1993) *Lactobacillus* GG fermented dairy products with documented health claims. *International Food Ingredients*, No. 1/2, 17–20.

Salminen, S. (1994) Healthful properties of *Lactobacillus* GG. *Dairy Industries International*, **59**(1), 36–7.

Salminen, S., Gorbach, S. and Salminen, K. (1991a) Fermented whey drink and yogurt-type product manufactured using *Lactobacillus* strain. *Food Technology*, **45**(6), 112.

Salminen, S., Salminen, K. and Gorbach, S. (1991b) *Lactobacillus* GG (Gefilac™) fermented whey drink and yoghurt. *Scandinavian Dairy Information*, **5**(3), 66–7.

Salminen, S., Deighton, M. and Gorbach, S. (1993) Lactic acid bacteria in health and disease, in *Lactic Acid Bacteria* (eds S. Salminen and S. von Wright), Marcell Dekker, New York, pp. 199–225.

Samuelsson, E.-G. and Ulrich, P. (1982) Processing of ymer based on ultrafiltration, in *Brief*

Communications of XXI International Dairy Congress, vol. 1, Book 1, Mir Publishers, Moscow, pp. 288–9.

Sandine, W.E., Radich, P.C. and Elliker, P.R. (1972) Ecology of lactic streptococci: a review. *Journal of Milk and Food Technology*, **35**, 176–84.

Savello, P.A. and Dargan, R.A. (1995) Improved yogurt physical properties using ultrafiltration and very-high temperature heating. *Milchwissenschaft*, **50**, 86–90.

Saxelin, M., Ahokas, M. and Salminen, S. (1993) Dose response on the faecal colonisation of *Lactobacillus* strain GG administered in two different formulations. *Microbial Ecology in Health and Disease*, **6**, 119–22.

Saxelin, M., Elo, S., Salminen, S. and Vapaatalo, H. (1991) Dose response colonisation of faeces after oral administration of *Lactobacillus casei* strain GG. *Microbial Ecology in Health and Disease*, **4**, 209–14.

Scardovi, V. (1986) Genus *Bifidobacterium*, in *Bergey's Manual of Systematic Bacteriology*, vol. 2 (eds P.H.A. Sneath, N.S. Mair, M.E. Sharpe and J.C. Holt), Williams & Wilkins, Baltimore, pp. 1418–34.

Schacht, E. and Syrazyrski, A. (1975) [Progurt[R] – a new cultured product: its manufacturing technology and dietetic value]. *Industria Lechera*, No. 646, 9–11 (in Spanish). Cited in *Dairy Science Abstracts* (1976), **38**, 741 [7013].

Schmidt, D.J. and Poll, J.K. (1986) Electrokinetic measurements on unheated and heated casein micelle systems. *Netherlands Milk and Dairy Journal*, **40**, 269–80.

Sellars, R.L. (1991) Acidophilus products, in *Therapeutic Properties of Fermented Milks* (ed. R.K. Robinson), Elsevier Applied Science Publishers, London, pp. 81–116.

Sgorbati, B., Biavati, B. and Palenzona, D. (1995) The genus *Bifidobacterium*, in *The Genera of Lactic Acid Bacteria*, vol. 2 (eds B.J.B. Wood, W.H. Holzapfel), Blackie Academic & Professional, London, pp. 279–306.

Shah, N.P., Spurgeon, K.R. and Gilmore, T.M. (1993) Use of dry whey and lactose hydrolysis in yogurt bases. *Milchwissenschaft*, **48**, 494–8.

Shalo, P.L. and Hansen, K.K. (1973) Maziwa lala – a fermented milk. *World Animal Review*, FAO publication No. 5, pp. 33–37.

Sharma, S.K. and Dalgleish, D.G. (1994) Effect of heat treatment on the incorporation of milk serum proteins into the fat globule membrane of homogenized milk. *Journal of Dairy Research*, **61**, 375–84.

Sharma, N. and Ghandi, D.N. (1981) Preparation of acidophilin: I. selection of starter cultures. *Cultured Dairy Products Journal*, **16**(2), 6–10.

Sharma, N. and Ghandi, D.N. (1983) Preparation of acidophilin: II. chemical, bacteriological and sensory evaluation. *Cultured Dairy Products*, **18**(3), 19–30.

Siitonen, S., Vapaatalo, H., Salminen, S., Gordin, A., Saxelin, M., Wikberg, R. and Kirkkola, A.-L. (1990) Effect of *Lactobacillus* GG yoghurt in prevention of antibiotic associated diarrhoea. *Annals of Medicine*, **22**, 57–9.

Silva, M., Jacobus, N.V., Deneke, C. and Gorbach, S.L. (1987) Antimicrobial substance from a human *Lactobacillus* strain. *Antimicrobial Agents and Chemotherapy*, **31**, 1231–3.

Simpson, W.J. and Taguchi, H. (1995) The genus *Pediococcus*, with notes on the general *Tetratogenococcus* and *Aerococcus*, in *The Genera of Lactic Acid Bacteria*, vol. 2 (eds N.J.B. Woods and W.H. Holzapfel), Blackie Academic & Professional, London, pp. 125–72.

Singh, H. (1993) Heat-induced interactions of proteins in milk in *Protein & Fat Globule Modifications by Heat Treatment, Homogenizations & Other Technological Means for High Quality Dairy Products*, IDF Special Issue No. 9303, International Dairy Federation, Brussels, pp. 191–204.

Singh, H. and Creamer, L.K. (1991) Influence of concentration of milk solids on the dissociation of micellar κ-casein on heating reconstituted milk at 120°C. *Journal of Dairy Research*, **58**, 99–105.

Singh, H. Sharma, R. and Taylor, M.W. (1993) Protein–fat interactions in recombined milk, in *Protein & Fat Globule Modifications by Heat Treatment, Homogenization & Other Technological Means for High Quality Dairy Products*, IDF Special Issue No. 9303, International Dairy Federation, Brussels, pp. 30–9.

Skorodumova, A.M. (1959) Acidophilus-yeast milk, a medical sour milk beverage, in *Proceedings of XV International Dairy Congress*, vol. 3, section 5, Richard Clay, Bungay, pp. 1418–22.

Sjollema, A. (1988) Specifications of dairy products used as raw materials for recombining. *Netherlands Milk and Dairy Journal*, **42**, 365–74.

Sneath, P.H.A., Mair, N.S., Sharpe, M.E. and Holt, J.G. (eds) (1986) *Bergey's Manual of Systematic Bacteriology*, vol. 2, Williams & Wilkins, Baltimore.

Soback, S. (1981) Growth of some lactic acid bacteria in milk containing sulfadiazine. *Acta Veterinaria Scandinavica*, **22**, 493–500.

Stansbridge, E.M., Walker, V., Hall, M.A., Smith, M.R., Bacon, C. and Chen, S. (1993) Effects of feeding premature infants with *Lactobacillus* GG on gut fermentation. *Archives of Disease in Childhood*, **69**, 488–92.

Subramanian, P. and Shankar, P.A. (1985) Commensalistic interaction between *Lactobacillus acidophilus* and lactose-fermenting yeasts in the preparation of acidophilus-yeast milk. *Cultured Dairy Products Journal*, **20**(4), 17–26.

Sundman, V. (1953) On the Microbiology of Finnish ropy sour milk, in *XIII International Dairy Congress*, vol. III, Mouton & Co., The Hague, pp. 1420–7.

Tahajod, A.S. and Rand, A.G. (1993) Bioprocess combinations for manufacturing cultured dairy products. *Cultured Dairy Products Journal*, **28**(1), 10–14.

Takizawa, S., Kojima, S., Tamura, S., Fujinaga, S., Benno, Y. and Nakase, T. (1994) *Lactobacillus kefirgranum* sp. nov. and *Lactobacillus parakefir* sp. nov., two new species from kefir grains. *International Journal of Systematic Bacteriology*, **44**, 435–9.

Tamime, A.Y. (1990) Microbiology of starter cultures, in *Dairy Microbiology – The Microbiology of Milk Products*, vol. 2, 2nd edn (ed. R.K. Robinson), Elsevier Applied Science Publishers, London, pp. 131–201.

Tamime, A.Y. (1993) Yoghurt-based products, in *Encyclopaedia of Food Science, Food Technology and Nutrition*, vol. 7 (eds R. Macrae, R.K. Robinson, M.J. Saddler), Academic Press, London, pp. 4972–7.

Tamime, A.Y. and Crawford, R.J.M. (1984) The microbiological quality of yoghurt cheese (known in the Lebanon as labneh anbaris) after one year storage at 20°C. *Egyptian Journal of Dairy Science*, **12**, 299–312.

Tamime, A.Y. and Deeth, H.C. (1980) Yogurt: technology and biochemistry. *Journal of Food Protection*, **43**, 939–77.

Tamime, A.Y. and Kirkegaard, J. (1991) Manufacture of Feta cheese – industrial, in *Feta and Related Cheeses* (eds R.K. Robinson and A.Y. Tamime), Ellis Horwood, London, pp. 70–143.

Tamime, A.Y. and Robinson, R.K. (1978) Some aspects of the production of a concentrated yoghurt (labneh) popular in the Middle East. *Milchwissenschaft*, **33**, 209–12.

Tamime, A.Y. and Robinson, R.K. (1985) *Yoghurt – Science and Technology*, Pergamon Press, Oxford.

Tamime, A.Y. and Robinson, R.K. (1988) Fermented milks and their future trends: II. technological aspects. *Journal of Dairy Research*, **55**, 281–307.

Tamime, A.Y., Kalab, M. and Davies, G. (1984) Microstructure of set-style yoghurt manufactured from cow's milk fortified by various methods. *Food Microstructure*, **3**, 83–92.

Tamime, A.Y., Davies, G. and Hamilton, M.P. (1987) The quality of yogurt on retail sale in Ayrshire: Part 1. chemical and microbiological evaluation. *Dairy Industries International*, **52**(6), 19–21.

Tamime, A.Y., Kalab, M. and Davies, G. (1989a) Rheology and microstructure of strained yoghurt (labneh) made from cow's milk by three different methods. *Food Microstructure*, **8**, 125–35.

Tamime, A.Y., Davies, G., Chehade, A.S. and Mahdi, H.A. (1989b) The production of 'labneh' by ultrafiltration: a new technology. *Journal of the Society of Dairy Technology*, **42**, 35–9.

Tamime, A.Y., Kalab, M. and Davies, G. (1991a) The effect of processing temperatures on the microstructure and firmness of labneh made from cow's milk by the traditional method or by ultrafiltration. *Food Structure*, **10**, 345–52.

Tamime, A.Y., Davies, G., Chehade, A.S. and Mahdi, H.A. (1991b) The effect of processing temperatures on the quality of labneh made by ultrafiltration. *Journal of the Society of Dairy Technology*, **44**, 99–103.

Tamime, A.Y., Kalab, M., Davies, G. and Mahdi, H.A. (1991c) Microstructure and firmness of labneh (high solids yoghurt) made from cow's, goat's and sheep's milks by a traditional method or by ultrafiltration. *Food Structure*, **10**, 37–44.

Tamime, A.Y., Barclay, M.N.I., Davies, G. and Barrantes, E. (1994) Production of low-calorie yogurt using skim milk powder and fat-substitutes: 1. a review. *Milchwissenschaft*, **49**, 85–8.

Tamime, A.Y., Marshall, V.M. and Robinson, R.K. (1995a) Microbiology and technological aspects of milk fermented by bifidobacteria. *Journal of Dairy Research*, **62**, 151–87.

Tamime, A.Y., Kalab, M., Muir, D.D. and Barrantes, E. (1995b) The microstructure of set-style, natural yogurt made by substituting microparticulate whey protein for milk fat. *Journal of the Society of Dairy Technology*, **48**, 107–11.

Tamime, A.Y., Barrantes, E. and Sword, A.M. (1996) The effect of starch-based fat substitutes on the microstructure of set-style yogurt made from reconstituted skimmed milk powder. *Journal of the Society of Dairy Technology*, **49**, 1–10.

Tannock, G.W. (1992) Genetic manipulation of gut microorganisms, in *Probiotics – the Scientific Basis* (ed. R. Fuller), Chapman & Hall, London, pp. 181–207.

Tantaoui-Elaraki, A. and El Marrakchi, A. (1987) Study of Moroccan dairy products: Iben and smen. *MIRCEN Journal*, **3**, 211–20.

Tantaoui-Elaraki, A., Berrada, M., El Marrakchi, A. and Berramou, A. (1983) Etude sur le Iben Marocain. *Lait*, **63**, 230–45.

Terzid-Vidojevid, A. (1991) Mikroflora fermentisanih pavlaka. *Review of Research Work at the Faculty of Agriculture, Belgrade*, **36**, 85–95 (in Slovenian). Cited in *Dairy Science Abstracts*, **57**, 492 [4028].

Teuber, M. (1995) The genus *Lactococcus*, in *The Genera of Lactic Acid Bacteria*, vol. 2 (eds B.J.B. Wood and W.H. Holzapfel), Blackie Academic & Professional, London, pp. 173–234.

Thompson, J.K. and Collins, M.A. (1989) A comparison of the plasmid profiles of strains of lactic streptococci isolated from commercial mixed strain starter culture with those from fermented milk. *Milchwissenschaft*, **44**, 65–9.

Thompson, J.K., Collins, M.A. and Johnston, D.E. (1988) From the Caucasu to Ulster – the Co. Fermanagh buttermilk plant. *Ulster Folk Life*, **34**, 54–9.

Thompson, J.K., Johnston, D.E., Murphy, R.J. and Collins, M.A. (1990) Characteristics of a milk fermentation from rural Northern Ireland which resembles kefir. *Irish Journal of Food Science & Technology*, **14**, 35–49.

Toba, T., Arihara, K. and Adachi, S. (1990a) Distribution of microorganisms with particular reference to encapsulated bacteria in kefir grains. *International Journal of Food Microbiology* **10**, 219–24.

Toba, T., Nakajima, H., Tobitani, A. and Adachi, S. (1990b) Scanning electron microscopy and texture studies on characteristic consistency of Nordic ropy sour milk. *International Journal of Food Microbiology*, **11**, 313–20.

Toba, T., Kotani, T. and Adachi, S. (1991a) Capsular polysaccharide of a slime-forming *Lactococcus lactis* ssp. *cremoris* LAPT 3001 isolated from Swedish fermented milk 'långfil'. *International Journal of Microbiology*, **12**, 167–72.

Toba, T., Uemura, H., Mukai, T., Fuji, T., Itoh, T. and Adachi, S. (1991b) A new fermented milk using capsular polysaccharide producing *Lactobacillus kefiranofaciens* isolated from kefir grains. *Journal of Dairy Research*, **58**, 497–502.

Ueda, M., Iwasawa, S., Miyata, N. and Ahiko, K. (1982) Hexose-phosphorylating activities of *Torulopsis holmii* KY-5 isolated from kefir grain. *Agricultural and Biological Chemistry*, **46**, 2637–43.

van Boekel, M.A.J.S. and Walstra, P. (1995) Effect of heat treatment on chemical and physical changes to milkfat globules, in *Heat-Induced Changes*, 2nd edn (ed P.F. Fox), IDF Special Issue No. 9501, International Dairy Federation, Brussels, pp. 51–65.

van Hooydonk, A.C.M., Hagendoorn, H.G. and Boerrigter, I.J. (1986) pH-induced physico-chemical changes of casein micelles in milk and their effect on renneting: 1. effect of acidification on physico-chemical properties. *Netherlands Milk and Dairy Journal*, **40**, 281–96.

van Vliet, T., Roefs, S.P.F.M., Zoon, P. and Walstra, P. (1989) Rheological properties of casein gels. *Journal of Dairy Research*, **56**, 529–34.

Vayssier, Y. (1978a) Le kefir: analyse qualitative et quantitative. *Revue Laitière Française*, No. 361, 73–5.

Vayssier, Y. (1978b) Le kefir: etude et mise au point d'un levian pour la préparation d'une boisson. *Revue Laitière Françaide*, No. 362, 131–4.

Vedamuthu, E.R. (1978) How to make better buttermilk. *Dairy & Ice Cream Field*, **161**(1), 66B–66H.

Vedamuthu, E.R. (1985) What is wrong with cultured buttermilk? *Dairy and Food Sanitation*, **5**, 8–13.

Visser, J., Minihan, A., Smits, P., Tjan, S.B. and Heertje, I. (1986) Effects of pH and temperature on the milk salt system. *Netherlands Milk and Dairy Journal*, **40**, 351–68.

Vlahopoulou, I., Bell, A. and Wilbey, A. (1994) Starter culture effects on caprine yogurt fermentation. *Journal of the Society of Dairy Technology*, **47**, 121–3.

Wahlgren, N.M., Dejmek, P. and Drakenberg, T. (1990) A ^{43}Ca and ^{31}P NMR study of the calcium and phosphate equilibra in heated milk solutions. *Journal of Dairy Research*, **57**, 355–64.

Walker, D.K. and Gilliland, S.E. (1987) Buttermilk manufacture using a combination of direct acidification and citrate fermentation by *Leuconostoc cremoris. Journal of Dairy Science*, **70**, 2055–62.

Walstra, P. and Jenness, R. (1984) *Dairy Chemistry and Physics*, Wiley & Sons, New York.

Walstra, P. and van Vliet, T. (1986) The physical chemistry of curd making. *Netherlands Milk and Dairy Journal*, **40**, 241–59.

Wanghof, J., Schulthess, W. and Struebi, P. (1992) Flavouring of the cultured milk product mala with fruits available in Kenya. *Milchwissenschaft*, **47**, 27–31.

Welch, C. (1987) Nutritional and therapeutic aspects of *Lactobacillus acidophilus* in dairy products. *Cultured Dairy Products Journal*, **22**(2), 23–6.

Whalen, C.A., Gilmore, T.M., Spurgeon, K.R. and Parsons, J.G. (1988) Yogurt manufacture from whey–caseinate blends and hydrolyzed lactose. *Journal of Dairy Science*, **71**, 299–305.

White, C.H. (1978) Manufacturing better buttermilk. *Cultured Dairy Products Journal*, **13**(1), 16–20.

White, C.H. (1995) Manufacture of high quality yogurt. *Cultured Dairy Products Journal*, **30**(2), 18–26.

Wilbey, R.A. (1994) Production of butter and dairy based spreads, in *Modern Dairy Technology – Advances in Milk Processing*, vol. 1, 2nd edn (ed. R.K. Robinson), Chapman & Hall, London, pp. 107–58.

Wilcek, A. (1990) Methods for classification of skim-milk powder, in *Recombination of Milk and Milk Products*, IDF Special Issue No. 9001, International Dairy Federation, Brussels, pp. 135–40.

Wilmsen, A. (1991) The use of milk proteins in the dairy industry. *Scandinavian Dairy Journal*, **5**(2), 50–3.

Xiong, Y.L., Dawson, K.A. and Wan, L. (1993) Thermal aggregation of β-lactoglobulin: effect of pH, ionic environment and thiol reagent. *Journal of Dairy Science*, **76**, 70–7.

Yondem, F., Ozilgen, M. and Bozoglu, T.F. (1989) Growth kinetics of *Streptococcus thermophilus* at sub-bacteriostatic penicillin G concentration. *Journal of Dairy Science*, **72**, 2444–51.

Zhang, Z.P. and Aoki, T. (1995) Effect of modification of amino groups on crosslinking of casein by micellar calcium phosphate. *Journal of Dairy Science*, **78**, 36–43.

Note from the Authors

It is evident that the old nomenclature of *Lactococcus lactis* biovar *diacetylactis* has been used in Chapters 3 and 4. The reason for this is that such starter culture (i.e. flavour producer) has to be differentiated from other closely related organisms within its genus, especially during the manufacture of certain fermented dairy products. However, the current taxonomic name is *Lactococcus lactis* subsp. *lactis,* and also the taxonomic position of *Lactobacillus paracasei* subsp. *paracasei* or *Lactobacillus casei* subsp. *casei* has not been clarified. For more information on dairy starter cultures the reader is referred to *Dairy Starter Cultures* (T. Cogan and J.-P. Accolas (eds) 1996 – VCH Publishers (UK) Ltd, Cambridge, UK).

4 Physiology and biochemistry of fermented milks

V.M.E. MARSHALL and A.Y. TAMIME

4.1 Introduction

The many fermented milks familiar across the world, which have traditionally been made by spontaneous growth and action of microorganisms in the special environment afforded by milk, are today carefully controlled microbial processes for which selected cultures have been developed. The technology required for production of large quantities of the now-familiar yoghurt, cheeses, fromage frais, etc. have developed from the knowledge of the physiology and biochemistry of the microorganisms involved. The diversity of fermented milks makes a discussion of starter physiology and biochemistry complex, particularly as there has been an increase in the types of fermented milks available since publication of the first edition of this book. Products such as Laban, 'bio-yoghurts' and the thickened and pourable (drinking) varieties of yoghurt have appeared widely in Europe. No longer can the physiology and biochemistry be confined to a general discussion of a few species of lactococci, leuconostocs, lactobacilli and streptococci. Over the last 10 years much work has been done on the biochemistry and molecular biology of these organisms. Not only is catabolism an important consideration for a successful fermentation to produce a product of good quality in terms of flavour and stability, but anabolic pathways also have a role in providing texture-modifying polysaccharides, and in providing other compounds with a role in preservation and health-promoting properties (bacteriocins and expression of cell surface components).

Although the lactic acid bacteria are widely distributed in nature, their nutritional requirements are complex. Table 4.1 shows the fermentation ability, G + C content of the DNA, and growth temperatures of the different bacteria that have been used to ferment milk. These are characteristics which may be used to differentiate the genera and species. The inability to synthesize a full complement of amino acids dictates the natural habitat of these organisms. Milk is a nutritionally rich medium which will support the growth of many micro-organisms, but the processing of milk enables us to control the type of growth to achieve a desirable product. Key attributes required by the microorganism include the ability to obtain energy from lactose and the ability to break down and utilize milk proteins – in so doing producing good flavour, aroma and texture. Other attributes, such as prolonged shelf-life, stability during storage, and possible health benefits are also dependent upon microbial activity and metabolism. Many fermented

Table 4.1 Selected differential characteristics of fermented milks starter cultures

Starter organisms	G+C[a] (mean %)	Growth 10°C	Growth 45°C	ESC	AMY	ARA	CEL	FRU	GAL	LAC	MAL	MAN	MNE	MLZ	MEL	RAF	RHA	RIB	SAL	SOR	SAC	TRE	XYL
Lac. lactis subsp. *lactis*	33.8–36.9	+	–	d	d	d	–	–	+	+	+	d	–	–	–	–	–	+	d	–	d	d	d
biovar *diacetylactis*	33.6–34.8	+	–	d	d	d	–	–	+	+	+	d	–	–	–	–	–	+	d	–	–	d	d
subsp. *cremoris*	35.0–36.2	+	–	d	–	–	–	–	+	+	–	–	–	–	–	–	–	+	d	–	–	d	–
Leu. mesenteroides																							
subsp. *mesenteroides*	37.0–39.0	+	–	d	d	+	d	+	+	(d)	+	d	+	–	d	d	–	+	d	–	+	+	d
subsp. *dextranicum*	37.0–40.0	+	–	d	d	–	d	+	d	+	+	d	d	–	d	d	–	–	d	–	+	+	d
subsp. *cremoris*	38.0–40.0	+	–	–	–	–	–	–	+	+	d	–	–	–	–	–	–	–	–	–	–	–	–
S. thermophilus	37.0–40.0	–	+	+	–	–	–	+	(d)	+	–	–	–	–	(d)	d	–	–	–	–	+	–	d
Lb. delbrueckii subsp. *delbrueckii*	49.0–51.0	–	+	+	–	d	d	+	d	–	d	–	+	–	–	–	–	–	–	–	+	d	–
subsp. *lactis*	49.0–51.0	–	+	–	–	–	d	+	d	–	d	–	+	–	–	–	–	–	+	–	+	+	–
subsp. *bulgaricus*	49.0–51.0	–	+	–	–	–	–	+	d	+	–	–	–	–	–	–	–	+	–	–	–	–	–
Lb. acidophilus	34.0–37.0	–	+	+	+	–	+	+	+	+	+	d	d	–	d	d	–	+	+	–	+	d	–
Lb. helveticus	38.0–40.0	–	+	–	–	–	–	d	+	+	d	–	–	–	–	–	–	–	–	–	–	d	–
Lb. kefranofaciens	34.0–35.0	–	–	+	+	–	–	–	+	+	+	–	–	–	+	+	–	–	+	–	+	–	–
Lb. paracasei subsp. *paracasei*	45.0–47.0	+	–	+	+	d	+	+	+	d	+	+	+	+	–	–	–	+	+	+	+	+	–
Lb. rhamnosus	45.0–47.0	+	+	+	+	d	+	+	+	+	+	+	+	+	+	–	+	+	+	+	+	+	–
Lb. kefir	41.0–42.0	+	–	+	–	–	–	+	+	+	+	–	–	–	+	+	–	–	–	–	+	–	–
Lb. reuteri	40.0–42.0	–	+	–	–	+	–	+	+	+	+	–	–	–	+	+	–	+	–	–	+	–	d
B. adolescentis	58.9			+	+	+	+	+	+	+	+	d	d	+	d	+	–	–	+	d	+	d	+
bifidum	60.8			–	–	–	–	+	+	+	+	–	–	d	+	–	–	+	+	–	d	–	–
breve	58.4			–	–	–	d	+	+	+	+	d	d	–	+	+	–	+	+	d	+	d	–
infantis	60.5			–	–	–	–	+	+	+	+	–	d	d	+	+	–	+	–	–	+	–	d
longum	60.5			–	–	+	–	+	+	+	+	d	d	+	+	+	d	+	–	–	+	–	d
P. acidilactici	38.0–44.0	–	+	d	d	d	+	+	+	d	–	d	+	–	–	d	d	+	d	–	(d)	d	+

[a]Mean % of guanine and cytosine in deoxyribonucleic acid (DNA).

[b]ESC, aesculin; AMY, amygdalin; ARA, arabinose; CEL, cellobiose; FRU, fructose; GAL, galactose; LAC, lactose; MAL, maltose; MAN, mannose; MLZ, melezitose; RAF, raffinose; RHA, rhamnose; RIB, ribose; SAL, salicin; SOR, sorbitol; SAC, saccharose/sucrose; TRE, trehalose; XYL, xylose.

+: >90% strains positive; –: >90% strains negative; d: positive reaction by 11–89% of strains; (d): delayed reaction.

Data compiled from Garvie (1986), Hardie (1986), Kandler and Weiss (1986), Scardovi (1986), Biavati *et al.* (1992), Hammes *et al.* (1992), Hardie and Whiley (1992, 1995), Holzapfel and Schillinger (1992), Teuber (1995), Weiss (1992), Dellaglio *et al.* (1995), Hammes and Vogel (1995), Sgorbati *et al.* (1995), Simpson and Taguchi (1995) and Teuber *et al.* (1992).

foods rely on a natural fermentation where a population of key organisms (often species of lactic acid bacteria or yeasts and sometimes both) develop at the expense of other (undesirable) types. Commercial processing of fermented milks, however, does not have to rely on such chance happenings. Milk is normally heated before fermentation in order to eliminate the undesirable types and is subsequently inoculated with a starter culture specifically formulated to provide the required attributes.

This chapter provides an overview of the biochemical pathways important in achieving fermented milks of the type and character demanded, and discusses some aspects of the physiology that are important for successful population development.

4.1.1 Milk as a medium for microbial growth

Milk contains the disaccharide lactose, milk proteins, casein and the whey proteins, fat, vitamins and minerals. There is little 'free' iron and few 'free' amino acids. The lack of free iron has been sufficient in some cases to limit those organisms that have a specific growth requirement for iron. Organisms must be able to metabolize lactose, or become associated with organisms that hydrolyse lactose and release a sugar that they can metabolize. This latter, associative growth is demonstrated for kefir where a non-lactose-utilizing yeast is closely associated with the lactose-metabolizing flora of the kefir grain (see below). The paucity of free amino acids in freshly drawn, good-quality milk means that organisms must be equipped with an effective proteolytic system where enzymes act in concert, first to break specific peptide bonds in the protein, to further hydrolyse the peptides, and then to transport di- or tri-peptides and amino acids into the cell for further transformations. The nature of these hydrolyses are of particular importance to the manufacture of good-quality dairy products (see Chapter 9). The structure of the milk protein matrix and its formation is different for each dairy product. For example, in cheese-making, there is a requirement for the casein micelles to become fused, to merge and contract in order to expel the whey. In yoghurt, expulsion of whey (syneresis) is an undesirable phenomenon: the requirement is for a protein matrix to form an open network in which whey is trapped. Kefir and buttermilk, on the other hand, are both pourable and consumed as beverages. Moreover, kefir has the added complexity of requiring a protein structure which will hold carbon dioxide gas.

4.1.2 Associative growth

(a) Thermophilic organisms. The associative growth between *Lactobacillus delbrueckii* subsp. *bulgaricus* and *Streptococcus thermophilus* described in the first edition of this book and in many textbooks concerned with dairy microbiology still holds for many of the strains in use today. This associ-

ation may be briefly described as follows. Each organism provides compounds which benefit the other. The *Streptococcus* benefits from the stronger proteolytic activity of the *Lactobacillus* and in return provides carbon dioxide and formate which stimulate the *Lactobacillus*. Study of associative growth of these two organisms continues. Ascon-Reyes *et al.* (1995) showed that the amount of carbon dioxide produced by *S. thermophilus* in skimmed milk was related to the amount of urea degraded and that in the presence of *Lb. delbrueckii* subsp. *bulgaricus* and *Lactobacillus acidophilus* the carbon dioxide disappeared. In another report (Amoroso and Manca de Nadra, 1990), mutual stimulation in milk is observed, while in LAPT medium (containing yeast extract, peptone, tryptone and Tween) with different sugars, only stimulatory effects of the *Streptococus* on the *Lactobacillus* were observed. This is an expected result as the nitrogen source in LAPT medium is readily available and not dependent on proteolytic activity (the mechanism for stimulation of the *Streptococcus*). Thus, the medium used could demonstrate only one side of the partnership, namely that dependent on the carbon metabolic pathways. This underlines the importance of understanding the special qualities of milk as a growth medium: it has an ample supply of a simple disaccharide and an ample, but complex, nitrogen source. It is also important to remember that both these organisms grow perfectly well in milk. Indeed, many of the mild 'bio-yoghurts' are prepared with mixed cultures some of which include *Lb. delbrueckii* subsp. *bulgaricus* or *S. thermophilus*, but which do not appear to be dependent on a mutual stimulation of growth and metabolism for a successful fermentation.

(b) Mesophilic cultures.   Leuconostocs, in spite of their widespread use by the dairy industry, do not grow well in milk (Goel and Marth, 1969); *Leuconostoc lactis*, for example, grows in litmus milk, producing acid but with no reduction of litmus. The most likely reason for this is a lack of proteolytic ability (Cogan and Jordan, 1994). Leuconostocs are normally grown in mixed culture with lactococci, which have good proteolytic ability. Thus, associative growth is likely and there is some evidence for this: Boquien *et al.* (1988) showed that growth rates of *Leu. lactis* were significantly greater when grown in the presence of *Lactococcus lactis* subsp. *lactis*, but there was no change in growth rate of the latter organism.

Kefir, a traditional fermented milk known in Eastern Turkey, Poland and as far west as Northern Ireland, is another example where associative growth is important. This fermented milk is dependent on a mixed-culture fermentation involving lactose fermenting and non-lactose-fermenting yeasts and lactic acid bacteria (*Lac. lactis* subsp. *lactis* and *Lactobacillus kefir*). The non-lactose-fermenting yeast (*Saccharomyces cerevisiae*) has a preference for galactose, even when glucose is present in the medium in addition to galactose (Hirota and Kekuchi, 1976), which may indicate that its carbon source is made available in milk as a consequence of lactic acid bacterial metabolism. However, *Lac. lactis* subsp. *lactis* and *Lactococcus lactis* subsp. *cremoris* may not release galactose when utilizing lactose, as they metabolize the two sugar moieties of lactose simultaneously; consequently, it may be the *Lb. kefir* which provides the carbon source. However, this latter organism, on primary isolation from the kefir grain, ferments only arabinose, although after sub-culture it will ferment additional sugars, including lactose (Marshall *et al.*, 1984; Thompson *et al.*, 1990). Thus, it may be that it is the lactose-fermenting yeast (*Candida kefyr*) that releases the galactose required for the non-lactose-fermenting yeast and the associative growth occurs between the yeasts and not between yeasts and lactic acid bacteria. Thompson *et al.* (1990) found some evidence in their study of the buttermilk plant of Northern Ireland, that the yeasts were able to stimulate growth of the lactobacilli, although these authors were cautious in this interpretation because of 'the apparent propensity of the bacteria to open up alternative pathways for fermentation'.

4.1.3 Non-traditional organisms

The inclusion of strains of *Lb. acidophilus*, *Lactobacillus paracasei* subsp. *paracasei* and certain species of *Bifidobacterium* in yoghurt is now common and these are marketed as 'bio' products. They are non-traditional as they are not dairy organisms in the true sense, their natural habitat being the mammalian gastrointestinal tract. Bifidobacteria, unlike *Lb. acidophilus*, *Lb. delbrueckii* subsp. *bulgaricus* and *S. thermophilus*, are heterofermentative organisms, producing acetate in addition to lactate. Milk fermented with *Bifidobacterium bifidum* at 37°C to pH 4.5 has a molar ratio of lactate to acetate of $2:3$ and the flavour therefore is different.

The ability of some species of bifidobacteria to ferment oligosaccharides such as inulin and oligofructose has led to conjecture about the value of this metabolic ability to preferentially favour growth of bifidobacteria in the colon (Wang and Gibson, 1993; Gibson *et al.*, 1994). Oligofructose specifically stimulates bifidobacteria in the presence of colonic anaerobes such as *Bacteroides* spp. and *Clostridium* spp.; inclusion of these non-digestible, 'pre-biotics' in the diet may therefore beneficially affect the host by improving its intestinal microbial balance.

The new 'bio-yoghurts' have been developed by the dairy industry and are successful in terms of flavour and texture. The dairy and the non-dairy cultures may be grown together, but any associative growth, if investigated, has not been widely published. Medina and Jordano (1995) looked at the population dynamics of fermented milks containing *Lb. acidophilus*, species of bifidobacteria and *S. thermophilus* and suggested that consumption of oxygen by the cocci was important to maintenance of the probiotic population.

4.2   Carbohydrate metabolism

The lactococci, lactobacilli, leuconostocs and bifidobacteria involved in fermenting milk do not have a tricarboxylic acid cycle, although some of its enzymes may be present. Nor is there a cytochrome system for harnessing energy from electrons of NADH. Energy is largely obtained via substrate-level phosphorylation and the ATPases of the cytoplasmic membrane. Carbohydrate is metabolized either through homofermentative or hetero-fermentative metabolic pathways.

4.2.1   Homolactic fermentation

Lactococcus spp., *Lb. delbrueckii* subsp. *bulgaricus*, *Lb. acidophilus* and *S. thermophilus* all ferment lactose homofermentatively (Figure 4.1). However, the lactococci and some strains of *Lb. acidophilus* differ from the other species mentioned in that both the glucose and galatose parts of the lactose molecule are catabolized at the same time. The key step in this pathway is at the entry point of lactose into the cell. Lactose is phosphorylated by phosphoenolpyruvate (PEP) during translocation by the PEP-dependent phosphotransferase system (PEP:PTS) described by McKay *et al.* (1969). The lactose, therefore, appears in the cell as lactose phosphate and requires a different enzyme, β-D-phosphogalactosidase (Laue and MacDonald, 1968), to split it into its monosaccharide components. The products of the reaction are glucose and galactose 6-phosphate. The galactose is then catabolized via the Tagatose pathway at the same time as the glucose is catabolized via the Embden–Meyerhof–Parnas pathways. Only if galactose phosphate is de-phosphorylated will galactose remain unmetabolized and be excreted from the cell. This has been observed in certain cases with *Lac. lactis* subsp. *cremoris* (Benthin *et al.*, 1994).

The pathways converge at dihydroxyacetone phosphate and glyceraldehyde 3-phosphate. The 3-carbon sugars become further oxidized to 3- and 2-phosphoglycerate, and then to phosphoenolpyruvate, yielding one ATP and one NADH. Phosphoenolpyruvate is energy-rich and will fuel further sugar uptake via the PTS, but also important is the pyruvate kinase enzyme which converts PEP into pyruvate to yield a further molecule of ATP, and provides the lactate dehydrogenase enzyme with its substrate to allow for oxidation of the NADH and the return of NAD for further oxidation of the incoming sugar. PEP and its enzyme pyruvate kinase occupy a pivotal role in regulation of lactose metabolism. All phosphorylated intermediates before 1,3-diphosphoglycerate are activators of pyruvate kinase (Thomas, 1976), while the immediate precursor of PEP does not activate the kinase. Control of pyruvate kinase therefore regulates the intracellular PEP 'pool' and hence couples sugar entry to sugar metabolism, providing a tight link between substrate entry and its subsequent metabolism.

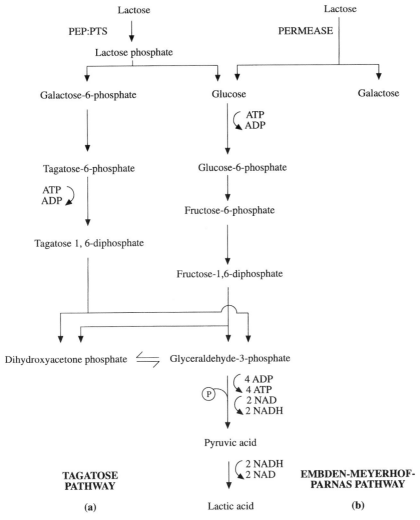

Figure 4.1 Homolactic fermentation of lactose after translocation, (a) by a PEP:PTS mechanism, and (b) by a Permease.

The lactococci are enabled in this mode of lactose metabolism through the possession of a plasmid which carries the genes for the PTS uptake and the β-D-phosphogalactosidase gene. There is some disagreement over the nature of lactose uptake in *Lb. acidophilus*, and Kanatami and Oshimura (1994) have reported on a plasmid-encoded PEP/PTS lactose uptake mechanism and a β-D-phosphogalactosidase gene in *Lb. acidophilus* LTF421 isolated from fermented milk. Loss of the plasmid resulted in a lactose negative genotype. The 1419 kb *pbg* gene (encoding 473 amino acid residues) had a deduced sequence which showed a 60% homology with β-D-phosphogalactosidase sequences in other Gram-positive organisms.

Homolactic fermentation by *Lb. delbrueckii* subsp. *bulgaricus, Lb. acidophilus* and *S. thermophilus* follows the Embden–Meyerhof–Parnas pathway for glucose catabolism. Lactose enters the cell via a permease and arrives as an unphosphorylated disaccharide. The lactose is split into its component sugars by the enzyme β-D-galactosidase which results in two non-phosphorylated sugars, glucose and galactose. The glucose is catabolized to pyruvate (Figure 4.2) and the galactose is excreted from the cell. When all the glucose has been utilized, *S. thermophilus* and *Lb. acidophilus* will utilize galactose via the Leloir pathway (Figure 4.2) where galactokinase is the first enzyme of the pathway. In media supplied with galactose as the only sugar some strains of *S. thermophilus* exhibit low growth; most strains of *Lb. delbrueckii* subsp. *bulgaricus* are unable to utilize galactose.

From genetic studies, the β-D-galactosidase enzymes of *S. thermophilus, Lb. delbrueckii* subsp. *bulgaricus* and *Leu. lactis* are similar (David *et al.,* 1992). The enzyme from *Lb. delbrueckii* subsp. *bulgaricus* may have a requirement for a magnesium ions for activity (Adams *et al.,* 1994) and a pH optimum of 6.5–7.0, although the enzyme is stable at a pH 5.8 (Gupta *et al.,* 1994).

Lactate dehydrogenase is also important in the control of carbohydrate metabolism. The lactococcal enzyme is activated by fructose 1,6-bisphosphate (Jonas *et al.,* 1972) and by tagatose 1,2-bisphosphate (Thomas, 1976). In homofermentative lactobacilli, this may not always be the case, as the enzyme from many of the species has been found to have constitutively high activity which is independent of the presence of fructose 1,6-bisphosphate (Gasser *et al.,* 1970). Two types of lactate dehydrogenase enzyme therefore exist, and both may be present in the actively metabolizing cell (Taguchi and Ohta, 1991, 1993). Sequencing of the lactate dehydrogenase gene from *S. thermophilus* (Ito and Sasaki, 1994) shows it to have 328 amino acid residues, and a reasonably high sequence homology with that of *Streptococcus mutans* (89.0%), *Lac. lactis* subsp. *lactis* (76.3%) and *Lactobacillus plantarum* (60%).

4.2.2   Heterolactic fermentation

Leuconostoc species are important in the production of cultured buttermilk, cottage cheese and sour cream. These lactic acid bacteria ferment lactose and glucose via a heterofermentative pathway (Figure 4.2). Lactose enters the cell via a permease and is split by a β-D-galactosidase enzyme. The enzyme is encoded by two overlapping genes (*lac*L and *lac*M), producing two proteins (72 kDa and 35 kDa, respectively) which are both required for enzyme activity (David *et al.,* 1992). Glucose 6-phosphate is oxidized to

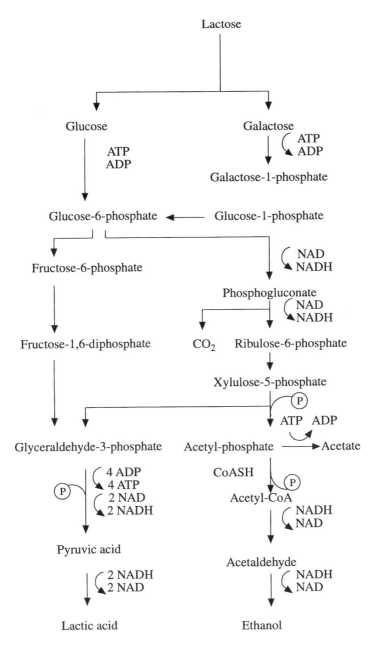

**LELOIR PATHWAYS   HEXOSE MONOPHOSPHATE
SHUNT PATHWAY**

Figure 4.2 Galactose metabolism by the Leloir pathway and heterolactate fermentation by the hexose monophosphate shunt.

phosphogluconate, then decarboxylated and the resulting pentose split into a 2-carbon compound and a 3-carbon sugar. This phosphoroclastic split is catalysed by a (pentose) phosphoketolase which gives the pathway its alternative name, the phosphoketolase pathway. The 3-carbon sugar (glyceraldehyde 3-phosphate) has the same fate as that found in the Embden–Meyerhof–Parnas pathway with the pyruvate finally being reduced to lactate. The 2-carbon moiety (acetyl phosphate) is an important energy-rich compound which may be converted to acetyl CoA, reduced to acetaldehyde, and further reduced to ethanol so that NAD can be re-generated to oxidize

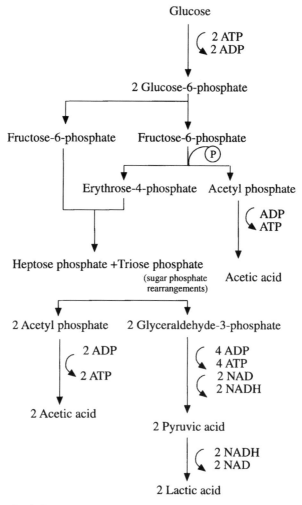

Figure 4.3 Heterolactic fermentation by bifidobacteria showing the phosphoroclastic split at fructose 6-phosphate. (After Stanier *et al.* (1987); reproduced courtesy of Prentice-Hall.)

more glucose molecules, or it may yield ATP and acetate. This latter reaction may be limited, however, by the need to generate the NAD by the reduction of acetate to ethanol. The most usual products of the fermentation are, therefore, equimolar quantities of lactate, carbon dioxide and ethanol. Only one mole of ATP is produced. A review paper by Cogan and Jordan (1994) provides a wider discussion of the metabolism of the *Leuconostoc* spp.

The bifidobacteria are also heterofermentative and catabolize glucose to mixed end-products (Figure 4.3). Unlike the leuconostocs, catabolism of glucose produces no carbon dioxide because there is no early step involving a decarboxylation. Hexoses are catabolized by a fructose 6-phosphate shunt and the pathway involves an alternative (hexose) phosphoketolase enzyme in the phosphoroclastic split. The enzyme in question is fructose 6-phosphate phosphoketolase, while the key enzyme for leuconostocs is xylulose 5-phosphate phosphoketolase (a pentose enzyme). The products of the bifido-pathway are lactate and acetate with more acetate than lactate being produced. The ratios of acetate : lactate are 3 : 2 as a result of catabolism of two moles of glucose.

In terms of flavour, milks fermented with heterofermentative species may be very different. The leuconostocs and the heterofermentative lactobacilli (such as *Lb. paracasei* subsp. *paracasei*) do not produce acetate, while bifidobacteria will produce substantial amounts.

4.3 Nitrogen metabolism

Most lactic acid bacteria require, or are stimulated by, amino acids and successful growth in milk is dependent on milk protein degradation by extracellular proteinases and peptidases. More is known of the lactococcal proteolysis systems than those of the leuconostocs, lactobacilli and streptococci (see Chapters 1, 7 and 9).

Stefanitsi *et al.* (1994) showed that *Lb. delbrueckii* subsp. *bulgaricus* had two cell-wall-associated proteinases whose activities were greatly enhanced when cells were grown in milk. Both were able to hydrolyse casein; one was rapidly inactivated by heating, inhibited by EDTA, and reactivated by addition of zinc ions, suggesting a metalloproteinase. The second enzyme was more stable to heat, inhibited by phenylmethylsulphonyl fluoride, insensitive to *N*-ethylmaleimide, and activated by calcium ions, suggesting a serine proteinase. No information is available as to the similarities with the lactococcal system where it is known, for example, that a maturation protein is required for active forms of the serine proteinase to be located in and secreted from the lactococcal cell envelope (Vos *et al.*, 1989; Haandrikman *et al.*, 1991).

Both *S. thermophilus* and *Lb. delbrueckii* subsp. *bulgaricus* possess a number of peptidases. The aminopeptidase from *S. thermophilus* NCDO 573 has been purified and characterized (Midwinter and Pritchard, 1994). It is able to hydrolyse a range of peptides, showing the highest activity towards lysyl derivatives. It was unable to hydrolyse peptides at the N-terminal side of a proline residue, and showed a high degree of N-terminal sequence homology with the aminopeptidase of *Lac. lactis* subsp. *cremoris*. A 'general aminopeptidase' from *S. thermophilus* CNRZ 302 was shown by Rul *et al.* (1994) to be a monomer of 97 kDa with maximum activity at pH 7 (36°C), and to have a preference for hydrophobic or basic amino acids at the N-terminal position.

A proline iminopeptidase of *Lb. delbrueckii* subsp. *bulgaricus* CNRZ 397 has also been purified and characterized by cloning it into *Escherchia coli* (Gilbert *et al.*, 1994). The enzyme is a trimer (three identical subunits) with a molecular weight of ∼ 100 kDa. It has a high specific activity for di- and tri-peptides, and is inhibited by 3,4-dichloroiso-coumarin, a serine proteinase inhibitor. It is unable to hydrolyse peptides with hydroxyproline at the N-terminus. There is evidence that this enzyme is located in the cell envelope (Atlan *et al.*, 1994). Other peptidases have also been characterized: x-prolyl dipeptidyl aminopeptidase from *Lb. delbrueckii* subsp. *bulgaricus* catalyses the hydrolytic removal of N-terminal dipeptidyl residues from peptides containing proline in the penultimate position (Meyer-Barton *et al.*, 1993) and leucyl aminopeptidase which showed similarities with other prolinases and iminopeptidases (Klein *et al.*, 1994, 1995).

The similarities of the amino acid sequences of the proline aminopeptidases of the lactic acid bacteria has led Guedon *et al.* (1995) to speculate that intergeneric transfer of insertion sequences could occur during co-culture. A chromosomal insertion sequence found in *S. thermophilus* has >99% homology with an insertion sequence found in *Lac. lactis* subsp. *lactis*.

4.4   Pathways leading to flavour compounds

A hundred years ago starter cultures were unknown, but throughout this century dairy microbiologists and molecular biologists have unravelled the mixes of organisms and determined their genetic maps. More is now known of the metabolic mechanisms which lead to production of flavour and aroma compounds. Table 3.3 (p. 62) lists the organisms associated with fermented milks, together with the important metabolic products. Figure 4.4 highlights the reactions which need to occur, and the types of organism required in order to obtain the particular flavours which are characteristic of each fermented product.

Organisms are selected and combined depending on their characteristics. For example, fast acid-producing, thermophilic organisms are used for yoghurt manufacture; buttermilk requires a slower rate of acid production

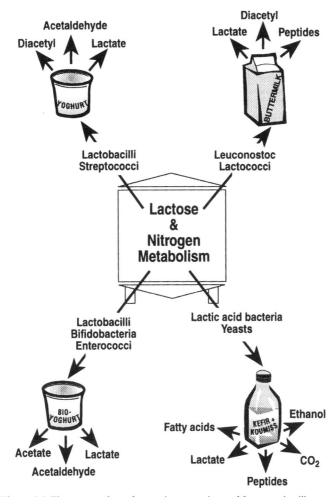

Figure 4.4 Flavour products from microorganisms of fermented milks.

and development of diacetyl, and for kefir a gentle release of carbon dioxide is required. 'Bio-yoghurts' use non-traditional bacteria which have different flavour attributes (acetate in addition to lactate and acetaldehyde).

4.4.1 Metabolic routes to flavour production

The utilization of food materials by any organism proceeds by specific sequences of enzymic reactions which fulfil two main functions: (i) they supply energy necessary for biosynthetic and other energy-dependent processes such as translocation mechanisms; and (ii) they maintain cellular physiology. Pathways giving rise to precursors are catabolic, such as those for lactose metabolism (section 4.2), while those which are synthetic such as those which produce polysaccharides (section 4.5) are anabolic, and there are central pathways and reactions which link the two. In the lactic acid bacteria the important catabolic pathways are dependent on enzymes which hydrolyse lactose and the milk proteins. These may also be responsible for the flavour and aroma of the fermented milks.

Lactose catabolism is essentially oxidative and generates reduced nucleotides. These are in limited supply and there is a need for re-cycling. The conversions adopted by the bacteria in re-generating the pyridine dinucleotides dictates the type of end-metabolites that are then excreted from the cell. It is these end-metabolites which are important to flavour. The principal end-product of the metabolic pathways shown in Figures 4.1 and 4.2 is lactic acid, and this compound is responsible for the characteristic sourness of all the fermented dairy products.

(a) Diacetyl. Diacetyl is the compound which gives buttermilk and some yoghurts a sweet, buttery aroma. It is produced by *Lac. lactis* subsp. *lactis*, biovar *diacetylactis* and leuconostocs when citrate is present. The co-metabolism of lactose and citrate has been studied by a number of workers (for reviews see Kempler and McKay, 1981; Cogan and Jordan, 1994). Citrate is not metabolized as an energy source, but is readily utilized in the presence of another fermentable carbohydrate. In both the lactococci and the leuconostocs, citrate uptake is plasmid-encoded and is coupled to translocation of protons in response to the proton-motive force generated by ATP hydrolysis (David *et al.*, 1992; Bellingier *et al.*, 1994). Within the cell it is catabolized by the routes outlined in Figure 4.5. Carbon dioxide is released from decarboxylation to the intermediate, acetaldehyde–thiamine pyrophosphate ([acetaldehyde–TPP]). This 'active aldehyde' is likely to remain associated with the enzyme pyruvate decarboxylase which requires thiamine pyrophosphate as its cofactor. Decarboxylation may be followed by a number of different transformations resulting in diacetyl, acetoin and/or 2,3-butanediol as shown in Figure 4.5. Citrate lyase, the first enzyme of the pathway, and acetolactate synthase which gives rise to acetolactate from which acetoin is formed, have been purified from *Lac. lactis* subsp. *lactis* variants which carry the *cit* plasmid (Bowien and Gottschalk, 1977; Snoep *et al.*, 1992). The citrate lyase enzyme is inducible in leuconostocs, but is constitutive in the biovar *diacetylactis* of *Lac. lactis* subsp. *lactis*.

There are two metabolic routes which yield diacetyl. Oxidative decarboxylation of acetolactate and condensation of acetaldehyde–TPP with acetyl CoA. The former may not be the major route as recent studies have shown that strains do not generally produce large amounts of acetolactate (Cogan and Jordan 1994). Degradation is dependent on pH and redox potential (Eh); if strains excrete acetolactate at the beginning of the

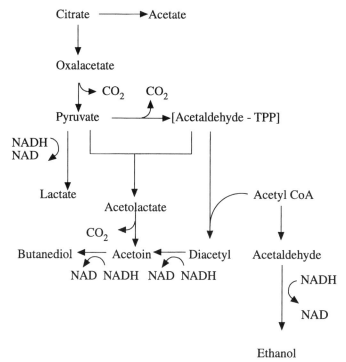

Figure 4.5 Fates of pyruvate and their implications on flavour in fermented milks.

fermentation in milk when the redox potential is high, it will be degraded to diacetyl and acetoin. If, however, acetolactate is present at the end of the fermentation when the redox potential is low, only acetoin will be formed (Monnet *et al.*, 1994). Co-metabolism of citrate and sugar results in increased growth rates, suggesting an advantage to the organism of metabolizing the citrate. This is accomplished by diverting pyruvate away from ethanol towards lactate production in order to re-generate NAD. The acetyl phosphate will then yield ATP and acetate with no necessity to send the acetyl phosphate down the acetaldehyde/ethanol 'leg' of the hexose monophosphate shunt pathway (Figure 4.2). The requirement for NAD is the driving force for many of the routes for pyruvate utilization. A conflict exists between the need for energy generation, ATP from acetyl phosphate, for example, and NAD recycling. Bellingier *et al.* (1994) found that in some *Leuconostoc* strains, co-metabolism of citrate with glucose resulted in the end-products of acetate and lactate; whereas only a few strains produced acetoin and diacetyl. These authors suggested that the increased levels of pyruvate which result from co-metabolism are reduced to lactate and used to re-oxidize NAD(P)H, so relieving the cell of the need to reduce acetaldehyde to ethanol; no strains were found to produce ethanol. Co-metabolism of lactose and citrate in cit$^+$ *Lactococcus* yields diacetyl, acetoin and 2,3-butanediol, with maximum production coinciding with disappearance of citrate from the medium (Crow, 1990). The role of acetolactate synthase in the lactococci has recently been investigated (Snoep *et al.*, 1992). The enzyme has a high K_M for pyruvate (50 mM) which means that the enzyme is active only when intracellular pyruvate levels are high (such as when co-metabolism of citrate and lactose occurs). Thus, in the lactococci, acetoin will be produced as a mechanism of preventing accumulation of pyruvate.

(b) Acetaldehyde. Acetaldehyde is important for yoghurt flavour and aroma. It may be reduced to ethanol by the enzyme alcohol dehydrogenase. The yoghurt bacteria, *Lb. delbrueckii* subsp. *bulgaricus* and *S. thermophilus*, however, do not have this enzyme, so that any intracellular acetaldehyde will be excreted as an end-metabolite. This may not be the case for non-traditional organisms such as *Lb. acidophilus* as many strains reduce acetaldehyde to ethanol via an alcohol dehydrogenase (Marshall and Cole, 1983). However, recent work (Gonzalez *et al.*, 1994) has suggested that when probiotic organisms such as *Lb. acidophilus* and *Lb. paracasei* subsp. *paracasei* are grown in milk, a phosphoketolase route to acetaldehyde can operate and that acetaldehyde will be excreted because the specific activity of enzymes synthesizing acetaldehyde are higher than that of the alcohol dehydrogenase.

The routes to acetaldehyde production may not necessarily be from carbohydrate degradation. Nitrogen metabolism is an important consideration. The amino acid threonine, when converted to another amino acid, glycine, results in equimolar generation of acetaldehyde – this mechanism may well be important to the lactobacilli.

4.4.2   Yeast and mould fermentation

Fungal metabolism results in three types of degradative change, all of which can potentially contribute to flavour. These are lipolysis and metabolism of triglycerides, proteolysis of casein, and oxidation of lactic acid. The moulds which colonize the surface of fermented milk products will have a predominantly oxidizing activity and the ability to metabolize lactic acid will have a local de-acidifying effect. Lipolysis may result in medium-chain fatty acids; saturated fatty acids of medium length ($C_5–C_{12}$) are found in the milk of most mammals and can be released by the action of lipases on milk. In addition to contributing their own flavour, short-chain fatty acids may be further metabolized to methyl ketones – volatiles important to flavour of soft cheeses.

All yeasts and moulds that are associated with fermented milks will degrade casein. Strains of *Geotrichum candidum* have different proteolytic abilities; some strains show weak proteolysis and a yeast-like appearance, while others (mould-ripened cheese isolates) are more strongly proteolytic (Gueguen and Lenoir, 1975). The intra- and extracellular proteolytic activities are similar, showing pH optima between 5.5 and 6.0, and maximum stability at pH 5.0–6.5. The peptidases are likely to be endopeptidases (Lenoir, 1984). The flavour profiles of these have not been fully investigated for the fermented milks and very little has been published. The contribution of a fungal microflora to flavour and aroma is better known for the mould-ripened cheeses (see Chapters 5 and 8).

Kefir owes its flavour characteristics not only to metabolism of lactic acid bacteria, but also to yeast. Ethanol has very little impact on flavour although it may contribute to aroma attributes. Kahala *et al.* (1993) examined the peptide content of a number of Finnish milk products and found greater rate of proteolysis and a greater number of peptides in kefir than in yoghurt. No comment was made on the contribution to flavour.

Before closing this part of a section concerned with the physiology of the organisms and the contribution to flavour, it should be noted that (and as pointed out in the previous edition of this book, where it was a recurrent theme to the chapter), the mechanisms which produce flavour compounds are a result of a metabolism which is necessary for the organism to grow efficiently and to achieve its maximum population. Knowledge of that physiology and the pressures which promote a certain (advantageous) metabolism enables the dairy technologist to provide consistent quality. It also provides a platform from which the molecular geneticist and the microbial physiologist can manipulate metabolic fluxes towards important end-metabolites.

4.5 Polysaccharide secretion

4.5.1 Exopolymers from mesophiles and thermophiles

A number of authors have investigated the viscous (ropy) nature of milk after fermentation with mesophilic and thermophilic lactic acid bacteria and it is generally accepted that the ropiness is related to the synthesis and secretion of exopolymers (Macura and Townsley, 1984; Manca de Nadra *et al.*, 1985; Cerning *et al.*, 1986, 1988, 1990, 1992; Pidoux *et al.*, 1988, 1990; Doco *et al.*, 1990; Nakajima *et al.*, 1990; Garcia-Garibay and Marshall, 1991; Gruter *et al.*, 1992, 1993). The chemical composition of the exopolymers of the thermophilic strains has long been controversial. Some authors suggested that the ropy characteristic which developed in milk was as a result of a glycoprotein of which 47% was proteinaceous. Schellhaas (1983), however, found that the exopolymer consisted of 85% carbohydrate composed of glucose and galactose. There is now agreement that the exopolymers are polysaccharides and that many different types are secreted. The mesophilic leuconostocs produce homopolysaccharides (dextrans) while the lactococci and the lactobacilli produce heteropolysaccharides. These latter polymers are branched and will differ in composition depending on the carbohydrate source on which they are grown (Cerning *et al.*, 1994). In addition, Marshall *et al.* (1995) have shown that two different polysaccharides can be produced concurrently from the same organism, and that culturing milk with mixed cultures of different polysaccharide-producing and non-producing strains results in products with different rheological (viscoelastic and/or thixotropic) properties. Although ropiness is thought to be attributable to the extracellular polysaccharide, Zoon *et al.* (1994) show that the macrostructure demonstrated an inverse relationship between ropiness and permeability of the gel and that the protein network is equally important. This was further explored by Rohm and Kovac (1994), where results of physical measurements of yoghurt gel implied a permanent network with attachment of ropy bacteria to the protein resulting in a decrease of yoghurt firmness.

The structure of polysaccharides published in the past 5 years from mesophilic and thermophilic strains of the dairy lactic acid bacteria is remarkably similar, the majority of workers agreeing that galactose, glucose and rhamnose are present in descending order, but with different ratios. Others have found a polysaccharide from *S. thermophilus* which is composed mainly of galactose residues (Doco *et al.*, 1990), and Gruter *et al.* (1992) also found a galactan from *Lac. lactis* subsp. *cremoris*. Marshall *et al.* (1995) showed the presence of two polysaccharides in milk fermented with a strain of *Lac. lactis* subsp. *cremoris*, one of which was a phosphopolysaccharide with composition similar to that reported by Nakajima *et al.* (1990). The second contained galactose, glucose and glucosamine. The possibility that more than one polysaccharide can be secreted by one strain may explain some of the different compositions. The media used will also influence composition.

Kefir also contains a water-soluble polysaccharide; Toba *et al.* (1987) and Yokoi *et al.* (1990) have shown it to have a high molecular weight in excess of 10^6 Da with glucose and galactose in equimolar ratios. Mukai *et al.* (1988) hydrolysed the kefiran with a cellulase and found free glucose and a hexapolysaccharide repeat unit where one galactose was substituted at the 6-*O*-position, which confirmed earlier work by Kooiman (1968).

4.5.2   *Microbial physiology and polysaccharide production*

Synthesis and secretion of exopolysaccharides occur during different growth phases and the type of polymer is influenced by growth conditions. Synthesis may be regulated by proteins located on the cell surface (Kontusaari and Forsèn, 1988). However, recent work has shown the presence of an enzyme 'switch', where the β anomer of glucose 1-phosphate can be utilized for polysaccharide production in maltose-assimilating cells of lactococci (Sjoberg and Hahn-Hagerdal, 1989). Polysaccharides are made by polymerizing precursors formed in the cytoplasm. Sugar nucleotides play an important role (derived from sugar 1-phosphates), and for polysaccharides other than dextrans, so does the lipid carrier located in the cell membrane. This latter carrier is also involved in the synthesis of cell wall polymers, (lipopolysaccharides, peptidoglycans, teichoic acids) so there is competition for this facilitating membrane component during different phases of growth (Sutherland, 1985). Thus, the nature and composition of exopolysaccharides as capsules or as disassociated polysaccharide are influenced by both medium composition, biosynthetic pathways, phase, and rate of microbial growth. These are important factors in the understanding of polysaccharide synthesis and secretion. In some cases the ropy character is associated with plasmid DNA. This has been reported for the ropy strains of *Lb. paracasei* subsp. *paracasei* NCIB 4114 (Vescovo *et al.*, 1989), and *Lb. paracasei* subsp. *paracasei* CG11 (Neve *et al.*, 1988). In other cases there appears to be no association with extra-chromosomal DNA. Sequence data of specific genes involved in polysaccharide production are not known for the thermophilic lactic acid bacteria, whereas major parts of the genes of, for example, xanthan and succinoglucan are now known.

The instability of this characteristic is acknowledged, but is yet unexplained. The loss on repeated sub-culture, or after prolonged incubation at high temperatures, has been attributed to loss of plasmids (Vedamuthu and Neville, 1986; von Wright and Tynkkynen, 1987; Neve *et al.*, 1988). For thermophilic bacteria such as *S. thermophilus* and *Lb. delbrueckii* subsp. *bulgaricus* this may not be the explanation, as these strains have not, to date, been shown to harbour plasmids. Instability of alginate production has recently been attributed to a particular chromosomal gene (*algT*) in *Pseudomonas aeruginosa* (Wozniac and Ohman, 1994) and a similar situation may prevail in the thermophilic lactics.

4.6 Therapeutic properties of fermented milks

The interest in the potential health benefits associated with lactic acid bacteria that are used to manufacture fermented milks dates back to the early part of this century. A Russian scientist suggested that regular consumption of yoghurt or closely related products would prolong life (Metchnikoff, 1910). The organism that has been identified by Metchnikoff was *Bacillus bulgaricus* (present designation is *Lb. delbrueckii* subsp. *bulgaricus*), and hence such fermented milk product has been the subject of much speculation and attracted wide scientific interest.

4.6.1 Antagonistic compounds

The souring of milk is primarily due to the fermentative conversion of lactose to organic acids, mainly lactic and acetic acid. Thus, with a concomitant lowering of the pH of milk from 6.8 to <4.6, this leads to an increased shelf-life and safety of fermented milk products against some food pathogens (e.g. *Staphylococcus aureus, Bacillus cereus, Salmonella* spp. and *Clostridium* spp.) without the addition of preservatives and/or the application of heat after the fermentation stage. The inhibitory activity of the organic acids is governed by the dissociation constant (pK_a) and the acid concentration at a given pH. Therefore, an organic acid of high pK_a value has more acid in the undissociated form and has a stronger antimicrobial activity. For example, the activity and pK_a values of some organic acids are: lactic (3.85) < acetic (4.73) < propionic (4.87) < benzoic (4.19) (Gould, 1991; De Vuyst and Vandamme, 1994), and it has been reported by Lindgren and Dobrogosz (1990) that acetic acid has up to four times more of the acid in the undissociated form at pH 4.0–4.6 when compared with lactic acid. Furthermore, the undissociated forms of lipophilic acids can penetrate a microbial cell, dissociate to produce hydrogen ions, interfere with metabolic function, and cause an inhibitory effect.

In the past few decades food-borne pathogens (e.g. *Listeria monocytogenes, Campylobacter jejuni* and *Yersinia enterocolitica*), which are capable of growing and surviving at low refrigerated temperature, have been the focus of food safety and received some scientific attention. It is well established that many inhibitory substances, produced in small amounts compared with lactic and acetic acid, are synthesized by a wide spectrum of lactic acid bacteria. It is clear that the antagonistic compounds produced are complex and numerous, and some have been identified: free fatty acids, carbonyl compounds, hydrogen peroxide, ethanol, ammonia, formic acid, carbon dioxide, benzoate, bacteriolytic enzymes, several unidentified inhibitory substances, and bacteriocins (De Vuyst and Vandamme, 1994). The latter inhibitory substances will be discussed in detail.

4.6.2 Bacteriocins

The discovery of antibacterial substances named colicins produced by *E. coli* dates back to the 1920s (De Vuyst and Vandamme, 1994). 'Colicin-like' inhibitory substances are also produced by Gram-positive bacteria including all dairy- and food-fermenting genera of lactic acid bacteria. As a consequence, the term 'bacteriocin' was proposed, and these antimicrobial substances have similar characteristics. Tagg *et al.* (1976) have suggested six criteria to characterize bacteriocins produced by Gram-positive bacteria after extensive studies on colicins produced by Gram-negative microorganisms. They are:

1. Proteinaceous is nature.
2. Bactericidal rather than just bacteriostatic.
3. Capable of having a specific binding site for pathogens, distinguishing them from the activity of other antimicrobial substances.

4. Plasmid-mediated.
5. Produced by lethal biosynthesis.
6. Active against closely related bacteria.

Although these criteria have been used by many researchers in this field to characterize bacteriocins produced by lactic acid bacteria, only a few antimicrobial proteins fit these six criteria. Montville and Kaiser (1993) have discussed separately each of these criteria and highlighted the limitation of Tagg *et al.'s* (1976) approach of starter cultures bacteriocins.

Microbial antagonism of certain strains of group N streptococci (now called lactococci) that inhibited the growth of lactobacilli was first reported by Rogers (1928) and later by Whitehead (1933) who identified the inhibitory compound to be proteinaceous in nature. Ten years later, the failure or slow acid development during the manufacture of Cheddar cheese was attributed to the starter cultures used rather than bacteriophage and/or presence of antibiotics in the milk (Hunter and Whitehead, 1944). The microbial compound was isolated, concentrated and found to exhibit an inhibitory activity against several pathogenic microorganisms where it was classified as an antibiotic called 'nisin' (Mattick and Hirsch, 1947). The name (nisin) was derived from: 'Group N streptococci (now called lactococci) Inhibitory Substance IN (the usual ending for the names of antibiotics)'.

At present, around 70 different types of bacteriocins have been identified and produced by lactic acid bacteria and *Bifidobacterium* spp. Table 4.2 summarizes some selected characteristics of these bacteriocins including the molecular mass, sensitivity and inhibitory spectrum against different microorganisms.

(a) Nomenclature and classification.   With the exception of nisin and very few inhibitory substances shown in Table 4.2, the bacteriocin nomenclature is somewhat similar to enzyme nomenclature. Whereas the suffix '-ase' is used to denote the proteolytic activity of an enzyme, the suffix '-cin' is normally appended to the genus or species name of the microorganism to denote the bacteriocin (Table 4.2). Not all the bacteriocins produced by lactic acid bacteria have been purified to homogeneity and genetic characterization is still in its infancy. Nevertheless, over the past decade the work on the genetics of recombinant DNA technology of bacteriocins produced by lactic acid bacteria has markedly increased in many laboratories worldwide, and for further information on this topic the reader is referred to Klaenhammer (1988, 1993), Schillinger (1990), Piard and Desmazeaud (1991, 1992), James *et al.* (1992), Hoover (1992, 1993), Nettles and Barefoot (1993), Hoover and Steenson (1993) and De Vuyst and Vandamme (1994).

According to Klaenhammer *et al.* (1993) and De Vuyst and Vandamme (1994) the classification of bacteriocins produced by lactic acid bacteria

Table 4.2. Some selected characteristics of bacteriocins produced by fermented milks starter cultures

Starter organisms and strain	Bacteriocin	Molecular mass (kDa)	Sensitivity	Comments
Lac. lactis subsp. *lactis* 10, 71 and 300	Lactostrepcins (Las) 1, 2, 3 and 4	>10	Pronase, α-chymotrypsin, trypsin, phospholipase D and pH 7 except Las 4	Las 1, 2, 3 and 4 are heat-resistant and only active in acidic conditions against *Lactococcus* spp; Las 1 displayed high activity against some streptococci of groups A, C and G.
ATCC 11454, 6F3, NIZ0 22186, N8, SIK-83 and NCFB 894	Nisin A and Z	~3.5	Erepsin, α-chymotrypsin, pancreatin, nisinase, elastase and subtilopeptidase	The difference between nisin A and Z is the histidine residue in position 27 which is replaced by asparagine; nisin is heat stable and active against lactococci, bacilli, clostridia, micrococci, pediococci, leuconostocs, pneumococci, streptococci, and actinomycetes; other nisin forms are B, C, D and E which differ only in biological activity but they have not been elucidated in detail.
6F3, 2F6, 5O8, 6F5 and 7C1	Bacteriocin (Bac) V, VI and VII	NR[a]	Trypsin for Bac V and VII	All Bac are heat-resistant (100°C for 10 min) at pH 4.5 and 7.0; type V exhibited relatively narrow inhibitory spectrum; types VI and VII are active towards Gram-positive bacteria and the former type is resistant to trypsin, suggesting a nisin identity; type VII is sensitive to nisin and inhibited nisin producers.
CNRZ 481	Lacticin 481	1.3–2.9	Pronase, α-chymotrypsin, pediococci, streptococci, proteinase k	Active against lactococci, leuconostocs, lactobacilli, propionic acid bacteria, *Clostridium tyrobutyricum*, and *Staphylococcus carnosus*; it is heat-resistant (100°C for 1 h) at pH 4.5 or 7.0.
DRC 1	Dricin	NR	Pronase, α-chymotrypsin, and trypsin	Heat-stable at 100°C only at pH 4.6; this bacteriocin is only active against *Lac. lactis* subsp. *lactis* HID 113.
ADRIA 85LO30	Lactococcin DR	2.3–2.4	Pronase and trypsin	Heat-stable at 100°C for 20 min at pH 7 and active from pH 4.0–11.0; exhibited bactericidal effect against strains of *C. tyrobutyricum*, *S. thermophilus* 13 and TAO 61 and *Lb. helveticus*.

Strain	Bacteriocin	pH	Enzyme sensitivity	Comments
Lac. lactis biovar *diacetylactis* 6F7 WM4	Bacteriocin (Bac) VIII	NR	Trypsin	Similar to lactococcin I (see section *Lac. lactis* subsp. *cremoris* AC1).
WM4	Bacteriocin (Bac) WM4	NR	Pronase, α-chymotrypsin and trypsin	Inhibitory only to lactococci and is heat-stable at 80°C for 60 min.
	Lactococcin A	3.4–5.8	Endoprotease glu-C and trypsin	Active against lactococci strains and clostridia.
S50	Bacteriocin S50	NR	Proteinase K, α-chymotrypsin, pronase E, trypsin and pepsin	Antibacterial spectrum is against *Lactococcus* spp; it is heat-stable (100°C for 60 min) and active from pH 2 to 11.
DPC 938	Lactocin D	NR	NR	It is similar to lactococcin A.
Lac. lactis subsp. *cremoris* 202	Lactostrepsin 5	20	Trypsin, α-chymotrypsin, lipase and pronase	Similar characteristics to Las 1, 2, 3 and 4, but also exhibit inhibitory spectrum against *Lb. helveticus*, *Bacillus cereus* and *Leuconostoc* spp.
AC1, 3A6, 9B4, 4G6, 1A1, 3E9, KC3, W3, 4E9 and 3C9	Bacteriocin (Bac) I, II, III and IV	NR	Trypsin, heat inactivates Bac II and III	Bac I to IV are active against lactococci and to a few of the genera *Leuconostoc*, *Pediococcus* and *Clostridum*; Bac II and III exhibited characteristics resembling diplococcin while Bac I may belong to lactococcins A, B and M.
9B4	Lactococcins A, B and M	3.4–5.8	NR	See sections above and below (Bac I, II, III, IV, and Lactococcin A).
LMG 2130	Lactocin A	3.4–5.8	Endoprotease glu-C and trypsin	Active against lactococci; no loss of activity when stored at −20°C in 60% ethanol containing 2.5 mM sodium phosphate.
AC1	Lactococcin I	6.0	Proteolytic enzymes	Heat-stable at 100°C for 30 min and at pH 4.5–7.0; active against lactococci and clostridia.
LMG 2081	Lactococcin G	~8.5	NR	This bacteriocin has two peptides known as α and β.
346	Diplococcin	3.5–5.3	Pronase, α-chymotrypsin, pepsin and trypsin	Inhibitory spectrum is against lactococci only; heat-stable at 100°C for 60 min.

Table 4.2. (*Continued*)

Starter organisms and strain	Bacteriocin	Molecular mass (kDa)	Sensitivity	Comments
Leu mesenteriodes subsp. *mesenteroides* UL5	Mesenterocin 5	4.5	Pronase and chloroform	It is heat-stable at 100°C for 30 min; this bacteriocin is active against *L. monocytogenes*, *Enterococcus faecalis*, *Brevibacterium linens*, *Pediococcus* spp. and *Pentosaceus* spp., but not dairy starter cultures; the amino acid sequence is similar to pediocin PA-1.
Y105	Mesentericin Y105	2.5–3.0	Pronase, proteinase K, trypsin, chymotrypsin and heat at 100°C for 120 min at pH 6.8	This bacteriocin is active against a wide range of *Listeria* spp. and *Clostridium* spp., but limited to other microorganisms including lactic acid bacteria; it is heat-stable at 60°C for 120 min at pH 4.5.
FR 52	Mesenterocin 52	<10	Protease	It is heat-stable at 80°C for 6 h at pH 4.5; this bacteriocin is active against other *Leuconostoc* strains and several of *Enterococcus* spp. and *Listeria* spp.
S. thermophilus	NR	<0.7	NR	The antimicrobial compound(s) is heat-stable (100°C for 10 min) and displayed inhibitory activity to Gram-negative and Gram-positive bacteria.
STB 40 and STB 78	Bacteriocin STB 40 and STB 78	10–20	Lipase, α-chymotrypsin, trypsin, and pronase	Both bacteriocins are stable between pH 2.0–12.0 and heat-resistant; they are active against *Enterococcus* spp. and *S. thermophilus* strains.
St 10	Bacteriocin St 10	>100	Proteolytic enzymes and α-amylase	This bacteriocin is only active against *S. thermophilus* and is heat-stable at 121°C for 15 min.
Sfi 13	Thermophilin 13	~4.0	NR	Thermophilin is heat stable (100°C for 1 h) and active in the pH range 1.6–10.0.
Lb. delbrueckii subsp. *bulgaricus* DDS 14	Bulgarican	NR	NR	This antimicrobial substance is thermostable (120°C for 60 min) and only active at acidic pH; it displayed a wide spectrum of inhibiting Gram-positive and Gram-negative bacteria.
Lb. delbrueckii subsp. *lactis*	Lactobacillin	NR	Catalase	It is active against staphylococci and clostridia; the inhibitory effect could be due to hydrogen peroxide activity.
JCM 1106 and JCM 1107	Lacticin A	NR	Trypsin, actinase E and heat at 60°C for 10 min	Lacticin A is active against *Lb. delbrueckii* spp.
JCM 1248	Lacticin B	NR	NR	Similar to lacticin A.

Lb. acidophilus		pH		
Lb. acidophilus	NR	NR	Lactocidin	Insensitive to catalase and has activity in a narrow pH range against closely related *Lb. acidophilus* strains as well as other Gram-positive and Gram-negative bacteria.
DDS 1	NR	NR	Acidophilin	It is thermostable at acidic pH and has a low molecular mass.
IFO 3205	Loss of activity during purification and concentration	3.5	Antibiotic-like	This bacteriocin has been shown to inhibit DNA biosynthesis in *E. coli*.
AC1	Trypsin, α-chymotrypsin, and heat at 50°C for 20 min	5.4	NR	It has inhibitory effect against *Salmonella typhosa, Shigella flexnerii, Pseudomonas aeruginosa* and *Staphylococcus aureus*; it is only active from pH 4.0 to 7.5.
N2	Proteinase K and pronase	6.2–8.1	Lactacin B	It is heat-stable (100°C for 1 h) and pH 5.0; the native protein size of this bacteriocin is >100 kDa; inhibitory spectrum is against lactobacilli.
VPI 11088	Proteinase K, subtilisin, trypsin and ficin	2.5–6.3	Lactacin F	It is heat-resistant (121°C for 15 min); activity is against lactobacilli and *E. faecalis*; the molecular size of native lactacin F is 180 kDa.
NCFM	Pepsin and trypsin	NR	NR	The inhibitory compound is heat-stable (121°C for 15 min) and active only against *Lb. acidophilus* strains.
M46	Trypsin	2.5	Bacteriocin M46	The molecular complex is >100 kDa, moderately heat-stable and active between pH 2.0–12.0; this substance strongly inhibits clostridia and listeriae.
LAPT 1060	Actinase E, trypsin and heat at 50–70°C for 10 min	NR	Acidophilucin A	It is inhibitory only towards *Lb. delbrueckii* subsp. *bulgaricus* and *Lb. helveticus*.
2181	NR	0.2	Acidolin	Acidolin has inhibitory effect against spore-formers, enteropathogens and only limited against lactic acid bacteria; it is thermostable at 121°C for 30 min.
TK 8912	Protease	NR	Acidocin 8912	This antibacterial substance is heat-stable at 120°C for 20 min; it is active against certain strains of *Lb. acidophilus, Lb. casei* subsp. *casei, Lactobacillus plantarum* and *Lac. lactis* subsp. *lactis*, but not active against *E. coli* and *Bacillus subtilis*.

Table 4.2. (*Continued*)

Starter organisms and strain	Bacteriocin	Molecular mass (kDa)	Sensitivity	Comments
Lb. helveticus LP27	Lactocin 27	12.4	Trypsin, pronase and ficin	Its inhibitory spectrum is against lactobacilli and it is a protein–lipopolysaccharide complex heat-stable at 100°C for 60 min.
481	Helveticin J	37	Wide range of proteolytic enzymes and heat	It is only active against few strains of lactobacilli; heat-stable at 100°C for 30 min.
1829	Helveticin V-1829	NR	Proteinase K, ficin, trypsin, pronase, pH > 7.0 and heat at 50°C for 30 min	It is active only against lactobacilli.
Lb. paracasei subsp. *paracasei* B80	Caseicin 80	40–42	Pronase E, trypsin, pepsin, α-chymotrypsin, proteinase K and heat at >60°C for 10 min	This bacteriocin is stable between pH 3.0 and 9.0; only active against *Lb. casei* subsp. *casei* B109.
LHS	Caseicin LHS	NR	NR	Caseicin LHS appears to be a glycoprotein and it displays a broad spectrum of activity.
Lb. rhamnosus GG	NR	<1.0	NR	It resists various proteases and active against wide range of bacteria including lactic acid bacteria but not against other lactobacilli; it is heat-resistant at 90°C for 1 h and antibacterial substance is short-chain fatty acids.
GR1	NR	12–14	Heat-labile	The inhibitory agent is lipophilic in nature and active against *E. coli*.
Lb. reuteri LA6	Reutericin 6	>200	Actinase E and trypsin	Such bacteriocin is active against *Lactobacillus* spp.; it is heat-stable at 100°C for 1 h and at pH 4.0–10.0.
Bifidobacterium spp. (commercial strain)	NR	NR	Trypsin, pepsin and pronase	This bacteriocin is heat-stable at 100°C for 30 min and active at pH between 2 to 10; its activity is against *Lactococcus* spp., *Bifidobacterium* spp., *Lactobacillus* spp. and *Clostridium* spp.

Bacteriocin / strains	pI	Sensitivity to enzymes	Properties
B. bifidum ATCC 29521 } B. breve ATCC 15700 } ATCC 15698 } B. infantis ATCC 15697 } ATCC 15702 } ATCC 25962 } B. longum ATCC 15707 } ATCC 15708 }	NR	As above	These strains have showed antimicrobial activity against S. thermophilus and Lb. acidophilus; only the B. breve strains have showed activity against Clostridium perfringens and B. infantis ATCC 25962 against Cl. tyrobutyricum.
Pediocin AcH P. acidilactici H, E, F, and M (commercial strain PC)	2.5–2.7	Chymotrypsin, papain, trypsin ficin and proteinase K	Pediocin AcH is heat-stable at 70–80°C for 15 min and at pH range from 2.5–9; it has antibacterial activity against many genera such as Pediococcus, Lactobacillus, Leuconostoc, Bacillus, Listeria, Enterococcus, Clostridium, Staphylococcus and Propionibacterium.
Pediocin PA-1 PAC 1.0, PO2	4.6	Protease, papain and α-chymotrypsin	It is stable to heating at 100°C for 10 min and between pH 4–7; pediocin PA-1 is inhibitory to lactic acid bacteria and Listeria spp.; the native molecular size is 16.5 kDa.
Pediocin JD Pediocin SJ-1 JD 1-23 SJ-1	NR 4	NR Trypsin, α-chymotrypsin, protease and α-amylase	NR Pediocin SJ-1 is heat-stable at 65–121°C and at pH 3.6, but its activity decreases significantly at pH 7; this antibacterial agent is active against selective Lactobacillus spp., Cl. perfringens and L. monocytogenes; the native bacteriocin exists in the form of monomers and aggregates of molecular weights ranging between 80–150 kDa.

a Not reported or not determined.
Data compiled from Hosono et al. (1977), Pulusani et al. (1979), Geis et al. (1983), Meghrous et al. (1990), Schillinger (1990), Thuault et al. (1991), Héchard et al. (1992), Kanatani et al. (1992), Vaughan et al. (1992), Mathieu et al. (1993), Nettles and Barefoot (1993), Schved et al. (1993) and De Vuyst and Vandamme (1994).

Table 4.3 Mode of action of some lactic acid bacteriocins

Bacteriocin	Comments
Lactostrepsins (Las 1-5)	Exert rapid bactericidal effect on susceptible strains cells (i.e. retard uptake of uridine, inhibit synthesis of DNA, RNA and protein, leakage of potassium ions and ATP, and ATP hydrolysis occurs); they adsorb rapidly to indicator cells and inactivate them within 3 min (e.g. Las 5); more than one molecule of Las is required to kill 1 cfu of susceptible cells, indicating a multi-hit mechanism.
Nisin A and Z	They are characterized by a strong bactericidal effect due to presence of dehydrated and lanthionine residues (e.g. Dha, Dhb, Ala-S-Ala, and Abu-S-Ala); it consists of elongated amphilic cationic peptides and contains several hydrophobic amino acids residues, it adsorbs to bacterial cells and creates pores affecting the cytoplasmic membrane similar to the action of cationic surface-active detergent so that intracellular components leak out; it inhibits murein biosynthesis by trapping lipid intermediates leading to lysis of phospholipid liposomes; another possible mechanism of nisin is that it inactivates the sulphydryl groups in the membrane.
Lacticin 481	This bacteriocin has a bactericidal and non-bacteriolytic effect, and acts in a similar way to nisin; in cheese-making, it is highly hydrophobic and has the tendency to bind to milk components.
Lactococcin A	The structure consists of a small hydrophobic polypeptide which permeates *only* the cytoplasmic membrane of sensitive lactococcal cells by the formation of pores; this indicates that a membrane-associated receptor specific in *Lactococcus* spp. is required for this bacteriocin to be effective; the mode of action of lactococcins A and B is similar sharing a common receptor, but not lactococcin M.
Lactococcin G	The activity of this bacteriocin is due to the presence of two peptides (α and β) where the N-terminal halves of each peptide may form amphilic helices, suggesting that these peptides are pore-forming toxins, and the C-terminal halves of α and β peptides are made up of polar amino acids.
Diplococcin	This inhibitory substance is rapidly adsorbed and bound to sensitive cells affecting the DNA and RNA biosynthesis which may suggest that the diplococcin action is targeting the cell membrane.
Lactacin B	Both the crude and purified forms of lactacin B are bactericidal to sensitive cells; it does not cause cellular lysis of host cells and adsorbs non-specifically to sensitive and insensitive lactobacilli because it is a highly hydrophobic peptide; the lethal effect to the cellular membrane may be similar to the mode of action of nisin and pediocin AcH.
Lactocin 27	It imparts bacteriostatic effect against *L. helveticus* LS18 by inhibiting protein synthesis and by interfering with potassium and sodium transport.
Pediocin AcH and PA-1	The bactericidal action of these two bacteriocins is to destabilize the cytoplasmic membrane; it involves adsorption of the pediocin molecule on specific receptors on the cell wall, entrance of molecule through cell wall and finally contacting the cytoplasmic membrane; the inability of pediocin AcH to adsorb to Gram-negative bacteria is due to lack of cell-wall lipotechoic acid; alternative mode of actions of pediocin AcH in Gram-positive sensitive cells probably produces conformational alteration in the three-dimensional configuration of the cell wall, leading to loss of its barrier functions.

Data compiled from De Vuyst and Vandamme (1994).

including some differentiation characteristics are subdivided into three general classes:

Class I: lantibiotics, small peptides containing 19–37 amino acids and heat stable (e.g. nisin and lacticin 481).

Class II: small non-lantibiotic or small hydrophobic peptides (< 15 kDa), heat stable (> 30 min at 100°C to 15 min at 121°C) and some examples include lactococcin A, lactacin B, lacticin F, lactocin 27 and pediocin PA-1.

Class III: large non-lantibiotics or large proteins (> 15 kDa) and heat labile (inactivation within 10–15 min at 60 to 100°C) (e.g. helveticin J, acidophilucin A, lacticin A and B, and caseicin 80).

Not all the bacteriocins shown in Table 4.2 have been fully characterized or extensively purified. Caution must be exercised when new bacteriocin activity is detected, and in particular to bacteriocins which were reported three to four decades ago. For example:

- lactobacillin, which is produced by *Lb. delbrueckii* subsp. *lactis*, has been identified as hydrogen peroxide
- Bac WM4 is similar to lactococcin A and lactocidin D where all these bacteriocins are produced by different strains of *Lac. lactis* biovar *diacetylactis*
- Bac VIII is similar to lactococcin I (De Vuyst and Vandamme, 1994).

It is evident, however, that one researcher's bacteriocin may well be another's under a different name, and efforts are being made to minimize such confusion regarding bacteriocins produced by lactic acid bacteria.

(b) Mode of action. Many of the bacteriocins are bactericidal, often as a result of a collapse in the proton motive force and loss of ability to retain the energy potential in the membrane. Only a few lactic acid bacteriocins have been fully studied; the mode of action of some selected bacteriocins is shown in Table 4.3. Knowledge of the structure has led to suggestions that a bacteriocin such as nisin can span the membrane forming a pore, which brings about the collapse of the membrane potential. Other bacteriocins, which have been categorized by a bactericidal mode of action, are bacteriocin St 10, mesentericin Y105, caseicin 80 (weak) and pediocin SJ-1.

4.6.3 Lactose malabsorption

The biochemical argument for the ability of fermented milks to relieve the discomfort experienced by malabsorbers is based on the ability of the fermenting organisms to catabolize lactose. The organisms possess enzymes which break down the sugar responsible for the symptoms. The relief may be afforded in three ways:

1. The lactose content of the product is reduced as a consequence of microbial action.
2. The enzymes of the bacteria are active in the gastrointestinal tract.
3. The organisms themselves continue to metabolize within the gastrointestinal tract.

The ability of the fermenting organism to reduce the lactose content in milk cannot be disputed. The yoghurt organisms will be able to utilize more that 50% of the lactose present on normal pasteurized skimmed milk, reducing the content from around $5\,g\,100\,ml^{-1}$ to around $2\,g\,100\,ml^{-1}$. However, if the milk is fortified by addition of skimmed milk powder or concentrated by evaporation so that the lactose content is increased to $7.5-8\,g\,100\,ml^{-1}$ (a practice common in yoghurt production), then the final lactose content after fermentation will be $4.5-5\,g\,100\,ml^{-1}$. Thus, the lactose content of the yoghurt will be very similar to the lactose content of regular pasteurized skimmed milk which is delivered to the doorstep. This is because the yoghurt organisms are only able to grow to a defined maximum population in milk which is dependent on a number of biochemical and physiological conditions. The lactose supply in milk is not limiting, and an increase in lactose content will not increase the amount of lactose consumed by the bacteria.

The bacteria, however, possess lactose-metabolizing enzymes, which could continue to act on the substrate if they were not confined to the bacterial cell. There are two lactose-hydrolysing enzymes associated with the lactic acid bacteria: β-D-galactosidase and β-D-phosphogalactosidase. The former acts on lactose while the latter acts on lactose phosphate (see section 4.2) which is not the naturally occurring sugar in milk and dairy products. The traditional yoghurt organisms and the leuconostocs possess β-D-galactosidase, while the lactococci have a β-D-phosphogalactosidase. The position with the bifidobacteria and *Lb. acidophilus* is unclear. Some strains may have both types of enzyme.

The ability of probiotic strains to persist in the gastrointestinal tract is pertinent to this discussion. The catabolism of lactose by both enzymes is possible if the bacteria are viable and actively metabolizing in the gastrointestinal tract. The bifidobacteria and *Lb. acidophilus* will survive stomach acid and bile salts, while there is some evidence that *S. thermophilus* and *Lb. delbrueckii* subsp. *bulgaricus* do not. If probiotic strains possess β-D-phosphogalactosidase, then it is important that the cells not only survive the environment of the gastrointestinal tract, but that they actively metabolize in this ecological niche, as intact cells of this type are required if they are to relieve the symptoms of lactose malabsorbtion. Fermented milks can help lactose malabsorbers, but only if attention is paid to the formulation and the types of organism used for the fermentation.

4.6.4 Bile salt resistance and cholesterol 'assimilation'

A high serum cholesterol concentration may be a contributing risk factor for development of cardiovascular disease. The majority of cholesterol is produced and conserved by the body and synthesized according to needs, the major site of metabolism being the liver, although the intestines also synthesize appreciable amounts. The rate of cholesterol formation is influenced by cholesterol availability from dietary sources and it has been suggested that fermented foods can reduce serum cholesterol concentration by reducing the intestinal absorption of dietary and endogenous cholesterol (Jaspers et al., 1984; Lin et al., 1989) or by inhibiting cholesterol synthesis in the liver (Kritchevsky et al. 1979; Grunewald, 1982). Imaizumi et al. (1992) examined extracts of milk fermented with 39 different strains of bifidobacteria and lactobacilli for effects on cholesterol synthesis and bile acid synthesis from cholesterol using cultured rat hepatocytes. Of these, the strains of bifidobacteria were more likely to suppress cholesterol synthesis from ^{14}C-labelled acetate and preparations from milks cultured with one strain of Lb. paracasei subsp. paracasei stimulated the rate-limiting enzyme of bile acid transformation from cholesterol. These authors, however, found many strain differences and no clear conclusions could be drawn on the general mechanism by which microbial activity within the gut might regulate serum cholesterol levels. Gilliland et al. (1985) and Walker and Gilliland (1993) suggested that a cholesterol lowering may be effected by an ability of bacteria to actively take up or assimilate cholesterol, and they reported that this occurred particularly in the presence of bile salts under anaerobic conditions. The bacteria which have received the most attention in this connection are a limited number of strains of Lb. acidophilus and some bifidobacteria. Recent work (Klaver and van der Meer, 1993), however, sheds doubt on the ability of these organisms to assimilate/metabolize cholesterol. These authors found that lactobacilli and B. bifidum deconjugate bile salts which co-precipitate with cholesterol as pH is lowered. De Smet et al. (1994) also reported on the possible lowering of cholesterol as a result of ability of bacteria to deconjugate bile salts, and suggested that because of the inter-relationship of bile salt/cholesterol metabolism, bile salt hydrolase-positive lactic acid bacteria could have potential in lowering serum cholesterol.

The bacterial strains that are used currently as dietary adjuncts because of their putative probiotic characteristics are few, and only in the Japanese study (Imaizumi, 1992), cited above, and that of Walker and Gilliland (1993) have studies been widened to include other strains of Lactobacillus spp. and bifidobacteria. In a recent paper, Marshall and Taylor (1995) studied a number of new isolates of Lb. acidophilus from human neonatal faecal samples. Klaver and van der Meer's findings of co-precipitation of cholesterol with deconjugated bile acids in acidic conditions were confirmed, but for one neonatal enteric strain not all the removal could be accounted for as a co-precipitation and cholesterol was 'removed' in the absence of bile. No evidence for a metabolic activity directed towards cholesterol was found, and the authors suggested an association of the cholesterol with the bacterial cell surface.

The co-precipitation may be a mechanism which can contribute to lowering a serum cholesterol, in that if a cholesterol–bile co-precipitate were to be excreted from the gut along with the faeces, then a loss of cholesterol and bile would occur. The net effect of such loss from the gastrointestinal tract would be to mobilize the enterohepatic cycle which synthesizes bile salts from cholesterol. The hypothesis would hold if the co-precipitate remained as a precipitate at the increased pH values (above pH 5.0) in the lower gut, and the cholesterol and bile were not re-absorbed. The probiotic organisms' ability to deconjugate bile may be helpful in the co-precipitation event, but the presence of deconjugated bile salts in the colon may contribute to risk of colon cancer.

4.6.5 Stimulation of the immune system

A number of authors (Vesely *et al.*, 1985; de Simone *et al.*, 1991; Marteau and Rambaud, 1993; Perdigon *et al.*, 1995) have discussed the possibility that lactic acid bacteria may 'prime' the immune system so that it is in a state of readiness. In such circumstances, should a serious pathogen be present, it could be quickly eliminated. The natural immune system comprises non-specific and specific defence mechanisms. Non-specific phagocytic cells (monocytes, macrophages and polynuclear cells) will internalize and kill pathogens. Macrophages, however, can present an antigen to the lymphocytes which are responsible for specific defence mechanisms (that is mechanisms directed at particular, 'non-self' entities) which elicit the secretory (circulating immunoglobulins) immune system.

For an antigen present in the gastrointestinal tract to elicit a secretory immune response, it must be transported to the Peyers patches (the gut associated lymphoid tissue or GALT). Studies with mice (de Simone *et al.*, 1987, 1988) have shown that when challenged with salmonellae, antibacterial activity of Peyers patches was increased in those animals fed live yoghurt (containing *Lb. delbrueckii* subsp. *bulgaricus* and *S. thermophilus*). In other studies with mice, Perdigon *et al.* (1990, 1995) showed that oral administration of *Lb. acidophilus*, *Lb. paracasei* subsp. *paracasei*, *Lb. delbrueckii* subsp. *bulgaricus* and *S. thermophilus* increased the systemic non-specific immune system (mononuclear phagocytes), and that *Lb. paracasei* subsp. *paracasei* was the most effective.

Kato *et al.* (1984) showed that mouse killer cell activity was promoted by cell wall antigens of a strain of *Lb. paracasei* subsp. *paracasei* and Nakajima *et al.* (1995) found that a slime-forming variant of *Lac. lactis* subsp. *cremoris* (which had a composition of 48.5% protein, 15.4% sugar and 1.1% phosphorus) enhanced antibody formation in mice. In other work by Namba *et al.* (1981), immunostimulation in guinea pigs was found on oral administration of proteinase and lysozyme-digested cell walls. Thus, the requirement for whole viable cells may be questioned. Fermented milks, however, are more than milk as a vehicle for carrying bacteria. Lactic acid bacteria degrade lactose and milk protein. The way that they degrade these milk components gives the milks their special flavours and properties (see Figure 4.4). Milk proteins may elicit an immune response in certain circumstances, and some peptides derived from milk proteins may be antigenic. The bacteria used to ferment milk have different proteinases which degrade milk protein in different ways. Moineau and Goulet (1991) have shown that the amount and type of milk proteolysis is an important consideration in assessing the immunostimulation found by administration of fermented milks.

Translation of results in mice to effects in humans must also be tempered with a consideration of the tolerances the human population and the mouse have for cows' milk and its (degraded) proteins.

4.7 Conclusions

Knowledge of the physiology and biochemistry of the lactococci in particular has been enhanced by the knowledge and experimental procedures of molecular biology. The complex metabolism involved in the microbial utilization of milk is now understood at the molecular, biochemical and genetic level. For example, catabolism of lactose, citrate and casein is understood at these basic levels for lactococci, some species of lactobacilli and fundamental work has begun on the leuconostocs. As a consequence, manipulation and choice of strains is made on an increasingly more possible rational basis by commercial enterprises. More information is still awaited, however, such as mechanisms for control of anabolic characteristics. What triggers a *Lactobacillus* to polymerize sugars instead of breaking them down for use as a energy source? What induces a *Lactobacillus* or a particular *Lactococcus* to cyclize and modify a peptide to create and secrete a bacteriocin? In the past decade our knowledge of the way lactic acid bacteria grow, utilize and behave in a milk environment is becoming deep and thorough. The use of lactic acid bacteria as probiotic organisms, however, depends on acquisition of a wider knowledge. How do these organisms behave in the constantly changing environment of the human gastrointestinal tract? What kind of microenvironments or niches will they have to adapt to? What kind of biochemical characteristics will the organism have to express to be of benefit to the human condition? These are the biochemical and physiological questions that need to be addressed if there is to be future successful exploitation of these types of organisms.

Acknowledgements

SAC, Research and Consultancy Services (A.Y.T.) receives financial support from the Scottish Office of Agriculture, Environment and Fisheries Department (SOAEFD).

4.8 References

Adams, R.M., Yoast, S., Mainzer, S.E., Moon, K., Dalombella, A.L., Estell, D.A., Power, S.D. and Schmidt, B.K. (1994) Characterization of two cold-sensitive mutants of the β-galactosidase from *Lactobacillus delbrueckii* subspecies *bulgaricus*. *Journal of Biological Chemistry*, **269**, 5666–72.

Amoroso, M.J. and Manca de Nadra, M.C. (1990) A new mixed culture of *Lactobacillus delbrueckii* subspecies *bulgaricus* and *Streptococcus salivarius* subspecies *thermophilus* isolated from commercial yoghurt. *Microbiologie, Aliments, Nutrition*, **8**, 105–13.

Ascon-Reyes, D.B., Ascon-Cabrera, M.A., Cochet, N. and Lebeault, J.M. (1995) Indirect conductance measurements of carbon dioxide produced by *Streptococcus salivarius* subspecies *thermophilus* JT 106 in pure and mixed cultures. *Journal of Dairy Science*, **78**, 8–16.

Atlan, D., Gilbert, C., Blanc, B. and Portalier, R. (1994) Cloning and sequencing of the *pep*Ip gene encoding a proline iminopeptidase from *Lactobacillus delbrueckii* subspecies *bulgaricus* CNRZ 397. *Microbiology*, **140**, 527–35.

Bellingier, P., Hemme, D. and Foucard, C. (1994) Citrate metabolism in 16 *Leuconostoc mesenteroides* subsp. *mesenteroides* and subspecies *dextranicum* strains. *Journal of Applied Bacteriology*, **77**, 54–60.

Benthin, S., Nielson, J. and Villadsen, J. (1994) Galactose expulsion during lactose metabolism in *Lactococcus lactis* subspecies *cremoris* FD1 due to dephosphorylation of intracellular galactose-6-phosphate. *Applied and Environmental Microbiology*, **60**, 1254–9.

Biavati, B., Sgorbati, B. and Scardovi, V. (1992) The genus of *Bifidobacterium*, in *The Prokaryotes – A Handbook on the Biology of Bacteria: Ecophysiology, Isolation, Identification, Applications*, vol. 1, 2nd edn (eds A. Balows, H.G. Trüper, M. Dworkin, W. Harder and K.-H. Schleifer), Springer Verlag, New York, pp. 816–33.

Boquien, C-Y., Corrieu, G. and Desmazeaud, M. (1988) Effect of fermentation conditions on growth of *Streptococcus cremoris* AM2 and *Leuconostoc lactis* CNRZ 1091 in pure and mixed cultures. *Applied and Environmental Microbiology*, **54**, 2527–31.

Bowien, S. and Gottschalk, G. (1977) Purification and properties of citrate lyase ligase from *Streptococcus diacetilactis*. *European Journal of Biochemistry*, **80**, 305–9.

Cerning, J., Bouillanne, C., Landon, M. and Desmazeaud, M. (1986) Isolation and characterization of exocellular polysaccharide produced by *Lactobacillus bulgaricus*. *Biotechnology Letters*, **8**, 625–8.

Cerning, J., Bouillanne, C., Desmazeaud, M. and Landon, M. (1988) Exocellular polysaccharide production by *Streptococcus thermophilus*. *Biotechnology Letters*, **10**, 255–60.

Cerning, J., Bouillanne, C., Landon, M. and Desmazeaud, M. (1990) Comparison of exocellular polysaccharide production by thermophilic lactic acid bacteria. *Sciences des Aliments*, **10**, 443–51.

Cerning, J., Bouillanne, C., Landon, M. and Desmazeaud, M. (1992) Isolation and characterization of exopolysaccharides from slime-forming mesophilic lactic acid bacteria. *Journal of Dairy Science*, **75**, 692–9.

Cerning, J., Renard, C.M.G.C., Thibault, J.F., Bouillanne, C., Landon, M., Desmazeaud, M. and Topisirovic, L. (1994) Carbon source requirements for exopolysaccharide production by *Lactobacillus casei* CG11 and partial structure analysis of the polymer. *Applied and Environmental Microbiology*, **60**, 3914–19.

Cogan, T.M. and Jordan, K.N. (1994) Metabolism of *Leuconostoc* bacteria. *Journal of Dairy Science*, **7**, 2704–17.

Crow, V.L. (1990) Properties of 2,3,-butanediol dehydrogenases from *Lactococcus lactis* ssp. *lactis* in relation to citrate fermentation. *Applied and Environmental Microbiology*, **58**, 1656–9.

David, S., Stevens, H., van Riel, M., Simons, M.G. and de Vos, W.M. (1992) *Leuconostoc lactis* β-galactosidase is encoded by two overlapping genes. *Journal of Bacteriology*, **174**, 4475–81.

Dellaglio F., Dicks, L.M.T. and Torriani, S. (1995) The genus *Leuconostoc*, in *The Genera of Lactic Acid Bacteria*, vol. 2 (eds B.J.B. Wood and W.H. Holzapfel), Blackie Academic & Professional, London, pp. 235–78.

de Simone, C., Vesely, R., Bianchi-Salvadori, B., Tzantzoglou, S., Cilli, A. and Lucci, L. (1987) Enhancement of immune response of murine Peyer's patches by a diet supplemented with yoghurt. *Immunopharmacology and Immunotoxicology*, **9**, 87–100.

de Simone, C., Tzantzoglou, S., Baldinelli, L., di Fabio, S., Bianchi-Salvadori, B., Jirillo, E. and Vesely, R. (1988) Enhancement of host resistance against *Salmonella typhimurium* infection by a diet supplemented with yoghurt. *Immunopharmacology and Immunotoxicology*, **10**, 399–415.

de Simone, C., Rosati, E., Moretti, S., Bianchi-Salvadori, B., Vesely, R. and Jirillo, E. (1991) Probiotics and stimulation of the immune response. *European Journal of Clinical Nutrition*, **45**, 32–4.

De Smet, I., van Hoorde, L., de Saeyer, N. vande Woestyne, M. and Verstraete, W. (1994) *In vitro* study of bile salt hydrolase (BSH) activity of BSH isogenic *Lactobacillus plantarum* 80 strains and estimation of cholesterol lowering through enhanced BSH activity. *Microbial Ecology in Health and Disease*, **7**, 315–29.

De Vuyst, L. and Vandamme, E.J. (eds) (1994) *Bacteriocins of Lactic Acid Bacteria*, Blackie Academic & Professional, London.

Doco, T., Wieruszeski, J-M., Fournet, B., Carcano, D., Ramos, P. and Loones, A. (1990) Structure of an exopolysaccharide produced by *Streptococcus thermophilus*. *Carbohydrate Research*, **198**, 313–21.

Garcia-Garibay, M. and Marshall, V.M. (1991) Polymer production by *Lactobacillus delbrueckii* spp. *bulgaricus*. *Journal of Applied Bacteriology*, **70**, 325–8.

Garvie, E.I. (1986) Genus *Leuconostoc*, in *Bergey's Manual of Systematic Bacteriology*, vol. 2 (eds P.H.A. Sneath, N.S. Mair, M.E. Sharpe and J.G. Holt), Williams & Wilkins, Baltimore, pp. 1071–5.

Gasser, F., Doudoroff, M. and Contopoulos, R. (1970) Purification and properties of NAD-independent lactic dehydrogenases of different species of *Lactobacillus*. *Journal of General Microbiology*, **62**, 241–50.

Geis, A., Singh, J. and Teuber, M. (1983) Potential of lactic streptococci to produce bacteriocin. *Applied and Environmental Microbiology*, **45**, 205–11.

Gibson, G.R., Willis, C.L. and van Loo, J. (1994) Non-digestible oligosaccharides and bifidobacteria implications for health. *International Sugar Journal*, **96**, number 1150.

Gilbert, C., Atlan, D., Blanc, B. and Portalier, R. (1994) Proline iminopeptidase from *Lactobacillus delbrueckii* subspecies *bulgaricus* CNRZ 397: purification and characterization. *Microbiology*, **140**, 537–42.

Gilliland, S.E., Nelson, C.R. and Maxwell, C. (1985) Assimilation of cholesterol by *Lactobacillus acidophilus*. *Applied and Environmental Microbiology*, **49**, 377–81.

Goel, M.C. and Marth, E.M. (1969) Growth of *Leuconostoc citrovorum* in skim milk at 22 and 30°C. *Journal of Dairy Science*, **52**, 1207–12.

Gonzalez, S., Morata de Ambrosisi, V., Manca de Nadra, M., Pesce de Ruiz Holgado, A. and Oliver, G. (1994) Acetaldehyde production by strains used as probiotics in fermented milks. *Journal of Food Protection*, **57**, 436–40.

Gould, G.W. (1991) Antimicrobial compounds, in *Biotechnology and Food Ingredients* (eds I. Goldberg and R. Williams), Van Nostrand Reinhold, New York, pp. 461–82.

Grunewald, K.K. (1982) Serum cholesterol levels in rats fed skim milk fermented by *Lactobacillus acidophilus*. *Journal of Food Science*, **47**, 2078–9.

Gruter, M., Leeflang, B.R., Kuiper, J., Kamerling, J.P. and Vleigenthart, J.F.G. (1992) Structure of the exopolysaccharide produced by *Lactococcus lactis* subspecies *cremoris* H414 grown in defined medium or skimmed milk. *Carbohydrate Research*, **231**, 273–91.

Gruter, M., Leeflang, B.R., Kuiper, J., Kamerling, J.P. and Vleigenthart, J.F.G. (1993) Structural characterization of the exopolysaccharide produced from *Lactobacillus delbrueckii* subspecies *bulgaricus* rr grown in skimmed milk. *Carbohydrate Research*, **239**, 209–26.

Guedon, G., Bourgloin, F., Pebay, M., Roussel, Y., Colmin, C., Simonet, J.M. and Decaris, B. (1995) Characterisation and distribution of two insertion sequences IS119 and iso-IS981 in

Streptococcus thermophilus: does intergeneric transfer of insertion sequences occur in lactic co-cultures? *Molecular Micribiology*, **16**, 69–78.

Gueguen, M. and Lenoir, J. (1975) Aptitude de l'espèce *Geotrichum candidum* à la production d'enzymes protéolytiques. *Lait*, **55**, 145–62.

Gupta, P.K., Mital, K.B., Garg, K.S. and Mishra, P.D. (1994) Influence of different factors on activity and stability of β galactosidase from *Lactobacillus acidophilus*. *Journal of Food Biochemistry*, **18**, 55–64.

Haandrikman, A.J., Meesters, H., Laan, S., Konings, W.M., Kok, J. and Venema, G. (1991) Processing of the lactococcal extracellular serine proteinase. *Applied and Environmental Microbiology*, **57**, 1899–904.

Hammes, W.P. and Vogel, R.F. (1995) The genus *Lactobacillus*, in *The Genera of Lactic Acid Bacteria*, vol. 2 (eds B.J.B. Wood and W.H. Holzapfel), Blackie Academic & Professional, London, pp. 19–54.

Hammes, W.P., Weiss, N. and Holzapfel, W. (1992) The genera *Lactobacillus* and *Carnobacterium*, in *The Prokaryotes – A Handbook on the Biology of Bacteria: Ecophysiology, Isolation, Identification, Applications*, vol. 2, 2nd edn (eds A. Balows, H.G. Trüper, M. Dworkin, W. Harder and K.-H. Schleifer), Springer Verlag, New York, pp. 1535–94.

Hardie, J.M. (1986) Genus *Streptococcus*, in *Bergey's Manual of Systematic Bacteriology*, vol. 2 (eds P.H.A. Sneath, N.S. Mair, M.E. Sharpe and J.G. Holt), Williams & Wilkins, Baltimore, pp. 1043–71.

Hardie, J.M. and Whiley, R.A. (1992) The genus *Streptococcus* – oral, in *The Prokaryotes – A Handbook on the Biology of Bacteria: Ecophysiology, Isolation, Identification, Applications*, vol. 2, 2nd edn (eds A. Balows, H.G. Trüper, M. Dworkin, W. Harder and K.-H. Schleifer), Springer Verlag, New York, pp. 1421–49.

Hardie, J.M. and Whiley, R.A. (1995) The genus *Streptococus*, in *The Genera of Lactic Acid Bacteria*, vol. 2 (eds B.J.B. Wood and W.H. Holzapfel), Blackie Academic & Professional, London, pp. 55–124.

Héchard, Y., Derijard, B., Letellier, F. and Cenatiempo, Y. (1992) Characterization and purification of mesentericin Y105, an anti-*listeria* bacteriocin from *Leuconostoc mesenteroides*. *Journal of General Microbiology*, **138**, 2725–31.

Hirota, T. and Kekuchi, T. (1976) Studies on kefir grains. Isolation and classification of micro-organisms from kefir grains and their characteristics. Reports of Snow Brand Milk Products Co. Laboratory, No. 74, pp. 63–82.

Holzapfel, W.H. and Schillinger, U. (1992) The genus *Leuconostoc*, in *The Prokaryotes – A Handbook on the Biology of Bacteria: Ecophysiology, Isolation, Identification, Applications*, vol. 2, 2nd edn (eds A. Balows, H.G. Trüper, M. Dworkin, W. Harder and K.-H. Schleifer), Springer Verlag, New York, pp. 1508–34.

Hoover, D.G. (1992) Bacteriocins: activities and applications, in *Encyclopaedia of Microbiology*, vol. 1 (ed J.L. Lederberg), Academic Press, New York, pp. 181–90.

Hoover, D.G. (1993) Bacteriocins with potential for use in foods, in *Antimicrobials in Foods*, 2nd edn (eds P.M. Davidson and A.L. Branen), Marcel Dekker, New York, pp. 409–40.

Hoover, D.G. and Steenson, L.R. (eds) (1993) *Bacteriocins of Lactic Acid Bacteria*, Academic Press, New York.

Hosono, A., Kastuki, K. and Tokita, F. (1977) Isolation and characterization of an inhibitory substance against *Escherichia coli* produced by *Lactobacillus acidophilus*. *Milchwissenschaft*, **32**, 727–30.

Hunter, G.J.E. and Whitehead, H.R. (1944) The influence of abnormal ('non-acid') milk on cheese starter cultures. *Journal of Dairy Research*, **13**, 123–6.

Imaizumi, K. Hirata, K., Zommara, M., Sugano, M. and Suzuki, Y. (1992) Effects of cultured milk products by *Lactobacillus* and *Bifidobacterium* species on the secretion of bile acids in hepatocytes and in rats. *Journal of Nutritional Science and Vitaminology*, **38**, 343–51.

Ito, Y. and Sasaki, T. (1994) Cloning and nucleotide sequencing of L-lactate dehydrogenase gene from *Streptococcus thermophilus* M-192. *Bioscience Biotechnology and Biochemistry*, **58**, 1569–73.

James, R., Lazdunski, C. and Pattus, F. (eds) (1992) *Bacteriocins, Microcins and Lantibiotics*, Springer Verlag, Berlin.

Jaspers, D.A., Massey, L.K. and Luedecke, L.D. (1984) Effect of consuming yoghurts prepared with three culture strains on human lipoproteins. *Journal of Food Science*, **49**, 1178–81.

Jonas, H.A., Anders, R.F. and Jago, G.R. (1972) Factors affecting the activity of the lactate dehydrogenase of *Streptococcus cremoris*. *Journal of Bacteriology*, **111**, 397–403.

Kahala, M., Pahkala, E. and Pihlanto-Leppala, A. (1993) Peptides in fermented Finnish products. *Agricultural Science Finland*, **2**, 379–85.

Kanatani, K. and Oshimura, K. (1994) Isolation and structural analysis of the phospho-β-galactosidase gene from *Lactobacillus acidophilus*. *Journal of Fermentation and Bioengineering*, **78**, 123–9.

Kanatani, K., Tahara, Y., Yoshida, K., Miura, H., Sakamoto, M. and Oshimura, M. (1992) Plasmid-associated bacteriocin production by and immunity of *Lactobacillus acidophilus* Tk8912. *Bioscience Biotechnology and Biochemistry*, **56**, 648–51.

Kandler, O. and Weiss, N. (1986) Regular, nonsporing Gram-positive rods, in *Bergey's Manual of Systematic Bacteriology*, vol. 2 (eds P.H.A. Sneath, N.S. Mair, M.E. Sharpe and J.G. Holt), Williams & Wilkins, Baltimore, pp. 1208–34.

Kato, I., Yokokura, T. and Mutai, M. (1984) Augmentation of mouse natural killer cell activity by *Lactobacillus casei* and its surface antigens. *Microbiology and Immunology*, **28**, 209–17.

Kempler, G.M. and McKay, L.L. (1981) Biochemistry and genetics of citrate utilisation in *Streptococcus lactis* ssp. *diacetylactis*. *Journal of Dairy Science*, **64**, 1527–39.

Klaenhammer, T.R. (1988) Bacteriocins of lactic acid bacteria. *Biochimie*, **70**, 337–49.

Klaenhammer, T.R. (1993) Genetics of bacteriocins produced by lactic acid bacteria. *FEMS Microbiology Review*, **12**, 39–85.

Klaenhammer, T.R., Fremaux, C., Ahn, C. and Milton, K. (1993) Molecular biology of bacteriocins produced by *Lactobacillus*, in *Bacteriocins of Lactic Acid Bacteria* (eds D.G. Hoover and L.R. Steenson) Academic Press, New York, pp. 151–80.

Klaver, F.M. and van der Meer, R. (1993) The assumed assimilation of cholesterol by lactobacilli and *Bifidobacterium bifidum* is due to their bile salt-deconjugating activity. *Applied and Environmental Microbiology*, **59**, 1120–4.

Klein, J.R., Schmidt, U. and Plapp, R. (1994) Cloning heterologous expression and sequencing of a novel proline iminopeptidase gene *pep*1 from *Lactobacillus delbrueckii* subspecies *lactis* DSM 7290. *Microbiology*, **140**, 1133–9.

Klein, J.R., Dick, A., Schick, J., Matern, H.T., Henrich, B. and Plapp, R. (1995) Molecular cloning and DNA sequence analysis of pepL, a leucyl aminopeptidase gene from *Lactobacillus delbrueckii* subspecies *lactis* DSM 7290. *European Journal of Biochemistry*, **228**, 570–8.

Kontusaari, S. and Forsèn, F. (1988) Finnish fermented milk viili: involvement of two cell surface proteins in production of slime by *Streptococcus lactis* subspecies *cremoris*. *Journal of Dairy Science*, **71**, 3197–202.

Kooiman, P. (1968) Chemical structure of kefiran, the water soluble polysaccharide of the kefir grain. *Carbohydrate Research*, **7**, 200–11.

Kritchevsky, D., Tepper, S.A. and Morrissey, R.B. (1979) Influence of whole or skim milk on cholesterol metabolism in rats. *The American Journal of Clinical Nutrition*, **32**, 597–600.

Laue, P. and MacDonald, R.E. (1968) Identification of thiomethyl-β-D-galactoside 6-phosphate accumulated by *Staphylococcus aureus*. *Journal of Biological Chemistry*, **243**, 680–2.

Lenoir, J. (1984) The surface flora and its role in the ripening of cheese. Bulletin 171, International Federation, Brussels, pp. 3–20.

Lindgren, S.E. and Dobrogosz, W.J. (1990) Antagonistic activities of lactic acid bacteria in food and feed fermentations. *FEMS Microbiology Reviews*, **87**, 149–64.

Lin, S.Y., Ayres, J.W., Winkler, W. Jr and Sandine, W.E. (1989) *Lactobacillus* effects on cholesterol *in vivo* results. *Journal of Dairy Science*, **72**, 2885–99.

Macura, D. and Townsley, P.M. (1984) Scandinavian ropy milk: identification and characterization of endogenous ropy lactic streptococci and their extracellular excretion. *Journal of Dairy Science*, **67**, 735–44.

Manca de Nadra, M.C., Strasser de Saad, A.M., Pesce de ruiz Holgado, A.A. and Oliver, G. (1985) Extracellular polysaccharide production by *Lactobacillus bulgaricus* CRL. *Milchwissenschaft*, **40**, 409–11.

Marshall, V.M. and Cole, W.M. (1983) Threonine aldolase and alcohol dehydrogenase activities in *Lactobacillus bulgaricus* and *Lactobacillus acidophilus* and their contribution to flavour production in fermented milks. *Journal of Dairy Research*, **50**, 375–9.

Marshall, V.M. and Taylor, E. (1995) Ability of neonatal human *Lactobacillus* isolates to remove cholesterol from liquid medium. *International Journal of Food Science and Technology*, **30**, 577–84.

Marshall, V.M., Cole, W.M. and Farrow, J.A.E. (1984) A note on the heterofermentative *Lactobacillus* isolated from kefir grains. *Journal of Applied Bacteriology*, **56**, 503–5.

Marshall, V.M., Cowie, E.N. and Moreton, R.S. (1995) Analysis and production of two exopolysaccharides from *Lactococcuc lactis* subspecies *lactis* LC 330. *Journal of Dairy Research*, **62**, 621–8.

Marteau, P. and Rambaud, J.-C. (1993) Potential of using lactic acid bacteria for therapy and immunomodulation in man. *FEMS Microbiology Reviews*, **12**, 207–20.

Mathieu, F., Suwandhi, I.S., Rekhif, N. Millière, J.B. and Lefebvre, G. (1993) Mesenterocin 52, a bacteriocin produced by *Leuconostoc mesenteroides* ssp. *mesenteroides* FR52. *Journal of Applied Bacteriology*, **74**, 372–9.

Mattick, A.T.R. and Hirsch, A. (1947) Further observations on an inhibitory substance (nisin) from lactic streptococci. *Lancet*, **2**, 5–7.

McKay, L.L., Walter, L.A., Sandine, W.E. and Elliker, P.R. (1969) Involvement of phosphoenol-pyruvate in lactose utilisation by group N streptococci. *Journal of Bacteriology*, **99**, 603–10.

Medina, L.M. and Jordano, R. (1995) Population dynamics of constitutive microbiota in BAT type fermented milk products. *Journal of Food Protection*, **58**, 70–6.

Meghrous, J., Euloge, P., Junelles, A.M., Ballongue, J. and Petitidemange, H. (1990) Screening of *Bifidobacterium* strains for bacteriocin production. *Biotechnology Letters*, **12**, 575–80.

Metchnikoff, E. (1910) *The Prolongation of Life: Optimistic Studies*, revised edition of 1907, Heinemann, London.

Meyer-Barton, E.C., Klein, J.R., Iman, M. and Plapp, R. (1993) Cloning and sequence analysis of the X-prolyl-dipeptidyl aminopeptidase gene (*pep*X) from *Lactobacillus delbrueckii* subspecies *lactis* DSM 7290. *Applied Microbiology and Biotechnology*, **40**, 82–9.

Midwinter, R.G. and Pritchard, G.G. (1994) Aminopeptidase N from *Streptococcus salivarius* subspecies *thermophilus* NCDO 573: purification and properties. *Journal of Applied Bacteriology*, **77**, 288–95.

Moineau, S. and Goulet, J. (1991) Effect of feeding fermented milks on the pulmonary macrophage activity in mice. *Milchwissenschaft*, **46**, 551–3.

Monnet, C., Schmitt, P. and Divies, C. (1994) Diacetyl production in milk by an α-acetolactic acid accumulating strain of *Lactococcus lactis* subspecies *lactis* biovar *diacetylactis*. *Journal of Dairy Science*, **77**, 2916–24.

Montville, T.J. and Kaiser, A.L. (1993) Antimicrobial proteins: classification, nomenclature, diversity and relationship to bacterioccus, in *Bacteriocins of Lactic Acid Bacteria* (eds D.G. Hoover and L.R. Steenson), Academic Press, San Diego, pp. 1–22.

Mukai, T., Toba, T., Itoh, T. and Adachi, S. (1988) Structural microheterogeneity of kefiran from kefir grains. *Japanese Journal of Zootechnical Science*, **59**, 167–76.

Nakajima, H., Toyodo, S., Toba, T., Ito, T., Mukai, T., Kitazawa, H. and Adachi, S. (1990) A novel phosphopolysaccharide from slime-forming *Lactococcus lactis* subspecies *cremoris* SBT0495. *Journal of Dairy Science*, **73**, 1472–7.

Nakajima, H., Toba, T. and Toyodo, S. (1995) Enhancement of antigen-specific antibody production by extracellular slime products from slime-forming *Lactococcus lactis* subspecies *cremoris* SBT 0495 in mice. *International Food Microbiology*, **25**, 153–8.

Namba, Y., Hidaka, Y., Taki, K. and Morimoto, T. (1981) Effect of oral administration of lysozyme or digested bacterial cell walls on immunostimulation in guinea pigs. *Infection and Immunity*, **31**, 580–3.

Nettles, C.G. and Barefoot, S.F. (1993) Biochemical and genetic characteristics of bacteriocins of food-associated lactic acid bacteria. *Journal of Food Protection*, **56**, 338–56.

Neve, H., Geis, A. and Tueber, M. (1988) Plasmid encoded functions of ropy lactic acid streptococcal strains from Scandinavian fermented milks. *Biochimie*, **70**, 437–42.

Perdigon, G. Alvarez, S., Nader de Macias, M., Roux, M.E. and de Ruiz Holgado, A.P. (1990) The oral administration of lactic acid bacteria increase the mucosal intestinal immunity in response to enteropathogens. *Journal of Food Protection*, **53**, 404–10.

Perdigon, G. Alvarez, S., Rachid, M., Aguero, G. and Gobbato, N. (1995) Probiotic bacteria for humans: clinical systems for evaluation of effectiveness. *Journal of Dairy Science*, **78**, 1597–606.

Piard, J.-C. and Desmazeaud, M. (1991) Inhibiting factors produced by lactic acid bacteria – 1. Oxygen metabolites and catabolism end products. *Lait*, **71**, 525–41.

Piard, J.-C. and Desmazeaud, M. (1992) Inhibiting factors produced by lactic acid bacteria – 2. Bacteriocins and other antibacterial substances. *Lait*, **72**, 113–42.

Pidoux, M., Brillouet, J.M. and Quemener, B. (1988) Characterization of the polysaccharides from a *Lactobacillus brevis* and from sugary kefir grains. *Biotechnology Letters*, **10**, 415–20.

Pidoux, M., Marshall, V.M., Zanoni, P. and Brooker, B.E. (1990) Lactobacilli isolated from sugary kefir grains capable of polysaccharide production and mini-cell formation. *Journal of Applied Bacteriology*, **69**, 311–20.

Pulusani, S.R., Rao, D.R. and Sunki, G.R. (1979) Antimicrobial activity of lactic cultures: partial purification and characterization of antimicrobial compound(s) produced by *Streptococcus thermophilus*. *Journal of Food Science*, **44**, 575–8.

Rogers, L.A. (1928) The inhibiting effect of *Streptococcus lactis* on *Lactobacillus bulgaricus*. *Journal of Bacteriology*, **16**, 321–5.

Rohm, H. and Kovac, A. (1994) Effects of starter cultures on linear viscoelastic and physical properties of yoghurt gels. *Journal of Texture Studies*, **25**, 311–29.

Rul, F., Monnet, V. and Gripon, J.C. (1994) Purification and characterization of a general aminopeptidase (St-pepN) from *Streptococcus salivarius* subspecies *thermophilus* CNRZ 302. *Journal of Dairy Science*, **77**, 2880–9.

Scardovi, V. (1986) Genus *Bifidobacterium*, in *Bergey's Manual of Systematic Bacteriology*; vol. 2 (eds P.H.A. Sneath, N.S. Mair, M.E. Sharpe and J.G. Holt), Williams & Wilkins, Baltimore, pp. 1418–34.

Schellhaass, S.M. (1983) *Characterisation of the exocellular slime produced by bacterial starter cultures used in the manufacture of fermented dairy products*, Thesis, University of Minnesota.

Schillinger, U. (1990) Bacteriocins of lactic acid bacteria, in *Biotechnology of Food Safety* (eds D.D. Bills and S. Kung), Butterworth-Heineman, Boston, pp. 55–74.

Schved, F., Lalazar, A., Henis, Y. and Juven, B.J. (1993) Purification, partial characterization and plasmid-linkage of pediocin SJ-1, a bacteriocin produced by *Pediococcus acidilactici*. *Journal of Applied Bacteriology*, **74**, 67–77.

Sgorbati, B., Biavati, B. and Palenzona, D. (1995) The genus *Bifidobacterium*, in *The Genera of Lactic Acid Bacteria*, vol. 2 (eds B.J.B. Wood and W.H. Holzapfel), Blackie Academic & Professional, London, pp. 279–306.

Simpson, W.J. and Taguchi, H. (1995) The genus *Pediococcus*, with notes on the genera *Tetratogenococcus* and *Aerococcus*, in *The Genera of Lactic Acid Bacteria*, vol. 2 (eds B.J.B. Wood and W.H. Holzapfel), Blackie Academic & Professional, London, pp. 125–72.

Sjoberg, A. and Hahn-Hagerdal, B. (1989) β-Glucose-1-phosphate, a possible mediator for polysaccharide formation in maltose assimilating *Lactococcus lactis* subspecies *lactis*. *Applied and Environmental Microbiology*, **55**, 1549–54.

Snoep, J.L., Teixeira-Mattos, M.J., Starrenburg, M.J.C. and Hugenholtz, J. (1992) Isolation, characterization and physiological role of pyruvate dehydrogenase complex and α-acetolactate synthase of *Lactococcus lactis* subsp. *lactis* bv. *diacetylactis*. *Journal of Bacteriology*, **174**, 4838–41.

Stanier, R.Y., Ingram, J.L., Wheelis, M.L. and Painter, P.R. (1987) *The Microbial World*, 5th edn, Macmillan Education, London, pp. 495–504.

Stefanitsi, D., Sakellaris, G. and Gavel, R. (1994) The presence of two proteinases associated with the cell wall of *Lactobacillus delbrueckii* subspecies *bulgaricus*. *FEMS Microbiology Letters*, **128**, 53–8.

Sutherland, I.W. (1985) Biosynthesis and composition of Gram-negative bacterial extracellular and wall polysaccharides. *Annual Review of Microbiology*, **39**, 243–70.

Tagg, J.R., Dajani, A.S. and Wannamaker, L.W. (1976) Bacteriocins of Gram-positive bacteria. *Bacteriology Reviews*, **40**, 722–56.

Taguchi, H. and Ohta, T. (1991). D-lactate dehydrogenase is a member of the D-isomer specific 2-hydroxyacid dehydrogenase family. Cloning, sequencing and expression in *Escherichia coli* of the D-lactate dehydrogenase gene of *Lactobacillus plantarum*. *Journal of Biological Chemistry*, **266**, 12588–94.

Taguchi, H. and Ohta, T. (1993) Histidine 296 is essential for catalysis in *Lactobacillus plantarum* D-lactate dehydrogenase. *Journal of Biological Chemistry*, **268**, 18030–4.

Teuber, M. (1995) The genus *Lactococcus*, in *The Genera of Lactic Acid Bacteria*, vol. 2 (eds B.J.B. Wood and W.H. Holzapfel), Blackie Academic & Professional, London, pp. 173–234.

Teuber, M., Geis, A. and Neve, H. (1992) The genus *Lactococcus*, in *The Prokaryotes – A Handbook on the Biology of Bacteria: Ecophysiology, Isolation, Identification, Applications*, vol. 2, 2nd edn (eds A. Balows, H.G. Trüper, M. Dworkin, W. Harder and K.-H. Schleifer), Springer Verlag, New York, pp. 1482–501.

Thomas, T.D. (1976) Activator specificity of pyruvate kinase from lactic streptococci. *Journal of Bacteriology*, **125**, 1240–2.

Thompson, J.K., Johnson, D.E., Murphy, R.J. and Collins, M.A. (1990) Characteristics of a milk fermentation from rural Northern Ireland which resembles kefir. *Irish Journal of Food Science and Technology*, **14**, 35–49.

Thuault, D., Beliard, E., Le Guern, J. and Bourgeois, C.-M. (1991) Inhibition of *Clostridium tyrobutyricum* by bacteriocin-like substances produced by lactic acid bacteria. *Journal of Dairy Science*, **74**, 1145–50.

Toba, T., Arahara, K. and Adachi, S. (1987) Comparative study of polysaccharide from kefir grains, an encapsulated homo-fermentative *lactobacillus* species and *Lactobacillus kefir*. *Milchwissenschaft*, **42**, 565–8.

Vaughan, E.E., Daly, C. and Fitzgerald, G.F. (1992) Identification and characterization of helveticin V-1829, a bacteriocin produced by *Lactobacillus helveticus* 1829. *Journal of Applied Bacteriology*, **73**, 299–308.

Vedamuthu, E.R. and Neville, J.M. (1986) Involvement of a plasmid in production of ropiness (mucoidness) in milk cultures by *Streptococcus cremoris* MS. *Applied and Environmental Microbiology*, **51**, 677–82.

Vescovo, M., Scolari, G.L. and Bottazzi, V. (1989) Plasmid-encoded ropiness in *Lactobacillus casei* ssp. *casei*. *Biotechnology Letters*, **11**, 709–12.

Vesely, R., Negri, R., Bianchi-Salvadori, B., Lavezzari, D. and de Simone, C. (1985) Influence of a diet additioned with yoghurt on the mouse immune system. *Immunology and Immunopharmacology*, **5**, 30–5.

von Wright, A. and Tynkkynen, S. (1987) Construction of *Streptococcus lactis* ssp *cremoris* strains with a single plasmid associated with mucoidy. *Applied and Environmental Microbiology*, **53**, 1385–6.

Vos, P., van Asseldonk, M., van Jeveren, F., Siezen, R., Simons, G. and de Vos, W.M. (1989) A maturation protein is essential for production of active forms of *Lactococcus lactis* subspecies *lactis* SK11 serine proteinase located in or secreted from the cell envelope. *Journal of Bacteriology*, **171**, 2795–802.

Walker, D.K. and Gilliland, S.E. (1993) Relationships among bile tolerance, bile salt deconjugation and assimilation of cholesterol by *Lactobacillus acidophilus*. *Journal of Dairy Science*, **76**, 956–61.

Wang, X. and Gibson, G.R. (1993) Effects of the *in vitro* fermentation of oligofructose and inulin by bacteria growing in the human large intestine. *Journal of Applied Bacteriology*, **75**, 373–80.

Weiss, N. (1992) The genera *Pediococcus* and *Aerococcus*, in *The Prokaryotes – A Handbook on the Biology of Bacteria: Ecophysiology, Isolation, Identification, Applications*, vol. 2, 2nd ed (eds A. Balows, H.G. Trüper, M. Dworkin, W. Harder and K.-H. Schleifer), Springer Verlag, New York, pp. 1502–7.

Whitehead, H.R. (1933) A substance inhibiting bacterial growth, produced by certain strains of lactic streptococci. *Biochemical Journal*, **27**, 1793–800.

Wozniac, D.J. and Ohman, D.E. (1994) Transcriptional analysis of the *Pseudomonas aeruginosa* genes algR, algB and algD reveals a hierarchy of gene expression. *Journal of Bacteriology*, **176**, 6007–14.

Yokoi, H., Watanabe, T., Fuji, Y., Toba, T. and Adachi, S. (1990) Isolation and characterization of polysaccharide-producing bacteria from kefir grains. *Journal of Dairy Science*, **73**, 1684–9.

Zoon, P., van Marte, M.E., Smith, M. and Kingma, F. (1994) Consistency of yoghurt: role of ropy bacteria. *Voedingsmiddelentechnologie*, **27**, 12–13.

5 Flavour and texture in soft cheese
J.-C. GRIPON

5.1 Introduction

Cheeses can be classified schematically according to their moisture content into three major categories: hard, semi-hard and soft cheeses. The higher the moisture content/casein ratio, the softer the casein matrix of cheese. Soft cheeses have a typical taste and aroma due to a surface flora. If the latter is of bacterial origin, they are called bacterial surface-ripened cheeses, such as 'Munster', 'Pont L'évêque' and 'Limburger'. However, most soft cheeses are surface mould-ripened cheeses, whose surface is covered with a felt of a white *Penicillium*, i.e. *P. camemberti*. Camembert is a typical example of this type of cheese. It originates from Normandy and it was produced only in farms until the end of the 19th century when industrial-type cheese factories appeared. The world production of soft cheeses is much lower than that of hard and semi-hard cheeses. Soft cheeses are mainly produced in France but now also in other countries eager to offer a wider range of aromas and flavours to their consumers.

The current review is dedicated to surface mould-ripened soft cheeses and will consider the biochemical modifications which occur during ripening as well as the aroma and textural development.

5.2 The flora of surface mould-ripened cheeses

This flora is particularly complex in the case of cheeses made from raw milk and undergoes considerable modifications throughout ripening. Twenty-four hours after the beginning of the cheese-making process it is essentially composed of lactococci used as starters (Lenoir, 1963a). As early as the second day, a fungal flora, capable of developing in an acid medium and composed of yeasts and *Geotrichum candidum* colonizes the surface (Schmidt and Lenoir, 1980). On the fifth or sixth day a felt due to germination of spores of *Penicillium camemberti* appears. This felt, which is more or less thick and fluffy according to the strains, entirely covers the surface of the cheese as early as the eighth or tenth day. The development of *Penicillium* is accompanied by a rapid increase in pH which enables the appearance of an acid-sensitive surface bacterial flora made of micrococci and coryneform bacteria whose colonies are yellow or orange-coloured (Richard and Zadi,

1983; Richard, 1984). The complexity and the evolution of the flora are necessary for the flavour development of the traditional product.

Cheeses made from pasteurized milk have a more simple flora and were initially only prepared by seeding lactic starters and spores of *P. camemberti*. The taste of the products obtained was much more neutral than that of raw-milk Camemberts. In order to diversify the organoleptic qualities and to obtain products which are more typical and closer to traditional products, many producers of thermized or pasteurized products currently add to the cheese milk beside lactococci and spores of *P. camemberti*, selected strains of *G. candidum*, coryneform bacteria and yeasts. Even if *P. camemberti* is the essential typical characteristic of surface mould-ripened cheeses, the other constituents of the ripening flora clearly play a complementary role in the acquisition of final organoleptic qualities (Mourgues et al., 1983; Molimard *et al.*, 1994).

The strains of *P. camemberti* offered by the producers of fungal starters have different growth rate, mycelium density and height as well as different lipolytic and proteolytic activities. The morphological and physiological properties of this mould have been described by Moreau (1979), Choisy *et al.* (1984) and Cerning *et al.* (1987). Recent reviews by Guegen and Schmidt (1992) and Bergère and Tourneur (1992) are dedicated to yeasts and surface bacteria, respectively.

5.3   Biochemical reactions involved in soft cheese ripening

Three major biochemical processes, i.e. glycolysis, proteolysis and lipolysis, occur during cheese ripening and are responsible for the appearance of the basic flavour and texture of the products. The secondary modifications undergone by the products resulting from these primary reactions are responsible for the aroma and taste compounds which give their typical characteristics to soft cheeses.

5.3.1   Glycolysis

During traditional Camembert manufacturing, milk undergoes maturation (by addition of a small level of lactococci) until it reaches an approximate pH of 6.4. Rennet and lactococci are then added. A strong acidification occurs during draining to reach a pH close to 4.6 in the curd after 24 h (Lenoir, 1963a). Lactose is essentially converted by lactococci to L-lactate whose concentration is about 1.5% after 24 h (Choisy *et al.*, 1984). During early ripening, the surface flora, i.e. yeasts, *Geotrichum* and especially *P. camemberti*, metabolizes lactate into CO_2 and H_2O which leads to an increase in the surface pH and to a diffusion of the internal lactate towards the outer part of the cheese. A pH gradient appears between the surface and the centre of the cheese to reach values of about 7.0 in surface and 5.5–6.0

in the centre at the end of ripening (Lenoir, 1984). The consequences of this drastic pH modification are important and are observed at four levels.

1. De-acidification of the body enables an aerophil and acid-sensitive bacterial flora made of micrococci and coryneform bacteria such as *Arthrobacter* and *Brevibacterium linens* to colonize the surface. *Brevibacterium linens* contributes to the flavour development and can only settle if the surface pH is higher than 5.8 (Boyaval and Desmazeaud, 1983).
2. The neutralization of the body also enhances the action of numerous ripening enzymes whose optimum action pH is often closer to neutrality than that of an acid cheese curd. In the case of plasmin (optimum $pH \cong 7.0$) an increase in γ-casein concentrations in the outer area of Camemberts is observed at the end of ripening (Trieu-Cuot and Gripon, 1982). It is also most probably the case for many microbial enzymes such as lipase and some exocellular peptidases of *P. camemberti* or such as peptidases of lactococci.
3. Curd content in calcium phosphate exceeds its solubility threshold at the external pH of cheeses, leading to its precipitation in surface. During *Penicillium* growth, a concentration gradient appears and calcium phosphate migrates to the surface where it accumulates as a deposit. Thus, the rind of surface mould-ripened cheeses reaches high contents in calcium and inorganic phosphate (17 and $9\,\mathrm{g\,kg^{-1}}$, respectively whereas these contents are much lower in the centre of the cheese) (Le Graet *et al.*, 1983).
4. Lastly, as we will see later, the increase in pH considerably modifies the rheological properties of the body and leads to a softening which is typical of this type of cheese.

5.3.2 Proteolysis

The proteolysis observed in the external area of surface mould-ripened cheeses is significant but it remains lower than that of blue-veined cheeses. Soluble nitrogen at a pH of 4.6 in the outer area of a raw-milk Camembert represents approximately 35% of total nitrogen and only 25% in the centre (Lenoir, 1963b; Mourgues *et al.*, 1983). Small peptides are abundant in the soluble fraction (nitrogen in trichloracetic represents 20% of total nitrogen). The high ammonia content (7–8% of surface total nitrogen) reveals a considerable deamination of free amino acids (Lenoir, 1963b).

Electrophoretic studies show a rapid degradation of α_{s1}-casein into α_{s11}-casein. β-Casein is highly hydrolysed in the external area but much less in the internal area.

This high proteolysis results from the action of three different agents: rennet, plasmin, i.e. the native milk protease, and microbial enzymes, including those of *P. camemberti*, which play a major role.

Part of the rennet added to manufacturing milk is retained in the curd after draining and contributes to ripening. The more acid the pH at draining, the higher the amount of rennet retained. In the case of Camembert, acidification takes place during draining and the amount retained is higher than in other cheeses. Vassal and Gripon (1984) and Garnot *et al.* (1987) have observed that approximately 50% of the rennet added remained in the curd whereas values in the range of 10–15% were measured in pressed cheeses (Stadhouders and Hup, 1975). The action of rennet appears very rapidly in Camembert, since 6 h after the beginning of manufacturing a considerable proportion of α_{s1}-casein is hydrolysed into the α_{s1I}-peptide. Noomen (1978b) observed that the action of general proteolysis of the rennet in the curds is optimal around pH 5.0. The pH of the external area of Camembert rapidly exceeds this value and the action of rennet probably diminishes during late ripening. Since the pH increase is less marked and rapid in the centre of cheese, its influence on the activity of rennet is probably lower.

As stated previously, the production of γ-caseins resulting from the action of plasmin on β-casein is higher in the rind than in the core and considerably intensifies during late ripening (Trieu-Cuot and Gripon, 1982). As suggested by Noomen (1977, 1978a) its contribution to proteolysis in the outer part of soft cheeses is higher than in other varieties.

Studies performed with aseptic curds in which only *P. camemberti* develops confirmed that this microorganism causes an intense proteolysis leading to the release of major quantities of both peptides and free amino acids (Desmazeaud et al., 1976). Two exocellular endopeptidases, one aspartyl protease with an optimum pH close to 3.5 and one metalloprotease with an optimal pH close to 6.0, are produced in appreciable quantities (Cerning *et al.*, 1987). The production of these enzymes seems to vary little from one strain to another (Lenoir et Choisy, 1970).

Variation of the proteolytic activity in the curd was measured during the ripening of Camembert. In the centre of the cheese, the activity is very low and virtually does not vary during ripening. In the external area it suddenly increases after 6–7 days of ripening, i.e. when *Penicillium* develops. Aspartyl protease and metalloprotease are both synthesized. Their activities are maximal after about 15 days and then slowly diminish; these two enzymes are thus relatively stable in cheese (Lenoir, 1970, 1984).

The action of these two enzymes in curds leads to the occurrence of new peptides, some of which, i.e. β-ap1 and β-mp1, can easily be identified by electrophoresis and are the markers of the aspartyl and metalloprotease activities, respectively (Trieu-Cuot and Gripon, 1982). The intensity of the β-ap1 band regularly increases during ripening and shows that this enzyme remains active despite the pH of the external area of the curd which does not help the development of its activity. In contrast, the intensity of the β-mp1 band decreases after 15 days, suggesting either that the metalloprotease activity diminishes or that β-mp1 is degraded by another protease. The activity of proteases of *P. camemberti* is only detectable in the external area of cheeses which shows that the migration of these enzymes towards the interior of the curd remains very limited. The presence of β-ap1 was detected more than 7 mm deep but it probably shows the migration of β-ap1 rather than that of aspartyl protease. The results of Lenoir (1970) suggest that peptides migrate from the outer area towards the centre of cheese. Images of ripened Camemberts obtained by electron microscopy reveal mycelium alterations and suggest the action of intracellular proteases (Rousseau,1984). However, electrophoresis gels do not reveal the appearance of new hydrolysis products and the action of these enzymes is probably much more limited than that of exocellular enzymes.

P. camemberti also produces large quantities of free amino acids in cheeses (Desmazeaud *et al.*, 1976) through the synthesis of several exocellular peptidases. An acid carboxypeptidase (of serine carboxypeptidase type, optimum pH $\cong$ 3.5) releases hydrophobic amino acids and has the property of diminishing the bitterness of casein hydrolysates (Ahiko *et al.*, 1981). At least one aminopeptidase with an optimum pH of 8–8.5 (Auberger et al., 1985; Matsuoka *et al.*, 1991) is also synthesized. An intracellular carboxypeptidase (optimum pH 6.5) might also be involved during late ripening (Auberger *et al.*, 1985). Fuka and Matsuoka (1993) have characterized an iminopeptidase with an optimum pH of 7.0 capable of liberating prolyl residues in the N-terminal position.

It is generally considered that *G. candidum* plays a much lesser role in proteolysis than *P. camemberti*. On the one hand the proteolytic activity in the outer area of cheese only increases during *Penicillium* growth and not during that of *Geotrichum*, and on the other hand soft cheeses seeded with *Geotrichum* show a lower proteolysis than those seeded with *Penicillium* only (Vassal and Gripon, 1984). The exocellular endopeptidase activity of *G. candidum* is maximal at pH 6.0 (Gueguen and Lenoir, 1976). An exocellular aminopeptidase activity (optimum pH 7.5–8.0) as well as intracellular aminopeptidase (optimum pH 8.0–8.5) and carboxypeptidase (optimum pH 4.5) activities were seen (Gueguen and Schmidt, 1992).

In a collection of 165 isolated yeasts of Camembert, Schmidt (1982) observed a global intracellular proteolytic activity on casein with an optimum pH of approximately 6.0 but did not detect any exocellular activity. Several studies characterized the presence of peptidases in *Saccharomyces cerevisiae* (Jones, 1991). It is difficult to determine the activity of these enzymes in the curd but the activity remains lower than that of *Penicillium*.

Brevibacterium linens synthesizes several exocellular serine proteases as well as an aminopeptidase which is also exocellular and could participate in the proteolysis during late ripening (Hayashi and Law, 1989; Hayashi *et al.*, 1990).

Although the ripening flora plays a major role in surface mould-ripened cheeses, the action of lactic acid bacteria which takes place, as in all the other types of cheese, should not be forgotten. The lysis of cells of lactococci leads to the release of intracellular peptidases capable of hydrolysing the peptides produced by rennet, plasmin or microbial proteinases. The peptidase system of lactococci is well known. A dozen peptidases were isolated and characterized (see Chapter 9). These enzymes have all an optimum pH close to neutrality and as mentioned earlier, the pH increase in the outer area of the cheese probably enhances their action.

5.3.3 Lipolysis

The degree of lipolysis in surface mould-ripened cheeses is higher than in other varieties of cheese but remains lower than in blue-veined cheeses. In a ripened raw-milk Camembert, lipolysis reaches 6–10% of total fatty acids (Kuzdal and Kuzdal-Savoie, 1966). Lower values ranging between 3–5% were observed in products which were probably less ripened (Van Belle *et al.*, 1978). Lipolysis is always higher in the outer area and results from the activity of *P. camemberti* and *G. candidum*, the other microorganisms of the surface flora having a much more limited lipolytic activity (Lamberet and Lopez, 1982). It is worth noting that this major lipolysis does not lead to a rancid taste, probably because the fatty acids are neutralized by a pH increase of the curd.

Lamberet and Lenoir (1976a, b) noted that *P. camemberti* produces only one extracellular lipase which has an optimal activity on tributyrin at pH 9.0 and 35°C. The level of production of this enzyme varies from 1 to 10 (relative scale), depending on the strain (Lamberet and Lenoir, 1972). At pH 6.0, this lipase retains 50% of its maximal activity and remains very active between 0 and 20°C. It is more active when calcium ions are present. The production of lipase has been studied during the ripening of raw-milk Camemberts (Lamberet and Lopez, 1982). The activity appears after 10 days of ripening, during or shortly after mycelium growth, is maximal at 16 days and then decreases slightly until the 30th day when it increases again probably because of the lysis of the mycelia. In 10 cheeses of different origins, the lipase activity in the outer region of the cheese varied from 1.2 to 4.45 units g^{-1} of cheese.

The particularly high proportion of free oleic acids in Camembert has been attributed to *G. candidum* lipases (Kuzdzal and Kuzdzal-Savoie, 1966). This fungus synthesizes two exocellular lipases with an optimum pH close to 6.5 (Veeraragavan *et al.*, 1990). These two lipases have similar properties and the nucleotide sequence of their genes shares 86% of homology (Nagoa *et al.*, 1993). Their specificity is, however, different: lipase I shows a higher specific activity on long-chain unsaturated fatty acids (which would explain the high oleic acid content of Camemberts) whereas lipase II releases the medium-chain saturated fatty acids better (C_8, C_{10}, C_{12} and C_{14}) (Bertolini *et al.*, 1995).

5.4 Aroma development

Surface mould-ripened soft cheeses have a typical aroma which makes them clearly different from other types of cheese. Analysis of their volatile fraction showed a great complexity; more than 100 different compounds were detected in an extract of Camembert (Moinas *et al.*, 1973; Dumont *et al.*, 1974, 1976). These volatile molecules essentially result from the action of the microbial flora and their enzymes on products released by proteolysis, lipolysis and glycolysis. We still do not know much about the metabolisms and enzymes that produce these compounds. In contrast, we know that some of the compounds themselves are responsible for the typical aromatic notes of these cheeses. The present paragraph summarizes our current knowledge about the nature of the detected compounds, their mode of formation and the estimation of their organoleptic impact. Detailed reviews of these compounds have recently been made by Molimard (1994) and Molimard and Spinnler (1996). A list of the main identified compounds in surface mould-ripened soft cheeses is given in Table 5.1.

Fatty acids are precursors of methyl ketones, alcohols, lactones and esters but they are also aromatic compounds. Long-chain fatty acids are generally considered to have a minor role in the flavour whereas short- and medium-chain fatty acids have characteristic aromatic notes and much lower perception thresholds. Acetic and propanoic acids have a 'vinegar' smell and the butyric acid a rancid smell. The detection threshold of the isovaleric

Table 5.1 Volatile compounds isolated from Camembert cheese[a]

1-Alkanols	C2, 3, 4, 6, 2-methylpropanol, 3-methylbutanol, oct-1-en-3-ol, 2-phenylethanol
2-Alkanols	C4, 5, 6, 7, 9, 11
Methyl ketones	C4, 5, 6, 7, 8, 9, 10, 11, 12, 13, 15
Aldehydes	C6, 7, 9, 2 and 3-methylbutanal
Esters	C2, 4, 6, 8, 10-ethyl, 2-phenylethylacetate
Phenols	phenol, *p*-cresol
Lactones	C_9, C_{10}, C_{12}
Sulphur compounds	H_2S, methyl sulphide, methyldisulphide, methanethiol, 2,4-dithiapentane, 3,4-dithiahexane, 2,4,5-trithiahexane, 3-methylthio2,4-dithiapentane, 3-methylthiopropanol
Anisoles	anisole, 4-methylanisole, 2,4-dimethylanisole
Amines	phenylethylamine, $C_{2,3,4}$, diethylamine, isobutylamine, 3-methylbutylamine
Miscellaneous	dimethoxybenzene, isobutylacetamide

[a]From Adda (1984).

acid, whose smell recalls that of sweat and feet, is especially low (0.07 p.p.m. in water, Brennand et al., 1989). 4-Methyloctanoic and 4-ethyloctanoic acids are present only in triglycerides of goats' milk and have particularly low detection thresholds (0.02 and 0.006 at 2.24 p.p.m., respectively; Brennand et al., 1989). These fatty acids are responsible for the typical smell of cheeses made from goat milks' (Brennand et al., 1989; Le Qéuré et al., 1996). They are predominantly released in cheese under the action of the milk lipoprotein lipase and probably less by microbial lipases (Delacroix-Buchet et al., 1996).

Long- or medium-chain fatty acids proceed from the lipolysis mediated by P. camemberti and Geotrichum candidum, one of the lipases of the latter which preferentially releases oleic acid. Short-chain fatty acids result from the action of microbial enzymes on lactose (acetic acid, propionic acid) or on amino acids (isovaleric acid). Kuzdzal and Kuzdzal-Savoie (1966) estimated that 5% of free fatty acids of raw-milk Camembert do not result from lipolysis.

Methyl ketones are the most abundant volatile compounds in surface mould-ripened cheeses (25–60 μg per 100 g fat), with the highest values observed for heptatone and nonanone. They result from the action of P. camemberti which metabolizes fatty acids by partial β-oxidation (Lamberet et al., 1982). The free fatty acid is oxidized into β-ketoacyl-CoA. A thiohydrolase releases the ketoacid which is then decarboxylated to give a methyl ketone with one carbon atom less than the initial fatty acid (Lamberet et al, 1982). High quantities of heptatone and nonanone are not only explained by the degradation of octanoic and decanoic acids present in milk fat. Studies using long-chain fatty acids labelled with ^{14}C showed that they undergo successive cycles of β-oxidation (Dartey and Kinsella, 1973) to give methyl ketones with shorter chains. These compounds give aromatic notes of fruity, mouldy and blue-cheese type (Rothe et al., 1982).

Secondary alcohols such as 2-heptanol and 2-nonanol are detected in abundant quantities (Dumont et al., 1974). These compounds are produced by P. camemberti through the action of a reductase on the corresponding methyl ketones.

Phenylethanol and its esters (phenylethylacetate and phenylethyl-propanoate) are produced during early ripening and their concentration stabilizes at approximately 1 p.p.m. at the end of ripening (Roger et al., 1988). These compounds have a floral smell. Their perception threshold is of approximately 9 p.p.m. However, some taste testers have a lower perception threshold and can detect a floral note in a ripened Camembert which is probably due to the cumulated effect of phenylethanol and its esters. These compounds result from the metabolism of yeasts which develop on the surface during early ripening (Lee and Richard, 1984).

Oct-1-en-3-ol is also present in large quantities (Dumont et al., 1974). This compound has a smell of mushrooms and, given its low perception threshold, it brings a characteristic note and is probably a key compound of the aroma of Camembert. Its production is due to the metabolism of P.

camemberti on linoleic and linolenic acids (Chen and Wu, 1984; Karahadian *et al.*, 1985a).

Numerous esters were detected in the volatile fraction of Camembert (Dumont *et al.*, 1974). Alcohols issued from lactose fermentation or from catabolism of amino acids are esterified with short- or medium-chain fatty acids. They are produced during early ripening and mainly result from the action of esterases produced by yeasts and *Geotrichum* (Kallel-Mhiri and Miclo, 1993; Jollivet *et al.*, 1994;). Most of them have fruit-like aromatic notes.

Several sulphur compounds were identified in the volatile fraction of Camembert (Dumont *et al.*, 1976). They are described as having a strong smell of garlic, cabbage or highly ripened cheese (Cuer *et al.*, 1979). Their perception threshold is generally very low and they represent a major component of the flavour of raw-milk Camemberts. They originate from the degradation of methionine by cleavage of the carbon–sulphur bond under the action of a demethiolase (Hemme *et al.*, 1982; Jollivet *et al.*, 1994; Collin and Law, 1989). The methanethiol released is very reactive and is a precursor of other sulphur compounds.

Several microorganisms, i.e. *P. camemberti*, *G. candidum* and *B. linens*, are capable of releasing methanethiol from methionine (Hemme *et al.*, 1982; Collin and Law, 1989; Jolivet *et al.*, 1994). Sulphur compounds are absent from young cheeses since the concentration of their precursor, i.e. methionine, is too low.

The decarboxylase activity of the surface flora leads to the release of amines (Ney and Wiromata, 1971; Adda and Dumont, 1974). The organoleptic role of amines is still not well known. They could originate compounds such as *N*-isobutylacetamide, which is regularly detected in Camemberts (Adda *et al.*, 1982).

Mention should also be made of ammonia, which originates from the deamination of amino acids and which largely contributes to the aroma of surface mould-ripened cheeses. In 1971, Do Ngoc *et al.* observed high levels of ammonia (7–8% of total nitrogen) in Camemberts which were probably very ripened. This production of ammonia results from the deaminating activity of *P. camemberti*, *G. candidum* and coryneform bacteria (Karahadian and Lindsay, 1987).

5.5 Textural development

The influence of pH modifications throughout ripening on the cheese texture is particularly obvious in Camembert. The initial pH of the curd is approximately 4.6–4.7 and its texture is firm and brittle. During ripening, the surface flora, and especially *Geotrichum* then *Penicillium*, metabolizes the lactic acid resulting in a pH gradient between the rind and the core of cheese. This gradient is amplified by the production of ammonia resulting from the deaminating action of surface microorganisms. At the end of ripening, the surface pH of a Camembert reaches values of about 7.0, and the pH of the central part values close to or higher than 5.5. These pH modifications of the curd lead to an increase in the net charge of caseins and in their water adsorption capacity. The texture is consequently modified and the body of the curd becomes more smooth and soft. This softening first occurs in the outer part and then progresses towards the centre as ripening progresses. It can be seen on a cross-section of cheese that the external part appears more homogeneous and yellow. According to Noomen (1977, 1983), three conditions which are met in Camembert are necessary for the appearance of the clear softening of the body: (i) a high water content, (ii) a pH higher than 5.2; and (iii) the degradation of α_{S1}-casein into α_{S1I}-casein by rennet (the bond Phe_{23}–Phe_{24} is cleaved). It is possible to simulate the influence of pH modifications on the texture by incubating young curds of Camembert with no *Penicillium* seeding in an atmosphere containing ammonia. Ammonia is dissolved in the outer area of the curds leading to a pH increase and the softening of the body (Vassal *et al.*, 1984). The direct role of the proteolysis caused by *Penicillium* is probably minor compared with that caused by pH modifications.

The influence of pH on the curd texture is used in the manufacturing processes of certain soft cheeses called 'stabilized cheeses'. In these cheeses the acidification of the curd is limited in order to obtain a curd with a final pH $\geqslant 5.2$. This is achieved by washing the curd or by using slightly acidifying starters (protease-negative strains of *Lactococcus lactis* or strains of *Streptococcus thermophilus* used at a temperature lower than that necessary for their optimal growth). The curd obtained is slightly demineralized and, given its pH and water content, it rapidly acquires a soft and homogeneous texture under the action of rennet on α_{S1}-casein. The curd rapidly looks like a ripened cheese. Stabilized cheeses, which are made from pasteurized milk, have a milder taste and can be kept longer than raw-milk cheeses.

5.6 Controlling the defects of surface-moulded cheeses

The previous paragraphs showed the complexity of the floras and the biochemical modifications occurring in surface mould-ripened cheeses. The microbiological control of the surface flora is essential for the quality of the flavour and the appearance of these products. As previously mentioned, *Geotrichum* develops before *Penicillium*. If its growth is too intense following an insufficient draining, it disturbs the colonization by *Penicillium* and the cheeses have a defective appearance called 'toad skin'. Salting limits this defect by regulating the balance between these two moulds because *P. camemberti* is much less sensitive to salt than *G. candidum*. An insufficient salting enhances *Geotrichum* to the detriment of *Penicillium*. Conversely, oversalting inhibits the growth of *Geotrichum* and leads to a less rich flavour. It was also shown that *G. candidum* produces volatile compounds capable of inhibiting the germination of spores of various filamentous fungi (Tariq and Campbell, 1991). This mechanism could also play a role in the competition between the two moulds.

Surface mould-ripened cheeses can sometimes have a celluloid taste due to high production of styrene by *P. camemberti* (Adda *et al.*, 1989). Tests during which phenylalanine labelled with ^{13}C were added to a culture of *P. camemberti* showed that the production of styrene was bound to the catabolism of phenylalanine by *Penicillium* (H. Spinnler and P. Chaulet, personnal communication). Ripening conditions of 'stabilized cheeses' (low carbohydrates content and high ripening temperatures) can help the development of defects (Spinnler *et al.*, 1992).

As in the other cheese varieties, the bitterness of surface mould-ripened cheeses results from too high a content in hydrophobic peptides. *P. camemberti* plays a major role in the development of the defect, which can appear if the intensity of the *Penicillium* development is too high. When its growth is limited either by the presence of *Geotrichum*, or by the incubation of young cheeses in an ammonia atmosphere, the proteolysis due to *Penicillium* is lower and the defect disappears (Vassal and Gripon, 1984; Molimard *et al.*, 1994). In general, the risk of bitterness is higher when the pH of the curd is too low after 24 hours (for example following the use of a lactic starter there is too high a proportion of prt$^+$ cells) perhaps because these conditions enhance the growth and the production of proteases by *P. camemberti*. Martley (1975) observed that the bitterness of Camembert is linked to a high population of lactococci and disappears when this population is reduced, for example in the presence of phages.

Texture defects are also observed in surface-moulded cheeses. If acidification is too high the curd remains too brittle. In contrast, an insufficient acidification leads to too high a moisture content and too soft a body that tends to flow when the ripe cheese is cut.

Acknowledgements

The author wishes to thank Mr L. Vassal for fruitful discussions and Miss Claire Gay for translating this manuscript into English.

5.7 References

Adda, J. (1984) La formation de la flaveur, in *Le Fromage* (ed. A. Eck), Lavoisier, Paris, pp. 330–40.

Adda, J. and Dumont, J.P. (1974) Les substances responsables de l'arôme des fromages à pâte molle. *Le lait*, **54**, 1–30.

Adda, J., Gripon, J.-C. and Vassal, L. (1982) The chemistry of flavour and texture generation in cheese. *Food Chemistry*, **9**, 115–29.

Adda, J., DeKimpe, J., Vassal, L. and Spinnler, H.E. (1989) Production de styrène part *Penicillium camemberti* Thom. *Le Lait*, **69**, 115–20.

Ahiko, K., Iwasawa, S., Ulda, M. and Nigata, N. (1981) Studies on acid carboxypeptidase from *Penicillium caseicolum*. II. Hydrolysis of bitter peptides by acid carboxypeptidase and large scale preparation of the enzyme. *Report of Research of Laboratory, Snow Brand Milk Products Co.*, no. 77, pp. 135–40.

Auberger, B., Lamberet G. and Lenoir, J. (1985) Les activités enzymatiques de *Penicillium camemberti*. *Sciences des Aliments*, **5**, 239–43.

Bergère, J.-L. and Tourneur, C. (1992) Les bactéries de surface des fromages, in *Les groupes microbiens d'intérêt laitier* (eds J. Hermier, J. Lenoir and F. Weber), Cepil, Paris, pp. 127–63.

Berner G. (1971) Degradation of lactic acid during ripening of Camembert. Enzymatic determination of D- and L-lactate. *Milchwissenschaft*, **26**, 685–7.

Bertolini, M.C., Schrag, J.P., Cygler, M., Ziomek, E., Thomas, D.Y. and Vernet, T. (1995) Expression and characterization of *Geotrichum candidum* lipase I gene. Comparison of specificity profile with lipase II. *European Journal of Biochemistry*, **228**, 863–9.

Boyaval, P. and Desmazeaud, M. (1983) Le point des connaissances sur *Brevibacterium linens*. *Le Lait*, **63**, 187–216.

Brennand, C.P., Ha, J.K. and Lindsay, R.C. (1989) Aroma properties and thresholds of some branched-chain and other minor volatile fatty acids occurring in milk fat and meat lipids. *Journal of Sensory Studies*, **4**, 105–20.

Cerning, J., Gripon, J.-C., Lamberet, G. and Lenoir, J. (1987) Les activités biochimiques des *Penicillium* utilisés en fromagerie. *Le Lait*, **67**, 3–39.

Chen, C.C. and Wu, C.M. (1984) Studies on the enzymatic reduction of 1-octen-3-one in mushroom. *Journal of Agricultural and Food Chemistry*, **32**, 1342–4.

Choisy, C., Gueguen, M., Lenoir, J., Schmidt, J.-L., and Tourneur, C. (1984) L'affinage du fromage: les phénomènes microbiens, in *Le Fromage* (ed. A. Eck), Lavoisier, Paris, pp. 259–90.

Collin, J.-C. and Law, B.A. (1989) Isolement et caractérisation de la L-méthionine-γ-déméthiolase de *Brevibacterium linens* NCDO 739. *Sciences des Aliments*, **9**, 805–12.

Cuer, A., Dauphin, G., Kergomard, A., Roger, S., Dumond, J.P., Adda, J. (1979) Flavour properties of some sulfur compounds isolate from cheese. *Lebensmittel-Wissenschaft und Technologie*, **12**, 258–61.

Dartey, K. and Kinsella, E. (1973) Metabolism of ^{14}C-lauric acid to methyl ketones by the spores of *Penicillium roqueforti*. *Journal of Agricultural and Food Chemistry*, **21**, 933–6.

Delacroix-Buchet, A., Degas, C., Lamberet, G. and Vassal, L. (1996) Influence des variants AA et FF de la caséine α_{s1} caprine sur le rendement et les caractéristiques sensorielles du fromage. *Le Lait*, **76**, 217–41.

Desmazeaud, M., Gripon, J.-C., Le Bars, D. and Bergère, J.-L. (1976) Etude du rôle des microorganismes et des enzymes au cours de la maturation des fromages; III. Influence des microorganismes. *Le Lait*, **56**, 379–96.

Do Ngoc, M., Lenoir, J. and Choisy, C. (1971) Les acides aminés libres des fromages affinés de camembert, Saint-paulin et Gruyère de Comté. *Revue Laitière Française*, **288**, 447–62.

Dumont, J.P., Roger, S., Cerf, P. and Adda, J. (1974) Etude des composés neutres volatils présents dans le camembert. *Le Lait*, **54**, 501–30.

Dumond, J.P., Roger, S. and Adda, J. (1976) L'arôme de camembert: autres composés mineurs mis en évidence. *Le Lait*, **56**, 595–9.

Fuka, Y. and Matsuoka, H. (1993) The purification and characterization of prolyl aminopeptidase from *Penicillium camemberti*. *Journal of Dairy Science*, **76**, 2478–84.

Garnot, P., Molle, D. and Piot, M. (1987) Influence of pH, type of enzyme and ultrafiltration on the retention of milk clotting enzymes in camembert cheese. *Journal of Dairy Research*, **54**, 315–20.

Gueguen, M. and Lenoir, J. (1976) Caractères du système protéolytique de *Geotrichum candidum*. *Le Lait*, **56**, 439–48.

Gueguen, M. and Schmidt, J.-L. (1992) Les levures et Geotrichum candidum, in *Les groupes microbiens d'intérêt laitier* (eds J. Hermier, J. Lenoir and F. Weber), Cepil, Paris, pp. 145–219.

Hayashi, K. and Law, B.A. (1989) Purification and characterization of two aminopeptidases produced by *Brevibacterium linens*. *Journal of General Microbiology*, **135**, 2027–34.

Hayashi, K., Cliffe, A.J. and Law, B.A. (1990) Purification and characterization of five serine proteinases produced by *Brevibacterium linens*. *International Journal of Food Science and Technology*, **25**, 180–7.

Hemme, D., Bouillanne, C., Metro, F. and Desmazeaud, M.J. (1982) Microbial catabolism of amino acids during cheese ripening. *Sciences des Aliments*, **2**, 113–23.

Jollivet, N., Chataud, J., Vayssier, Y., Bensoussan, M. and Belin, J.M. (1994) Reduction of volatile compounds in model milk and cheese media by eight strains of *Geotrichum candidum*. *Journal of Dairy Research*, **61**, 241–8.

Jones, E.W. (1991) Three proteolytic systems in the yeast *Saccharomyces cerevisiae*. *Journal of Biological Chemistry*, **266**, 7963–6.

Kallel-Mhiri, H. and Miclo, A. (1993) Mechanism of ethyl acetate synthesis by *Kluayceromyces fragilis*. *FEMS Microbiology Letters*, **111**, 207–12.

Karahadian, C. and Lindsay, R.C. (1987) Integrated role of lactate, ammonia and calcium in texture development of mold surface-ripened cheese. *Journal of Dairy Science*, **70**, 909–18.

Karahadian, C., Josephson, D.B. and Lindsay, R.C. (1985a) Volatile compounds from *Penicillium* sp. contributing to musty-earthy notes to brie and camembert cheese flavors. *Journal of Agricutural Food Chemistry*, **33**, 339–43.

Karahadian, C., Josenhan, D.B. and Lindsay, R.C. (1985b) Contribution of *Penicillium* sp. to the flavour of brie and camembert cheese. *Journal of Dairy Science*, **68**, 1865–77.

Kuzdzal, N. and Kuzdzal-Savoie, S. (1966) Etude comparée des acides gras non volatils libres et estérifiés dans les fromages. *Proceeding of the 17th International Dairy Congress, Munich*, Vol. D2, pp. 335–42.

Lamberet G., and Lenoir, J. (1972) Aptitude de l'espèce *Penicillium caseicolum* à la production d'enzymes lipolytiques. *Le Lait*, **52**, 175–92.

Lamberet, G. and Lenoir, J. (1976a) Les caractères du système lipolitique de l'espèce *Penicillium caseicolum*. Nature du système. *Le Lait*, **56**, 119–34.

Lamberet, G. and Lenoir, J. (1976b) Les caractères du système lipolitique de l'espèce *Penicillium caseiocolum*. Purification et propriétés de la lipase majeure. *Le Lait*, **56**, 622–44.

Lamberet G. and Lopez, M. (1982) Lipolytic activity in camembert type cheeses. *Proceeding of the 21st International Dairy Congress, Moscow*, Vol. 1 (1), p. 499.

Lamberet, G., Auberger, B., Canteri, C. and Lenoir, J. (1982) L'aptitude de *Penicillium caseicolum* à la dégradation oxydative des acides gras. *Revue Laitière Française*, **406**, 13–19.

Lee, C.W. and Richard, J. (1984) Catabolism of L-phenylalanine by some microorganisms of cheese origin. *Journal of Dairy Research*, **51**, 461–9.

Le Graet, Y., Lepienne, A., Brulé, G. and Ducruet, P. (1983) Migration du calcium et des phosphates inorganiques dans les fromages à pâte molle de type camembert au cours de l'affinage. *Le Lait*, **63**, 317–32.

Le Quéré, J.L., Demaizières, D., Negrello, C., Lesschaeve, I., Issanchou, S. and Salles, C. (1996) Goat cheese flavour. Identification of the character flavour impact compounds. *Proceeding of the IDF Symposium on ripening and quality of cheeses*, Besançon, p. 4.

Lenoir, J. (1963a) La flore microbienne du camembert et son évolution au cours de la maturation. *Le Lait*, **43**, 262–70.

Lenoir, J. (1963b) Note sur la composition en matières azotées des fromages affinés de camembert, Saint-Paulin et gruyère de Comté. *Annales de Technologie Agricole*, **12**, 51–7.

Lenoir, J. (1970) L'activité protéasique dans les fromages à pâte molle de type camembert. *Revue Laitière Française*, **275**, 231–6.

Lenoir, J. (1984) The surface flora and its role in the ripening of cheese. *Bulletin of International Dairy Federation*, no. 171, pp. 3–20.

Lenoir, J. and Choisy, C. (1970) Aptitude de l'espèce *Penicillium caseicolum* à la production d'enzymes protéolytiques. *Le Lait*, **51**, 138–57.

Martley, F.G. (1975) Comportement et rôle des streptocoques lactiques du levain en fabrication de camembert. *Le Lait*, **55**, 310–23.

Matsuoka, H., Fuka, Y., Kaminogawa, S. and Yamauchi, K. (1991) Purification and debittering effect of aminopeptidase II from *Penicillium caseicolum*. *Journal of Agricultural and Food Chemistry*, **39**, 1392–5.

Moinas, M., Groux, M. and Hormun, I. (1973) La flaveur des fromages: une méthodologie nouvelle d'isolement de constituants volatils. Application au roquefort et au camembert. *Le Lait*, **53**, 601–9.

Molimard, P. (1994) Etude de la coopération entre *Geotrichum candidum* et *Penicillium camemberti*: impact sur le profil aromatique et sur les qualités organoleptiques d'un fromage de type camembert. *Theis*, University of Dijon, France.

Molimard, P. and Spinnler, H.E. (1996) Compounds involved in the flavor of surface mold-ripened cheeses: origins and properties. *Journal of Dairy Science*, **79**, 769–84.

Molimard, P., Lesschaeve, I., Bouvier, I., Vassal L., Schlich, P., Issanchou, S. and Spinnler, H.E. (1994) Amertume et fractions azotées de fromages à pâte molle de type camembert: rôle de l'association de *Penicillium camemberti* avec *Geotrichum candidum*. *Le Lait*, **74**, 361–4.

Moreau, C. (1979) Nomenclature des *Penicillium* utiles à la préparation du camembert. *Le Lait*, **59**, 219–33.

Mourgues, R., Bergère, J.-L. and Vassal, L. (1983) Possibilités d'améliorer les qualités organoleptiques des fromages de camembert grâce à l'utilisation de *Geotrichum candidum*. *La Technique Laitière*, **978**, 11–15.

Nagao, T., Shimada, Y., Sugihara, A. and Tominaga, Y. (1993) Cloning and sequencing of two chromosomal lipase genes from *Geotrichum candidum*. *Journal of Biochemistry*, **113**, 776–80.

Ney, K.H. and Wiromata, I.P.G. (1971) Aliphatic monosamine in German and French camembert. *Lebensmittel-Wissenschaft und Technologie*, **146**, 343–4.

Noomen, A. (1977) Noordhollandse meshanger cheese: a model for research on cheese ripening. 2. The ripening of the cheese. *Netherlands Milk and Dairy Journal*, **31**, 75–102.

Noomen, A. (1978a) Activity of proteolytic enzymes in simulated soft cheeses (Meshanger type). 1. Activity of milk protease. *Netherlands Milk and Dairy Journal*, **32**, 26–48.

Noomen, A. (1978b) Activity of proteolytic enzymes in simulated soft cheeses: (Meshanger type). 2. Activity of calf rennet. *Netherlands Milk and Dairy Journal*, **32**, 49–68.

Noomen, A. (1983). The role of the surface flora in the softening of cheeses with a low initial pH. *Netherlands Milk and Dairy Journal*, **37**, 229–32.

Pyysalo, H. and Suihko, M. (1976) Odour characterization and threshold values of some volatile compounds in fresh mushroom. *Lebensmittel-Wissenschaft und Technologie*, **9**, 371–8.

Richard, J. (1984) Evolution de la flore microbienne à la surface des camemberts fabriqués avec du lait cru. *Le Lait*, **64**, 496–520.

Richard, J. and Zadi, H. (1983) Inventaire de la flore bactérienne dominante des camemberts fabriqués avec du lait cru. *Le Lait*, **63**, 25–42.

Roger, S., Degas, C. and Gripon, J.-C. (1988) Production of phenylethyl alcohol and its esters during ripening of traditonal camembert. *Food Chemistry*, **28**, 129–40.

Rothe, M., Engst, W. and Erhardt, V. (1982) Studies on characterization of blue cheese flavour. *Die Nahrung*, **26**, 591–602.

Rousseau, M. (1984) Study of the surface flora of traditional camembert cheese by scanning electron microscopy. *Milchwissenschaft*, **39**, 129–35.

Schmidt, J.L. (1982) Proteolytic activity of yeast isolated from camembert. *Proceedings of the 21st International Dairy Congress*, Moscow, Vol. 1 (2), p. 365.

Schmidt, J.L. and Lenoir, J. (1980) Contribution à l'étude de la flore levure du fromage de camembert. *Le Lait*, **60**, 272–82.

Spinnler, H.E., Grosjean, O. and Bouvier, I. (1992) Effect of culture parameters on the production of styrene (vinyl benzene) and 1-octene-3-ol by *Penicillium caseicolum*. *Journal of Dairy Research*, **59**, 533–41.

Stadhouders and Hup, G. (1975) Factors affecting bitter flavour in gouda cheese. *Netherlands Milk and Dairy Journal*, **29**, 335–53.

Tariq, V.M. and Campbell, V.M. (1991) Influence of volatile metabolites from *Geotrichum candidum* on other fungi. *Mycology Research*, **95**, 891–3.

Trieu-Cuot, P. and Gripon, J.-C. (1982) A study of proteolysis during camembert cheese ripening using isoelectric focusing and two-dimensional electrophoresis. *Journal of Dairy Research*, **49**, 501–10.

Van Belle, M., Vervack, W. and Foulon, M. (1978) Composition en acides gras supérieurs de quelques types de fromages consommés en Belgique. *Le Lait*, **58**, 246–60.

Vassal, L. and Gripon, J.-C. (1984) L'amertume des fromages à pâte molle de type camembert: rôle de la présure et de *Penicillium caseicolum*, moyens de la contrôler, *Le Lait*, **64**, 397–417.

Vassal, L., Monnet, V., Le Bars, D., Roux, C. and Gripon, J.-C. (1984) Relation entre le pH, la composition chimique et la texture des fromages de type camembert. *Le Lait*, **66**, 341–51.

Veeraragavan, K., Colpitts, T. and Gibbs, B.F. (1990) Purification and characterization of two distinct lipases from *Geotrichum candidum*. *Biochimica et Biophysica Acta*, **1044**, 26–33.

6 Flavour and texture in low-fat cheese
Y. ARDÖ

6.1 Introduction

In many countries several cheese varieties are made with reduced fat content to meet a general wish to decrease fat intake among the population. A cheese may be classified as low-fat only if the fat content is considerably reduced. A demand often expressed is a fat content half or less than that of normal-fat cheese. If the fat content is two-thirds or less of a corresponding normal-fat cheese, it is commonly referred to as a reduced-fat cheese. Decreasing the fat content in cheese has over time been surrounded by many problems that have been successfully overcome for some varieties but not for others. Well-known problems are firm and elastic or doughy texture as well as blend flavour or pronounced bitterness. The firmness and elasticity, likely to be more related to casein breakdown than to fat, may occur even in reduced-fat cheeses with the same moisture in non-fat solids (MNFS) as a normal-fat cheese (Emmons, 1980). In the reduced-fat cheese more protein matrix is present, requiring more effort of cutting or deformation in texture assessment.

A mild-flavoured Cheddar with a fat content of 25% may be produced by altering the holding times and temperatures during manufacture to establish similar MNFS as in normal-fat Cheddar (33% fat) (Banks *et al.*, 1989, 1994). Some chemical properties of the cheeses are changed in these experiments. Cheese with lower fat content contains lower concentrations of free short-chain fatty acids, C_4, C_6 and C_8, but not of the longer free fatty acids, C_{10}, C_{12}, C_{14} and C_{16}, as normal-fat Cheddar. The level of methyl ketones is considerably reduced in the lower-fat cheeses. Cheddar-type cheese with only 16% fat content made with similar MNFS fails to develop flavour.

Some different Swedish varieties of semi-hard, reduced-fat cheese containing two-thirds of the normal fat content and still fairly similar to traditional mild cheeses are successfully made by altering the production procedures to achieve similar MNFS, salt content and pH in the reduced-fat (17%) as in normal-fat (28%) cheeses (Ardö, 1993; Ardö and Gripon, 1995). The cheeses develop an attractive flavour with most of the notes present in the normal-fat cheese varieties. However, the texture of the reduced-fat cheeses, and especially of those with a fat content as low as 10%, easily develops too quickly and may become over-ripened before any sharp or intense flavour appears. Methods to overcome this include the addition of ripening cultures or enzymes. These additions are likely to introduce flavours uncharacteristic of the cheese varieties studied and are useful only in the development of new cheese varieties with reduced fat content. New Swedish cheese varieties with a fat content of 10% or less and with its own characteristic flavour has been developed this way (Ardö et al., 1989).

Reduced-fat Mozzarella is successfully made with retained stretch and melt characteristics by using homogenized milk, lowered preparation temperature and refrigerated storage (Thunick et al., 1993a, b). The stretchability and meltability depend on casein breakdown and especially the rennet cleavage of α_{s1}- to α_{s1}-I-casein. A lowered temperature (about 32°C instead of 46°C) causes a higher rennet activity in cheese. Increasing peptidolysis by changing starter also enhances the textural properties (Merril et al., 1994).

Smear bacteria-ripened cheese varieties are successfully made with reduced fat content (Hargrove et al., 1966; Ramanauskas, 1978). The surface bacteria produce thioesters that strongly influence flavour, and also ammonia-ions that soften the body by competing with calcium in the casein network, and also by increasing the pH.

6.2 Role of fat in cheese

The fat has a crucial role in preventing the casein network of the cheese body from shrinking into a tough, inedible structure. To obtain an attractive texture of a reduced-fat cheese, most commonly, water and/or whey are used to replace the fat in filling up the three-dimensional casein network. Sometimes different water-holding agents like salt or whey proteins are added to stabilize the structure. However, the cooking temperature has a greater influence on the moisture content of cheese than an addition of 6% concentrate of denatured whey protein to the cheese milk (Nes, 1980). The changes in the reduced-fat cheese manufacturing technique that have been needed are developed mostly within industry and kept as in-house knowledge that is not easily accessible. However, some significant material has been published, for instance, the effects of adding fresh buttermilk, reconstituted buttermilk or reconstituted skim milk to low-fat cheese milk (Madsen et al., 1966). Cheese moisture and yield increased with the additions but unfortunately a bitter flavour developed, making this technique limited for use in cheese varieties that need only a short period of ripening. Increasing the content of whey proteins in low-fat cheese using ultrafiltration (UF) techniques mediates a smoother consistency of the low-fat cheese, an increase in cheese yield and an accurate control of the moisture content in the cheese (de Boer and Nooy 1980). Native whey proteins are resistant to the proteolytic enzymes of rennet and starter, which prevent development of bitter peptides from them in cheese (de Koning et al., 1981). However, casein breakdown in cheese decreases with increasing concentration of the milk as a result of interactions between whey proteins and proteolytic enzymes, e.g. plasmin inhibition by beta-lactoglobulin (Bech, 1993). This may be overcome by adding plasmin or plasmin activator to cheese milk (c.f. Chapter 1) or introducing their genes into a cheese starter. UF is a promising technique for the production of new varieties of reduced-fat cheese.

Lipolysis of milk-fat in cheese produces flavour compounds among which some free fatty acids are the main contributers to the flavour of mould-ripened and hard Italian cheese varieties (Chapter 5). The flavour quality of these cheese varieties does not have to be lost when the fat content is reduced (Jameson, 1990). There is still enough fat for a relevant production of these flavour compounds in a reduced-fat cheese. The situation is likely to be the same in other reduced-fat cheese varieties where the products of lipolysis are only required in very low concentrations to give characteristic flavour.

Milk-fat is thought to contribute to cheese flavour by dissolving flavour compounds that may be produced from the hydrolysis of fat or protein, or in other ways. The fat is also beneficial for masking flavours, bitterness being an example. The amino acid composition of many bitter peptides comprises negatively charged amino acids at one end and mainly hydrophobic amino acids at the other and their arrangement at the interface between fat and water could hide bitterness. This emphasizes the concept that the less fat in a cheese, the greater the tendency for it to taste bitter.

Normal-fat cheese contains a greater area of lipid–water interface than does reduced-fat cheese. This specific part of the cheese could be important both for our flavour perception and for the microbial and enzymatic activities in the cheese. The interface contains the fat membrane materials; e.g. water-holding biphasic compounds such as phospholipids and phos-phopeptides, as well as enzymes and several other bioactive compounds. Experiments which introduced more fat membrane material (buttermilk preparations) to low-fat cheese have, however, only marginally improved flavour. Homogenization of the fat together with the membrane materials was needed to achieve any effect, indicating that an increase in the fat/protein/water interface area is the most important improvement (Mayes et al., 1994). The enzyme content of the buttermilk preparations used in these experiments was not determined.

The water is more easily evaporated from cheese containing less fat, and the body may become too dry during ripening. To overcome this problem, cheese varieties which normally are made with higher water content and are ripened in plastic films are chosen to be made with a reduced fat content. However, if this technology is adopted, it must be taken into consideration that high water activity in cheese stimulates bacterial activities and may lead to failure in flavour development due to retarded autolysis among starter bacteria, as well as increased growth of contaminant bacteria and moulds.

6.3 Consequences of measures taken to slow down syneresis

Controlling the syneresis process for the purpose of retaining sufficient water in the curds is essential for the production of reduced- and low-fat cheese. This is promoted by decreasing holding-times and temperatures during production, adding less rennet and calcium, and also by cutting and draining the curd at a higher pH. These changes in the cheese production procedure may influence several enzymes and bacteria that mediate cheese maturation, i.e. rennet, plasmin, starter and non-starter bacteria.

Higher pH at draining and/or less rennet added to cheese milk decrease the amount of rennet incorporated into a reduced-fat cheese. A lower temperature in cheese during the first 24 hours decreases the early activity of both rennet and plasmin. Lower plasmin activity than in normal-fat cheese may also be the result of less activation of plasmin during cooking at a lower temperature (Noomen, 1975; Richardson, 1983; Farkye and Fox, 1992). A lower activity on casein from rennet as well as from plasmin produces smaller amounts of large and medium-sized peptides that are in turn substrates for starter and non-starter enzymes. Rennet added to milk stimulates amino acid production mediated by non-starter bacteria (Hickey *et al.*, 1983). Due to the different enzymatic specificities of rennet and plasmin, the small peptides produced from the medium-sized to large peptides generated by plasmin activity on casein will be different, certainly more hydrophilic, than those peptides that are produced from rennet-

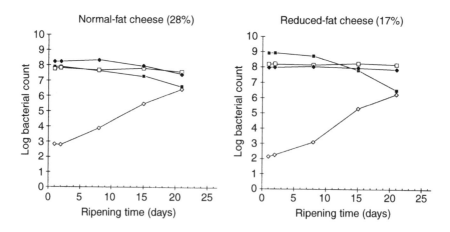

Figure 6.1 Bacterial growth in round-eyed, semi-hard cheese with normal (28%) and reduced (17%) fat content. The cheeses were made with a mesophilic mixed-strain DL-starter repeated five times. (From Ardö, 1993.). ■, *Lactococcus lactis* subsp. *lactis* and *cremoris* (starter); □, *Lactococcus lactis* subsp. *lactis* biovar. *diacetylactis* (starter); ◆, *Leuconostoc mesenteroides* subsp. *cremoris* (starter); ◇, Lactobacilli (non-starter).

generated, medium-sized peptides. These differences in peptide composition of ripened cheese should influence both flavour and texture.

Cooking the curd is not only done to obtain optimal syneresis, but also to restrict growth of starter bacteria and slow down the pH-development in the fresh cheese. It may also induce autolysis of the starter bacteria occuring later during maturation. The lower temperatures often used for production of reduced-fat cheese are closer to optimal growth temperatures of meso-philic starter bacteria than the temperatures used for normal-fat cheese. Bacterial growth in a reduced-fat cheese made with a mesophilic, mixed strain, aroma-producing (DL) starter differs in some ways from that in a normal-fat cheese (Figure 6.1). Up to ten times higher bacterial counts can be obtained in fresh reduced-fat cheeses as compared with normal-fat cheeses (Ardö, 1993). Starter bacteria rapidly growing in fresh cheese are likely to produce high amounts of proteolytic enzymes. Cheeses containing high amounts of lactococci are susceptible to develop bitterness (Lowrie *et al.*, 1972). A greater variety of peptides are produced by starter proteases (recently reviewed by Tan *et al.*, 1993) including bitter peptides of different size that contribute to cheese flavour, but also cause defects if they accumulate in high concentrations. A larger amount of small and medium-sized peptides similar to those produced by starter proteases has been found in the reduced- as compared with normal-fat cheeses (Ardö, 1994).

6.4 Casein breakdown in reduced-fat cheese

The fat in cheese as such does not influence the proteolytic activities (Rank, 1985; Banks *et al.*, 1989; Ardö, 1993, Ardö and Gripon, 1995). The fat content, however, has (as described above) an important indirect role for the ripening process due to changes in the production procedure. Some prob-lems resulting from the changes occur to different extents in reduced-fat cheese of different varieties. An example of this is that young Cheddar has lower pH, higher salt content, lower water content and lower temperature than many varieties of young semi-hard, round-eyed or open-texture cheese varieties. All these parameters influence primary proteolysis. Problems that may occur are:

- Insufficient casein breakdown;
- detrimentally increased activity of mesophilic starter proteases;
- insufficient autolysis of starter bacteria leading to a shortage of amino acids; and
- stimulated and uncontrolled growth of non-starter bacteria.

Firmer and more elastic texture of reduced-fat cheese has been explained by insufficient breakdown of the casein matrix and especially hydrolysis of α_{s1}-casein. This conclusion has been drawn from knowledge about the softening process occuring during maturation of normal-fat cheese (Creamer and Olson, 1982). The few results allowing comparisons between proteolysis in normal- and reduced-fat cheeses that have been published show a decreased casein breakdown during early ripening in cheeses with reduced fat-content (Banks *et al.*, 1989; Ardö, 1993), whereas a proportionally higher amount of α_{s1}-casein in reduced-fat cheese than in a normal-fat cheese would need increased casein breakdown for the development of an attractive, smooth cheese body.

As described above, the increased activity of mesophilic starter proteases may result from decreased cooking temperatures; conditions closer to optimum for these bacteria during cheese production are reached, so they grow to higher numbers and produce more proteases, attacking mainly medium-sized to large peptides in the cheese that have been produced by rennet and/or plasmin activity on casein. These activities may lead to increased amounts of bitter peptides.

The amino acids and small peptides are considered responsible for the background flavour found in most cheese varieties (Mulder 1952; Richardson and Creamer, 1973; Guigoz and Solms, 1974; Visser *et al.*, 1983; Cliffe *et al.*, 1993). Relatively low amounts of amino acids are found in some semi-hard reduced-fat cheeses, which could be a consequence of decreased autolysis due to lower cooking temperatures and/or a high content of MNFS of these cheese varieties (Ardö and Gripon, 1995). A balanced ripening process includes the hydrolysis of peptides to an extent that is greater than effected by normal bacterial metabolism. Peptides that are not able to enter the bacteria cells must be degraded directly in the cheese matrix. A high activity of intracellular peptidases on substrates outside the bacterial cells may occur first after cell lysis and enzyme leakage (c.f. Chapter 7). The amino acids resulting from these peptidolytic activities accumulate more or less in cheese during ripening. Amino acids are catabolized to a variable extent in different cheese varieties (Manning, 1979; Hemme *et al.*, 1982), which contribute to the characteristic cheese flavours and specific amino acid patterns for each cheese variety.

Higher redox potential, higher pH and higher water activity of reduced-fat cheeses may stimulate growth of several contaminants, e.g. clostridia (Ardö, 1993; Urbach, 1993).

6.5 Use of enzymes and bacteria to improve quality of reduced-fat cheese

Many problems in reduced-fat cheese technology may be overcome by altering cheese-making procedures as described above. To obtain the stronger flavours, or to make cheese with a fat content less than about two-thirds of normal, however, this is not sufficient. Four methods have been followed in attempts to enhance the flavour and control the quality of low-fat cheese: (i) to use free or microencapsulated ripening enzymes; (ii) to use attenuated bacterial cells which release enzymes to act freely in the cheese; (iii) to use adjunct starters growing in cheese; or (iv) to use specially designed, reduced-fat starters.

Activities needed to improve flavour and texture development in reduced-fat cheese are closely related to those being studied extensively for accelerating cheese ripening (these are dealt with in Chapter 7). Addition of ripening enzymes to the cheese-milk usually leads to the loss of part of the enzymes in the whey. A technique using enzymes in microcapsules made of milk-fat has been developed for production of Cheddar (Magee and Olson, 1981). It cannot, however, be applied to the production of semi-hard cheese varieties with cooking temperatures of 35–40°C or more, because the milk-fat would melt in the cheese vat. Multilamellar liposomes containing proteolytic enzymes may be added to a variety of cheeses in order to increase proteolysis and this technique may be applied to cheese production in the future (Law and King, 1985; Kirby et al., 1987; Skeie, 1994).

Bacterial cells may be attenuated by freezing, heating, chemical treatment and/or drying for the purpose of temporarily inactivating them metabolically, but not actually killing them. The technique of using bacterial cells attenuated by heat treatment to preserve the desired enzyme activities and inactivate undesired enzymes is used successfully for reduced-fat cheese (Ardö et al., 1989; Ardö, 1994; Skeie et al., 1995). The bacterial enzymes are kept entrapped in their heat-treated membranes during cheese manufacture and leak into the cheese during early ripening. Not only is the acidification activity controlled by the heat treatment but so also are unwanted proteolytic activities. A high production of desired, broad-specificity aminopeptidases can be found in thermophilic lactobacilli that do not grow to significant numbers at the low temperatures used in reduced-fat cheese production. However, their properties are excellent as heat-treated cultures added as pre-grown biomass, where growth at cheese temperatures is not necessary, or even not wanted. Rapid growth on cheap milk- or whey-based substrates in a fermenter is relatively easy to produce such biomass. (Ardö et al., 1989; Lopez-Fandiño and Ardö, 1991; Ardö and Jönsson, 1994). The results of adding a preparation of heat-treated Lb. helveticus to semi-hard, low-fat (10%) cheese are enhanced flavour, elimination of bitterness and improved texture.

A special advantage with the heat-treated lactobacilli is that the first step of proteolysis in cheese is hardly affected. The main function of the addition is to accelerate the breakdown of peptides produced in cheese by ordinary activities of rennet, plasmin and starter proteases. Consequently, the effect of this addition will be controlled by normal activities in cheese. Casein breakdown, measured as soluble N at pH 4.6 in the cheese, is only marginally increased by the addition of the heat-treated lactobacilli. However, the peptide composition of the water-soluble N fractions of cheese will be different, as demonstrated by HPLC (Ardö, 1994). Of special interest is that the amount of medium-sized to small, rather hydrophobic peptides is considerably decreased in the cheeses with the addition of attenuated lactobacilli; instead, larger amounts of amino acids and hydrophilic peptides are present. An explanation of these results is that the peptidases of the added lactobacilli hydrolyse the medium-sized peptides and produce smaller peptides and amino acids. Substantial changes in the peptide composition of the water-soluble cheese fraction has the potential to influence water-binding capacity of cheese, as well as flavour. Aminopeptidolytic enzymes are released from the lactobacilli in the cheese early during ripening. As a consequence, the concentration of amino acids and small peptides becomes twice as high in cheeses with added lactobacilli as in those without (Figure 6.2). A significant increase is achieved also in acetaldehyde content (Figure 6.3). Some other volatile compounds develop similarly, indicating further breakdown of the amino acids mediated by the added lactobacilli and/or stimulation of starter activities (Skeie *et al.*, 1995). The addition of heat-treated lactobacilli may stimulate the starter bacteria activities; results of several experiments indicate that activities of *Lactococcus lactis* subsp. *lactis* biovar. *diacetylactis* are stimulated by the heat-treated lactobacilli. These bacteria, which consume citrate and produce CO_2, are added to cheese-milk as one of the species of a mesophilic DL-starter, also containing *Lactococcus lactis* subsp. *lactis* and *cremoris* and *Leuconostoc mesenteroides* subsp. *cremoris*. The pH decreases slightly faster, gas production is increased, and

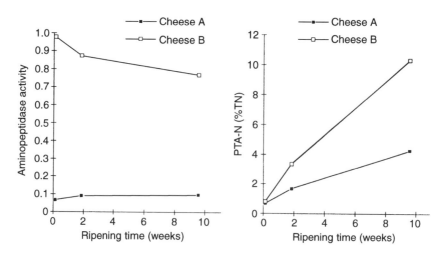

Figure 6.2 Phosphotungstic acid-soluble nitrogen (PTA-N) content and aminopeptidolytic activity (on leucine-*p*-nitroanilide) in the matrix of reduced-fat cheese without (cheese A) and with (cheese B) the addition of heat-treated lactobacilli. (PTA-N includes amino acids and small peptides.) (From Ardö, 1994.)

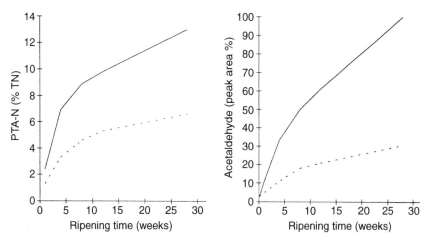

Figure 6.3 Influence of heat-treated *Lactobacillus helveticus* on the amount of small peptides and amino acids measured as phosphotungstic acid-soluble nitrogen (PTA-N) and acetaldehyde in low-fat cheese during ripening. (PTA-N includes amino acids and small peptides.) ——, cheese with lactobacilli; ....., control cheese. (From Ardö, 1994.)

in some cheeses with retarded starter activity on citrate, this was stimulated by the addition which by itself does not consume any citrate (Ardö and Pettersson, 1988). Adding heat-treated lactobacilli to reduced-fat Drabant cheese (17%), mediates a low but significant increase in the number of *Lc. diacetylactis*, but no changes of the total amount of *Lc. lactis* subsp. *lactis* and *cremoris* (Ardö, 1994).

Several bacterial strains have been tested as adjunct starters to enhance flavour in reduced-fat cheese. Experiments with addition of pediococci induced a sharp, aged, Cheddar flavour in Cheddar with 23% fat, while micrococci introduced bitterness (Bhowmik *et al.*, 1990). Lactobacilli have been used successfully as adjuncts to control adventitious metabolizing bacteria in Cheddar cheese (Broome *et al.*, 1990). Flavour development in low-fat Cheddar cheese has also been accelerated by the addition of lyophilized extracts from several cheese-related bacteria (Chen *et al.*, 1990; El Soda *et al.*, 1991). Many of these bacteria grow naturally in cheese but only poorly in milk and have special nutritional requirements that could be available in cheese after some time of ripening. This makes them interesting as viable adjunct starters when added in low numbers to the milk for the purpose of slow growth in cheese. They must be antagonistic to detrimental bacteria, tolerate pH, redox potential and salt content of cheese and they must grow at cheese-ripening temperatures. They should produce large amounts of aminopeptidases with broad specificity while growing in cheese. Some other important properties are the possibility for them to be cultivated in a fermenter, the absence of detrimental activities like decarboxylation (producing gas and amines) and, of course, the presence of desired flavour-producing activities such as amino acid catabolism. Their tendency to autolyse may also be important. Further research is needed to find the best strain or combination of strains for each cheese variety in question.

Bacterial strains to be used as starters in reduced-fat cheese should be carefully selected on the basis of their temperature sensitivity, autolytic properties and proteolytic/peptidolytic activities. The ideal situation is that the starter bacteria chosen should be influenced by the production temperatures used for reduced-fat cheese in a similar way as normal starters are during the production of normal-fat cheese, e.g. the autolysis should be induced at a lower temperature for a reduced-fat cheese starter. Another desired property is an increased tendency to autolysis and leak enzymes into the cheese matrix with relatively high MNFS. The stimulation of salt addition on early autolysis in Cheddar cannot be achieved in a semi-hard, brine-salted cheese. Some strains have been developed to increase proteolysis and enhance flavour of reduced-fat Cheddar cheese by increasing the number of starter bacteria without interfering with acidification of the curd, i.e. cultures of *Lactococcus lactis*, selected among natural mutants which are highly peptidolytic and lactase-negative for addition to cheese vat or Cheddar curd (Bech, 1992; Coulson *et al.*, 1992). The concept of eliminating only acidification activity (lactase-negative strains) to make it possible to increase the number of starter bacteria in cheese is hardly adequate for semi-hard, reduced-fat cheese. Also, common proteolysis must be restricted in those cheeses, while high and broad peptidolytic properties are desired.

Future research and development is expected to bring out the use of the variety of possibilities that exist to improve quality and should increase the number of low-fat cheese varieties on the market.

6.6   References

Ardö, Y. (1993) Characterizing ripening in low-fat, semi-hard round-eyed cheese made with undefined mesophilic DL-starter. *Int. Dairy J.*, **3**, 343–57.

Ardö, Y. (1994) *Proteolysis and its impact on flavour development in reduced-fat semi-hard cheese made with mesophilic undefined DL-starter*, PhD Thesis, Pure and Applied Biochemistry, Lund University, Sweden.

Ardö, Y. and Gripon, J.-C. (1995) Comparative study of peptidolysis in some semi-hard round-eyed cheese varieties with different fat contents. *J. Dairy Res.*, **62**, 543–7.

Ardö, Y. and Jönsson, L. (1994) A study of chromatographic profiling of aminopeptidolytic activities in lactobacilli as a tool for strain identification. *J. Dairy Res.* **61**, 573–9.

Ardö, Y. and Pettersson, H.-E. (1988) Accelerated cheese ripening with heat-treated cells of *Lactobacillus helveticus* and a commercial proteolytic enzyme. *J. Dairy Res.*, **55**, 239–45.

Ardö, Y., Larsson, P.-O., Lindmark Månsson, H. and Hedenberg, A. (1989) Studies of peptidolysis during early maturation and its influence on low-fat cheese quality. *Milchwissenschaft*, **44**, 485–90.

Banks, J.M., Brechany, E.Y. and Christie, W.W. (1989) The production of low fat Cheddar-type cheese. *J. Soc. Dairy Technol.*, **42**, 6–9.

Banks, J.M., Hunter, E.A. and Muir, D.D. (1994) Sensory properties of Cheddar cheese: effect of fat content on maturation. *Milchwissenschaft*, **49**, 8–12.

Bech, A.-M. (1992) Enzymes for the acceleration of cheese ripening. *Bull. I.D.F.*, **269**, 24–8.

Bech, A.-M. (1993) Characterizing ripening in UF-cheese. *Int. Dairy J.*, **3**, 329–42.

Bhowmik, T., Riesterer, R., Van Boekel, M.A.J.S. and Marth, E.H. (1990) Characteristics of low-fat Cheddar cheese made with added *Micrococcus* or *Pediococcus* species. *Milchwissenschaft*, **45**, 230–5.

Broome, M.C., Krause, D.A. and Hickey, M.W. (1990) The use of non-starter lactobacilli in Cheddar cheese manufacture. *Aus. J. Dairy Technol.*, November, 67–73.

Chen, C., El Soda, M., Riesterer, B. and Olson, N. (1990) Acceleration of low-fat Cheddar cheese ripening using lyophilized extracts from several cheese-related microorganisms. *Brief Communications of the XXII International Dairy Congress, Montreal*, Vol. II, p. 332.

Cliffe, A.J., Marks, J.D. and Mulholland, F. (1993) Isolation and characterization of non-volatile flavours from cheese: peptide profile of flavour fractions from Cheddar cheese, determined by reverse-phase high-performance liquid chromatography. *Int. Dairy J.*, **3**, 379–87.

Coulson, J., Pawlett, D. and Wivell, R. (1992) Accelerated ripening of Cheddar cheese. *Bull. I.D.F.*, **269**, 29–35.

Creamer, L.K. and Olson, N.F. (1982) Rheological evaluation of maturing Cheddar cheese. *J. Food Sci.*, **47**, 631–6, 646.

de Boer, R. and Nooy, P.F.C. (1980) Low-fat semi-hard cheese from ultrafiltrated milk. *Nordeuropaeisk mejeri tidsskrift*, **3**, 52–61.

de Koning, P.J., Boer, de R., Both, P. and Nooy, P.F.C. (1981) Comparison of proteolysis in a low-fat semi-hard type of cheese manufactured by standard and by ultrafiltration techniques. *Neth. Milk Dairy J.*, **35**, 35–46.

El Soda, M., Chen, C., Riesterer, B. and Olson, N. (1991) Acceleration of low-fat Cheddar cheese ripening using lyophilized extracts of freeze shocked cells of some cheese-related microorganisms. *Milchwissenschaft*, **46**, 358–60.

Emmons, D.B., Kalab, M., Larmond, E. and Lowrie, R.J. (1980) Milk gel structure. X. Texture and microstructure in Cheddar cheese made from whole milk and from homogenized low-fat milk. *J. Texture Studies*, **11**, 15–34.

Farkye, N. and Fox, P.F. (1992) Contribution of plasmin to Cheddar cheese ripening: effect of added plasmin. *J. Dairy Res.*, **59**, 209–16.

Guigoz, Y. and Solms, J. (1974) Isolation of a bitter tasting peptide from "Alpkäse", a Swiss mountain-cheese. *Lebensmittel-Wissenshaft und-Technologie*, **7**, 356–7.

Hargrove, R.E., McDonough, F.E. and Tittsler, R.P. (1966) New type of ripened low-fat cheese. *J. Dairy Sci.*, **49**, 796–9.

Hemme, D., Bouillanne, C., Métro, F. and Desmazeaud, M.J. (1982) Microbial catabolism of amino acids during cheese ripening. *Sciences des Aliments*, **2**, 113–23.

Hickey, M.W., Hillier, A.J. and Jago, G.R. (1983) Peptidase activities in lactobacilli. *Aus. J. Dairy Technol.*, **38**, 118–23.

Jameson, G.W. (1990) Cheese with less fat. *Aus. J. Dairy Technol.*, **45**, 93–8.

Kirby, C., Brooker, B.E. and Law, B.A. (1987) Accelerated ripening of cheese using liposome-encapsulated enzymes. *Int. J. Food Sci. Technol.* **22**, 355–75.

Law, B.A. and King, J.S. (1985) Use of liposomes for proteinase addition to Cheddar cheese. *J. Dairy Res.*, **52**, 183–8.

López-Fandiño, R. and Ardö, Y. (1991) Effect of heat treatment on the proteolytic/peptidolytic enzyme system of a *Lactobacillus delbrückii* subsp. *bulgaricus* strain. *J. Dairy Res.*, **58**, 469–75.

Lowrie, R.J., Lawrence, R.C., Pearce, L.E. and Richards, E.L. (1972) Cheddar cheese flavour. III. The growth of lactic streptococci during cheesemaking and the effect on bitterness development. *N.Z. J. Dairy Sci. Technol.*, **7**, 44–50.

Madsen, F.M., Reinbold, W. and Warren S.C. Jr. (1966) Low-fat cheese. *Manufactured Milk Products Journal*, **57**, 18–22.

Magee, E.L. and Olson, N.F. (1981) Microencapsulation of cheese ripening systems: formation of microcapsules. *J. Dairy Sci.*, **64**, 600–10.

Manning, D.J. (1979) Chemical production of essential Cheddar flavour compounds. *J. Dairy Res.*, **46**, 531–7.

Mayes, J.J., Urbach, G. and Sutherland, B.J. (1994) Does addition of buttermilk affect the organoleptic properties of low-fat Cheddar cheese. *Aus. J. Dairy Technol.*, **49**, 39–41.

Merill, R.K., Oberg, C.J. McMahon, D.J. (1994) A method for manufacturing reduced-fat Mozzarella cheese. *J. Dairy Sci.*, **77**, 1783–9.

Mulder, H. (1952) Taste and flavour forming substances in cheese. *Neth. Milk Dairy J.*, **6**, 157–68.

Nes, Å.M. (1980) Mager ost av goudatype ysta med tilsetjing av myseproteiner til ystemjolka. *Meieriposten*, **2**, 35–43.

Noomen, A. (1975) Proteolytic activity of milk protease in raw and pasteurized cow's milk. *Neth. Milk Dairy J.*, **29**, 153–61.

Ramanauskas, R. (1978) Characteristics of the ripening of new types of semi-hard cheeses with smeary rind. *Brief communications of the XX International Dairy Congress, Paris*, Vol. E, pp. 777–8.

Rank, T. (1985) Proteolysis and flavour development in low-fat and whole milk Colby and Cheddar-type cheeses. *Dissertation Abstracts International*, **62**, 132–3.

Richardson, B.C. (1983) The proteinases of bovine milk and the effect of pasteurization on their activity. *N.Z. J. Dairy Sci. Technol.*, **18**, 233–45.

Richardson, B.C. and Creamer, L.K. (1973) Casein proteolysis and bitter peptides in Cheddar cheese. *N.Z. J. Dairy Sci. Technol.*, **8**, 46–51.

Skeie, S. (1994) Developments in microencapsulation science applicable to cheese research and development. A review. *Int. Dairy J.* **4**, 573–95.

Skeie, S., Narvhus, J.A., Ardö, Y. and Abrahamsen, R. (1995) Influences of liposome-encapsulated Neutrase and heat-treated lactobacilli on the quality of low fat Gouda type cheese. *J. Dairy Res.*, **62**, 131–9.

Tan, P.S.T., Poolman, B. and Konings, W.N. (1993) Proteolytic enzymes of *Lactococcus lactis*. *J. Dairy Res.*, **60**, 269–86.

Thunick, M.H., Mackay, K.L., Shieh, J.J., Smith, P.W., Cooke, P. and Malin, E.L. (1993a) Rheology and microstructure of low-fat Mozzarella Cheese. *Int. Dairy J.*, **3**, 649–62.

Thunick, M.H., Malin, E.L., Smith, P.W., Shieh, J.J., Sullivan, B.C., Mackay, K.L. and Holsinger, V.H. (1993b) Proteolysis and rheology of low fat and full fat Mozzarella cheeses prepared from homogenized milk. *J. Dairy Sci.*, **76**, 3621–8.

Urbach, G. (1993) Relations between cheese flavour and chemical composition. *Int. Dairy J.*, **3**, 389–422.

Visser, S., Slangen, K.J., Hup, G. and Stadhouders, J. (1983) Bitter flavour in cheese. 3. Comparative gel-chromatographic analysis of hydrophobic peptide fractions from twelve Gouda-type cheeses and identification of bitter peptides isolated from a cheese made with *Streptococcus cremoris* strain HP. *Neth. Milk Dairy J.*, **37**, 181–92.

7 Control and enhancement of flavour in cheese
M. EL SODA

7.1 Introduction

Cheese manufacture can be defined in relatively simple terms as the removal of moisture from a curd obtained enzymatically or isoelectrically. The process involves the concentration of the main milk components, casein and fat by 6- to 12-fold. This concentration process is regulated by different factors including pH, temperature, time, agitation, etc. The resulting cheese is rubbery and essentially flavourless.

The next manufacturing step begins when the cheese is placed in the curing rooms. It is during ripening that each type of cheese develops its characteristic flavour and texture due to a complex series of biochemical reactions (Figure 7.1) resulting from the action of the following agents:

- The proteolytic agents: plasmin, the indigenous milk proteinase, the residual coagulant, as well as the proteinase and peptidase systems of the starter and non-starter microflora.
- The lipolytic agents: most of the changes occurring in the fat can be referred to the action of the lipases and esterases of the starter and non-starter microorganisms. Milk lipase probably has a very limited role.
- The degradation of lactose and citrate is mainly due to the enzyme systems from the added starter cultures or the microorganisms present in the cheese environment.

Cheese ripening time varies with each cheese type from 4 weeks for soft varieties like Camembert, to up to 4 years for the very hard types like Parmesan (Table 7.1). It is also during ripening that special characteristics can develop, i.e. blue veining in Roquefort, the holes in Swiss varieties, the red smear on Limburger, or the white mould on Camembert. During ripening, the conditions of temperature and humidity are carefully controlled to promote the development of the desirable microbial flora and the secretion of the enzymes responsible for the biochemical changes taking place during ripening. As a general rule, a temperature ranging from 4 to 12°C and a relative humidity of 80–90% are used for most cheeses.

The ripening of cheese is a slow and consequently an expensive process that is still not fully controllable. The cost of cheese ripening and aging is quite high; it was reported that the aging time of Cheddar cheese adds significantly to product costs ranging from 1.5 to 3 cents per month per

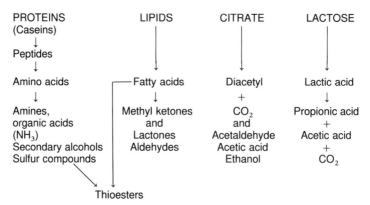

Figure 7.1 Metabolic pathways leading to flavour compounds formation in cheese.

pound of cheese aged (Freund, 1986). The development of an efficient way to reduce the aging time would allow substantial cost savings to the cheese industry.

Due to the economical and academic importance of controlling and enhancing the flavour of cheese, extensive work was published on the subject. An analysis of the publications (El Soda, 1993) indicated that work in this area was rather limited up to the early seventies, since then, more than 300 articles were published. The general trend for these publications indicate that in the 1980s most of the work was directed towards the addition of free enzymes obtained from various sources to the cheese milk or curd. In the late 1980s and early 1990s, interest was in favour of methods using whole cells of lactic acid bacteria in an attenuated form. This situation is likely to continue.

Table 7.1 The ripening time for different varieties of cheese

Cheese variety	Ripening time
Parmesan	2–4 years
Cheddar	9–12 months
Provolone	12 months
Kashkaval	10 months
Romano	5–12 months
Gruyère	8–12 months
Danablu	6–8 months
Cheshire	6–9 months
Wensleydale	6–12 months
Beaufort	6–8 months
Stilton	6–12 months
Brie	4–5 months
Tilster	3–5 months
Ras	3–4 months

7.2 Strategies used for the enhancement of cheese flavour

The methods used to reduce the ripening of cheese and/or enhance flavour formation will be described, and the advantages and disadvantages of each method will be considered.

7.2.1 Elevated ripening temperature

This method offers the simplest approach to reduce cheese ripening and was attempted for the first time when Sanders *et al.* (1946) obtained a fully matured cheese in 3–4 months if the temperature was raised to 16°C.

More recently, Law *et al.* (1979) demonstrated that an increase of the ripening temperature from 6°C to 13°C led to a 50% increase in flavour formation.

Studies in Australia have further contributed to the use of elevated ripening temperature as a means of accelerated ripening. Aston *et al.* (1985), in an intensive study where different temperature regimes were adopted, concluded that the minimum temperature at which cheese could be stored for 32 weeks without reduction in quality was 15°C. The combined action of elevated ripening temperature and increased mutant starter population (lactose-negative; Lac⁻), (proteinase-negative; Prt⁻), and addition of Neutrase to cheese curd was studied by Fedrick *et al.* (1986b) as an approach to accelerate the maturation. According to that study, the elevated temperature (16°C versus the usual 8°C) accounted for more than 50% of the reduction in the maturation time. The authors speculated that the caseinolysis, peptidolysis, and flavour development from the amino acid precursors all could have been affected by the increase in temperature.

The use of Neutrase, FlavourAge and increased rennet levels in combinations with elevated ripening temperature were evaluated by Guinee *et al.* (1991) and Wilkinson *et al.* (1992) for their effect on proteolysis and flavour development in Cheddar cheese. The obtained results could be summarized as follows:

- The addition of freeze-shocked mutant starter gave moderate increase in proteolysis.
- A synergistic effect on proteolysis was noted between Neutrase and freeze-shocked starter.
- At a temperature higher than 10°C and in the presence of Neutrase or FlavourAge the obtained cheese exhibited a soft texture in addition to a bitter and/or soapy flavour.

The conclusion reached by most authors in this area indicates that, although increasing the ripening may offer to the cheese-maker a technologically simple method to speed up flavour-forming reactions in a cheese system, a great deal of attention should be given to the quality of the milk and the hygienic conditions used for cheese production to avoid flavour defects and the potential growth of pathogens.

7.2.2 Slurry systems

A cheese curd slurry has been recently defined as a semi-solid paste containing about 40% solids and possessing the characteristic flavour of a particular cheese used in its preparation (Thakar and Upadhyay, 1992). The process involves the addition of water and salt to cheese. Trace elements and reduced glutathione are also used as additives. The slurry is incubated under anaerobic conditions at 30°C for 4–5 days with agitation. The method was developed by Kristoffersen et al. (1967); since then, several additives were evaluated to intensify typical flavour development, the most effective being glutathione. Other additives included citrate, manganese sulphate, riboflavin and cobalt. Glutathione was reported to enhance the release of peptides in the slurries, probably through the activation of certain enzymes (Harper et al., 1971). It was also reported that glutathione-treated slurries exhibited higher esterase activity (Harper et al., 1980). The mechanism of flavour enhancement in the slurries is not fully understood, but would probably be due to the enzymes of lactic acid bacteria present in the cheese used in their manufacture. The role of lactic acid bacteria in the flavour development of slurries was clearly demonstrated by Singh and Kristoffersen (1971, 1972). They suggested that if curd prepared by a direct acidification method is to be used for slurry preparation, it is necessary to add 5% of an active mixed culture for the development of a typical Cheddar-like flavour in 8 days at 30°C. More recently it was demonstrated that a slurry prepared by direct acidification contained only carbonyl sulphide, whereas the use of lactic culture in slurry preparation also produced hydrogen sulphide, methanethiol and dimethyl-sulphide. These sulphur compounds are important for the development of the typical flavour of Cheddar cheese (Ponce-Trevino et al., 1987; see also Chapter 8 of this volume).

In addition to their role in the development of typical cheese flavour in the slurries, Cliffe and Law (1990) demonstrated that lactic acid bacteria can also act as debittering agents in slurries. The authors noticed the presence of many very late-eluting peaks using reverse-phase high performance liquid chromatography, indicating hydrophobicity and therefore bitterness in a Cheddar cheese slurry treated with Neutrase. The bitterness disappeared when the slurry was treated further with a peptidase-containing extract from Lactococcus lactis.

A recent development to the slurry system is the attempt by Revah and Lebeault (1989) to produce blue cheese flavour in granular curds fermented with Penicillium roqueforti to which lipase or lipolysed cream was added. The authors report an 8- to 10-fold higher concentration of carbonyls than in traditional cheese. Sterilized cream and butter oil were also used as a substrate for the screening of six commercial lipolytic enzyme preparations with the aim of producing a blue cheese concentrate (Tomasini *et al.*, 1993). The results obtained indicate that UHT cream was the best substrate for the production of free fatty acids necessary for the preparation of the flavour concentrate, while the lipase of *P. roqueforti* was shown to be the best enzyme preparation tested for this purpose.

Cheese slurries have seen a wide range of applications. They are used in processed cheese formulations, snacks, crackers and imitation dairy products. They are also an ingredient in the production of enzyme-modified cheese (Moskowitz, 1980) as well as being a screening medium to select proteinases and/or peptidases to be used in the accelerated ripening of cheese (Law, 1990).

The major disadvantage of this technology is the difficulty of controlling the process. Incubation at 30°C can induce the growth of contaminants.

7.2.3 Addition of enzymes

The biochemical reactions leading to the formation of flavour compounds in cheese are thought to result from the combined action of the ripening agents. It was therefore logical to see that more than 45% of the work on accelerated cheese ripening accomplished before 1990 was directed towards the evaluation of the influence of one or more enzyme on speeding-up the ripening process.

Table 7.2 lists the different commercial enzymes evaluated, and their sources. Attention was also drawn towards the enzymes extracted from cheese-related microorganisms (intracellular as well as extracellular enzymes) in a crude or partially purified form. These attempts are listed in Table 7.3.

The conclusions reached could be summarized as follows:

- For *β*-galactosidase it is difficult to determine the precise role of the enzyme in reducing the maturation time. It is also more likely that the positive effect of *β*-galactosidase was due to the presence of contaminant proteinases in some of the commercial preparations tested (Hemme *et al.*, 1978).
- Rennet paste (containing pre-gastric esterases) as well as commercial lipase preparations are successfully used in many countries around the world for the production of cheese varieties known for their characteristic piquant flavour. Provolone, Caciocavallo, some blue-veined varieties, the Egyptian Ras cheese and Feta cheese made by ultrafiltration are examples for such cheeses.
- Proteinase addition was in most cases accompanied by bitter flavour development, a softer body and a marked reduction in cheese yield. The problem was overcome to some extent by the use of a mixture of proteinases and peptidases (Law and Wigmore, 1983).

Table 7.2   Enzymes used to enhance cheese flavour

Enzyme	Source or composition	Manufacturer
β-galactosidase		
Maxilact	*Kluyveromyces lactis*	Gist Brocades
Lactozym	*Kluyveromyces fragilis*	NOVO
Proteinases		
Neutrase	*Bacillus subtilis*	NOVO
Maxatase	*Bacillus subtilis*	Gist Brocades
Maxazyme	*Bacillus subtilis*	Gist Brocades
Rulactine[a]	*Micrococcus caseolyticus*	Roussel Uclaf
Corolase	*Aspergillus* sp.	Rohm Tech
Prozyme	*Aspergillus* sp.	Amano
Acid proteinase	*Aspergillus oryzea*	
Lipases		
Italase	Animal	Dairyland-Sanofi
Capalase	Animal	Dairyland-Sanofi
Kid lipase	Animal	Hansen's Laboratories
Lamb lipase	Animal	Hansen's Laboratories
Palatase 750 L	*Aspergillus niger*	NOVO
Palatase 200	*Mucor miehei*	NOVO
Piccantase	*Mucor miehei*	Gist Brocades
Enzyme mixtures		
Naturage[a]	Protease + peptidase + culture	Miles
FlavourAge	Protease + lipase from *Aspergillus oryzea*	Hansen's Laboratories
Accelase	Protease + peptidases	Imperial Biotechnology

[a]Production discontinued.

Plasmin was also considered as an accelerated ripening agent: Farkye and Fox (1991) reported that cheese with added plasmin inhibitor, 6-aminohexanoic acid, showed different electrophoretic protein patterns as well as different levels of water-soluble nitrogen when compared with the control cheese, suggesting that plasmin plays a role during maturation. The same authors (Farkye and Fox, 1992) evaluated the effect of adding plasmin to cheese-milk on the rate of maturation. It was concluded that, when added to cheese milk, most of the plasmin is retained in the curd, and no activity was found in the whey. Plasmin addition also led to a 20% increase in the level of soluble N in cheese, which had better organoleptic properties and was not bitter. Comparable results were also obtained by Farkye and Landkammer (1992). However, the cost of plasmin represented an obstacle for industrial application and the authors suggested the activation of milk plasminogen to substitute for added plasmin.

Enzyme mixtures seem to be more effective than the addition of a single proteinase due to the fact that the addition of a single enzyme very often leads to a disturbance of the flavour components in a cheese system and

Table 7.3 Cheese-related bacterial enzymes used to enhance flavour development in cheese

Microorganism	Mode of addition	Effect on cheese ripening	References
Lc. lactis	Cell homogenate	Liberation of amino acids in cheese	Gripon et al. (1977)
Lb. casei	Cell-free extract	Accelerated ripening Bitter flavour defect	El Soda et al. (1981)
Lb. delbruckii subsp. bulgaricus Lb. delbruckii subsp. lactis Lb. helveticus	Cell-free extract	Acelerated ripening Bitter flavour defect	El Soda et al. (1982)
Lc. lactis subsp. lactis	Cell-free extract + Neutrase	Accelerated ripening No flavour defect	Law and Wigmore (1983)
Lb. casei	Cell-free extract	Acceleration of Domiati cheese ripening (50%)	Mashaly et al. (1986)
Lb. casei Lb. plantarum Lb. brevis	Cell free extract	Acceleration of Domiati cheese ripening	El Soda et al. (1986)
Brevibacterium linens	Partly purified extracellular aminopeptidase + Neutrase	Accelerated ripening of Cheddar cheese No flavour defect	Hayashi et al. (1990a)
Brevibacterium linens	Partly purified extracellular serine proteinase	Accelerated ripening Flavour enhancement	Hayashi et al. (1990b)
Propionibacterium shermanii Brevibacterium linens Lb. delbruckii subsp. bulgaricus	Crude extract + Neutrase	Bitter flavour defect in Ras cheese	Ezzat (1990)
Lb. casei Lb. helveticus	Cell free extract + Neutrase or Rulactine	Accelerated of Ras cheese ripening Flavour enhancement	El Soda et al. (1990)
Lb. casei	Homogenized cells	Increase in amino nitrogen of Cheddar cheese Debittering effect	Trépanier et al. (1991)
Lb. casei, Pediococcus Propionibacterium sp. Leu. mesenteroidis	Lyophilized extract	No effect on Cheddar cheese ripening	El Soda et al. (1991)

causes flavour defects. Enzyme manufacturers have developed enzyme mixtures as the basis for several products, obtained from selected strains of microorganisms; Arbige *et al.* (1986) prepared a lipase–proteinase mixture from *Aspergillus oryzae* and the product is commercialized under the brand name FlavourAge. Accelase is another commercial preparation available in the market containing several aminopeptidases from *Lc. lactis* subsp *lactis*.

Different formulations of the products, either for the acceleration of cheddar cheese ripening or for the generation of the flavour and texture of traditional cheddar types in low-fat cheeses, are offered by Imperial Biotechnology, and both preparations were referenced and evaluated by Wilkinson (1990). The authors concluded that proteolysis was highest with added Flavourage, intermediate in Accelase-treated cheese, and lowest in the control. The same trend was found for concentrations of low-molecular weight peptides and amino acids. Sensory evaluation of the cheeses indicated little acceleration of flavour development by any of the treatments. In the case of Flavourage, off-flavours were detected.

One of the major limitations in the use of free enzymes is the method of enzyme addition. If the enzymes are added to milk, only a very small portion is retained in the curd, which increases costs. Addition of proteinases to milk also results in a reduction in cheese yield and flavour defects from proteolysis during manufacture and in the early stages of maturation. The resultant cheese whey is contaminated with active enzymes and may be unsuitable for certain food applications.

Addition of enzymes to curd is efficient only in the case of Cheddar-type cheeses which enable the addition of enzymes with the salt during milling of the curd. Even with Cheddar cheese, it is often difficult to ensure an equal distribution of the salt–enzyme mixture in the curd without 'hot spots' where excessive proteolysis and lipolysis occur. The use of entrapped enzymes and attenuated bacterial cells has been suggested to overcome these problems.

7.2.4   Entrapped enzymes

Two methods were adopted for enzyme entrapment and are discussed here. The first is liposome technology. Enzyme entrapment in liposomes has been used to overcome the problems associated with the use of free enzymes. The advantages of this approach can be summarized as follows:

- Phospholipid vesicles protect the milk protein substrates from enzymatic action until after the cheese has been formed and pressed. They could therefore avoid bitter flavour formation during ripening and losses in yield which results from the direct addition of an exogenous proteinase to cheese milk.
- The resulting cheese whey is free from exogenously added enzymes.
- The substances from which they are normally composed are found in high concentrations in most foods, which make them food-acceptable.
- Because liposome technology offers the possibility of preparing a wide range of vesicles varying in size, net charge, sensitivity to pH and/or temperature, liposomes of variable stability can be obtained.
- Methods are also available for their industrial preparation.

An overview of liposome utilization for the accelerated ripening of cheese is given in Table 7.4. Several types of liposomes have been engineered for the acceleration of cheese ripening with encapsulation efficiencies varying from 1% for small unilamillar vesicles to 55% for dehydration rehydration

Table 7.4 Liposome-entrapped molecules for flavour enhancement in cheese

Entrapped molecule	Type of liposome and encapsulation efficiency	Effect on cheese ripening	References
Rulactine	7% (MLV), 1% (SUV), 7% (REV)	Significant degradation of β-casein in Saint-Paulin cheese	Piard et al. (1986)
Neutrase	1–2% (MLV)	Protection of milk proteins during curd formation	Law and King (1985)
	17% (MLV), 30% (REV)	Bitter-quality cheese obtained with REV	Alkhalaf et al. (1988)
	37% (DRV)	Accelerated ripening of Cheddar cheese No flavour defects	Kirby et al. (1987)
	22% (DRV)	Bitter flavour defect in Taleggio cheese	Scolari et al. (1993)
Corolase PN	11% (MLV), 14% (REV), 4% (SUV)	Bitter flavour defect in Gouda cheese made from ultrafiltered milk	Spangler et al. (1989)
Trypsin	10% (MLV), 14% (ML/MF)	Enhanced proteolysis	Larivière et al. (1991)
Chymosin	13% (DRV)	Enhanced degradation of α_s-casein and β-casein during ripening	Picon et al. (1994)
Lb. casei extract Lb. helveticus extract	1–2% (SUV)	Little effect on flavour development in Cheddar cheese	El Soda et al. (1983, 1984)
Lb. helveticus extract	19% (MLV)	Little effect on flavour development in Gouda cheese made from ultrafiltered milk	Spangler et al. (1989)

SUV, small unltilamellar vesicles; MLV, multilamellar vesicles; REV, reverse-phase evaporation vesicles; DRV, dehydrated–rehydrated vesicles; MLV/MF, microfluidized MLV; MF, microfluidized.

vesicles. The retention of the vesicles in the cheese (not shown in the table) varies (from 17% to 90%) and seems to be related to liposome size (Piard et al., 1986).

Concerning the impact of liposome addition on the flavour and texture of the cheese, in most studies, the authors reported that proteolysis was significantly enhanced in the liposome-treated cheese when compared with the control (Law and King, 1985; Piard et al., 1986; Kirby et al., 1987; Alkhalaf et al., 1988, 1989; Spangler et al., 1989; Larivière et al., 1991; Scolari et al., 1993; Shehata et al., 1995). When Alkhalaf et al. (1988) compared the release of Neutrase from multilamellar vesicles (MLV) and reverse-phase evaporation vesicles (REV), they suggested that REV were more stable in the cheese environment, thus leading to a slower release of Neutrase which in turn led to a better protection of the milk proteins during coagulation and draining. Sensory evaluation of the cheese also indicated that REV prevented the development of bitterness. On the other hand, MLV-treated cheeses were more crumbly and showed a pronounced bitter flavour. Spangler *et al.* (1989) succeeded in improving the performances of MLV by using a combination of liposome-entrapped Corolase PN and freeze-shocked *Lactobacillus helveticus* to enhance the flavour of Gouda cheese made from ultrafiltered milk. This study shows that the combination of a gross proteolytic agent (Corolase PN) and a source of exopeptidases (the frozen cells of *Lb. helveticus*) capable of releasing low-molecular weight peptides and amino acids, which are precursors of flavour compounds, appears to be an efficient means of reducing the ripening time of cheese made from ultrafiltered milk.

Several studies were undertaken for a better understanding of the behaviour of the vesicles in a cheese system.

Law and King (1985) clearly demonstrated that liposomes protected the milk proteins from proteinase attack during curd formation. They also showed that the enzyme was released into cheese during ripening and led to an increased hydrolysis of β-casein.

Alkhalaf *et al.* (1989) studied the influence of liposome surface charge on the rate of encapsulation of Neutrase and on the rate of liposome retention in Saint-Paulin cheese. The encapsulation efficiency of ^{14}C-labelled Neutrase in positively charged REV was 32%, while it was 26% and 19% respectively for negatively charged and neutral liposomes. Proteinase adsorption on the liposome surface was found to be very low, i.e. less than 0.5% in the case of positive liposomes, and about 2% for negative liposomes. The level of liposome retention in Saint-Paulin cheese was 42, 31 and 24% respectively for positive, negative and neutral liposomes. The higher retention of the positively charged liposomes was attributed to the formation of electrostatic bonds with casein micelles at pH 6.5. Monitoring the rate of protein breakdown during ripening indicated a faster degradation of β-casein in the case of the cheese treated with negatively charged liposomes. This is in agreement with the finding that negatively charged vesicles are less stable than positively charged liposomes. The authors also determined the stability of their liposome preparations as a function of pH, temperature and sodium chloride concentration. The results indicated a faster release of ^{14}C-labelled Neutrase when pH, temperature or NaCl concentration were increased.

An electron microscopy study accomplished by Kirby *et al.* (1987) revealed that liposomes are grouped in clusters in areas between the globules and the casein matrix.

The controlled release of enzymes into cheese was attempted by El-Soda *et al.* (1989) using temperature-sensitive liposomes made from dipalmitoyl phosphatidylcholine. This lipid has a transition temperature of 41°C and is stable at temperatures below 35°C. It was observed that the temperature-sensitive vesicles were not ideal for this purpose, because of the relatively high temperature used during the preparation of the vesicles (45°C) which led to enzyme denaturation. The obligation to use a low coagulation and manufacturing temperature during the cheese-making process (25–30°C) to avoid leakage of the entrapped substance from the vesicles is also a serious limitation for their application. Alternatively, liposomes engineered to have a phase transition at a pH that is reasonably well defined (pH 5.5 to 5.0) will probably be more appropriate for enzyme delivery under cheese-making conditions. For example, incorporation of pH-sensitive lysolipids such as N-palmitoyl-L-homocysteine into vesicle membranes, made mainly of phospholipids, yielded pH 6.5-sensitive liposomes (Yatvin et al., 1987).

Despite the relatively high number of publications describing the use of liposome-entrapped enzymes as a means to reduce the ripening time of cheese and to enhance flavour development, conclusions are still difficult to reach. This is probably due to the fact that result reproducibility, especially for the encapsulation efficiency of the enzyme in the vesicles and liposome retention in the cheese, seems questionable. Work in this area is therefore still needed using radioactively labelled derivatives of the molecule to be entrapped (Piard et al., 1986; Alkhalaf et al., 1988, 1989; Laloy, 1996) rather than the conventional markers used for this purpose (i.e. carboxyfluorescein), which in most cases are smaller than the enzyme to be encapsulated, and may also behave differently in a cheese system. An economical evaluation of the process at this point is also becoming inevitable.

The second method is the encapsulation of the cell-free extract from bacteria or whole bacterial cells with appropriate substrates in milk-fat capsules. This technology was developed at the University of Wisconsin, Madison. Microencapsulation was accomplished by extruding a water/oil emulsion, consisting of an aqueous solution dispersed in a molten mixture of milk-fat and emulsifier, under high pressure through an orifice submerged in a chilled dispersion fluid. The aqueous solution (the carrier) contained the enzyme, cell-free extracts or microorganisms (Magee and Olson, 1981a).

Work on capsule stability, the influence of emulsions, and the evaluation of milk fat fractions for microcapsule preparation has also been reported (Magee and Olson, 1981b; Braun, 1984).

Several applications were described; Magee et al. (1981b) entrapped the cell-free extract of Lc. lactis subsp. lactis biovar diacetylactis with substrate and cofactors for the generation of diacetyl and acetoin. Cheese containing the capsules produced about eight times more diacetyl than did the control cheese. A system generating 3-methyl-butanal and 3-methyl-butanol from leucine (Braun et al., 1983), as well as systems with cofactor recycling were evaluated (Braun and Olson, 1986). Their application in low-fat cheese was also described (Braun et al., 1982).

More recently, Pannell and Olson (1989) have suggested the use of microcapsules in the production of blue cheese flavours based on their in vitro experiments. The microcapsules contained spores of P. roqueforti, phosphate buffer, and a low temperature melting fraction of butter. The butter fraction was encapsulated either after a pre-lipolysis or along with a pancreatic lipase. The spores were the source for the enzymes that transform the free fatty acids to methyl ketones, contributing to the cheese flavour. The free fatty acids originating from the pre-lipolysed butter were consumed in about 50 h of incubation at 10°C and methyl ketones concentration peaked at this time and sharply declined thereafter, perhaps being transformed into other, secondary products. The capsules containing the lipase, however, maintained the concentration of methyl ketones for a much longer time, directly affected by the lipase concentration. Thus, it appears that a regulated production of methyl ketone could be achieved during cheese ripening by using microcapsules containing a careful balance of the fat, lipase and spores.

Industrial application of microencapsulated flavour-generating systems should lead to the enhancement of low-fat cheese flavour and possibly to the production of a flavour-enhanced cheese to be used in the snack industry.

7.2.5   Modified bacterial cells

In the early attempts to reduce the ripening time of cheese, researchers increased the number of starter cells added to the cheese-milk. Two strategies were used for this purpose; using a higher inoculum (Stadhouders, 1961) or by adding stimulants to favour starter growth. The stimulants used included: trace elements (Hofi et al., 1973a), protein hydrolysates (Hofi et al., 1973b) and autolysed cells (Nassib, 1974). The cheese produced in most cases exhibited a lower pH, a higher moisture content and, very often, flavour defects.

Different methods were therefore attempted to modify the desired micro-organism in such a way that it would not produce lactic acid, would not compete with the normal starter, and will still deliver active enzymes into the cheese matrix. Figure 7.2 illustrates the different approaches undertaken in that respect.

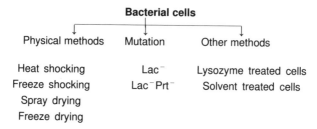

Figure 7.2 Different methods used to decrease acid production in adjunct cells.

(a) Physical methods.   The basic idea behind these methods is to expose the cells to conditions that will result in a significant decrease in their ability to produce lactic acid with only a limited reduction of their cellular proteinase, peptidase and esterase activities. Several methods were evaluated:

Heat shocking.   Addition of moderate amount of heat-treated *Lb. helveticus* (69°C for 20 s) combined with untreated mixed-strain starters accelerated the ripening of a Swedish hard cheese (Bie and Sjöstrom, 1975). Acid production by the *Lb. helveticus* culture appeared to have been minimized, while its proteolytic activity remained unaffected by this treatment.

Petterson and Sjöstrom (1975) revealed that heating at 59°C and 69°C for 15 s was optimal for mixed mesophilic and lactobacilli cultures respectively. When concentrates of heat-shocked cultures were added to cheese milk at a level of 2% (v/v), 90% of the added cells were entrapped in the curd but entrapment efficiency decreased at higher levels of addition. Proteolysis in Swedish household cheese was increased and flavour score improved by addition of the heat-shocked cells to the cheese milk, *Lb. helveticus* being the most effective. The extent of proteolysis did not increase with the level of heat-shocked mesophilic culture added, but did with the *Lb. helveticus* culture, suggesting some limiting factor in the former. Bitterness was not observed in any of the cheeses.

Abdel Baky *et al.* (1986) made cheese from curd incorporated with a heat-shocked culture of either *Lb. casei* subsp. *casei* or *Lb. helveticus*. These treatments did not greatly affect cheese composition but influenced flavour intensity. The heat-shocked lactobacilli-treated cheeses showed the desirable flavour 1 to 2 months earlier than control cheese.

When comparing the influence of *Lb. helveticus*, *Lb. bulgaricus* and *Strep. thermophilus* heat-shocked cells on the ripening of Gouda cheese, Bartels *et al.* (1987a) showed that *Lb. helveticus* cells gave the best results. A reduction of 25% on the ripening period of Gouda cheese was also measured by Exterkate *et al.* (1987) as a result of the addition of thermoshocked cultures of the mixed-strain starter.

Evaluation of the synergistic effect of extracellular proteolytic enzymes and the heat-shocked cells of *Lb. helveticus* was undertaken by Ardö and Petterson (1988). The authors demonstrated that heat-shocked cells accelerated the breakdown of peptides rather than casein hydrolysis. The addition of attenuated cells also eliminated the bitter flavour caused by Neutrase addition. In addition to the previously described cheeses, Vafopoulou *et al.* (1989) successfully used heat-shocked cells of *Strep. thermophilus* and *Lb. delbruckii bulgaricus* to shorten the aging of Feta cheese.

In an extensive study accomplished at Université Laval, Canada, El Abboudi *et al.* (1990) concluded that the development of typical flavor in Cheddar cheese was accelerated by the addition of homogenized cells of *Lb.*

casei subsp. *casei*. This treatment did not increase gross proteolysis as measured by trichloroacetic acid-soluble nitrogen (TCA-SN), but accelerated the breakdown of peptides, thus increasing the amount of amino acid N in the cheese and also reducing bitterness. Comparison between different methods for homogenate addition indicated that addition in the lyophilized form at renneting was the method of choice. In the study, the authors also concluded that live cells of *Lb. casei* subsp. *casei* seemed to accelerate methyl ketone production and improve the overall quality of the cheese. Combining the neutral proteinase Neutrase to viable and heat-shocked lactobacilli was also considered by El Abboudi *et al.* (1990, 1991a, b), who concluded that the best results were obtained by adding Neutrase at 1.0×10^{-5} Anson units g^{-1} of cheese and 2.0% (vol/vol) heat-shocked cells.

A continuous flow pasteurization method was developed for heat-shocking of *Lactobacillus* strains. The optimum conditions recommended were 64°C for 18 s. An increase of 2°C drastically inactivated the cell wall proteinase and aminopeptidase activities. Saint-Paulin cheese made with an extra inoculum of 1% (v/v) heat-shocked cell suspension accelerated the production of free amino acids during ripening. The cheese developed an increased mature flavour and overall quality after 15 or 30 days of ripening and showed no texture modification, bitterness or other flavour defect. Cheeses made with the same amount of untreated cells developed considerable flavour and texture defects (Castaneda *et al.*, 1990).

Freeze-shocking. Freeze-shocking at sub-optimal temperature leads to damage of the cell wall and membrane which results in a partial loss of the cell ability to produce acid, and induces cell lyses. The method was therefore adopted to attenuate several organisms (Table 7.5). Comparisons between the intracellular activities of frozen and unfrozen cells revealed that several factors are important in that respect; the microorganism used, the enzymatic activity measured, the freezing temperature, the storage time, the temperature and time of thawing, and finally the growth medium. In general, it was reported that *Lactobacillus* sp. are more susceptible to cryo-injury (Ray and Speck, 1973). It seems that aminopeptidases are more resistant to freezing when compared with protease, proline iminopeptidase and dipeptidase (Frey *et al.*, 1986; Kamaly and Marth, 1989). Khalid *et al.* (1991) however detected minor inhibition of the different peptidase activities due to the freezing of *Lb. helveticus*. No reports are available on the effect of freezing on the cell wall associated proteinase activity.

Freezing and holding the cells at −96°C for 1 week caused a 50% reduction in the number of *Lb. helveticus* grown in MRS broth. A 75% reduction was measured for cells grown in skim milk (Khalid *et al.*, 1990). Storing the cells in the form of a pellet or resuspended in buffer as well as the storage temperature and the thawing temperature and time also seems

Table 7.5 Conditions used to prepare freeze-shocked cells

Microorganism	Freeze-shocking conditions	Impact on cheese ripening	References
Lb. helveticus	−24°C/24 h	Substantial increase in water soluble nitrogen Debittering effect	Bartels *et al.* (1987b)
Lb. helveticus	−24°C/24 h + liposome-entrapped Corolase PN	Enhancement of Gouda cheese flavour made from UF milk	Spangler *et al.* (1989)
Lb. casei and *Lb. helveticus*	−20°C/20 h	50% reducton of Ras cheese ripening with *Lb. helveticus*	Aly (1990)
Micrococcus sp.	−96°C	Brothy and unclean flavour after 6 months of ripening	Bhowmik *et al.* (1990)
Pediococcus sp.	−96°C	Sharp, aged Cheddar flavour after 6 months	Bhowmik *et al.* (1990)
(Lac⁻) *Lc. lactis* subsp. *lactis* biovar *diacetylactis*	−20°C/24 h refrozen at 40°C stored at −20°C + Neutrase	Mutant starter gave moderate increase in proteolysis (1–2%)	Guinee *et al.* (1991)
Lb. casei and *Pediococcus* sp.	−20°C/20 h	Enhanced proteolysis of Cheddar cheese Acid flavour defect with both microorganisms and calcium lactate crystals with Pediococci	El Soda *et al.* (1991)
Lac⁺ and Lac⁻ *Lb. casei*	−20°C/20 h	Ras cheese made with Lac⁻ mutant obtained higher scores	El Soda *et al.* (1992)
Pediococcus halophilus	−30°C/36 h	Enhanced proteolysis and lipolysis in Ras cheese	El Shafei *et al.* (1992)
Lb. helveticus	−15°C/96 h	Enhanced proteolysis Debittering effect	Kim *et al.* (1994)
Leuconostoc sp.	−20°C/24 h	Decreased bitterness in Ras cheese	El Shafei (1994a)
Bifidobacterium sp.	−20°C/24 h	Enhanced proteolysis and lipolysis in Ras cheese	El Shafei (1994b)

to affect cell viability and intercellular enzyme activity (Frey *et al.*, Kamaly and Marth, 1989; Khalid *et al.*, 1990, 1991).

Comparative studies on the impact of heat-shocking and freeze-shocking bacterial cells on cheese ripening is limited and to some extent contradictory. Although Ezzat and El-Shafei (1991) suggested that heat-shocked cells are a better candidate for the enhancement of Ras cheese flavour, Aly (1994) reported no significant differences between the two treatments on low-fat Kashkaval cheese quality.

Spray-drying. Spray-drying was recently proposed as an attenuation method and compared with freezing and freeze drying. Johnson and Etzel (1995) subjected *Lb. helveticus* to spray-drying at an outlet temperature of 82°C and 120°C, freeze-drying, and freezing processes. Changes in viability, intracellular enzyme activity, lactic acid production, permeability, and autolysis were measured and compared for each of the various treatments.

Freeze-dried and frozen cells had the highest residual viabilities with averages of 49% and 55%, respectively. Cells spray-dried at an outlet air temperature of 82°C retained 16% of their initial viability whereas only 0.078% survived spray-drying at an outlet air temperature of 120°C. Aminopeptidase and β-galactosidase activities decreased with each processing method whereas dipeptidase and proteolytic activity increased. Decreases and increases in residual enzyme activities were less severe for cells spray-dried at lower temperatures than for those spray-dried at higher tempertures. Acid production rates for the freeze-dried and frozen cells were 0.09% and 0.08% acid h^{-1}, respectively. The corresponding figures for spray-dried cells ranged from 0.12 to 0.15% acid h^{-1}. Finally, freeze-dried and frozen cells autolysed more easily than spray-dried cells when suspended in 0.01 M sodium phosphate buffer containing 10^{-4} M EDTA at 41°C.

Low-fat Cheddar cheese was then made with the previously described adjuncts (Johnson *et al.*, 1995). The authors concluded from their evaluation of the cheese that because distinct flavour differences were noted among cheeses made with adjuncts, spray-dried starter culture adjuncts might be ideal for the production of specialty cheeses.

(b) Mutant strains (Lac⁻ and/or Prt⁻). The use of Lac⁻ mutants also provides a source of 'ripening enzyme bags' that will not interfere with acid production during cheese-making.

The use of Lac⁻ Prt⁻ mutant of *Lc. lactis* subsp. *lactis* C2 was described by Grieve and Dully (1984). The authors report an acceleration of proteolysis in the experimental cheese, as well as advancement of flavour development by up to 12 weeks over the control. Experimental cheeses received superior flavour scores and did not exhibit any flavour defect.

The same mutant was also evaluated by Fedrick *et al.* (1986a) and led to accelerated ripening.

Oberg *et al.* (1986) manufactured Cheddar cheese using Prt⁻ as suggested by Richardson *et al.* (1983) to reduce bitterness. The quality of the experimental and control cheeses were similar with only lower values for flavour noted for Prt⁻ cheese. Scores for body were higher for the Prt⁻ cheese after 180 days of ripening.

Stadhouders *et al.* (1988) in the Netherlands investigated the possibilities of adding either Prt⁺ or Prt⁻ strains to the normal starter during Gouda cheese making. The authors' observations led to the conclusion that cultures of exclusively Prt$^+$ variants were not useful as starters for making cheese with a good flavour development. However, the presence of 20% Prt$^-$ variants was sufficient to give maximal proteolysis and flavour development.

In another attempt by Kamaly *et al.* (1989), acceleration of Cheddar cheese ripening was made by adding the Lac$^-$ and Prt$^-$ mutants of *Lc. lactis* and *Lc. lactis* subsp. *Cremoris* to cheese-milk. During storage for 6 months, mutant-containing cheese developed higher levels of phosphotungstic acid-soluble amino N than did control cheese. Levels of TCA-SN in mutant-containing and control cheeses were similar. Mutant-containing cheese had a slightly softer body and more intense flavour than control cheese; a slight off-flavour (described as 'unclean') appeared in experimental cheese but was absent from control cheese. Bitterness did not develop in mutant-containing cheese during 6 months. Results of this work suggest that incorporation of the mutant Lac$^-$ Prt$^-$ lactic acid streptococci into cheese-milk as a source of enzymes could be useful in accelerated ripening of cheese.

Reduction of the ripening period using Lac$^-$ mutants of lactococci has also been reported by Abrahamsen *et al.* (1989). The authors recommended the selection of Lac$^-$ strains which would autolyse rapidly in the cheese to assure a high level of peptidases during the early stages of the ripening process.

In Ireland, Farkye *et al.* (1990) followed proteolysis and flavour developments in Cheddar cheese made with Prt$^+$ *Lc. lactis* subsp. *cremoris* UC-317 or its Prt$^-$ variant UC-041. The conclusion reached in this work was to some extent different from those previously described. The levels of water-soluble N in Prt$^+$ Prt$^-$ cheeses of the same age were not significantly different. However, the Prt$^+$ cheese had a significantly higher level of amino acid N throughout ripening. Proteinase-positive cheeses received slightly higher scores for flavour and body and texture than Prt$^-$ cheese of the same age. Nevertheless, the overall quality of all the cheeses was good, suggesting that starter peptidase activity may be more important than starter proteinases in flavour development in cheese during ripening. This is indeed in agreement with the experiment of Oberg *et al.* (1986) who found that the rate of proteolysis in cheese made with Prt$^-$ starter was not different from the control cheese.

(c) Other methods

Lysozyme-treated cells.   Addition of lysozyme-treated cells to cheese was proposed by Law (1978). The author (Law, 1980) however considers the method to be too expensive for industrial applications.

Solvent-treated cells.   Addition of starter-cells treated with organic solvents led to a slight acceleration of the ripening process and a decrease in bitterness (Stadhouders *et al.*, 1983). The method is, however, impractical because of the legal barriers.

7.2.6   Genetically modified starter cells

Genetically modified microorganisms or mutant strains lacking certain key enzymes were recently investigated for a better understanding of the role of lactic acid bacterial enzymes in cheese ripening. El-Abboudi *et al.* (1992) using an X-prolyl dipeptidyl aminopeptidase (XPDP)-deficient mutant of *Lb. casei* demonstrated that no significant differences could be detected in either the rate of ripening or the taste of the cheese made with the parent strain and the (XPDP) deficient strain, which suggest that enzyme has a minor role in the development of the characteristic flavour of the cheese. On the other hand, aminopeptidase N seems to play a key role in the hydrolysis of bitter peptides in cheese. In fact, Baankreis (1992) obtained a very bitter off-flavour when cheese was manufactured with an aminopeptidase N-negative strain of lactococcus which was not present in the cheese made with the parent strain. The possible role of aminopeptidase as a debittering agent was also confirmed by Prost and Chamba (1994). Three vats of Emmental cheese were made with *Lb. helveticus* strain L_1 (a strain exhibiting high aminopeptidase activity), strain L_2 or strain L_3 (clones selected for their lack of aminopeptidase activity); the starter culture, milk composition and all the manufacturing parameters were the same for the three treatments. The physicochemical analyses of cheeses showed no differences in acidification, moisture of curd, and casein degradation. Differences between ripened cheeses which were observed only by sensorial analysis revealed that cheeses made with the aminopeptidase-deficient lactobacilli were more bitter than those made with L_1. The authors suggest that the mutants should not be considered as bitter peptide-producing clones, but rather as allowing an accumulation of these peptides in cheeses. They also explained the bitterness in ripened cheeses made with *Lb. delbrueckii* subsp. *lactis* to be due to their very low aminopeptidase activity.

The contribution of lactococcal starter proteinases degradation in Cheddar cheese was investigated by Law *et al.* (1993) using *Lc. lactis* subsp. *lactis* UC317, its proteinase-negative derivative FH041, and variants of UC317 modified in proteinase production, location, and specificity. *Lc. lactis* subsp. *lactis* FH041 was transformed by electroporation with plasmids pC13601, pC13602, or pNZ521. Plasmids pC13601 and pC13602 harbour the cloned proteinase genes of *Lc. lactis* subsp. *lactis* UC317 on a high copy number vector and, as such, encode an increased concentration of cell wall-associated and secreted enzymes, respectively. Plasmid pNZ521 contains the cloned proteinase genes from *Lc. lactis* subsp. *cremoris* Sk11. Assessment of proteolysis and flavour development in Cheddar cheese made with these strains revealed that starter proteinases are required for the accumulation of small peptides and free amino acids in Cheddar cheese. The strain in which the proteinase remained attached to the cell wall appeared to contribute more to proteolysis than the strain that secreted the enzyme. Water-soluble peptides unique to *Lc. lactis* subsp. *cremoris* SK11 and *Lc. lactis* subsp. *lactis* UC317 were detected by PAGE and HPLC, respectively. Sensory evaluation showed that the flavours of all cheeses made with proteinase-positive starters were similar, but cheeses made with proteinase-negative starter lacked flavour. More recently, McGarry *et al.* (1994) compared the performances of different genetically modified lactococci during Cheddar cheese ripening. The following derivatives of *Lc. lactis* subsp. *lactis* UC317 were evaluated: L3601, a proteinase-negative derivative of UC317 transformed with high-copy-number plasmid pC13601 containing the cloned proteinase gene complex from UC317; AM312, a proteinase-negative derivative of UC317 transformed with plasmid pMG36enpr containing the neutral proteinase gene from *B. subtilis*; AC322, JL3601 transformed with pMG36enpr; AC311, UC317 transformed with plasmid pNZ1120, which contains the aminopeptidase N (*pep N*) gene from *Lc. lactis* subsp. *lactis* MG1363; and AC321, JL3601 transformed with pNZ1120. Organoleptic and chemical analyses indicated that the control cheeses, which were made with UC317, were of the highest quality; cheeses made with strains harbouring pC13601 in addition to either pMG36enpr (AC322) or pANZ1120 (AC321) did not ripen in a significantly different manner than cheeses made with AM312 (containing only pMG36enpr) or AC311 (containing only pNZ1120), respectively; cheeses made with strains that overproduce *pep N* did not have improved body, texture, and flavour characteristics; and cheeses with strains harbouring the neutral proteinase from *B. subtilis* (AM312 and AC322) underwent greatly accelerated proteolysis.

Despite the fact that direct application of the previously described results is still not possible, several conclusions can be reached from the utilization of genetically modified starter:

- X-Prolyl dipeptidyl dipeptidase may not have a direct impact on flavour development.
- Although the precise role of aminopeptidase in flavour development is not totally elucidated, their debittering action is now well recognized.
- Starter peptideses and proteinases produce small peptides and amino acids in cheese, but may not have a direct impact on flavour.
- Cloning of exogenous proteinases in starter cells leads to enhanced proteolysis.

7.2.7 Other cheese flavour enhancement methods

In addition to what may be call the 'conventional' accelerated ripening methods that were described in this chapter, other information is also available in patents describing cheese in an electric field (Japanese patent N. 63-502877 (502877/1988)) or subjecting the cheese to a high pressure (US Patent 5, 180, 596/1993) varying from 100 to 2500 kg cm^{-2}. The mechanistic impact on flavour development is, however, not described in this literature.

7.3   Enzymes release from cheese microorganisms

The increase in levels of free amino acids and free fatty acids in cheese during ripening is partly due to activities of peptidases and esterases released after the lysis of cheese microorganisms.

Autolysis could be defined as the spontaneous disintegration of the bacterial cell as the result of age or unfavourable physiological conditions which activate autolysin(s), enzymes found in the cell and capable of hydrolysing the cell wall peptidoglycan structure. The physiological functions of these enzymes is not fully understood, but they probably play a role during cell division, wall growth and wall turnover.

This process is of great importance during cheese ripening because it leads to the release of the intracellular enzymes that are now known to play a key role in protein and fat hydrolysis.

Studies on the autolytic properties of cheese-related bacteria which started in the 1940s (Hansen, 1941) were not extensive up to the last five years but a growing interest has emerged more recently.

The work on the autolytic properties of lactic acid bacteria was recently reviewed by El Soda et al. (1995a) and will not be reported here. The impact of autolysis on the release of intracellular enzymes during cheese ripening will however be considered.

Release of intracellular XPDP from Lc. lactis subsp. cremoris was first demonstrated by Wilkinson et al. (1989) in a cheese system. Monitoring of XPDP release over a period of 120 days was then accomplished (Wilkinson et al., 1994). Very little enzyme release could be detected during the first 20 days of cheese ripening. This is then followed by a gradual increase in XPDP activity which then reaches a peak after 60 days. A decline in activity was then noticed and no XPDP activity was measurable after 120 days of ripening. Increased of XPDP activity was also lower in cheeses ripened at 4°C when compared with cheese ripened at 10°C.

Intracellular enzyme release was also studied in two Lac^- Prt^+ mutant strains Lc. lactis subsp. cremoris and Lc. lactis subsp. lactis which autolyse rapidly at cheese-ripening temperatures and at ionic strengths found in cheese (Birkeland et al., 1992). DNA as well as LDH and proline aminopeptidase activities were found to be higher in the experimental cheese when compared to the control where only the starter organism was present. Proteolysis assessment during ripening using either amino nitrogen determination or FPLC of the peptide fractions also revealed higher values of amino nitrogen in the experimental cheeses.

In a rather detailed study Chapot-Chartier *et al.* (1994) monitored the autolysis of two lactococcal strains in Saint-Paulin-type cheese. The authors also determined the viability of the organisms in cheese, the morphological changes in the bacteria observed by electron microscopy, as well as the release of intracellular peptidases. For one of the strains (*Lc. lactis* subsp. *cremoris* AM2) lysis occurred from the first week of ripening leading to a decrease of cell viability which was also confirmed by electron microscopy observation. The authors also demonstrated that the aminopeptidase activity was measurable from the first day of ripening and remained active during the 60 days of the experiment.

A 30% decrease in activity could however be measured at the end of the ripening period. Maximum aminopeptidase activity was detected after 20 days of ripening. The levels of measurable XPDP activities in the cheese seems to be lower than the aminopeptidases, and no XPDP activity could be measured after 40 days of ripening. In the cheese made with strain *Lc. lactis* subsp. *lactis* NCDO 763, an opposite situation could be observed. High cell viability, and no release of enzymes could be monitored during the first 3 weeks of ripening. This work that highlighted the differences in the autolytic properties of the lactococci was confirmed (Wilkinson *et al.*, 1994) for three *Lc. lactis* subsp. *cremoris* strains.

El Soda *et al.* (1993) also demonstrated that the release of aminopeptidase activity from a highly autolytic strain of *Lb. casei* as compared with a poorly autolytic strain of the same species was significantly higher in the case of the highly autolytic strain. Aminopeptidase activity was then monitored in Cheddar-type cheese where both strains were evaluated separately. Acidification was accomplished using gluconodeltalactone to eliminate interference from the starter peptidases. The enzyme could be detected in the cheese manufactured with the highly autolytic strain after 48 h of ripening, while a week was necessary to detect aminopeptidase activity in the cheese made with the strain showing little autolysis. No aminopeptidase activity was measurable in a control cheese made under the same experimental conditions.

Experiments accomplished in a buffer system (Dako *et al.*, 1994) confirm the results described in cheese systems. Protein, aminopeptidase and XPDP release from *Lb. casei* were monitored at 7°C in a buffer system. As far as enzyme release was concerned, little activity could be measured during the first 3 h of incubation; this was then followed by a gradual increase during the first 9 h, after which both activities reached a plateau. On the other hand, protein release seems to be increasing during the 48 h of the experiment, which was also the case for cell autolysis.

Such results show beyond doubt that the pre-selection of strains based on their autolytic properties is becoming an important criterion which starter manufacturers should consider for their ripening strains.

7.4 Flavour enhancement in non-conventional cheeses

7.4.1 Cheese made from recombined milk

Due to the shortage of milk supplies, the dairy industries of several countries rely either partially or totally on milk powder for cheese making. Attempts were therefore made to improve the quality of the cheese made from recombined milk using different lipase preparations (Abd El Salam et al., 1983; Ashour et al., 1986).

The rate of ripening of Ras cheese made from fresh milk or recombined milk treated with proteolytic and lipolytic enzyme preparations were compared (El Soda et al., 1988). Enzymes were added to the curd in both cases. The cheeses made from recombined milk showed higher values for soluble nitrogen and fatty acids than the fresh-milk cheese. This was explained by the fact that the partial denaturation of milk protein during milk powder manufacture and the alteration of the milk-fat globule membrane during the production of butteroil make the resulting recombined milk cheese more accessible to proteolytic and lipolytic enzymes.

7.4.2 Cheese made by ultrafiltration

Cheese obtained through ultrafiltration (UF) technology was described as resisting ripening which was attributed to several factors: a reduced level of residual rennet (Green et al., 1981; Green, 1985) and/or the concentration of proteinase and peptidase inhibitors by ultrafiltration (Hickey et al., 1983). The presence of β-lactoglobulin, which inhibits plasmin activity (Chen and Ledford 1971; Kaminogawa et al., 1972), may be a contributing factor. Some attempts have been made to overcome this problem. Spangler et al. (1989) concluded that the combined use of liposome-entrapped proteinase and freeze-shocked Lactobacillus cells may be an efficient means of reducing the curing time of cheese made from UF milk. El Shibiny et al. (1991) obtained high-quality cheese using different combination of starters.

Rapidly ripened ultrafiltered retentates were recently described by Aly et al. (1995) and used to manufacture a processed cheese spread. The authors report that good quality flavour could be obtained by using a combination of 0.05% sodium dodecylsulphate and 0.005% of a commercial lipase preparation. For UF cheeses in which proteolysis is not the major event during ripening (Feta, Blue and Camembert), the addition of lipases seems to mask many flavour defects (Lelièvre and Lawrence, 1989).

7.4.3 Cheese made from ovine, caprine or buffalos' milk

In different areas of the world, including India, Spain, France, Greece, Italy and Egypt, cheeses are made from milks other than cows' milk. The water buffalo, ewe and goat are the main animals used for that purpose. Since the milks of these animals differ in composition from cows' milk their behaviour towards accelerated ripening agents could be expected to differ.

Fernandez-Garcia *et al.* (1990) made Manchego-type cheese from ewes' or cows' milk. Different concentrations of Neutrase were added to the curd. The rates of proteolysis and lipolysis in both cheeses were then compared. It was concluded that bovine caseins were more susceptible to proteolysis than ovine caseins. This was confirmed by electrophoresis of the protein fraction. The degradation of bovine β-casein was 90%, while it was only 42% in ewes' milk cheese. The values for α-caseins were 50% and 37% for cows' and ewes' milks, respectively. It is of interest that the cows' milk cheese developed a strong bitter taste that was not detectable in the ewes' milk cheese.

Nuñez *et al.* (1991) compared the influence of two commercial preparations of neutral proteinase from *B. subtilis* (Neutrase L and Novozym 257) on the ripening of Manchego cheese made from ewes' milk. The conclusions reached from this work indicate that Manchego cheese made from milk to which 0.004 Anson units L^{-1} of Neutrase L proteinase was added would develop, in 35 days at 16°C, the same flavour intensity as control Manchego cheese held for 90 days at 12°C.

The influence of lipases (Palatase 200 from *Mucor miehei* and Palatase 750 from *A. niger*) alone or in combination with a proteinase (MKC from *A. oryzae* and Neutrase) on the maturation of Manchego cheese made from 70% ewes' and 30% cows' milk were described by Fernandez-Garcia *et al.* (1994). Addition of both lipases alone, or in combination with the proteases, increased lipolysis. The *M. miehei* lipase affected mainly the long-chain fatty acids of cheese, and the *A. niger* lipase affected the short- and long-chain fatty acids equally. The influence of the proteinases on the rate of lipolysis was different; the *Bacillus subtilis* protease preparation increases lipolysis significantly, either by contamination of the enzymatic solution or by a synergistic effect with the added *A. niger* lipase. On the other hand an opposite effect was noticed with the protease from *A. oryzae*; the combination of proteases and lipases increased proteolysis as measured by indexes of soluble N and HPLC of free amino acids while the exclusive addition of lipases decreased proteolysis. Sensory analysis of the cheese showed a slight but significant acceleration of ripening (16%) for the cheeses treated with *A. niger* lipase at the beginning of aging. The *M. miehei* lipase produced a soapy flavour in the cheeses because of the excessive release of long-chain fatty acids. The cheeses containing the *B. subtilis* protease developed bitterness and sticky and crumbly texture because of the intense breakdown of β-casein. The authors recommend that enzymes should be carefully selected according to their lipolytic activity and the ratios of free fatty acid release.

For goats' milk, no work on the acceleration of traditional goats' milk cheese is available in the literature. However, some attempts were made for the acceleration of Ras cheese ripening made from goats' milk (El Shazly *et al.*, 1993) or from a mixture of goats' and cows' milk (Ammar *et al.*, 1994). Addition of cheese slurries seems to be effective in that respect.

In the case of buffalos' milk, most work has focused on lipase addition. The results indicate that flavour development in buffalos' milk cheese is considerably slower than cows' milk cheese. Reduction in maturation time was made possible by the addition of lipases, proteases and viable *Lb. casei* cells to the milled curd (Kanawjia and Singh, 1990). The addition of higher levels of carefully selected starters seems to have a positive effect on the enhancement of Gouda cheese flavour made from buffalos' milk (Rajesh and Kanawya, 1990).

7.4.4 Low-fat cheese

The general understanding of the health risks related to high fat intake is increasing throughout the world and has led cheese manufacturers to develop low-fat products.

A cheese may be classified as low-fat only if the fat content is subtractically reduced (see Chapter 6 of this volume). A demand often expressed is that the fat should be half or less than of normal-fat cheese. If the fat content is two-thirds or less of normal-fat cheese it is commonly referred to as reduced-fat cheese. Most commonly, water is used as the substitute for fat to obtain an attractive texture of a reduced-fat cheese. Sometimes fat replacers are added to stabilize the structure and improve the texture of the cheese.

Milk fat contributes to cheese flavour partially by dissolving flavour compounds. These may be produced from the hydrolysis of fat or protein, or in other ways. The fat is also beneficial for masking off-flavours, bitterness being an example. The amino acid composition of many bitter peptides comprises basic amino acids at one end and mainly hydrophobic ones at the other. The arrangement of molecules with such properties at the interface between fat and water might hide their bitterness. This stresses the impression that less fat in the cheese increases its tendency to taste bitter. Full-fat cheese contains a greater area of lipid–water interface than reduced-fat cheese.

Some attempts aiming for the enhancement of flavour in low-fat cheese were described in the literature.

Banks *et al.* (1989) developed a technique for the production of Cheddar-type cheese of 25% and 16% fat content. The development of flavour and texture in these cheese was monitored over a 6-month maturation time. At the lower fat content the cheese did not develop adequate Cheddar flavour. The cheese also showed an over-fine texture, and mild Cheddar flavour; a better texture was found in the cheese with 25% fat.

Ardö, in Sweden, has been active in this area and has developed a system using heat-treated lactobacilli to give a desirable aroma in low-fat cheese (Ardö et al., 1989; Ardö and Mansson, 1990). The heat-shocked lactobacilli released their aminopeptidase early during ripening. The enzyme exhibited high debittering activity and the low-fat cheese containing the heat-shocked lactobacilli developed a desirable flavour. The concept was shown to be applicable both to cheeses with round eyes and to those with open texture, which offers possibilities for increasing the number of attractive low-fat varieties.

Acceleration of the ripening of low-fat cheese has also been achieved using lyophilized extracts of Lb. casei, Lb. helveticus, Propionibacterium shermanii, Pediococcus sp., or Brevibacterium linens (El Soda et al., 1991). Cheeses treated with bacterial extract showed significantly higher values of phosphotungstic acid (PTA)-soluble nitrogen. Flavour defects, including bitterness, were not detected in the cheeses but none exhibited a pronounced Cheddar flavour.

Coulson et al. (1992) suggested the use of Accelase or cultures based on aminopeptidase systems. These authors indicate that Accelase, which contains six different aminopeptidases from Lc. lactis, reduces the maturation time for full-fat cheese and improves the flavour and texture of low-fat cheeses.

The use of cultures for low fat flavour improvement was also reported by Banks et al. (1993). The authors concluded that low-fat Cheddar cheese manufactured using a modified procedure (1.8% salt in the presence of culture adjuncts) led to improved Cheddar flavour development. Textural differences between the low-fat and the full-fat product remained.

Attempts were also made to enhance low-fat cheese flavour using protease and lipase enzymes derived from A. oryzae (Brandsma et al., 1993). The authors reported marked rancidity in the enzyme treated cheese after 8 to 12 weeks ripening and recommend the use of enzyme preparations with balanced proteases and lipases.

Attenuated Lb. helveticus, either in a frozen (Aly, 1994) or spray-dried (Johnson et al., 1995) form, were reported to improve the flavour of low-fat Kashkaval and Cheddar cheese respectively.

Laloy et al. (1996) have recently demonstrated that full-fat cheese contained 4 to 10 times more starter cells when compared with cheese made from skimmed milk. The number of colony-forming units in a cheese with 50% less fat was intermediate. These results may to some extent explain the lack of flavour in low-fat cheese and clarify the possible role of the so-called low-fat cheese adjuncts.

Very little work on the impact of fat substitute on flavour enhancement has been reported. Lucey and Gorry (1993) studied the effect of adding Simplesse 100 in the manufacture of Cheddar cheese. The resulting cheese had a higher moisture content, higher yield, and a softer body than Cheddar manufactured without Simplesse incorporation. Both cheeses exhibited an acid flavour defect. Paquin *et al.* (1993) developed a fat substitute from whey using microfluidization. Full-fat Cheddar cheese obtained by the addition of the ingredient possessed the same textural and flavour characteristics as the conventional products. The colour of the cheese was, however, whiter and the yield on a moisture basis higher. No commercial trials in low-fat cheese manufacture were reported however.

7.5   Economic aspects

Despite the fact that the whole area of accelerated ripening and flavour enhancement was developed for economic purposes, very limited economic analysis of the strategies proposed have been made available.

Fedrick (1987) presented a cost/benefit analysis of cheese maturation. His study included the following ripening agents: curing at 13°C as well as the addition of the following enzymes, FlavourAge, Naturage, Accelase, Neutrase or Werribee maturation system. The author concluded that the cost of FlavourAge, Naturage and Accelase are relatively high compared with the calculated benefits. The use of higher ripening temperature resulted in the higher net benefit per ton, taking into consideration the risks discussed in section 7.2.1.

Ezzat *et al.* (1991) evaluated the economics of using a mixture of Rulactine (an enzyme from *Micrococcus* sp.) and Piccantase in Ras cheese production and revealed significant decrease in production costs. The major drawback of this work is the discontinuation of Rulactine production.

With the new trends in flavour enhancement strategies, cost/benefit analysis of attenuated cells and entrapped enzymes is becoming an urgent necessity.

7.6   Future perspectives

From the published literature in the area of accelerated ripening and flavour enhancement, and from the trends taken from recent patents, it seems that improving the flavour or reducing the ripening of a traditional cheese will not be reached through the addition of an exogenous agent. This goal can only be achieved by using organisms isolated from the cheese environment. One would therefore expect that future research and applications in the area of flavour enhancement of traditional cheese will be directed towards improving the performances of cheese-related bacteria. More work in the area of cell attenuation is to be expected.

A comparison between the different methods described in the literature (heat-shocking, freezing and spray-drying) using the same organism on the same type of cheese is needed, especially since recent findings have shown that heat-shocked cells lysed at a much slower rate when compared with freeze-shocked cells. El Soda *et al.* (1995a) on the other hand, suggested that scaling-up the freezing process may be a problem for industrial applications, especially if more than one freeze and thawing cycle is needed. The information available on spray-drying was derived from one strain of *Lb. helveticus*; data on other strains and species are therefore required.

Despite the efforts reported by the different genetic engineering groups (Kok *et al.*, 1985; Van de Guchte *et al.*, 1990; Alen-Boerrigter *et al.*, 1991; Bruinenberg *et al.*, 1992; Riepe and McKay, 1994), the recent work illustrating the impact of genetically modified bacteria on cheese ripening is more on the negative side (McGarry *et al.*, 1994), which indicates the need for further research.

The isolation and selection of microorganisms from their natural habitat should not be neglected; currently, starter manufacturers are offering to cheese-makers adjunct cultures in a frozen or freeze-dried form. Some cultures seem to have positive effects on the quality of conventional full-fat cheese, while others are more appropriate for the enhancement of low-fat cheese. In most cases, the information offered by the manufacturer is limited to the levels of aminopeptidase activity of the culture. More academic work on these commercial cultures is needed for a better understanding of their role in flavour enhancement.

Basic research on cheese ripening and the agents involved are still needed for a better control of the process. The mechanism of enzyme release from cheese-related bacteria and their migration in the cheese matrix, as well as their stability in the cheese environment, are important questions to which very few answers are currently available (*c.f.* Law and Mulholland, 1995).

A recent study by Laloy *et al.* (1996) using electron microscopy indicated that more than 85% of the starter cells are located in the peripheral region of fat globules during the early stages of the ripening process, the cells appearing to be either in direct contact with the fat globule membrane or in a limited area around it. This suggests an active interaction between the starter cells and the fat globule membrane. After 1 and 2 months of ripening, this interaction became more intimate and the fat globule membrane appeared altered at the area in contact with the starter cells. Observations of more than 100 ghost cells led to the conclusion that the cells were either included into the fat globule membrane or appeared directly in contact with the interior of the globules.

These results, which were expected (Kiely *et al.*, 1993) but never confirmed, provide some evidence on the location of the cells in the cheese matrix. Laloy (1996) was also able to follow the release of an aminopeptidase from *Lb. casei* subsp. *pseudoplantarium* in a cheese system using an immunological assay. Such a technique may be used in the future to determine the migration of the enzymes released from bacterial cells.

Evidence for stability of aminopeptidase (El Abboudi *et al.*, 1991a) and esterase (El Soda *et al.*, 1995b) in a cheese system were recently demonstrated. This work was however accomplished on a *Lb. casei* strain and should therefore be demonstrated in other lactic acid bacteria, since previous results indicate that the aminopeptidase of lactococci could not be detected in ripened cheese (Desmazeaud and Vassal, 1979).

A recent review on the lipolytic activity of cheese-related microorganisms (El-Soda *et al.*, 1995b) reveals the lack of information in this area, and the need of further efforts, especially since the role of fat and the products resulting from its hydrolysis are probably as important as those of proteolysis in determining the flavour bouquet of a wide variety of cheeses.

7.7   References

Abdel Baky, A., El-Neshawy, A., Rabie, A. and Ashour, M. (1986) Heat-shocked lactobacilli for accelerating flavour development of Ras cheese. *J. Food Chem.*, **21**, 301–13.

Abd El Salam, M., El Shibiny, S., El Koussey, L. and Haggag, H. (1983) Domiati cheese made with ultrafiltered reconstituted milk and lipolysed recombined cream. *J. Dairy Res.*, **50**, 237–40.

Abrahamsen, R., Birkeland, S. and Langsrud, T. (1989) Acceleration of cheese ripening by the use of Lac⁻ mutants of group N-streptococci. *Acta Aliment. Pol.*, **15**, 123–31.

Alen-Boerrigter, I., Baankreio, R. and de Vos, W. (1991) Characterization and overexpression of the *Lactococcus lactis* Pep N. gene and localization of its products, aminopeptidase N. *Appl. Environ. Microbiol.*, **57**, 2555–61.

Alkhalaf, W., Piard, J., El-Soda, M., Gripon, J., Desmazeaud, M. and Vassal, L. (1988) Liposome a proteinase carrier for cheese ripening acceleration. *J. Food Sci.*, **53**, 1674–9.

Alkhalaf, W., El-Soda, M., Gripon, J. and Vassal, L. (1989) Acceleration of cheese ripening with liposome entrapped proteinase. Influence of liposome net charge. *J. Dairy Sci.*, **72**, 2233–8.

Aly, M. (1990) Utilization of freeze-shocked lactobacilli for enhancing flavour development of Ras cheese. *Die Nahrung*, **34**, 329–35.

Aly, M. (1994). Flavour-enhancement of low-fat kashkaval cheese using heat- or freeze-shocked *Lactobacillus delbrueckii* var. *helveticus* cultures. *Die Nahrung*, **38**, 504–10.

Aly, M., Abdel-Baky, A., Farahat, S. and Hana, U. (1995) Quality of processed cheese spread made using ultrafiltered retentates treated with some ripening agents. *Int. Dairy Journal*, **5**, 191–209.

Ammar, E., El-Shazly, A., Nasr, M. and El-Saadany, M. (1994) Effect of using autolyzed starter and cheese slurry on acceleration of Ras cheese ripening made from mixture of goat's and cow's milk. *Egyptian J. Dairy Sci.*, **22**, 67–80.

Arbige, P., Freund, S., Silver, S. and Zelko, J. (1986) Novel lipase for Cheddar cheese flavor development. *Food Technol.*, **40**, 91.

Ardö, T. and Mansson, H. (1990) Heat treated lactobacilli develop desirable aroma in low-fat cheese. *Scan. Dairy Info.*, **4**, 38–40.

Ardö, T. and Petterson, H. (1988) Accelerated cheese ripening with heat-treated cells of *Lactobacillus helveticus* and a commercial proteolytic enzyme. *J. Dairy Res.*, **55**, 239–45.

Ardö, T., Larsson, P., Mansson, L. and Hedenberg, A. (1989) Studies on peptidolysis during early maturation and its influence on low-fat cheese quality. *Milchwissenschaft*, **44**, 485–90.

Ashour, M., Abd El-Baky, A. and El-Neshawy, A. (1986) Improving the quality of Domiati cheese made from recombined milk. *Food Chem.*, **20**, 85–96.

Aston, J., Giles, J., Durwar, I. and Dulley, J. (1985) Effect of elevated ripening temperatures on proteolysis and flavour development in Cheddar cheese. *J. Dairy Res.*, **52**, 565–72.

Baankreis, R. (1992) *The role of lactococcal peptidases in cheese ripening*, PhD Thesis, University of Amsterdam.

Banks, J., Brechany, E. and Christie, W. (1989) The production of low fat Cheddar-type cheese. *J. Soc. Dairy Technol.*, **42**, 6–9.

Banks, J., Hunter, E. and Muir, D. (1993). Sensory properties of low fat Cheddar cheese. Effect of salt content and adjunct culture. *J. Soc. Dairy Technol.*, **46**, 119–23.

Bartels, H., Johnson, M. and Olson, N. (1987a) Accelerated ripening of Gouda cheese. I–Effect of heat-shocked thermophilic lactobacilli and streptococci on proteolysis and flavor development. *Milchwissenschaft*, **42**, 83–8.

Bartels, H., Johnson, M. and Olson, N. (1987b) Accelerated ripening of Gouda cheese. II–Effect of freeze shocked *Lactobacillus helveticus* on proteolysis and flavor development. *Milchwissenschaft*, **42**, 139–44.

Bhowmik, T., Riesterer, R., Van Boekel, M. and Marth, E. (1990) Characteristics of low-fat Cheddar cheese made with added *Micrococcus* or *Pediococcus* species. *Milchwissenschaft*, **45**, 230–5.

Bie, R. and Sjöstrom, G. (1975) Autolytic properties of some lactic acid bacteria used in cheese production. Part II–Experiments with fluid substrates and cheese. *Milchwissenschaft*, **30**, 739–47.

Birkeland, S., Abrahamsen, R. and Langsrud, T. (1992) Accelerated cheese ripening: use of lac⁻ mutants of lactococci. *J. Dairy Res.*, **59**, 389–400.

Brandsma, R., Mistry, V., Anderson, D. and Baldwin, K. (1993) Reduced fat Cheddar cheese from condensed milk. 3–Accelerated ripening. *J. Dairy Sci.*, **77**, 897–906.

Braun, S. (1984) *Microencapsulated multi-enzyme systems to produce flavours and recycle cofactors*, PhD Thesis, University of Wisconsin, Madison, WI.

Braun, S. and Olson, N. (1986) Microencapsulation of cell-free extracts to demonstrate the feasibility of heterogenous enzyme systems and cofactor recycling for development of flavor in cheese. *J. Dairy Sci.*, **69**, 1202–8.

Braun, S., Olson, N. and Lindsay, R. (1982). Microencapsulation of bacterial cell-free extract to produce acetic acid for enhancement of cheese flavor. *J. Food Sci.*, **47**, 1803–7.

Braun, S., Olson, N. and Lindsay, R. (1983) Production of flavor compounds: aldehydes and alcohols from leucine by microencapsulated cell-free extracts of *Streptococcus lactis* var. *maltigenes. J. Food Biochem.*, **7**, 23–41.

Bruinenberg, P., Vos, P. and De Vos, W. (1992) Proteinase over production in *Lactococcus lactis* strains: regulation and effect on growth and acidification in milk. *Appl. Environ. Microbiol.*, **58**, 78–84.

Castaneda, V., Gripon, J. and Rousseau, M. (1990). Accelerated ripening of a saint-poulin cheese variant by addition of a heat-shocked *Lactobacillus* suspension. *Neth. Milk Dairy J.*, **44**, 49–62.

Chapot-Chartier, M., Deniel, C., Rousseau, M., Vassal, L. and Gripon, J. (1994) Autolysis of two strains of *Lactococcus lactis* during cheese ripening. *Int. Dairy J.*, **4**, 251–69.

Chen, J. and Ledford, R. (1971) Purification and characterisation of milk protease. *J. Dairy Sci.*, **54** (suppl. 1), 763 (abstract).

Cliffe, A. and Law, B. (1990) Peptide composition of enzyme treated Cheddar cheese slurries, determined by reverse phase high performance liquid chromatography. *Food Chem.*, **36**, 73–80.

Coulson, J., Pawlett, D. and Wivell, R. (1992) Accelerated ripening of cheese. *Bull. Int. Dairy Fed.*, **262**, 29–35.

Dako, E., El-Soda, M., Vuillemard, J. and Simard, R. (1994) Autolytic properties of lactic acid bacteria. *Fod. Res. Int.*, **28**, 503–9.

Desmazeaud, M. and Vassal, L. (1979) Activité protéque intracellulaire de streptocoques lactiques mésophiles "rôle au cours de l'affinage des fromages". *Le Lait*, **59**, 327–44.

El-Abboudi, G., Trépanier, S., Pandian, S. and Simard, R. (1990) Utilisation de cellules atténuée et d'extraits cellulaires de *Lactobacillus* pour accélérer la maturation du fromage Cheddar. *Can. Inst. Food Sci. Technol. J.*, **22**, 407 (abstract).

El-Abboudi, M., El-Soda, M., Pandian, S., Barreau, M., Trepanier, G. and Simard, R. (1991a) Peptidase activities in debittering and nondebittering strains of lactobacilli. *Int. Dairy J.*, **1**, 55–64.

El-Abboudi, M., Pandian, S., Trepanier, G., Simard, R. and Lee, B. (1991b) Heat-shocked lactobacilli for acceleration of Cheddar cheese ripening. *J. Food Sci.*, **56**, 948–9.

El-Abboudi, M., El-Soda, M., Johnson, M., Olson, N., Simard, R. and Pandian, S. (1992) Peptidase deficient mutants, a new tool for the study of cheese ripening. *Milchwissenschaft*, **47**, 625–8.

El-Shafei, H. (1994a) Accelerated ripening of Ras cheese using cell free extract, freeze and heat-shocked *Leuconostoc* spp. cultures. *Die Nahrung*, **38**, 606–11.

El-Shafei, H. (1994b) Manufacture of Ras cheese with cell free extract, freeze and heat shocked strains of *Bifidobacterium* spp. *Indian J. Dairy Sci.*, **47**, 774–9.

El-Shafei, H., Hantira, A., Ezzat, N. and El-Soda, M. (1992) Characteristics of Ras cheese made with freeze-shocked *Pediococcus halophilus*. *Food Sci. Technol. Lebensm. Wiss.*, **25**, 438–41.

El-Shazly, A., Ammar, E., Nasr, M. and El-Saadany (1993) Acceleration of Ras cheese ripening made from goat's milk as affected by using autolyzed starter and cheese slurry. *Egyptian J. Appl. Sci.*, **8**, 423.

El-Shibiny, S., Mahran, G., Haggag, H., Mahfouz, M. and El-Shiekh, M. (1991) Accelerated ripening of UF Ras cheese. *Egyptian J. Dairy Sci.*, **19**, 25–34.

El-Soda, M. (1993) Accelerated maturation of cheese. *Int. Dairy J.*, **3**, 513–44.

El-Soda, M., Desmazeaud, S., Abou-Donia, S. and Kamal, K. (1981) Acceleration of cheese ripening by the addition of whole cells or cell-free extracts from *Lactobacillus casei* to the cheese curd. *Milchwissenschaft*, **36**, 140–2.

El-Soda, M., Desmazeaud, S., Abou-Donia, S. and Badran, A. (1982) Acceleration of cheese ripening by the addition of extracts from *Lactobacillus helveticus*, *Lactobacillus bulgaricus* and *Lactobacillus lactis* to the cheese curd. *Milchwissenschaft*, **37**, 325–7.

El-Soda, M., Fathallah, S. and Ezzat, N. (1983) Acceleration of Cheddar cheese ripening with liposome trapped extracts from *Lactobacillus casei*. *J. Dairy Sci.*, **66** (suppl. 1), 78 (abstract).

El-Soda, M., Korayem, M. and Ezzat, N. (1984) Acceleration of Cheddar cheese ripening with liposome trapped extracts from *Lactobacillus helveticus*. Proceedings of 2nd Egyptian Conference on Dairy Science Technology, p. 28.

El-Soda, M., Ezzat, N., El-Deeb, S., Mashaly, R. and Moustapha, F. (1986) Acceleration of Domiatti cheese ripening using extracts from several lactobacilli. *Le Lait*, **66**, 177–84.

El-Soda, M., Ezzat, N., Salam, A. and Khamis, A. (1988) Accelerated ripening of cheese made from recombined milk. *International Dairy Fed. Special Issue No. 9001*, Recombination of milk and milk products, pp. 290–7.

El-Soda, M., Johnson, M. and Olson, N. (1989) Temperatures sensitive liposomes, a controlled release system for the acceleration of cheese ripening. *Milchwissenschaft*, **44**, 213–14.

El-Soda, M., Ezzat, N., El-Abassy, F., Wahba, A. and Hassanein, S. (1990) Acceleration of Ras cheese ripening. II. Combination of a gross proteolytic agent with the cell free extract of some lactobacilli. *Egyptian J. Dairy Sci.*, **18**, 183–93.

El-Soda, M., Chen, B., Riesterer, B. and Olson, N. (1991) Acceleration of low-fat cheese ripening using lyophilized extracts of freeze-shocked cells of some cheese related microorganisms. *Milchwissenschaft*, **46**, 358–60.

El-Soda, M., Hantira, A., Ezzat, N. and El-Shafei, H. (1992) Accelerated ripening of Ras cheese using freeze-shocked mutants strains of *Lactobacillus casei*. *Food Chem.*, **44**, 179–84.

El-Soda, M., Kim, L. and Olson, N. (1993) Autolytic properties of several *Lactobacillus casei* strains. *J. Dairy Sci.*, **76** (suppl. 1), 130 (abstract).

El Soda, M., Farkye, N., Vuillemard, J., Simard, R., Olson, N., El Kholy, W., Dako, E., Medrano, E., Gaber, M. and Lim, L. (1995a) Autolysis of lactic acid bacteria: Impact on flavour development in cheese, in *Food Flavors: Generation Analysis and Process Influence*, (ed. G. Charalambous), Elsevier Science B.V., Amsterdam, pp. 2205–23.

El Soda, M., Law, J., Tsakalidou, E. and Kalantzopoulos, G. (1995b) Lipolytic activity of cheese related microorganisms and its impact on cheese flavour, in *Food Flavors: Generation Analysis and Process Influence*, (ed. G. Charalambous), Elsevier Science B.V., Amsterdam, pp. 1823–48.

Exterkate, F., Deveer, G. and Stadhouders, J. (1987) Acceleration of the ripening process of Gouda cheese by using heat-treated mixed strain. *Neth. Milk Dairy J.*, **41**, 307–20.

Ezzat, N. (1990) Accelerated ripening of Ras cheese with a commercial proteinase and intracellular enzymes from *Lactobacillus bulgaricus*, *Propionibacterium freudenreichii* and *Brevibacterium linens*. *Le Lait*, **70**, 459–66.

Ezzat, N. and El-Shafei (1991) Acceleration ripening of Ras cheese using freeze and heat-shocked *Lactobacillus helveticus*. *Egyptian J. Dairy Sci.*, **19**, 347–58.

Ezzat, N., Hawary, M., El Soda, M. and Abdel Naby, E. (1991) Economical evaluation of Ras cheese ripening using commercial enzymes (in Arabic). *Ann. Agric. Sci. Mostohor.*, **29**, 1–12.

Farkye, N. and Fox, P. (1991) Preliminary study on the contribution of plasmin to proteolysis in Cheddar cheese. Cheese containing plasmin inhibitor 6-aminohexanoic acid. *J. Agric. Food Chem.*, **39**, 766–8.

Farkye, N. and Fox, P. (1992) Contribution of plasmin to Cheddar cheese ripening: effect of added plasmin. *J. Dairy Res.*, **59**, 209–16.

Farkye, N. and Landkammer, C. (1992) Contribution of plasmin to Cheddar cheese ripening: effect of added plasmin. *J. Dairy Res.*, **59**, 209–16.

Farkye, N., Fox, P., Fitzgerald, G. and Daly, C. (1990) Proteolysis and flavour development in Cheddar cheese made exclusively with single strain proteinase-positive or proteinase-negative starters. *J. Dairy Res.*, **73**, 874.

Fedrick, I. (1987) Technology and economics of the accelerated ripening of Cheddar cheese. *Aust. J. Dairy Technol.*, **42**, 33–6.

Fedrick, I., Aston, W., Nottingham, S. and Dulley, J. (1986b) The effect of neutral fungal protease on Cheddar cheese ripening. *N.Z.J. Dairy Sci. Technol.*, **21**, 9–19.

Fedrick, I., Cromie, S., Dulley, J. and Giles, J. (1986a) The effects of increased starter populations, added neutral proteinase and elevated temperature storage in Cheddar cheese manufacture and maturation. *N.Z.J. Dairy Sci. Technol.*, **21**, 191–203.

Fernandez-Garcia, E., Lopez-Fandino, R., Olans, A. and Ramos, M. (1990) Comparative study of the proteolytic activity of a *Bacillus subtilis* neutral protease preparation during early stages of ripening of cheese made of cow and ewe milk. *Milchwissenschaft*, **45**, 428–31.

Fernandez-Garcia, E., Lopez-Fandino, R., Alonso, L. and Ranos, M. (1994) The use of lipolytic and proteolytic enzymes in the manufacture of Manchego cheese from Ovine and Bovine milk. *J. Dairy Sci.*, **77**, 2139–49.

Freund, P. (1986) A unique system for enhancing Cheddar cheese ageing. Seventh Biennial Cheese Industry Conference, Logan, USA.

Frey, J., Johnson, M. and Marth, E. (1986) Peptidases and proteases in barley extract a potential source of enzymes for use in cheese ripening. *Milchwissenschaft*, **41**, 488.

Green, M. (1985) Effect of milk pretreatment and making conditions on the properties of Cheddar cheese from milk concentrated by ultrafiltration. *J. Dairy Res.*, **52**, 555–64.

Green, M., Glover, F., Scurlock, E., Marshall, R. and Hatfield, D. (1981) Effect of use of milk concentrated by ultrafiltration on the manufacture and ripening of Cheddar cheese. *J. Dairy Res.*, **48**, 333–41.

Grieve, P. and Dulley, J. (1984) Use of *Streptococcus lactis* lac⁻ mutants for accelerating Cheddar cheese ripening. 2–Their effect on the rate of proteolysis and flavor development. *Aust. J. Dairy Technol.*, **38**, 49–54.

Gripon, J., Desmazeaud, M., Le Bars, D. and Bergère, J. (1977) Role of proteolytic enzymes of *Streptococcus lactis*, *Penicillium roqueforti* and *Penicillium caseicolum* during cheese ripening *J. Dairy Sci.*, **60**, 1532–8.

Guinee, T., Wilkinson, M., Mulholland, E. and Fox, P. (1991) The influence of ripening temperature, added commercial enzyme preparations and attenuated mutant (Lactose negative) *Lactococcus* starter bacteria on the proteolysis and maturation of Cheddar cheese. *J. Food Sci. Technol.*, **15**, 27–52.

Hansen, A. (1941) A study in cheese ripening. The influence of autolyzed cells of *Streptococcus cremoris* and *Streptoccus lactis* on the development of *Lactobacillus casei*. *J. Dairy Sci.*, **24**, 969–75.

Harper, W., de Carmona, A. and Kristoffersen, T. (1971) Protein degradation in Cheddar cheese slurries. *J. Food Sci.*, **36**, 503–6.

Harper, W., de Carmona, A. and Chen, J. (1980) Esterases of lactic streptococci and their stability in cheese slurry systems. *Milchwissenschaft*, **35**, 129–32.

Hayashi, K., Revell, D. and Law, B. (1990a) Effect of partially purified extracellular serine proteinases produced by *Brevibacterium linens* on the accelerated ripening of Cheddar cheese. *J. Dairy Sci.*, **73**, 579–83.

Hayashi, K., Revell, D. and Law, B. (1990b) Accelerated ripening of Cheddar cheese with the aminopeptidase of *Brevibacterium linens* and a commercial neutral proteinase. *J. Dairy Res.*, **57**, 571–7.

Hemme, D., Vassal, L. and Auclair, J. (1978) Stimulation of *Streptococcus thermophilus* by the addition of lactase or extracts of lactobacilli to milk. *Proc. 20th International Dairy Congress*, I.E.: 513–14.

Hickey, M., Van Leeuwen, A., Hiller, A. and Jago, G. (1983) Amino acid accumulation in Cheddar cheese manufactured from normal and ultrafiltered milk. *Aust. J. Dairy Technol.*, **38**, 110.

Hofi, A., Mahran, G., Abdel-Salam, M. and Riffaat, I. (1973a) Acceleration of cephalotyre (Ras) cheese ripening by using trace elements. II–Optimum conditions. *Egyptian J. Dairy Sci.*, **1**, 45–55.

Hofi, A., Mahran, G., Ashour, M., Khorshid, A. and Farahat, S. (1973b) The use of casein and whey protein hydrolysates in Ras cheese making. *Egyptian J. Dairy Sci.*, **1**, 79–83.

Johnson, J. and Etzel, M. (1995) Properties of *Lactobacillus helveticus* CNRZ-32 attenuated by spray-drying, freeze-drying, or freezing. *J. Dairy Sci.*, **78**, 761–8.

Johnson, J., Etzel, M., Chen, C. and Johnson, M. (1995) Accelerated ripening of reduced-fat Cheddar cheese using four attenuated *Lactobacillus helveticus* CNRZ-32 adjuncts. *J. Dairy Sci.*, **78**, 769–76.

Kamaly, K. and Marth, E. (1989) Enzyme activities of cell-free extract from mutant strains of lactic streptococci subjected to sublethal heating or freeze-thawing. *Cryobiology*, **26**, 496–500.

Kamaly, K., Johnson, M. and Marth, E. (1989) Characteristics of Cheddar cheese made with mutant strains of lactic streptococci as adjunct sources of enzymes. *Milchwissenschaft*, **44**, 343–6.

Kaminogawa, S., Mizobuchi, H. and Jamauchi, K. (1972) Comparison of bovine milk protease with plasmin. *Agric. Biol. Chem.*, **36**, 2163–7.

Kanawjia, S. and Singh, S. (1990) Effect of lipase addition on enhancement of flavor and biochemical changes in buffalo milk Cheddar cheese. *Indian J. Dairy Sci.*, **43**, 1–8.

Khalid, N., El-Soda, M. and Marth, E. (1990) Esterases of *L. helveticus* and *L. delbruckii* ssp. *bulgaricus. J. Dairy Sci.*, **73**, 2111.

Khalid, N., El-Soda, M. and Marth, E. (1991) Peptide hydrolases of *Lactobacillus helveticus* and *Lactobacillus delbrueckii* ssp. *bulgaricus. J. Dairy Sci.*, **74**, 29–45.

Kiely, L.J., Kindstedt, P.S., Hendricks, G.M., Levis, J.E., Yun, J.J. and Barbano, D.M. (1993) Age related changes in the microstructure of Mozarella cheese. *Food Struct.*, **12**, 13–20.

Kim, M., Kim, S. and Olson, N. (1994) Effect of commercial fungal proteases and freeze-shocked *Lactobacillus helveticus* CDR. 101 on accelerating cheese fermentation. 1. Composition. *Milchwissenschaft*, **49**, 56–9.

Kirby, C., Brooker, B. and Law, B. (1987) Accelerated ripening of cheese using liposome encapsulated enzyme. *Int. J. Food Sci. Technol.*, **22**, 355–75.

Kok, J., van Dijl, J., van der Vossen, M. and Venema, G. (1985) Cloning and expression of a *Streptococcus cremoris* proteinase in *Bacillus subtilis* and *Streptococcus lactis. Appl. Environ. Microbiol.*, **50**, 94.

Kristoffersen, T., Mikolajeik, E. and Gould, I. (1967) Cheddar cheese flavour. IV. Directed and accelerated ripening process. *J. Dairy Sci.*, **50**, 292–7.

Laloy, E. (1996) *Etude de l'incorporation de liposomes chargés d'enzymes d'affinage dans un caillé de fromage Cheddar*, PhD Thesis, Université Laval, Québec, Canada.

Laloy, E., Vuillemard, J., El-Soda, M. and Simard, R. (1996) Influence of the fat content of Cheddar cheese on retention and localization of starters. *Int. Dairy J.* (in press).

Lariviere, B., El-Soda, M., Soucy, Y., Trepanier, G., Paquin, P. and Vuillemard, J. (1991) Microfluidized liposomes for the acceleration of cheese ripening. *Int. Dairy J.*, **1**, 111–24.

Law, B. (1978) The accelerated ripening of cheese by the use of non conventional starters and enzymes. A preliminary assessment. *Bull. Int. Dairy Fed.*, **108**, 40–8.

Law, B. (1980) Accelerated ripening of cheese. *Dairy Ind. Int.*, **48**, 15.

Law, B. (1990) The application of biotechnology to the ripening of cheese. *Proc. 23rd Internat. Dairy Congress, Vol 2*, Mutual Press, Ottowa, Canada, pp. 1616–24.

Law, B. and King, J. (1985) Use of liposome for proteinase addition to Cheddar cheese. *J. Dairy Res.*, **52**, 183–8.

Law, B. and Mulholland, F. (1995) Enzymology of lactococci in relation to flavour development from milk proteins. *Int. Dairy J.*, **5**, 833–54.

Law, B. and Wigmore, A. (1983) Accelerated ripening of Cheddar cheese with a commercial proteinase and intracellular enzymes from starter streptococci. *J. Dairy Res.*, **50**, 519–26.

Law, B., Hosking, Z. and Chapman, H. (1979) The effect of some manufacturing conditions on the development of flavour in Cheddar cheese. *J. Soc. Dairy Technol.*, **32**, 87–90.

Law, J., Fitzgerald, G., Uniackelowe, T., Daly, C. and Fox, P. (1993) The contribution of lactococcal starter proteinases to proteolysis in Cheddar cheese. *J. Dairy Sci.*, **76**, 2455–67.

Lelièvre, J. and Lawrence, R. (1988) Manufacture of cheese from milk concentrated by ultrafiltration. *J. Dairy Res.*, **55**, 465–78.

Lucey, J. and Gorry, C. (1993) Effect of Simplesse 100 on manufacture of low-fat Cheddar cheese, in *Cheese yield and factors affecting its control*, International Dairy Federation, 9402, pp. 439–47.

Magee, E. and Olson, N. (1981a) Microencapsulation of cheese ripening systems: formation of microcapsules. *J. Dairy Sci.*, **64**, 600–10.

Magee, E. and Olson, N. (1981b) Microencapsulation of cheese ripening systems: stability of microcapsules. *J. Dairy Sci.*, **64**, 611–15.

Magee, E., Olson, N. and Lindsay, R. (1981) Microencapsulation of cheese ripening production of diacetyl and acetoin cheese by encapsulated bacteria cell-free extract. *J. Dairy Sci.*, **64**, 616–21.

Mashaly, R., Ezzat, N., El-Soda, M., El-Deeb, S. and Mostafa, F. (1986) Acceleration of Domiati cheese ripening by the addition of cell-free extracts from *Lactobacillus casei* to the cheese curd. *Indian J. Dairy Sci.*, **39**, 426–30.

McGarry, A., Law, J., Coffey, A., Daly, C., Fox, P. and Fitzgerald, G. (1994) Effect of genetically modifying the lactococcal proteolytic system on ripening and flavor development in Cheddar cheese. *Appl. Environ. Microbiol.*, **60**, 4226–33.

Moskowitz, G. (1980) Flavor development in cheese, in *The Analysis and Control of less Desirable Flavors in Foods and Beverages* (ed. G. Charalambous), Academic Press, New York, pp. 53–70.

Nassib, T. (1974) Acceleration of Ras cheese ripening by autolysed starter. *Assiut J. Agric. Sci.*, **5**, 123–9.

Nuñez, M., Guillen, A., Rodriguez-Marin, M., Marcilla, A., Gaya, P. and Medina, M. (1991) Accelerated ripening of ewes' milk manchego cheese: the effect of neutral proteinases. *J. Dairy Sci.*, **74**, 4108–18.

Oberg, C., Davis, L., Richard, G. and Ernstrom, C. (1986) Manufacture of Cheddar cheese using proteinases-negative mutants of *Streptococcus cremoris*. *J. Dairy Sci.*, **69**, 2975–81.

Pannell, L. and Olson, N. (1989) Production of methyl ketones in milk fat-coated microcapsules containing free fatty acids and *Penicillium roqueforti* spores. *J. Dairy Sci.*, **72** (suppl. 1), 117 (abstract).

Paquin, P., Lebeuf, Y., Richard, J. and Kalab, M. (1993) Microparticulation of milk proteins by high pressure homogenization to produce a fat substrate, in *Protein and Fat Globule modifications by heat treatment, homogenization and other technological means for high quality dairy products*, International Dairy Federation, Special Issue 9303, pp. 389–96.

Petterson, H. and Sjöstrom, G. (1975) Accelerated cheese ripening a method of increasing the number of lactic starter without detrimental effect to the cheese making process and its effect on cheese ripening. *J. Dairy Res.*, **42**, 313–26.

Piard, J., El-Soda, M., Alkhalaf, W., Rousseau, M., Desmazeaud, M., Vassal, L., and Gripon, J. (1986) Acceleration of cheese ripening with entrapped proteinase. *Biotechnol. Lett.*, **8**, 241–6.

Picon, A., Gaya, P., Medina, M. and Nunez, M. (1994) The effect of liposome encapsulation of chymosin derived by fermentation on Manchego cheese ripening. *J. Dairy Sci.*, **77**, 16–23.

Ponce-Trevino, R., Richter, R. and Dill, C. (1987) Influence of lactic acid bacteria on the production of volatile sulfur-containing compounds in Cheddar cheese slurries. *J. Dairy Sci.*, **70** (suppl. 1), 59.

Prost, F. and Chamba, J. (1994) Effect of aminopeptidase activity of thermophilic lactobacilli on Emmental cheese characteristics. *J. Dairy Sci.*, **77**, 24–33.

Rajesh, P. and Kanawya, S. (1990) Effect of starter cultures on yield, composition and sensory characteristics of buffalo milk Gouda cheese using hannilase rennet. *Indian J. Dairy Sci.*, **43**, 608–13.

Ray, B. and Speck, M. (1973) Freeze-injury in bacteria. *Crit. Rev. Clin. Lab. Sci.*, **4**, 161–7.

Revah, S. and Lebeault, J. (1989) Accelerated production of blue cheese flavors by fermentation of granular curd with lipase addition. *Le Lait*, **69**, 281–9.

Richardson, G., Ernstrom, C., Kim, J. and Daly, C. (1983) Proteinase negative variants of *Streptococcus cremoris* for cheese starters. *J. Dairy Sci.*, **66**, 2278–86.

Riepe, H. and McKay, L. (1994) Oversecretion of the neutral protease from *Bacillus subtilis* in *Lactococcus lactis* spp. *lactis* JF254. *J. Dairy Sci.*, **77**, 2150–9.

Sanders, G., Tittsler, R. and Walter, H. (1946) The rapid ripening of Cheddar cheese made from pasteurized milk. US Dept. Agr. Bur. Dairy Ind., BDIM-Inf-29.

Scolari, G., Vescovo, M., Sarra, P.G., Bottazzi, V. (1993) Proteolysis in cheese made with liposome entrapped proteolytic enzymes. *Le Lait*, **73**, 281–92.

Shehata, A., Shehata, T., Hagras, A. Ali, A. (1995) Acceleration of Ras cheese ripening with liposomes-entrapped *Bacillus subtilis* protease. *Proceedings of the 6th Egyptian Conference for Dairy Science and Technology*, 41 (abstract).

Singh, S. and Kristoffersen, T. (1971) Accelerated ripening of direct acidified curd. *J. Dairy Sci.*, **54**, 756–9.

Singh, S. and Kristoffersen, T. (1972) Cheese flavor development using direct acidified curd. *J. Dairy Sci.*, **55**, 744–9.

Spangler, P., El-Soda, M., Johnson, M., Olson, N., Amundson, C. and Hill, C. (1989) Accelerated ripening of Gouda cheese made from ultrafiltered milk using liposome entrapped enzyme and freeze-shocked lactobacilli. *Milchwissenschaft*, **44**, 199–203.

Stadhouders, J. (1961) The hydrolysis of protein during the ripening of cheese. Some methods to accelerate the ripening. *Neth. Milk Dairy J.*, **15**, 151–64.

Stadhouders, J., Hup, G., Exterkate, F. and Visser, S. (1983) Bitter flavour in cheese. I–Mechanism of the formation of the bitter flavour defect in cheese. *Neth. Milk Dairy J.*, **37**, 157–67.

Stadhouders, J., Toepoel, L. and Wouters, J. (1988) Cheese making with prt⁻ variants of N-streptococci and their miturex. Phage sensitivity, proteolysis and flavour development during ripening. *Neth. Milk Dairy J.*, **42**, 183–93.

Thakar, P. and Upadhyay, K. (1992) Cheese and slurry, a review. *Cultured Dairy Prod. J.*, **27**, 9–12.

Tomasini, A., Bustillo, G. and Lebeault, J. (1993) Fat lipolyzed with a commercial lipase for the production of Blue cheese flavour. *Int. Dairy J.*, **3**, 117–27.

Trépanier, G., Simard, R. and Lee, B. (1991) Lactic acid bacteria relation to accelerated maturation of Cheddar cheese. *J. Food Sci.*, **56**, 1238–40.

Vafopoulou, A., Atichanidis, E. and Zerfiridis, G. (1989) Accelerated ripening of Feta cheese with heat-shocked cultures or microbial proteinases. *J. Dairy Res.*, **56**, 285–96.

van de Guchte, M., Kadde, J., Vossen J., Kok, J. and Venema, G. (1990) Heterologous gene expression in *Lactococcus lactis* subsp. *lactis*: synthesis, secretion, and processing of the *Bacillus subtilis* neutral proteinase. *Appl. Env. Microbiol.*, **56**, 2606–11.

Wilkinson, M. (1990) Acceleration of cheese ripening. *Proceedings 2nd Moorepark Cheese Conference*, pp. 111–19.

Wilkinson, M., O'Keeffe, A. and Fox, P. (1989) Factors affecting lysis of starter bacteria in Cheddar and Cheshire cheese. *Irish J. Food Sci. Technol.*, **13**, 158 (abstract).

Wilkinson, M., Guinee, T., O'Callaghan, D. and Fox, P. (1992) Effect of commercial enzymes on proteolysis and ripening in Cheddar cheese. *Le Lait*, **72**, 449–59.

Wilkinson, M., Guinee, T. and Fox, P. (1994) Factors which may influence the determination of autolysis of starter bacteria during Cheddar cheese ripening. *Int. Dairy J.*, **4**, 141–60.

Yatvin, M., Tegmo-Larsson, I., Dennis, W. (1987) Temperature and pH sensitive liposomes for drug targeting, in *Methods in Enzymology*, Vol. 149. (ed. R. Green and W.H. Dennis), Academic Press, Inc., San Diego, CA, USA, pp. 77–87.

8 The chemical and biochemical basis of cheese and milk aroma

G. URBACH

8.1 Introduction

Flavour is defined as taste plus aroma. All aroma compounds are volatile but by no means all volatiles contribute to aroma. As taste is discussed in Chapter 9, the present chapter is restricted to volatile compounds. Taste is perceived on the tongue but aroma is perceived in the nose where it enters either from the outside by sniffing, or from the back of the throat when food is masticated.

8.2 Isolation and identification of volatiles from dairy products

The first step in the identification of aromas must be the separation of volatiles from the food matrix, usually in a form in which they can be subjected to gas chromatography. The choice then is either to attempt to identify all the volatiles, usually by a combination of gas chromatography and mass spectrometry (GC/MS), or to sniff the effluent from the gas chromatograph and identify only the areas of interest. The second method is used most often where off-flavours need to be identified. This method fails, when applied to desirable aroma, if the desirable aroma is produced by the physiological interaction of two or more compounds or if the compound responsible for the desirable aroma is very unstable. This is the case with Cheddar cheese where methanethiol appears to be essential for the aroma but, on its own, smells like cooked cabbage and needs to be mixed with one or more (as yet unidentified) acidic compounds to produce the typical Cheddar odour (Wijesundera and Urbach, 1993). In all cases the total effluent from the GC needs to be collected to check whether it reproduces the aroma of the injected sample. A critical overview of methods used in aroma chemistry is given by Maarse and Belz (1981).

Methods used for aroma isolation and identification at the Swiss Federal Dairy Research Institute (FAM) are described by Bosset and Lavanchy (1991); these are high-vacuum distillation using tower extraction, steam distillation using a rotary evaporator, direct dynamic headspace collection and injection into the gas chromatograph, direct coupling between super-critical fluid extraction and capillary fluid chromatography. Bosset *et al.*

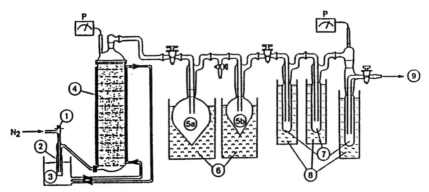

Figure 8.1   Apparatus used for high-vacuum distillation (tower extraction). 1–3, Pressure, flow and moisture regulators for the stripping gas (nitrogen); 4, Extraction tower filled in layers with grated cheese; 5a, 5b, Cold traps ($-77°C$); 6, Coolant (alcohol and dry ice); 7, Cold traps ($-196°C$); 8, Coolant (liquid nitrogen); 9, Diffusion pump. P, pressure gauges. (Reproduced with permission from *Lebensmittel-Wissenschaft und -Technologie*; Liardon *et al.*, 1982.)

(1996) have since updated their survey of methods of isolation of volatiles and have compared the effectiveness of four methods of isolating volatiles from Swiss Emmental cheese, namely high-vacuum distillation, steam distillation using a Rotavapor, and two systems of dynamic headspace analysis, MWS-1[R] from Rektorik and LSC-2000[R] purge-and-trap equipment from Tekmar. The authors conclude that, although there is no ideal method of extraction, the dynamic headspace analysis using the LSC-2000[R] is the most promising technique due to the small sample size (3.5 g), the short analysis time (90 min, including sample preparation) and the large number of compounds obtainable (98 peaks).

8.2.1   Tower extraction

Tower extraction is illustrated in Figure 8.1. It is carried out at 0–10°C under high vacuum (Liardon *et al.*, 1982) and has the advantage of avoiding artefacts resulting from heat or oxidation.

8.2.2   Steam stripping

Figure 8.2 illustrates a steam stripping system which is operated at 60°C with a rotary evaporator (Imhof and Bosset, 1989). This system is considerably faster than the tower extraction and requires only 250 g compared with 2.5 kg cheese, but could be subject to heat-induced artefacts.

8.2.3   Supercritical fluid extraction

Figure 8.3 illustrates the apparatus developed at FAM for supercritical fluid extraction (SFE) with CO_2 with direct coupling to a capillary column in a

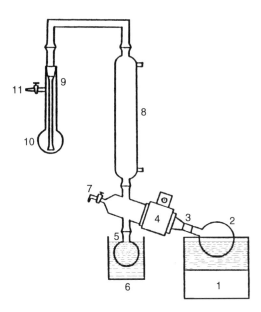

Figure 8.2 Steam distillation using the Rotavap rotary evaporator. 1. Water bath (60°C); 2, Round-bottomed flask containing the sample dispersed in water; 3, 4, 7, Rotary evaporator; 5, 6, Preliminary trap cooled with ice water; 8, Low-temperature cooling (about − 15°C); 9, 10, Liquid nitrogen trap; 11, Water pump. (Reproduced with permission from *Mitteilungen aus dem Gebiete der Lebensmitteluntersuchung und Hygiene*; Imhof and Bosset, 1989.)

supercritical fluid chromatograph (SFC) (Gmür *et al.*, 1987a, b, c). Gmür *et al.* (1986) found that the optimal conditions for the extraction of free fatty acids, which are important in the aroma of many cheese varieties, were 100 bar and 40°C. If the supercritical fluid extract is to be examined by gas chromatography/mass spectrometry, the extract first has to be subjected to high-vacuum distillation to separate the small amount of non-volatile material which SFE extracts together with the volatiles.

8.2.4 *Molecular distillation and concentration*

Since heating often produces undesirable chemical changes, distillations in the Dairy Research Laboratory (now renamed Melbourne Laboratory) of the CSIRO Division of Food Science and Technology are generally performed under vacuum. With products such as butter oil, which have a strong affinity for lipophilic volatiles and hence reduce their volatility, molecular distillation is used. Here, high-boiling compounds such as C12 lactones and acids can be collected from a food matrix at 35°C onto a cold-finger filled with liquid nitrogen (Stark *et al.*, 1973). The apparatus is shown diagramatically in Figure 8.4.

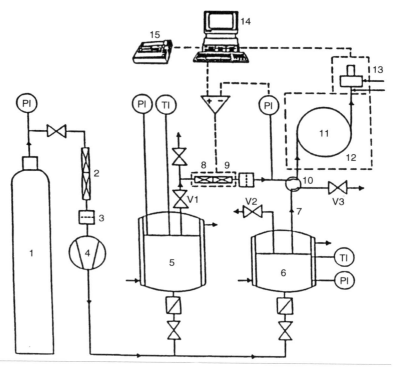

Figure 8.3 Apparatus with direct coupling between supercritical fluid extraction (SFE) and supercritical capillary fluid chromatography (SFC). 1–5, Preparation of the supercritical gas; 6, Extraction autoclave (SFE); 7, Capillary connection (SFE–SFC); 8, Pressure programming cylinder (SFC); 9, Gas filter (SFC); 10, Sample inlet (SFC); 11, Chromatographic capillary column (SFC); 12, Programmable oven (SFC); 13, Flame ionization detector (SFC); 14, Control computer (SFC); 15, Printer (SFC). (Reproduced with permission from the *Journal of Chromatography*; Gmür *et al.*, 1987b.)

When cheese is analysed it is grated into liquid nitrogen with a kitchen grater and the frozen cheese is powdered using mortar and pestle (Dimos, 1992). In the analysis of cheese, an aqueous distillate (about 30 ml from 100 g cheese) is collected in trap F on the first day and a cold-finger distillate is collected on the second day. Actually, in the case of Cheddar, the aroma collects in the aqueous fraction and the higher-boiling compounds which collect on the cold-finger do not contribute to Cheddar aroma.

The aqueous distillate is very dilute and must be concentrated. One method is solvent extraction with concentration of the extract. The disadvantages of this method are: (i) interference by the solvent in subsequent analysis; (ii) the solvent does not extract highly polar materials; and (iii) low-boiling compounds are lost during concentration, even if a very efficient reflux condenser is used. Other methods available for concentration of the

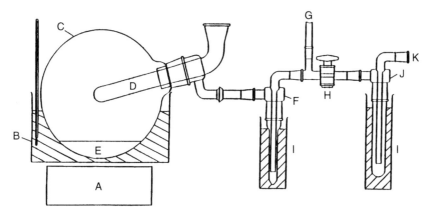

Figure 8.4 Apparatus used for cold-finger molecular distillation. A, Magnetic stirrer and heater (in the case of cheese the stirrer is not used); B, water bath; C, 5-l flask with B55 neck; D, cold finger; E, powdered cheese; F, trap (in the case of cheese this trap is replaced with a 100-ml capacity trap); G, to vacuum gauge; H, stopcock, 9-mm bore; I, liquid N_2 vessels; J, trap; K, to 2-way stopcock with capillary and thence to mechanical and diffusion pumps. (Reproduced with permission from the *Journal of Dairy Research*; Stark *et al.*, 1973.)

aqueous distillate are freeze concentration (Kepner *et al.*, 1969), distillation at 0 to 20 torr through a vertical condenser held at 0°C (concentration under reflux), and vacuum sublimation (Forss *et al.*, 1967).

(a) Freeze concentration. An apparatus which can be used for freeze concentration is shown in Figure 8.5. The beaker containing the aqueous solution is cooled in a mixture of salt and ice or other suitable method of cooling. The aqueous solution is constantly stirred, the stirrer being lifted as the water freezes onto the walls of the beaker. If the rate of freezing and the rate of stirring have been correctly adjusted, 5–40-fold concentration of solutes can be obtained in the last few millilitres of liquid which remains unfrozen.

(b) Concentration under reflux. The apparatus for concentration under reflux is shown in Figure 8.6. Concentration factors of 400 can be achieved with 70–100% recovery for compounds with up to 10 carbon atoms, depending on their chemical nature.

(c) Vacuum sublimation. Both freeze concentration and concentration under reflux can be followed by vacuum sublimation, or vacuum sublimation can be used as the sole means of concentration. The apparatus for vacuum sublimation is illustrated in Figure 8.7.

   A 5-ml portion of aqueous distillate or concentrate, in a 25-ml round-bottom flask, is frozen in liquid nitrogen and the system evacuated through

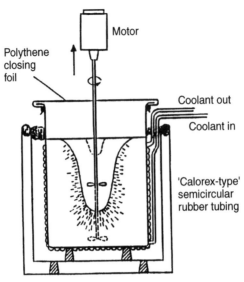

Figure 8.5 Apparatus for freeze-concentration. (Reprinted with permission from *J. Agric. Food Chem.*, Copyright 1967, American Chemical Society; Kepner *et al.*, 1969.)

tap D to a pressure of less than 10 micron. Tap D is then closed, the liquid nitrogen is removed from flask A, and the sample is allowed to sublime through trap C, which is cooled in a bath of dry-ice and acetone (to a suitable temperature between −40°C and −60°C) into another trap G which is cooled in liquid nitrogen. Most of the water remains in trap C, producing a concentrate in trap G which can be analysed directly by GC/MS. Up to 200-fold concentration can be achieved. When reflux concentration and vacuum sublimation are combined, up to 40 000-fold concentration can be achieved. The main disadvantage of vacuum sublimation is that it takes a whole day to concentrate a very small volume (5 ml).

8.2.5   Headspace analysis

(a) Static headspace analysis.   Figure 8.8 illustrates the cheese headspace sampler of Price and Manning (1983) which is used in the Dairy Research Laboratory for static headspace analysis of cheese. A 15-g sample is taken with the trier and allowed to equilibrate at 30°C. It is then shredded by being pushed through the perforated plate into the 50-ml chamber from where the headspace is sampled via the septum after exactly 5 min. Samples (5 ml) are analysed either by gas chromatography with a flame photometric detector (FPD) for sulphur compounds or by GC/MS for other volatiles (Dimos, 1992).

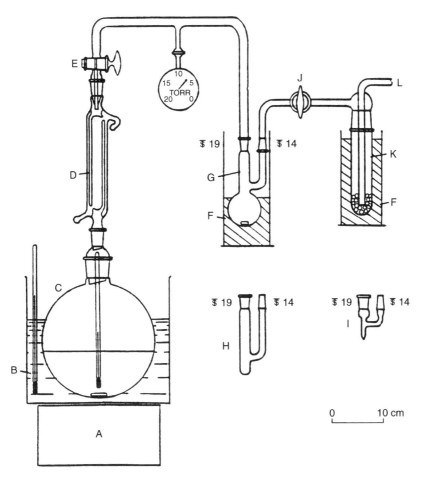

Figure 8.6 Apparatus for concentration under reflux (Forss *et al.*, 1967). A, Magnetic stirrer and heater; B, water bath; C, 5-l flask with B34 neck; D, double-surface (Davies) condenser with B24 joints; E, stopcock, 2-mm; F, liquid nitrogen vessel; G, collecting flask, volume 100 ml; H, collecting traps for volumes 2 to 10 ml; I, collecting trap for volumes below 1 ml; J, stopcock, 4-mm; K, trap containing glass beads; L, connection to mechanical pump, pressure 10^{-2} torr. (Reprinted with permission from *J. Agric. Food Chem.*, Copyright 1967, American Chemical Society.)

For the analysis of cheese curd slurries (Roberts *et al.*, 1995) 10 ml cheese curd slurry, in a 20-ml serum bottle, is deoxygenated by flushing with N_2 (purified by passage through a Midisart 2000 filter). The bottles are then sealed with Teflon lined silicone rubber caps, and stored for up to 2 months at 30°C. Samples (5 ml) of the headspace gases from separate bottles are analysed by GC/FPD and GC/MS at regular intervals. This technique is

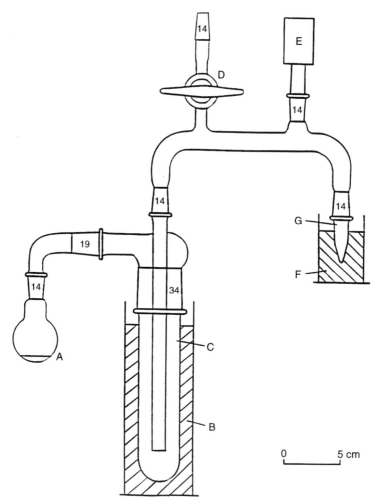

Figure 8.7 Apparatus for concentration by vacuum sublimation (Forss *et al.*, 1967). A, 25-ml round-bottomed flask containing 5 ml of aqueous distillate; B, dry-ice/acetone coolant; C, trap through which sample is allowed to sublime; D, stopcock (6-mm) to diffusion pump; E, Pirani gauge; F, liquid nitrogen vessel; G, trap in which concentrated sample is collected (less than 100 μl). Numbers indicate joint sizes. (Reprinted with permission from *J. Agric. Food Chem.*, Copyright 1967, American Chemical Society).

currently being used in an attempt to determine the flavour-producing potential of different starter cultures.

(b) Dynamic headspace analysis. For dynamic headspace analysis of cheese (Horwood, 1989) or milk (Urbach, 1987) the matrix, in a 1- to 3-l flask, is purged with ultrapure helium via a special washing bottle head with

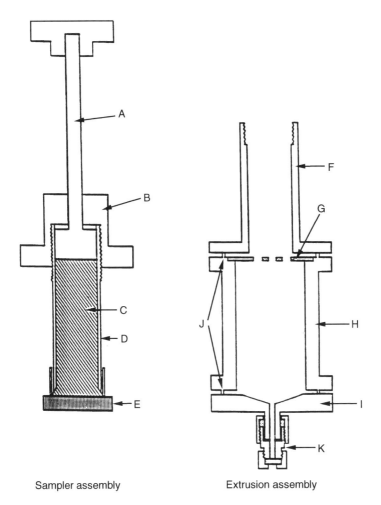

Figure 8.8 Cheese headspace sampler of Price and Manning (1983). A, plunger; B, handle; C, borer; D, cheese bore; E, cap; F, barrel; G, extrusion plate; H, main chamber; I, end-plate; J, sealing lips; K, septum port. (Reproduced with permission from the *Journal of Dairy Research.*)

a perforated glass ball at the end. The effluent is collected either on oxidized Porapak Q (Horwood, 1989) or NIOSH charcoal (Urbach, 1987) or Tenax or in dry-ice/acetone or in liquid nitrogen (Wijesundera and Urbach, 1993). Porapak Q, NIOSH charcoal and Tenax traps are heat-desorbed onto the chromatographic columns.

Imhof and Bosset (1994) used dynamic headspace analysis with standard addition of the compounds to be determined for the quantitative determination of volatile aroma compounds in pasteurized milk and fermented milk products.

8.3   Quantitative estimations

8.3.1   Estimation of acids

Bouzas *et al.* (1991a, b) developed a method for the simultaneous estimation of citric, orotic, pyruvic, lactic, formic, acetic, propionic, butyric and hippuric acids, all of which they found in 60-day-old commercial Cheddar cheese. In a simple, rapid isocratic HPLC method, sugars and organic acids were separated on an Aminex HPX-87 column in the H^+ form and detected using ultraviolet and refractive index detectors in series. Bouzas *et al.* (1991a) point out that orotic acid has been reported to be readily utilized by various bacteria used in the fermentation of dairy products.

8.3.2   Estimation of α-dicarbonyls

The α-dicarbonyls, diacetyl, glyoxal and methylglyoxal, are regarded as being important for the development of cheese aroma (Griffith and Hammond, 1989). McDonald (1992) used the reaction with 2,3-diaminonaphthalene and subsequent high-pressure liquid chromatography on a reverse-phase C18 column with fluorescence detection to determine these α-dicarbonyls in the culture supernatants from a wide variety of lactic acid bacteria and in Cheddar cheese manufactured with selected lactic acid bacterial adjunct cultures. However, because of the very reactive nature of the α-dicarbonyl compounds with amino compounds in media and cheese, only surplus concentrations, and not total concentrations were measured (Lindsay, 1994). Using bisulphite as a competitive binding compound for amino acids, it became possible to add sufficient levels of this compound to 'tie-up' the α-dicarbonyls and still permit growth of lactic acid ripening organisms. In this manner it became possible to study the rates of actual production of α-dicarbonyls by lactic organisms.

de Revel and Bertrand (1993) analysed the very volatile (*O*-2,3,4,5,6-pentafluorobenzyl) hydroxylamine hydrochloride derivatives of the α-dicarbonyls by GC/MS or GC/ECD.

8.3.3   Estimation of glutathione (GSH)

Glutathione (γ-glutamylcysteinylserine) appears to play an important role in the production of cheese aroma, although its exact functions in cheese have not been clearly defined.

Fernándes and Steel (1993) estimated glutathione in deproteinated bacterial cultures by reverse-phase high pressure liquid chromatography. Thiol compounds were derivatized with 5,5′-dithiobis-(2-nitrobenzoic acid) reagent and separated isocratically on a Hibar[R] LiChrosorb[R] RP-18 column (E. Merck, Darmstadt, Germany) with a UV detector set at 280 nm.

Enzymatic assays using glyoxylase or isomerases from a strain of *Pseudomonas*, or glutathione reductase have also been described (Jocelyn, 1972). Harper and Kristoffersen (1970) used glutathione reductase to measure oxidized glutathione, and glyoxylase to measure reduced glutathione. Fahey and Newton (1983) used reverse-phase, high-performance liquid chromatography of the monobromobimane derivatives of thiols, with fluorometric detection, for the identification and quantitation of all bacterial thiols, including glutathione, coenzyme-A, H_2S, SSO_3^{2-}, cysteine, 2-mercaptoethanesulphonic acid, 2-mercaptopyridine, pantetheine, 4'-P-pantetheine, homocysteine, γ-glutamylcysteine, cysteinylglycine as well as three unidentified thiols. Fahey and Newton (1983) found H_2S in all 12 species of bacteria which they examined. Glutathione was found mainly in facultative and aerobic Gram-negative bacteria, but even there it was not always the main thiol present, this being CoA in *Beneckea alginolytica*. In lactic acid bacteria the identities of thiols, other than glutathione and CoA, do not appear to have been investigated.

Recently, Gotti *et al.* (1994) described a selective high-pressure liquid chromatography determination of reduced glutathione based on pre-chromatographic derivatization with 4-(6-methylnaphthalen-2-yl)-4-oxo-2-butenoic acid. UV and fluorescence detection (l_{em} 450 nm, l_{exc} 300 nm) were both used.

8.3.4 Determination of redox potential

Kristoffersen and Gould (1959) used both indicator dyes and embedded platinum electrodes to determine the oxidation–reduction potentials of cheese. They obtained negative but different values in both cases and in both cases the initial low (i.e. negative) value increased and then fell again on further maturation (average values $-104\,mV$ initially, $-90\,mV$ after 1 month and $-217\,mV$ after 6 months), presumably reflecting the initial dying-off of the starter microflora to be replaced later by adventitious secondary microflora. Where these trends were significantly different from those described, the cheeses developed acidy, fermented, rancid or oxidized aromas. Horwood *et al.* (1994) incubated cheese curd slurries with three different levels of O_2 and also found that these slurries developed fermented aromas compared with anaerobically incubated slurries, which developed normal Cheddar aroma. The O_2 also raised the acetic acid content of the slurries, which these workers attributed to the oxidative decarboxylation of pyruvic acid; this has been shown to increase in aerated slurries (Harper, 1959). However, in cheese films with different O_2 transmission rates, the appearance of acetate is associated with the oxidation of lactate by cheese non-starter lactic acid bacteria, such as *Pediococcus pentosaceus* (Thomas, 1987). This could also explain the acetate levels of the oxygenated slurries.

Samples (1985) used platinum electrodes and a saturated calomel electrode to measure the oxidation–reduction potential in Cheddar cheese curd slurries. He was careful to eliminate oxygen and to have good contact between the slurries and the electrodes. Galesloot (1960) sealed the electrodes into the cheese to measure the oxidation–reduction potential of Gouda cheese. However, workers at the CSIRO Dairy Research Laboratory found that when they used this method for Cheddar cheese they had difficulty with making good contact between the electrodes and the cheese. When the electrodes were slightly moved the oxidation–reduction potential reading changed. It is doubtful how much meaning can be attatched to the oxidation–reduction potential of a heterogeneous medium such as cheese, although Kristoffersen et al. (1964b) do appear to have obtained reproducible readings using embedded platinum electrodes.

8.4 Aroma compounds in cheese and fermented milks

8.4.1 General considerations

In cultured milks, cultured butter and fresh cheese, which is ready to be eaten immediately after manufacture, the main aroma compounds are usually assumed to be diacetyl and acetaldehyde with the acid background provided largely by lactic acid. Recently, however, 2,3-pentanedione was also shown to be a major contributor to the aroma of numerous single strain and mixed-strain starters used for cultured milk products. Some 102 peaks were found by gas chromatography/mass spectrometry, of which 36 were identified (Imhof et al. , 1994a, 1995).

The aroma of matured cheese is the result of the interaction of starter bacteria, enzymes from the milk, enzymes from the rennet and accompanying lipases, and secondary flora. The starter culture, method of manufacture of the cheese and the secondary flora determine the cheese variety. The effect of enzymes from non-starter bacteria in the raw milk, and of enzymes from the milk itself (Samples, 1985), is considerably reduced by pasteurization or microfiltration (Bouton and Grappin, 1995) and pasteurization also affects the protein structure (McSweeney et al., 1993). Decrease in the number of non-starter bacteria and inactivation of native milk enzymes accounts for the fact that cheese from pasteurized milk lacks the full aroma of traditional cheese made from unpasteurized milk.

The pathway of protein breakdown to amino acids has been well established (Chapters 1 and 9) and it is generally assumed that this breakdown is essential for aroma formation in matured cheeses. However, increasing the level of free amino acids does not always produce an increase in aroma intensity (Law et al., 1976; Olson, 1994). Obviously, an increase in aroma intensity requires further metabolism of the amino acids (Ardö,

1994). Results from recent research suggest that good flavour-producing starters lyse more rapidly than starters which do not produce such intense flavour (Wilkinson *et al.*, 1994). The enzymes which are released by lysis thus appear to be important in flavour production. The fast-lysing bacteria also produce a higher level of amino acids than the more slowly lysing bacteria. Although amino acids themselves do not contribute to volatile aroma, they appear to be the major precursors. Just how aroma in cheese is produced from amino acids has not been established. Amino acids, and probably peptides, also act as substrates for the secondary flora. Protein breakdown products also contribute to the background savoury, non-specific cheesy flavour of all cheeses. The metabolites of citric acid, particularly diacetyl, have also been shown to play a major role in aroma production although it is not clear how these compounds influence flavour. It has been suggested that some essential aroma compounds are formed by the purely chemical interaction of carbonyl compounds and amino acids (Griffith and Hammond, 1989) and by other purely chemical reactions (Green and Manning, 1982). Although, on the face of it, this is an attractive theory, it does not explain the reason for the differences between cheese varieties. In a medium such as cheese, enzymatic reactions are much more likely. Thus, it has recently been shown that methanethiol is produced enzymatically by the action of a cell-free extract of starter bacteria on methionine (producing α-ketobutyric acid) (P. Bruinenberg, unpublished data).

The biochemical transformations involved in the ripening of cheese include not only the catabolism of proteins, milk fat, lactose and citrate but also shunts between these pathways. For example, Harper and Kristoffersen (1970) incorporated radioactive glucose into cheese curd slurry systems and recovered radioactivity in lactic acid, acetaldehyde, diacetyl, carbonyl compounds and amino acids. Lin *et al.* (1979) incorporated uniformly labelled ^{14}C-glucose in slurries of fresh Cheddar curd with and without added reduced glutathione. During ripening, lactic, pyruvic, α-ketoglutaric, formic, acetic, propionic and to a lesser extent butyric acids, acetaldehyde and diacetyl, were derived from carbohydrate at least in part. A greater proportion of the acids other than lactic and of acetaldehyde was derived from carbohydrate in the presence of added reduced glutathione than in its absence. Radioactive CO_2 was incorporated into malic and aspartic acids by *Lactobacillus casei* and into numerous amino acids by commercial cultures (Schormller, 1962). Harper and Wang (1980a, b, 1981) also showed that ^{14}C-alanine was catabolized in a cheese curd slurry to pyruvate, α-ketoglutarate, CO_2, acetaldehyde and diacetyl and ^{14}C-U-glutamic acid was also catabolized to CO_2, α-ketoglutaric acid, pyruvate and traces of acetaldehyde and diacetyl. Because Harper and Wang found no evidence of transamination, whereas other workers (Holden *et al.*, 1951; Schormüller, 1968) did report transaminations by lactic acid bacteria grown in synthetic cultures, they suggested that ripening processes in slurries appear to differ from metabolic studies with pure cultures. Although slurries produce correct cheese flavours, characteristic of the cheese type from which the slurry was made, even here, care must be taken when extrapolating from slurries to cheese. Harper and Kristoffersen (1970) did find that headspaces from Swiss cheese and free fatty acids and free peptides from Cheddar cheese were practically identical with these compounds found in the corresponding slurries. However, in later work (Harper *et al.*, 1979) nearly normal flavour and lower-molecular weight free fatty acid patterns were obtained by the slurry approach for fat-modified Cheddar and Romano, whereas, for cheese, there was less flavour development in fat-modified Cheddar and practically none in fat-modified Romano. Harper *et al.* (1979) attributed this finding to a difference in diffusivity of substrate molecules and higher microbial enzyme activity in the slurry system.

Reduced-fat cheese generally lacks flavour. It is not clear whether this is due to the lack of a precursor from the fat, the lack of the solvent power of the fat thus allowing essential aroma compounds to escape, or the different physical structure of the reduced-fat cheese which inhibits certain enzyme reactions that are essential for the formation of aroma compounds. Escape of esters, as well as a certain amount of oxidation, was shown to be responsible for the reduction in aroma in grated Parmesan when compared with Parmesan in a block (Dumont *et al.*, 1974b).

A high proportion of volatile compounds identified in foods is common to nearly all foods. For example, wine and cheese have many volatiles in common (Maarse *et al.*, 1989). Therefore the important aroma compounds must be sought among the relatively small number of aroma character-impact compounds, i.e. compounds which, on their own, are reminiscent of the product. Nonetheless, patterns of volatiles, many of which may not contribute to flavour, have been shown to be characteristic of cheese varieties (Palo and Hollá, 1982, 1983) and even distinguish between mild, tasty and vintage Cheddar (Lloyd and Ramshaw, 1985).

8.4.2 Ketones and alcohols

In semi-hard, mould-ripened cheese, such as blue cheese, heptan-2-one is an important aroma character impact compound, whereas in soft-type, mould-ripened cheeses nonan-2-one predominates (Adda *et al.*, 1982). The methyl ketones are formed by the action of the moulds, *Penicillium roqueforti* and *Penicillium caseicolum* whose enzymes lipolyse the fat and oxidize the resulting saturated fatty acids to β-ketoacids via the β-ketoacyl-coenzyme A. The β-ketoacids are then decarboxylated by decarboxylase to methyl ketones (Kinsella and Hwang, 1976; Okumura and Kinsella, 1985). The methyl ketones are in turn partially reduced to alkan-2-ols which are also important in the aroma of blue-vein cheese, having similar but heavier aroma notes to those of methyl ketones (Forss, 1972).

The volatile compounds that characterize the aroma of surface-ripened cheeses have been attributed collectively to the action of residual starter bacteria, *Penicillium* moulds, and lactate-utilizing yeasts which raise the pH and provide growth factors for coryneform bacteria such as *Brevibacterium linens*. Oct-1-en-3-ol provides the characteristic mushroom-like aroma to Camembert cheeses and imparts a masking effect on the blue-cheese aroma notes of methyl ketones. 1,5-Octadien-3-ol, 1,5-octadien-3-one, 3-octanol, 3-octanone (all green-plant-like), 8-nonen-2-one (blue-vein cheese-like; Yamamoto, 1976), 2-methylisoborneol (musty, mouldy), 2-methoxy-3- isopropylpyrazine (earthy, raw potato) are important contributors to the heavy earthy notes of Camembert and Brie aroma (Karahadian *et al.*, 1985). The C8 compounds are produced by the action of lipoxygenase from *Penicillium caseicolum* on linoleic and linolenic acids. The microflora of the surface smear is also responsible for the formation of the major methyl ketones of Swiss Gruyère (heptan-2-one, nonan-2-one and acetophenone) which, as well as the ketoalcohols (butan-3-ol-2-one, 3-methylbutan-3-ol-2-one, pentan-3-ol-2-one, pentan-2-ol-3-one, pentan-2-ol-4-one), occur in the surface smear (Bosset and Liardon, 1984). Methyl ketones, particularly heptan-2-one, contribute to the aroma of Parmesan (Parmigiano) (Meinhart and Schreier, 1986) and methyl ketones and the corresponding alcohols also play a major role in the aroma of Gorgonzola (Martelli, 1989).

The significance of methyl ketones in Cheddar aroma is not established. Significant quantities of pentan-2-one and heptan-2-one in the headspace of Cheddar is an indication of contamination with mould (Horwood and Urbach, 1990). Butanone is generally found in increasing quantities (accompanied by butan-2-ol) in maturing Cheddar. However, the headspace from premium quality mature Cheddar does not necessarily contain significant quantities of butanone and/or butan-2-ol, indicating that these compounds do not contribute to the aroma of Cheddar (Urbach, 1993).

8.4.3   Acids

Fatty acids mostly originate from the lipolysis of the milk-fat. The lipase responsible for their production probably originates from the milk itself, since milk-fat must be partially hydrolysed before the starter culture lipases become effective (Stadhouders and Veringa, 1973). A high proportion of the volatile fatty acids (Parodi, 1979) have been shown to be in the α-position of the triglyceride molecule. This is attacked preferentially by the lipoprotein lipase from the milk (Deeth and Fitz-Gerald, 1983) but not by the lipases from the starter bacteria. Reiter and Sharpe (1971) consistently found that the free fatty acid content of cheese-milk varied according to the method of milking, feeding regime and herd, quite independently of the lipolytic bacterial count; this was also reflected in the cheese. Although the lactic acid bacteria were lipolytic, their contribution to the free fatty acid levels in cheese were less important than that of the native milk lipase. The milk lipase was not permanently inactivated by heating and was reactivated in 24 h at 30°C. El Soda et al. (1986) confirmed earlier observations indicating that lactic acid bacteria have limited ability to hydrolyse triglycerides from milk-fat.

Low-molecular weight fatty acids (C_2–C_{10}) in cheese and cheese curd slurries are not necessarily formed by the lipolysis of milk-fat but also occur in systems where the milk-fat has been replaced by vegetable oil which does not contain these acids (Harper et al., 1978). Indeed, these authors suggest that such acids are synthesized from acetyl CoA since it was shown that small amounts of butyric acid are formed from glucose carbon in Cheddar cheese slurries made with milk-fat. When Harper et al. (1978) substituted various vegetable lipids for milk-fat in Romano and Cheddar cheese which they then used in slurry systems, up to 90% as much low-molecular weight free fatty acids were formed in slurries made with partially hydrogenated soya bean oil as were produced in the milk-fat control systems.

Fatty acids also occur as conjugates with amino acids or as 1-O-glycosidylesters in the milk of cows, sheep and goats (Lopez and Lindsay, 1993).

These conjugates can thus also act as precursors of free fatty acids.

Barlow et al. (1989a) plotted concentration of butyric and hexanoic acids against flavour and found that Cheddar flavour rose and then declined as butyric and hexanoic acid concentrations increased. The best Cheddar flavour was associated with 45–50 mg kg^{-1} butyric acid and 20–25 mg kg^{-1} hexanoic acid.

Perret (1978) calculated the extent of fat lipolysis in Cheddar cheese from the palmitic acid (C_{16}) value and found that good-quality cheeses lay in the range less than 0.52% lipolysis at 0 months to less than 1.6% at 20 months. Cheeses which lay above this range had off-flavours. In Perret's work the proportion of acids, other than acetic acid, was similar to the fatty acid composition of butterfat and therefore appeared to be the result of lipolysis. The levels of acetic, butyric and caproic acids as determined in commercial Cheddar were well above reported threshold values (Siek et al., 1969, 1971) and therefore seemed certain to contribute to the background flavour of the cheese.

Perret (1978) observed that the addition of low-molecular weight fatty acids to commercial Cheddar curd, followed by a period of maturation, had a much greater effect on flavour than similar additions to previously matured cheese. It is thus possible that it is metabolites of fatty acids rather than the fatty acids themselves which are important for Cheddar flavour.

Dimos and colleagues (1992; Dimos et al., 1996), in their comparison of full-fat (33%) and reduced-fat (7%) cheeses, found that the levels of butyric, hexanoic and octanoic acids in the reduced-fat cheeses were of the order of one-twelfth those in the full-fat cheese and not one-quarter as might have been expected from the fat ratios. This may have been partially due to the fact that the cheeses were distilled at their natural pH of 5.2 and therefore only the non-ionized portion of the acids was collected. Because of the higher level of protein in the reduced-fat cheese more free fatty acids could have been bound to amino groups in the protein. It is also possible that there was less lipolysis in the reduced-fat cheese due to the different ratios of fat globule surface areas and other differences in physical environment. Although the absolute amount of free fatty acids in the full-fat cheeses was considerably higher than in the reduced-fat cheeses, the rate of increase, i.e. the extent of lipolysis, was small even in the full-fat cheeses. This confirms the much more delicate free fatty acid aroma of Cheddar cheese when compared with Italian varieties such as Romano, Parmesan (Meinhart and Schreier, 1986) and Provolone where fatty acids C_4–C_8 largely contribute to character impact (Woo and Lindsay, 1984). Of the three Italian varieties, Romano has the highest level of free fatty acids and Parmesan the lowest. Free fatty acids also contribute to the aroma of Brie, Camembert, Roquefort and blue (Woo et al., 1984).

In cheeses where there is extensive protein degradation some acids are produced by breakdown of amino acids, e.g. 2- and 3-methylbutyric acids in Livarot and Pont-l'Évêque are probably produced from leucine and isoleucine (Stark and Adda, 1972). The origin of 3-methylpentanoic acid in these cheeses is not obvious, but is probably the result of microbial synthesis. With a flavour threshold of 3.2 p.p.m. and 0.07 p.p.m. for 2- and 3-methylbutyric acids respectively (Brennand et al., 1989), these acids impart a sweaty note to the cheeses in which they occur. They are not normally found in Cheddar (Barlow et al., 1989b) although Ha and Lindsay (1991a) did find 15.4 p.p.m. 2-methylbutanoic acid in typical American medium-aged Cheddar and 2.36 p.p.m. 3-methylbutanoic acid in rancid American Cheddar. 3-Methylpentanoic acid has a flavour threshold of 0.15 p.p.m. and is sheep-like and reminiscent of wet pine-wood. Ney (1985) regarded iso-butyric, isovaleric and isocaproic acids, together with methanethiol and H_2S as the key compounds in the aroma of Tilsit. Lamparsky and Klimes (1981) also reported the presence of 2-hydroxy-3-methyl-butyric acid, 2-hydroxy-4-methyl-pentanoic acid, 2-hydroxy-3-methyl-pentanoic acid and 2-hy-droxy-hexanoic acid in mild Cheddar and suggested that these acids are formed enzymatically where possible from the corresponding amino acids.

Ha and Lindsay (1991a, b) investigated the effect of branched-chain fatty acids on the aroma of cows', sheep's and goats' milk cheeses and found levels of free 2-ethylbutanoic, 2-ethylhexanoic, 2-ethylheptanoic and 4-methyloc-tanoic acids of 1.58, 0.08, 0.05 and 0.11 mg kg^{-1} respectively in typical Cheddar cheese. With flavour thresholds of 5.7 and 82.4 mg kg^{-1} for 2-ethylbutanoic and 2-ethylhexanoic acids and 0.6 mg kg^{-1} for 4-methyloc-tanoic acid, none of these branched-chain free fatty acids contributes to Cheddar aroma (Brennand et al., 1989) unless there is some sub-threshold interaction between aroma compounds in Cheddar. However, branched-chain fatty acids are the character-impact compounds of goat and sheep cheeses. Ha and Lindsay (1991a) found 10–50 p.p.b. ($\mu g\,kg^{-1}$) of the potent goaty 4-ethyloctanoic acid (threshold 6 p.p.b.; Brennand *et al.*, 1989) in goats' milk cheese. The concentrations of 4-methyloctanoic acid in the same cheeses ranged from 20–260 p.p.b. When evaluated singly at concentrations below 100 p.p.b., 4-methyloctanoic acid exhibited a mutton-aroma, but blended easily with the goaty aroma of 4-ethyloctanoic acid to give the overall impression of goatiness. The presence of notable quantities of other branched-chain and n-chain fatty acids provided a range of blended heavy, goaty–muttony aromas in either the goats' or sheep's milk cheeses. Ha and Lindsay (1991a, b) did not find the very goaty 4-ethyloct-2-enoic acid which workers at Givaudan isolated from goats' milk (Smith *et al.*, 1984).

In a preliminary investigation into the chiral ratios of branched-chain acids in cheese, Karl *et al.* (1994) found a 2:1 ratio favouring the S-isomer of 2-methylbutanoic acid in the headspace from Parmesan cheese. There are differences between the aromas and aroma intensities of S and R isomers of branched-chain acids but the differences for 2-methylbutanoic acid were not reported.

Ney (1981) reported α-keto-acids corresponding to almost every amino acid in Cheddar (Ney and Wirotama, 1971), Emmental, German blue-mould cheese (Ney and Wirotama, 1972), Italico, Tilsit (Ney, 1985), Fontina (Ney and Wirotama, 1978), Manchego, Parmesan, Gouda, processed cheese, Provolone and Camembert (Ney and Wirotama, 1973), (Table 8.1). According to Ney and Wirotama (1978) α-keto-3-methylbutyric acid and α-keto-3-methylvaleric acid have an intense cheese odour. Tanaka and Obata (1969) also reported that α-keto-isocaproic acid has a cheese-like aroma. The concentrations, and hence the possible contribution to aroma, of these acids varies largely between cheese types. That no one other than Ney and

Table 8.1 α-Keto-acids found in various cheeses. (From Ney and Wirotama, 1971, 1973, 1978.)

α-Keto-acid	Corresponding amino acid
2-Keto-succinic acid	Aspartic acid
2-Keto-3-hydroxy-propionic acid	Serine
2-Keto-3-hydroxy-butyric acid	Threonine
2-Keto-glutaric acid	Glutamic acid
2-Keto-acetic (glycolic) acid	Glycine
2-Keto-propionic (pyruvic) acid	Alanine
4-Methylthio-2-keto-butyric acid	Methionine
2-Keto-3-methyl-valeric acid	Leucine
3-*p*-Hydroxyphenyl-pyruvic acid	Tyrosine
Phenylpyruvic acid	Phenylalanine
Imidazolylpyruvic acid	Histidine
2-Keto-5-amino-caproic acid	Lysine

Wirotama has reported these α-keto-acids may be due to the fact that these compounds are difficult to handle by gas chromatography, whereas Ney and Wirotama formed the 2,4-dinitrophenylhydrazone derivatives which they then reduced to the corresponding amino acids. The amino acids were determined by the Moore and Stein method (Ney, 1973). The function of α-keto-acids in cheese flavour urgently needs to be re-investigated.

Benzoic acid also regularly occurs in cheese where it is formed by the enzyme-mediated breakdown of hippuric acid (Sieber *et al.*, 1990) or phenylalanine (Bosset *et al.*, 1990). It probably does not contribute to flavour but may act as a natural preservative.

8.4.4 *Phenolic compounds*

Ramshaw (1985) reviewed the occurrence and aroma quality of phenol, methylphenols and *p*-ethylphenol in milk and cheeses (Cheddar, Gouda, Roquefort, Camembert, Vacherin, Livarot) and pointed out that phenol was particularly important in Vacherin and surface-ripened cheeses such as Livarot. Phenols may be detected organoleptically in the range of p.p.m. or lower, the aroma strength being in the order 4-methylphenol > 4-ethylphenol ≫ phenol and the aroma strength in the medium being in the order water > lipid > cheese. Phenolic compounds appear to make a positive contribution to aroma at about threshold concentration, but tend towards an unpleasant note as their concentration rises. Their sensory quality ranges from sharp, medicinal through sweet, aromatic to smoky, charred, caramel, unpleasant and 'sheep yard'. Phenolic compounds also occur in milk conjugates as glucuronides, sulphates or phosphates and can be liberated by β-glucuronidase, aryl sulphatase or acid phosphatase which all occur in milk (Kitchen, 1985). Phenols in sheep's milk are mostly bound as phosphate and sulphate conjugates with lesser amounts of glucuronides. Phenols in cows' and goats' milks are mostly bound as sulphates with smaller amounts of glucuronides and no phosphates (Lopez and Lindsay, 1993).

In Cheddar, levels of phenol, *p*-methylphenol and *p*-ethylphenol up to 20 p.p.b. appeared to be normal but, at higher levels, *p*-methylphenol and *p*-ethylphenol caused 'cow-yard-like' off-flavours (Ramshaw *et al.*, 1990).

Ha and Lindsay (1991a) found that Pyrenees sheep's milk cheese contained significant amounts of methyl and ethyl substituted phenols (*m*- and *p*-methylphenol 29 and 66 p.p.m.; *o*-, *m*- and *p*-ethylphenol 21, 29 and 29ppb; and 3,4-dimethylphenol 26 p.p.b.) which contributed characterizing sheep-like aroma notes to this cheese variety. 4-Methyloctanoic and 4-ethyloctanoic acids along with *p*-cresol, *m*-cresol and 3,4-dimethylphenol appeared responsible for sheepy notes in sheep's milk Romano cheese. Phenol and cresols (*o, m, p*) contributed strong phenolic and medicinal aroma notes to smoked Provolone cheese. Low concentrations of volatile branched-chain fatty acids and phenols appeared to provide desirable background aromas to Parmesan cheeses (Ha and Lindsay, 1991b). Lopez and Lindsay (1993) also found thiophenol bound as conjugate in sheep's but not cows' or goats' milks.

8.4.5 Sulphur compounds

(a) *Methanethiol and other sulphydryl compounds.* The presence of methanethiol (CH_3SH) in Cheddar cheese was first confirmed by Libbey and Day (1963) although there were earlier reports of its presence in soft and semi-soft cheeses such as Tilsit, Limburger, Dutch-type cheeses and Romadur (Jarczynski and Kiermeier, 1956; Kiermeier and Jarczynski, 1962; Tsugo and Matsuoka, 1962).

The decomposing cabbage or decomposing protein odour of methanethiol is characteristic of surface-smear cheeses such as Limburg (Manning and Nursten, 1985) and methanethiol has been identified as the compound responsible for an unpleasant odour in Grana cheese (Battistotti *et al.*, 1985). However, in combination with an as yet unidentified compound or compounds it completely looses its cabbagey odour and is responsible for the typical odour of Cheddar (Wijesundera and Urbach, 1993). The interaction between methanethiol and the other compound(s) is probably a response of the odour receptors in the nose rather than being a chemical reaction between methanethiol and the other compound(s) because exact proportions do not seem to matter. A similar effect was found by Guth and Grosch (1994) for the volatiles from beef stew juice. When methanethiol was omitted from a mixture of 12 volatiles, which *in toto* completely reproduced the odour of beef stew juice, the odour of the remaining 11 compounds was completely different from that of beef stew juice. Lindsay and Rippe (1986) also showed that when methioninase plus methionine was incorporated into Cheddar cheese during manufacture, the cheese became noticeably sulphury, cooked-vegetable-like and toasted-cheese-like in character soon after manufacture. As ripening at 10°C progressed the flavour became more intensely Cheddar-like, showing that the methanethiol which was initially formed interacted, either chemically or physiologically, with another compound or compounds which was being formed during maturation.

Dimos *et al.* (1996), in a comparison of the volatiles from full-fat and reduced-fat Cheddar, showed that the level of methanethiol in the cheeses is highly correlated with the flavour grade, indicating that the lack of aroma in reduced-fat Cheddar is likely to be mainly due to lack of methanethiol. This was later confirmed by Wijesundera (unpublished data) when he added methanethiol to a bland slurry of reduced-fat Cheddar and thus produced a strong Cheddar aroma. It is not clear whether this lack of methanethiol in reduced-fat Cheddar is due to lack of solvent power to retain all the methanethiol which has been formed or whether less methanethiol is produced in the reduced-fat Cheddar because of its different physical structure.

Manning (Green and Manning, 1982) recognized the importance of methanethiol in the aroma of Cheddar and Stilton (Manning and Moore, 1979) and advanced the hypothesis that methanethiol is produced by a purely chemical decomposition of methionine at the low (-150 to $-200\,\mathrm{mV}$) redox potential of the cheese. Although a contribution to the pool of methanethiol by such a reaction cannot be entirely ruled out, the methanethiol is much more likely to be formed by the methioninase which Bruinenberg et al. (1996) have shown to be present in many lactic acid bacteria. Negative redox potential is probably a necessary condition to prevent the oxidation of methanethiol once it has been formed.

Kristoffersen and co-workers also recognized the importance of -SH (thiol) and -S-S- (disulphide) groups in the development of Cheddar flavour. Kristoffersen (1985) stored strictly fresh milk at 37°C and manufactured cheese at intervals from portions of the milk. Cheese made from milk stored for less than 3 h failed to develop full flavour, while cheese made from milk stored 3.5–5 h did develop full flavour. Holding of milk beyond 5 h also resulted in full-flavoured cheese, but defects such as unclean and bitter were prominent. High flavour was correlated with high levels of sulphydryl groups. In a separate experiment Kristoffersen (1985) made cheese from milk which had been treated as follows: (i) no treatment (raw); (ii) 61.5°C for 5 min, (iii) 61.5°C for 30 min; and (iv) 68.5°C for 15 min. The rate of formation and concentration of active sulphydryl groups produced in the cheese during maturation was inversely related to the intensity of heat treatment of the milk, as was the development of characteristic flavour. Kristoffersen (1973) advanced the hypothesis that milk-protein-based sulphydryl groups (-SH) serve as a reservoir for hydrogen in biological oxidation–reduction processes during manufacture and ripening of cheese. At least a portion of the masked -SH groups of fresh milk must be oxidized before or during cheese manufacture to produce -S-S- groups which, in turn, must be available for reduction to active -SH groups during cheese ripening. Bacterial growth may contribute to this transition. It is not clear from Kristoffersen's work how these sulphydryl groups are produced during cheese maturation and which are the compounds whose -SH groups are being measured. If one considers the fact that, of the proteins in cheese, α_{s1}-casein, β-casein and the γ-caseins contain no cysteine/cystine residues at all and α_{s2}- and κ-casein contain only two cysteine/cystine residues each in chains of 207 and 169 amino acids respectively (Swaisgood, 1982), it becomes questionable whether cysteine from cheese caseins could be the precursor of the -SH groups. However, Kristoffersen et al. (1964a) point out that there is sufficient whey protein in Cheddar cheese to account for the level of active -SH groups which they found. The fact that reduced glutathione hastens the formation of Cheddar flavour in Cheddar cheese curd slurries (Singh and Kristofferesen, 1970) and that it is oxidized immediately upon addition to cheese curd slurries suggests that the disulphide linkages that form in fresh milk on standing may be due to the formation of oxidized glutathione.

Green and Manning (1982) pointed out that cheese in which H_2S fails to appear early in ripening usually contains only low levels of methanethiol and they regarded this as support for their theory that methanethiol in ripening cheese is produced by the chemical action of H_2S on methionine. However, Roberts *et al.* (1995) have shown that, in the cell-free extract of the one starter which they examined, cysteine inhibited the action of the enzyme which produced methanethiol from methionine. The reduced production of methanethiol, when H_2S is not produced first, can thus also be explained as follows: when H_2S is produced, this is due to the breakdown of the inhibitor of the methanethiol-producing enzyme, thereby stimulating the production of methanethiol. This could also explain the fact that Cheddar cheese, when made from ultrafiltered milk, produces less methanethiol than conventionally made cheese from the same milk (Dimos *et al.*, 1996). The ultrafiltered cheese contains much higher levels of whey protein, and thus -SH groups, which could act to partially inhibit the production of methanethiol.

Samples (1985) used Cheddar cheese curd slurries to investigate the mechanism of volatile sulphydryl compound production. His results are summarized in Table 8.2. Heating the skim milk (75°C/15 min) stopped production of H_2S and reduced production of methanethiol, indicating that in the skim milk either an enzyme of the thiol-producing pathway was inactivated or a thiol precursor was destroyed. (The precursor for H_2S is

Table 8.2 Effect of various treatments on the production of H_2S and methanethiol (MeSH) from cheese curd slurries. (Summarized from Samples, 1985.)

Treatment	H_2S	MeSH
Untreated cheese milk	+ +	+ +
60°C/15 min of cheese skim milk	+	+
75°C/15 min of cheese skim milk	−	+
75°C/15 min of cheese cream	+ +	+ +
Glutathione	+ + + +	+ + + + E_h not lowered
Glutathione + AZA[a]	+ + +	+ + +
Cysteine	+ + + +	+ + + +
AZA	+	+ +
75°C/15 min of cheese milk + cysteine or glutathione	−	+
Methionine	+ +	+ +
Rennet	+ + +	+ +

[a]Azaserine, O-diazoacetyl-L-serine, $N_2CHCO-OCH_2CH(NH_2)COOH$, γ-glutamyltranspeptidase inhibitor.

likely to be cysteine or a cysteine containing peptide, and the precursor for methanethiol is likely to be methionine or a methionine-containing peptide.) However, the presence of either added glutathione or added cysteine in a slurry from heated (75°C/15 min) skim milk did not restore the production of the two thiols, H_2S and methanethiol, suggesting that heating inactivated the -SH-producing enzyme system (neither glutathione nor cysteine could be expected to act as a precursor for methanethiol). Addition of the γ-glutamyltranspeptidase inhibitor, AZA, (azaserine, O-diazoacetyl-L-serine, $N_2CHCOOCH_2CH(NH_2)COOH$) reduced production of H_2S in control milk slurries, but had no effect on methanethiol. γ-Glutamyltranspeptidase removes the glutamyl group from glutathione and thus makes the -SH on the cysteine more accessible for production of H_2S. Addition of reduced glutathione or cysteine to curd slurry from control cheese increased production of both H_2S and methanethiol although oxidation–reduction potential was not reduced, indicating that lowering of oxidation–reduction potential is not the initiator of thiol production. Addition of glutathione plus AZA decreased the production of both thiols from the level when only glutathione was added, again indicating that the -SH group on the glutathione was made less available by the presence of the γ-glutamyltranspeptidase inhibitor. These experiments indicate that γ-glutamyltranspeptidase is involved in the production of H_2S and possibly of methanethiol, but that another enzyme (or enzymes) is also involved in producing H_2S from -SH. Lack of -SH appears to be a limiting factor for the production of H_2S. Lack of methionine does not appear to be a limiting factor for the production of methanethiol, as addition of methionine does not increase H_2S or methanethiol levels. Increased production of H_2S appears to increase methanethiol, probably by preventing it from being oxidized to dimethyl disulphide. Kim (1985) and Kim et al. (1990) showed that in milk-fat-coated microcapsules containing Brevibacterium linens cysteine and methionine produced H_2S and methanethiol respectively. The presence of H_2S stabilized levels of methanethiol, which disappeared rapidly without the H_2S. The investigations of Samples (1985) showed that native milk enzymes and perhaps enzymes from the microflora in the cheese milk are responsible for the production of H_2S.

However, Bruinenberg et al. (1996) have shown that cell-free extracts from commonly occurring cheese lactic acid bacteria produce methanethiol directly from methionine and other S-containing substrates without producing H_2S from either methionine or cysteine. Samples (1985) commented that while negative redox potential may be a necessary condition for the production of volatile sulphydryl compounds, it is not a sufficient one, and that glutathione probably functions as something other than a reducing agent in the enhanced production of H_2S and methanethiol. Glutathione is recognized as an ubiquitous material which serves multiple functions in cell biochemistry. These include the reduction of sulphide groups in proteins, protection of sulphydryl enzymes and cofactor relationships, e.g. the glyoxylase enzyme system converting methyl glyoxal to lactic acid requires reduced glutathione as an essential coenzyme. In a cheese slurry system, reduced glutathione is oxidized immediately upon addition to the system (Harper and Kristoffersen, 1970) where it disaggregates the protein, thus allowing increased proteolysis. Glutathione also influences the rate of degradation of the caseins in the cheese slurry. Decay of soluble protein in slurries with glutathione is very rapid, and only when the degradation is more than 85% complete is the β-casein attacked. Characteristic flavour in the slurry has no relation to the α-casein content but is directly proportional to β-casein degradation. It thus appears that products derived from the breakdown of β-casein are more desirable for flavour than products derived from α-casein. However, addition of 5% β-casein to cheese curd slurries causes a disruption of normal fermentation. More recent work (Papoff et al., 1995) has shown that β-casein exists as three variants, A, B and C, and that these variants are hydrolysed by the plasmin from the milk. Papoff et al. (1995) found that for β-casein C and β-casein A[1], hydrolysis by plasmin leads to the same types of peptide products in different ratios. Consequently, they consider this to be the reason for the different cheese (Beaufort) flavours obtained from milk with different haplotypes.

Addition of glutathione also slightly accelerates total α-esterase activity and inhibits the degradation of esterase activity during later stages of the ripening process, i.e. it protects enzymes in the slurry system. Harper and Kristoffersen (1970) also found that where slurries without added glutathione gave good flavour development, glutathione was detectable for the first few days. Slurries that failed to develop normal flavour also failed to show detectable levels of glutathione after the first day of fermentation. In cheese-making, glutathione can be introduced into the cheese in the cells of the starters (Fernándes and Steele, 1993; Wiederholt and Steele, 1994). Some strains of Lactococci have been shown to be capable of accumulating reduced glutathione from the environment, but not of synthesizing it from its constituent amino acids.

Ponce-Trevino et al. (1987, 1988) made cheese slurries from cheese curd prepared (i) by culturing, or (ii) by acidification. Volatile sulphur compounds detected in the slurries made from the cultured cheese curd were H_2S, COS (carbonyl sulphide), methanethiol and dimethyl sulphide. Addition of reduced glutathione to such slurries did not affect the oxidation–reduction potential, but increased production of all the above sulphur compounds. When antibiotics were added to the cultured cheese curd slurries no H_2S or dimethyl sulphide were produced. When the cheese was made by acidification, only COS was detected but when the oxidation–reduction potential was lowered by the addition of NADH, methanethiol was also produced.

Methanethiol is present in Camembert, together with other sulphur compounds such as 2,4-dithiapentane, 3,4-dithiahexane, 2,4,5-trithiahexane and 3-methylthio-2,4-dithiapentane. They are responsible for the garlic note which may be found in well-ripened Camembert (Adda et al., 1988).

Hydrogen sulphide has been suggested as an important contributor to Cheddar aroma (Green and Manning, 1982) but Wijesundera and Urbach (1993) have shown that Cheddar aroma can be produced in the absence of H_2S. Cysteine, which is generally regarded as the precursor of H_2S, has only recently been reported as a free amino acid in cheese and there appears to be only one publication (Wood et al., 1985) where the presence of free cystine in cheese is reported (in Cheddar, Edam and Jarlsberg). A.F. Wood (private communication) also failed to find cysteine in Gouda, Marring Red, Gruyère, Emmental, Chester, Camembert, Maribo, Danbo and Samsoe cheeses. Wilkinson (1992) reported values between 50 and 100 μg of free cysteine per ml juice expressed from Cheddar cheese at 42 days' ripening at 10°C and of the order of 100 μg g^{-1} Cheddar ripened for 120 days at 4°C. Wilkinson (1992) also found about 100 μg cysteine ml^{-1} Cheddar cheese juice from cheese made with L. lactis subsp. cremoris AM2 when the cheese was cooked at normal temperatures and matured for 70 days at 10°C, compared with about double that amount when the cheese was cooked at low temperatures. Cooking at low temperatures had the effect of reducing the amount of most of the other amino acids compared with cooking at normal temperatures, presumably due to reduced lysis and hence lower amounts of amino acid-producing enzymes available in the low-temperature cooked cheese. It is thus possible that this reduced lysis also reduced the availability of the enzyme which decomposes cysteine. Normal cooking temperature produced a good-quality, non-bitter cheese compared with a bitter, off-flavoured cheese when low cooking temperatures were used. Morgan et al. (1995), in their attempt to increase lysis in the bitter strain HP by accompanying it with the bacteriocin-producing strain DPC3286, found 2–3 mg cysteine ml^{-1} in juice from 6-month-old cheese made either with HP only or a mixture of HP and DPC3286. Addition of DPC3286 increased the lysis of HP and produced non-bitter cheese. Recently, Darwish et al. (1994) reported values from 0 to 9.89 mg cystine per 100 g Egyptian Ras cheese. Swiatek and Poznanski (1959) determined the H_2S production and the loss of cystine/cysteine (the sum of cysteine/cystine bound in the protein plus free cysteine/cystine) in Trappist cheese during maturation and found that the amount of sulphur resulting from cystine/cysteine loss was much lower than the sulphur present in the developing H_2S. Therefore, the source of H_2S was not only free plus bound cystine/cysteine. On the other hand, Kristoffersen et al. (1964b) found a correlation between loss of -SH groups and increase in pyruvic acid, suggesting the oxidative loss of H_2S and NH_3 from cysteine as the source of the pyruvic acid. However, as pointed out earlier (Kristoffersen et al., 1964a), there is sufficient whey protein in Cheddar cheese to account for the level of active -SH groups. During maturation of Gouda cheese, β-lactoglobulin, a cysteine-containing whey protein, was shown to break down almost completely within 56 days (de Koning et al., 1981). The source of H_2S in cheese still has not been clarified.

Perret (1978), who was one of the few researchers who actually measured total volatile sulphur compounds in the cheese rather than relative amounts in the headspace, found that the best quality commercial Cheddar cheese contained $20-30\,\mu g$ H_2S, $7-8\,\mu g$ methanethiol per 100 g, plus a trace of disulphide (unspecified but presumably dimethyl disulphide). Fruity and fermented cheeses generally showed lower values. Cheeses with sulphury, eggy flavours contained $50-440\,\mu g$ H_2S. Cheeses with offensive oniony odours contain $40-600\,\mu g$ H_2S, and up to $4000\,\mu g$ methanethiol. With time, this methanethiol was oxidized to dimethyl disulphide and the cheese developed a distinct mustard odour and taste. Considerable decarboxylation of amino acids occurred in both types of defective cheeses although no abnormal populations of microorganisms were found. It is possible that the off-flavours resulted from the action of heat-resistant enzymes from psychrotrophic bacteria.

Thiol (-SH) groups, from proteins, glutathione, coenzyme-A, etc. and enzymes reacting with them in cheese originate from the milk, the rennet, starters and adjuncts. Their interaction is a complicated process which is of major importance in the formation of cheese aroma. Much more research is required in this field.

(b) Other sulphur compounds.   Dimethyl sulphide occurs in many cheeses but Manning et al. (1976) and Dimos (1992) showed that its occurrence is related to season, i.e. presumably feed, rather than aroma development.

Methylthioesters, CH_3-SCOR, have been identified in surface-ripened cheese (Dumont and Adda, 1979) and methylthioacetate, which has an aroma of cooking cauliflower, was found in the volatiles from Limburger cheese (Parliment et al., 1982). Cuer (1982) described methylthiopropionate as cheesy; since Cuer is French, the cheese is presumably Camembert.

Methional (3-methylthiopropanal) which, on its own, has an odour of cooked potato, has been identified in Cheddar (Day et al., 1960).

Methional and the two α-dicarbonyl compounds, 4-hydroxy-2,5-di-methyl-3(2H)-furanone (furaneol) and 5-ethyl-4-hydroxy-2-methyl-3(2H)-furanone belong to the key compounds of the aroma of Emmental cheese. Most likely the two furanones cause the sweet note in the aroma of this type of cheese (Preininger and Grosch, 1994; Preininger et al., 1994). Other compounds which also make some contribution to the aroma of Emmental, as determined by aroma-extract-dilution-analysis, are diacetyl, 3-methyl-butanal, ethyl butanoate, ethyl 3-methylbutanoate, heptan-2-one, oct-1-en-3-one, ethyl hexanoate, 2-sec-butyl-3-methoxypyrazine, skatole and δ-decalactone (Grosch, 1994).

3-Methythio-1-propanol (methionol), which has been reported to have the odour of raw potatoes (Muller *et al.*, 1971) has been found among the volatiles from premium quality Cheddar (Wijesundera and Urbach, 1993) and in Camembert (Dumont *et al.*, 1976). Ethyl 3-methylthiopropanoate was found as a trace component among the volatiles from Parmesan (Meinhart and Schreier, 1986).

Gallois and Langlois (1990) examined the volatiles from French blue cheeses and showed that high levels of methanethiol, dimethyl sulphide ('cowy') and dimethyl disulphide (cabbagey, like methanethiol) distinguished Bleu de Causses from Roquefort and Bleu d'Auvergne.

Sloot and Harkes (1975) identified 2,4-dithiapentane, with an odour threshold of 0.003 mg kg^{-1} above oil and $0.0003 \text{ mg kg}^{-1}$ above water, among the volatile components of Gouda cheese.

Eckert *et al.* (1964) synthesized *N*-dimethylmethioninol (*N*-dimethyl-2-amino-4-methylthio-butan-1-ol) and found that this compound had an intense odour of Camembert; however, it has not been found in cheese.

8.4.6 Terpenes

Vacherin, also known as Mont d'Or and fromage de boîte, is a cheese which has a ring of spruce wood put around it during ripening. This is held responsible for the high level of terpenes which this cheese contains. Dumont *et al.* (1974c) found 12 terpenes in this cheese, the major ones being terpineol, isoborneol and linalool.

Bosset *et al.* (1994) examined 14 Swiss Gruyère and Étivaz cheeses produced in alpine pastures and 20 others produced in the lowlands and confirmed the findings of Dumont and Adda (1978) and Dumont *et al.* (1981) for French Gruyère de Comté that alpine cheeses contained more terpenes than lowland cheeses. The alpine cheeses contained on average about three times as much limonene and four times as much nerol as the lowland cheeses. Pinene, whose concentration in the alpine cheeses was 1.3 times higher than that of nerol, was not detected in any of the lowland cheeses. Such findings are important where cheese producers want to claim particular naming rights for cheeses from a small specified area, as is the case for wines.

Guichard *et al.* (1987) showed that in the samples of Gruyère de Comté which they examined an unsaturated C_{16} hydrocarbon and the sesquiterpenes β-caryophyllene and an unidentified sesquiterpene (of molecular weight 206) were most characteristic of summer production. They also found the terpenes camphene, cymene and limonene and the sesquiterpene α-humulene.

Wilson (1989) isolated the terpenes, α- and β-pinene, D-limonene, linalool, α-terpineol and caryophyllene from New Zealand but not from Finnish milk-fat and showed that 1 p.p.m. of D-limonene, in particular, was responsible for the green/grassy aroma present in New Zealand milk-fat at certain times of the year.

8.4.7 α-Dicarbonyls and related compounds

Diacetyl is involved in the development of desirable cheesy aroma in matured cheese (Singh and Kristoffersen, 1971, 1972; McDonald, 1992) but does not appear to form part of this aroma because, by 6 months, when a good full Cheddar flavour has been developed, it has practically disappeared (Barlow et al., 1989a).

The α-dicarbonyls, glyoxal and methylglyoxal (the latter also known as α-ketopropanal or pyruvaldehyde), although flavourless themselves, are also involved in the production of cheesy aroma (McDonald, 1992). Methylglyoxal production by adjunct cultures was associated with strong cheesy aromas but it was also linked to meaty-brothy aromas in low-fat cheeses. Selection of adjunct cultures based on α-dicarbonyl production did not always result in desirable cheesy aromas because of concomitant production of unclean aroma compounds resulting from the breakdown of aromatic amino acids. Griffith and Hammond (1989) investigated the aroma-producing potential of amino acids with methylglyoxal, glyoxal, dihydroxyacetone and acetaldehyde and found that the amino acids important in aroma-generating reactions were valine, leucine, isoleucine, methionine, cysteine, phenylalanine, proline and lysine, which all produced the corresponding Strecker aldehydes. In addition, the reactions of methylglyoxal with amino acids produced benzaldehyde and acetophenone from phenylalanine, dimethyldisulphide and dimethyltrisulphide from methionine, and 2-acetyl-thiazole from cysteine. Alkylpyrazines were produced from lysine and dihydroxyacetone. 2-Acetyl-1-pyrroline was produced from certain carbonyls with proline and lysine; 2,5-dimethyl-4-hydroxy-3(2H)-furanone from methylglyoxal; 2-methylbenzaldehyde from proline and acetaldehyde; δ-valerolactam from the lysine–carbonyl combination. The reaction products accounted for most of the aromas previously noted in aqueous cheese extracts from American Swiss cheese and cultures of cheese microorganisms, although the fact that these reactions occurred in vitro does not prove that the aroma compounds were produced by the same purely chemical mechanism in cheese and cultures.

8.4.8   Esters

Esters are common constituents of cheese volatiles where they impart a fruity aroma. In Cheddar, this is regarded as a defect by professional cheese graders although consumers are prepared to pay a premium for fruity Cheddar. Imhof and Bosset (1994) found 14 different esters in Swiss

Emmental and esters are also important contributors to the aroma of Parmigiano (Meinhart and Schreier, 1986), the most abundant of the 38 esters identified being ethyl butyrate, ethyl hexanoate, ethyl acetate, ethyl octanoate, ethyl decanoate and methyl hexanoate. Ethyl esters may also play a part in the flavour of Manchego (a hard or semi-hard Spanish ewe's milk cheese which originates from La Mancha) where they reach total levels in the range of $8-18\ \mu g\ g^{-1}$ after 300 days of ripening (Martínez-Castro *et al.*, 1991).

8.4.9   Amines, amides and other nitrogen-containing compounds

Golovnya and her group used chromatography to identify 21 amines (primary, secondary and tertiary) from one strain of *Streptococcus lactis* (Golovnya *et al.*, 1969), 27 amines in Chanakh brine pickled cheese, Rossiiskii cheese and Dutch cheese (Magak'yan *et al.*, 1976), and 35 nitrogenous components in the volatiles from Chanakh Cheese (Table 8.3) (Dilanyan and Magak'yan, 1978). Some 14 amines (primary and secondary) were also found in Cheddar cheese (Ney and Wirotama, 1971) and 8 amines in German blue-mould cheese (Ney and Wirotama, 1972). No pathway for the formation of these amines has been shown or even suggested, and their contribution to flavour is also unknown.

Ney (1981) reported the presence of acetamide, propionamide, butyramide, isobutyramide and isovaleramide in Cheddar, Emmental, Edelpilzkäse (German blue-mould cheese) and Manchego cheeses. Neither the contribution to flavour nor the method of formation of these amides appears to have been reported. Dumont and Adda (1979) identified *N*-isobutylacetamide as a volatile compound with a bitter taste and a pungent odour in Pont-l'Évêque and Camembert. Its presence can be explained by the acetylation of isobutylamine (Adda *et al.*, 1982) in analogy with acetylation occurring in wine.

Table 8.3 Nitrogenous compounds identified from Chanakh cheese. (From Dilanyan and Magak'yan, 1978.)

Diethylamine	*iso*-Butylamine	Pyridine
Dimethylamine	Di-n-butylamine	α-Picoline
Trimethylamine	*iso*-Amylamine	β- or γ-Picoline
Triethylamine	Di-*iso*-amylamine	2,4-Dimethylpyridine
iso-Propylamine	Pyrrolidine	2,6-Dimethylpyridine
n-Propylamine	Pipridine	4-Ethylpyridine
Di-n-propylamine	n-Methylpiperidine	Methylamine (?)
Tri-n-propylamine	n-Ethylpiperidine	Ethylamine (?)
	Di[*iso*]propylamine (?)	

(?), Tenative identification.

Liardon *et al.* (1982) found 2,5-dimethylpyrazine, 2,6-dimethylpyrazine, ethylpyrazine, 2,3-dimethylpyrazine, ethylmethylpyrazine, trimethylpyrazine, tetramethylpyrazine, ethyltrimethylpyrazine, as well as 10 other nitrogen-containing compounds in the outer layer (including the smear) of Swiss Gruyère cheese where they were presumably formed by the microorganisms of the smear. Pyridine, pyrazine, 2,3-dimethylpyrazine, 2,6-dimethylpyrazine, 2-ethyl-3,5(6)-dimethylpyrazine and 2,3-diethyl-5-methylpyrazine were also found in Parmesan (Meinhart and Schreier, 1986) and six alkylpyrazines were found in Swiss Emmental where the fraction which contained them had a burnt potato odour (Sloot and Hofman, 1975).

8.4.10 Lactones

Lactones occur universally in milk-fat (Urbach, 1990) and therefore also occur in all cheeses, where their contribution to aroma is marginal. However, the sweet-flavoured γ-dodecanolactone and γ-dodec-*cis*-6-enolactone occur at much higher levels in milk from grain-fed than from pasture-fed cows (Urbach, 1990) and are thus responsible for the sweeter aroma of, for example, European cheeses as compared with Australian and New Zealand varieties (Wilson, 1989).

8.5 Some less-common cheese varieties

Artisanal cheeses from the Asturias region (a mountainous area in northern Spain) are highly prized and have intense flavour. Gamonedo is a semi-hard cheese made from a mixture of cow, goat and sheep milk, it has some greenish-blue veins and is salted, smoked and wrapped in fern leaves. Cabrales is a semi-hard cheese made from cows' milk; it has blue veins and is salted and wrapped in chestnut leaves. La Peral is a semi-hard cheese made from cows' milk, it has blue veins of *Penicillium roqueforti*. Peñamellera is a semi-hard cheese made from a mixture of cow, goat and sheep milk; it is white with salt added. Los Beyos is a semi-hard white cheese made of cows' milk; it is salted and smoked. Afuega'l pitu is a fresh cheese made from cows' milk; it is white or red depending on whether the curd is mixed with salt and/or red pimento. The three blue varieties are richer in free fatty acids, methyl ketones and 2-alkanols than the other cheeses. They all contain α-pinene, γ-terpinene and 1,8-cineole (Frutos *et al.*, 1991).

8.6 Cheese with high linoleic acid content

Feed supplements for ruminants in which the lipid has been protected against biohydrogenation in the rumen (Scott *et al.*, 1970) can be used to produce milk and meat with at least 20% linoleic acid in the fat. When such milk was used to make Cheddar, Cheedam and Gouda the flavour of these cheeses was more bland than that of cheese from conventional milk, even after 9 months' storage (Czulak *et al.*, 1974a). In the headspaces, acetic acid and ethanol were at an extremely low level in the polyunsaturated Cheddar cheese; compared with the control cheese and acetoin, pentanal and pentanol were significantly lower. These differences are unlikely to explain the flavour differences but probably point to differences in biochemical pathways. In subsequent work Broome *et al.* (1979) showed that linoleic acid inhibited the growth of all species of group N streptococci, as well as glycolysis and amino acid uptake in *Streptococcus lactis* C10. The metabolism of pyruvate by resting cells of *Str. lactis* was altered in the presence of linoleic acid in a manner which was consistent with the inhibition of the pyruvate dehydrogenase system. When Cheddar was made from high-linoleic acid milk using *Streptococcus cremoris* 1609/P plus YB (a yoghurt culture) the cheese had a strong flavour similar to Continental Swiss cheese (Czulak *et al.*, 1979). When the high-linoleic acid Cheddar was made with the addition of Capalase K (a pre-gastric kid lipase) the resultant cheese had a flavour resembling that of Romano cheese (Czulak *et al.*, 1974b) due to increased levels of C_4, C_6, C_8 and C_{10} acids. The flavour of Brie, Camembert and cream cheeses was not affected by high levels of linoleic acid in the milk (Czulak *et al.*, 1974b).

8.7 Enzyme-modified cheese

Enzyme-modified cheese is derived from cheese by enzymatic means. Enzymes may be added during manufacture or after aging. An incubation period under controlled conditions is required for proper flavour development. Commercially available enzyme-modified cheese flavours include mild, medium and sharp Cheddar, Colby, Swiss, Provolone, Romano, Mozzarella, Parmesan and Brick. Enzyme-modified cheeses are generally added to foods at a level of 0.1 to 2.0%, although they can be used at 5% to add dairy or cheesy notes to foods and to reduce the requirement for aged cheese in the formulation (Moskowitz and Noelck, 1987). There appear to be no reports on the flavour volatiles from enzyme-modified cheeses.

8.8 Off-flavours

Sorbic acid is widely used as a mould inhibitor, but a kerosene-flavoured taint due to *trans*-1,3-pentadiene (Horwood *et al.*, 1981) has been shown to result from the catabolism of sorbic acid by sorbate-resistant strains of moulds and yeasts (Daley *et al.*, 1986; Sensidoni *et al.*, 1994).

Furaneol, which has been described as having a burnt pineapple aroma, has been found in cultures of *Lactobacillus helveticus* (Kowalewska *et al.*, 1985), in low-fat Cheddar (McDonald, 1992) and, in combination with 4,5-dimethyl-3-hydroxy-2(5H)-furanone (sugar furanone) and several unidentified pyrazine compounds is claimed to be at least partially responsible for the brothy off-flavour of mature reduced-fat cheese (Johnson and Chen, 1991; Ha and Lindsay, both references as quoted by McDonald, 1992).

The malty aroma defect that can develop in raw milk is due to the metabolic activity of *Lactococcus lactis* subsp. *lactis* biovar. *maltigenes* (Morgan, 1976) and *Lactobacillus maltaromicus* (Miller *et al.*, 1974). Branched-chain amino acids are converted to keto-acids by transamination in the presence of pyridoxal phosphate. In the presence of thiamine pyrophosphate the keto-acids lose CO_2 and the aldehyde with one carbon atom less is formed. This is reduced by $NADH_2$ to the corresponding alcohol. Dunn and Lindsay (1985) detected compounds of this type in 20 samples of Cheddar cheese which had an unclean-type aroma and McDonald (1992) showed that these compounds are produced by a variety of *Lactobacillus*-type organisms. 3-Methyl butanal/2-methyl butanal and 2-methyl propanal cause the malty aromas, *p*-cresol causes an 'unclean-barny' aroma.

$$RCHCOOH + HOOCCH_2CH_2CCOOH \rightarrow RCCOOH + HOOCCH_2CH_2CHCOOH$$

$$
\begin{array}{ccccc}
| & & \| & \| & | \\
NH_2 & & O & O & NH_2
\end{array}
$$

$$RCCOH \rightarrow RCH + CO_2$$

$$
\begin{array}{cc}
\| & \| \\
O & O
\end{array}
$$

Strecker aldehyde

$$RCH \rightarrow RCH_2OH$$

$$
\begin{array}{c}
\| \\
O
\end{array}
$$

where R = $(CH_3)_2CH$-
$(CH_3)_2CHCH_2$-
$CH_3CH_2CH(CH_3)$-
$CH_3SCH_2CH_2$-
$(C_6H_5)CH_2$-

Pseudomonas fragi hydrolyses milk fat and esterifies certain of the lower fatty acids with ethanol, producing fruity aromas. A similar esterase is also present in certain lactic cultures used in the manufacture of Cheddar cheese (Morgan, 1976).

The Celluloid taste in some mould-ripened cheeses is due to styrene (Adda *et al.*, 1989). It is produced by some strains of *Penicillium camemberti* because of the deregulation of the oxidative metabolism in these strains. However, it can also enter foods by migration from the packaging material (Imhof *et al.*, 1994b). The same pathway also explains the presence of phthalate esters in most dairy products.

Although oct-1-en-3-ol is essential for the aroma of Camembert, a sample which had an excessive mushroom odour was found to contain an uncharacteristically large amount of this alcohol (Dumont et al., 1974d). 2-Methylpentane-2-thiol-4-one has been generated in cheese by the reaction of mesityl oxide (acetone dimer) with H_2S. It has an aroma of 'tom cat's urine' (Badings, 1967; Prante and Duus, 1986). 2-Methoxy-3-isopropyl-pyrazine was found to be resposible for the potato-like off-flavour in smear-coated cheese (Dumont et al., 1983; Gallois et al., 1988).

Some pollutants can enter dairy products from the environment (Imhof et al., 1994b), although these substances normally do not impart off-flavours (dichloromethane, chloroform, 1,1,1-trichloroethane, trichloroethylene, tetrachloroethylene, 1,4-dichlorobenzene). They have been detected in many cheese varieties. However, chlorophenols impart a phenolic flavour. These can be absorbed from wood which has been treated with pentachlorphenol. 2,4,5-Trichlorophenol has been found in the milk from cows whose feed included the herbicide 2,4,5-trichlorophenoxyacetic acid (Bjerke et al., 1972). Bosset et al. (1993) found that a chemical–foreign flavour in the rind from Emmental cheese was due to (1-methoxy-2-propyl)-acetate which originated from the epoxy resin with which the floor of the maturing room had been treated.

8.9 Conclusions

There has been a great tendency among flavour researchers to label indiscriminantly all volatile compounds as flavours or aromas. At the beginning of flavour research the need to identify as many volatiles as possible was justified. However, the time has now come when these volatiles need to be examined critically for their contribution to aroma. With the techniques currently available for identification and quantification, these procedures are relatively simple compared with the painstaking work of establishing flavour significance (Chapter 10), particularly in view of the fact that the only suitable instrument for this is a panel of humans who may be neither available nor willing. In Table 8.4 an attempt has been made to highlight those volatiles which possibly play a role in the aromas of the particular cheeses. However, the flavour of any particular cheese has not been fully identified until it can be reproduced with a synthetic mixture.

Some cheeses have flavours which can be readily related to certain groups of compounds, e.g. the most important contributors to the flavour of mould-ripened cheeses are the methyl ketones, the flavour of Camembert and related cheeses is in a large measure dependent on methyl ketones in combination with the mushroom-flavoured oct-1-en-3-ol, fresh cheeses depend largely on diacetyl and acetaldehyde for their flavour, and Italian cheeses are characterized by high concentrations of free fatty acids. Even with those cheeses, however, the complete flavour requires the bouquet of

Table 8.4   Important aroma compound in various cheeses

Cheese	Compound(s)
Cheddar	Methanethiol
Camembert	Nonan-2-one, oct-1-en-3-ol, N-isobutylacetamide, 2-phenyl ethanol, β-phenylethyl acetate, 2-heptanol, 2-nonanol, NH_3, isovaleric acid, isobutyric acid, hydroxybenzoic acid, hydroxyphenylacetic acid
Emmental	Methional, 4-hydroxy-2,5-dimethyl-3(2H)-furanone (furaneol) and 5-ethyl-4-hydroxy-2-methyl-3(2H)-furanone
Romano	Butyric acid, hexanoic acid, octanoic acid
Parmesan	Butyric acid, hexanoic acid, octanoic acid, ethyl butyrate, ethyl hexanoate, ethyl acetate, ethyl octanoate, ethyl decanoate, methyl hexanoate
Provolone	Butyric acid, hexanoic acid, octanoic acid
Goat's milk cheese	4-Methyloctanoic acid, 4-ethyloctanoic acid
Sheep's milk Romano cheese	4-Methyloctanoic and 4-ethyloctanoic acids, p-cresol, m-cresol, 3,4-dimethylphenol
Limburg	Methanethiol, methylthioacetate
Surface-ripened cheeses	Methylthioesters
Pont-l'Évêque	N-isobutylacetamide, phenol, isobutyric acid, 3-methyl-valeric acid, isovaleric acid, heptan-2-one, nonan-2-one, acetophenone, 2-phenyl ethanol, indole
Vacherin	Acetophenone, phenol, dimethyl disulphide, indole, terpineol, isoborneol, linalool
Roquefort	Oct-1-en-3-ol, methyl ketones
Livarot	Phenol, m- and p-cresol, dimethyl disulphide, isobutyric acid, 3-methylvaleric acid, isovaleric acid, benzoic acid, phenylacetic acid, nonan-2-one, acetophenone, 2-phenylethanol, 2-phenylethyl acetate, dimethyl disulphide, indole
Munster	Dimethyl disulphide, isobutyric acid, 3-methylvaleric acid, isovaleric acid, benzoic acid, phenylacetic acid
Trappiste	H_2S, methanethiol
Blue cheeses	Heptan-2-one, nonan-2-one, methyl esters of C4, 6, 8, 10, 12 acids, ethyl esters of C1, 2, 4, 6, 8, 10 acids
Brie	Isobutyric acid, isovaleric acid, methyl ketones, S-compounds, oct-1-en-3-ol
Carré de l'Est	Isobutyric acid, isovaleric acid, 3-methylvaleric acid
Epoisses	2-Phenylethanol
Maroilles	2-Phenylethanol, nonan-2-one, acetophenone, phenol, indole
Langres	2-Phenylethanol, dimethyl disulphide, styrene, indole
Buffalo Mozzarella	Oct-1-en-3-ol, nonanal, indole, component RI 975 with odour of truffles
Cow Mozzarella	Ethyl isobutanoate, ethyl 3-methylbutanoate
Melted cheese flavour	3-Hydroxy-5-methyl-2-hexanone

Data from Dumont *et al.* (1974a,c); Groux and Moinas (1974); Adda and Dumont (1974); Moio *et al.* (1993, 1994); and references listed in this chapter.

many more compounds. On the other hand, no compounds have so far been identified which are reminiscent of cheeses such as Cheddar, Gouda and Emmental, although many volatiles have been isolated from them. Their characteristic flavours could be due to: (i) traces of as yet unidentified compounds; (ii) very unstable compounds which are decomposed by the methods currently used for isolation and identification; or (iii) the physiological interaction of compounds which have already been identified but which need to be smelled at the same time in order to produce the characteristic odour. Undoubtedly, taste sensations such as slight bitterness in the case of Cheddar and sweetness in the case of Emmental also play a part. Perhaps the complex chemistry and biochemistry of cheese aromas and flavours will remain a mystey until an experimental basis is found to test the Component Balance Theory.

8.10   References and bibliography

References

Adda, J. and Dumont, J.P. (1974) [Substances responsible for the aroma of soft cheese]. *Le Lait*, **54**, 1–21.

Adda, J., Gripon, J.C. and Vassal, L. (1982) The chemistry of flavour and texture generation in cheese. *Food Chem.*, **9**, 115–29.

Adda, J., Czulak, J., Mocquot, G. and Vassal, L. (1988) Cheese, in *Meat Science, Milk Science and Technology*, Elsevier Science Publishers, Amsterdam, Netherlands, pp. 373–92.

Adda, J., Dekimpe, J., Vassal, L. and Spinnler, H.E. (1989) [Styrene production by *Penicillium camemberti* Thom.] *Le Lait*, **69**, 115–20.

Ardö, Y. (1994) *Proteolysis and its impact on flavour development in reduced-fat semi-hard cheese made with mesophilic undefined DL-starter*, PhD Thesis, Lund University.

Badings, H.T. (1967) Causes of Ribes flavor in cheese. *J. Dairy Sci.*, **50**, 1347–51.

Barlow, I., Lloyd, G.T., Ramshaw, E.H., Miller, A.J., McCabe, G.P. and McCabe, L. (1989a) Correlations and changes in flavour and chemical parameters of Cheddar cheeses during maturation. *Aust. J. Dairy Technol.*, **44**, 7–18.

Barlow, I., Lloyd, G.T., Ramshaw, E.H., Miller, A.J., McCabe, G.P. and McCabe, L. (1989b) Raw data for 'Correlations and changes in flavour and chemical parameters of Cheddar cheeses during maturation'. *Aust. J. Dairy Technol.*, **44**, 7–18. CSIRO Dairy Research Report No. 43.

Battistotti, B., Bosi, F., Scolari, G.L., Bottazzi, V., Chiusa, P. and Gonzaga, E. (1985) Development of an unpleasant odour in Grana cheese, resulting from the formation of methyl-mercaptan. *Scienza e Tecnica Lattiero-Casearia*, **36**, 77–97.

Bjerke, E.L., Hermann, J.L., Miller, P.W. and Wetters, J.H. (1972) Residue studies of phenoxy herbicides in milk and cream. *J. Agric. Food Chem.*, **20**, 963–7.

Bosset, J.O. and Lavanchy, P. (1991) Flavour research by the Swiss Federal Dairy Research Institute (FAM): retrospective, trends and future prospects. *Lebensmittel-Technologie*, **24**, 190–202.

Bosset, J.O. and Liardon, R. (1984) The aroma composition of Swiss Gruyère cheese. II. The neutral volatile components. *Lebensm.-Wiss. u. -Technol.*, **17**, 359–62.

Bosset, J.O., Bütikofer, U. and Sieber, R. (1990) Decomposition of phenylalanine – another path for the natural formation of benzoic acid in smeared cheeses. *Schweiz. Milchw. Forschung*, **19**, 46–50.

Bosset, J.O., Biedermann, R., Gauch, R. and Pfefferli, H. (1993) Off-flavour in the rind of Swiss Emmental cheese due to volatile compounds from the epoxy resin coated surface in a cheese ripening cellar. *Schweiz. Milchw. Forschung*, **22**, 8–11.

Bosset, J.O., Bütikofer, U., Gauch, R. and Sieber, R. (1994) Caractérisation de fromages d'alpages subalpins suisses: mise en évidence par GC–MS de terpènes et d'hydrocarbures aliphatiques lors de l'analyse par 'Purge and Trap' des arômes volatils de ces fromages. *Schweiz. Milchw. Forschung*, **23**, 37–41.

Bosset, J.O., Gauch, R., Mariaci, R. and Klein, B. (1996) Comparison of various sample treatments for the analysis of volatile compounds by GC/MS: Application to Swiss Emmental cheese. *Mitt. Gebiete Lebensm Hyg.* (in press).

Bouton, Y. and Grappin, R. (1995) Comparison of the final quality of a Swiss-type cheese made from raw or microfiltered milk. *Le Lait*, **75**, 31–44.

Bouzas, J., Kantt, C.A., Bodyfelt, F. and Torres, J.A. (1991a) Simultaneous determination of sugars and organic acids in Cheddar cheese by high-performance liquid chromatography. *J. Food Sci.*, **56**, 276–8.

Bouzas, J., Bodyfelt, F.W. and Torres, A. (1991b) A potential analytical assessment of Cheddar cheese flavor defects. *Int. Dairy J.*, **1**, 263–71.

Brennand, C.P., Ha, J.K. and Lindsay, R.C. (1989) Aroma properties and thresholds of some branched-chain and other minor volatile fatty acids occurring in milkfat and meat lipids. *J. Sensory Studies*, **4**, 105–20.

Broome, M.C., Thomas, M.P., Hillier, A.J., Horwood, J.F. and Jago, G.R. (1979) The effect of linoleic acid on the growth and metabolism of *Streptococcus lactis*. *Aust. J. Dairy. Technol.*, **34**, 163–8.

Bruinenberg, P.G., van den Ban, E., de Roo, G. and Limsowtin, G.K.Y. (1996) *Proc. V LAB Symposium, Veldhoven, Netherlands* (in press).

Cuer, A. (1982) [The aroma of cheese.] *Parfums, Cosmétiques, Arômes*, no. 44, 88–92.

Czulak, J., Hammond, L.A. and Horwood, J.F. (1974a) Cheese and cultured dairy products from milk with high linoleic acid content. I. Manufacture and physical and flavour characteristics. *Aust. J. Dairy Technol.*, **29**, 124–8.

Czulak, J., Hammond, L.A. and Horwood, J.F. (1974b) Cheese and cultured dairy products from milk with high linoleic acid content. II. Effect of added lipase on the flavour of the cheese. *Aust. J. Dairy Technol.*, **29**, 128–31.

Czulak, J., Horwood, J.F. and Hammond L.A. (1979) Cheese and cultured dairy products from milk with high linoleic acid content. III. The use of *Leuconostoc* and *Lactobacillus* starters for ripening the cheese. *CSIRO Dairy Research Report no. 27*, July 1979.

Daley, J.D., Lloyd, G.T., Ramshaw, E.H. and Stark, W. (1986) Off-flavours related to the use of sorbic acid as a food preservative. *CSIRO Food Research Quaterly*, **46**, 59–63.

Darwish, S.M., El-Difrawy, E.A., Mashaly, R. and Aiad, E. (1994) An assay for bitter peptides, amino acids, biogenic amines, glycerides, and fatty acids, in the bitter Ras cheese on local market. *Egyptian J. Dairy Sci.*, **22**, 1–10.

Day, E.A., Bassette, R. and Keeney, M. (1960) Identification of volatile carbonyl compounds from Cheddar cheese. *J. Dairy Sci.*, **43**, 463–74.

Deeth, H.C. and Fitz-Gerald, C.H. (1983) Lipolytic enzymes and hydrolytic rancidity in milk and milk products, in *Developments in Dairy Chemistry – 2 Lipids* (ed. P.F. Fox), Applied Science Publishers, London, New York, pp. 195–239.

de Koning, P.J., de Boer, R. and Nooy, P.F.C. (1981) Comparison of proteolysis in a low-fat semi-hard type of cheese manufactured by standard and by ultrafiltration techniques. *Neth. Milk Dairy J.*, **35**, 35–46.

de Revel, G. and Bertrand, A. (1993) A method for the detection of carbonyl compounds in wine. Analysis of glyoxal and methylglyoxal. *J. Sci. Food Agric.*, **61**, 267–72.

Dilanyan, Z.Kh. and Magak'yan, D.T. (1978) Volatile components of "Chanakh" pickled cheese. *Proc. 20th Int. Dairy Congr. Paris*, 1978. pp. 295–6.

Dimos, A. (1992) *A comparative study by GC/MS of the flavour volatiles produced during the maturation of full-fat and low-fat Cheddar cheese*, M.Sc. Thesis, Latrobe University, Bundoora, Victoria, Australia.

Dimos, A., Urbach, G.E. and Miller, A.J. (1996) Changes in flavour and volatiles of full-fat and low-fat Cheddar cheeses during maturation. *Int. Dairy J.* (accepted)

Dumont, J.P. and Adda, J. (1978) Occurrence of sesquiterpenes in mountain cheese volatiles. *J. Agric. Food Chem.*, **26**, 364–7.

Dumont, J.P. and Adda, J. (1979) Flavour formation in dairy products, in *Progress in Flavour Research* (ed. D.G. Land and H.E. Nursten) Applied Science Publishers Ltd, London, pp. 245–62.

Dumont, J.P., Roget, S. and Adda, J. (1974a) [Neutral volatile compounds in soft cheeses and surface ripened cheeses.] *Le Lait*, **54**, 31–43.

Dumont, J.P., Roget, S. and Adda, J. (1974b) Composés volatils du fromage entier et du fromage râpé: exemple du Parmesan. *Le Lait*, **54**, 386–96.

Dumont, J.P., Roget, S., Cerf, P. and Adda, J. (1974c) Etude de composés volatils neutres présents dans le Vacherin. *Le Lait*, **54**, 243–51.

Dumont, J.P., Roget, S., Cerf, P. and Adda, J. (1974d) [Neutral volatiles in Camembert cheese]. *Le Lait*, **54**, 501–16.

Dumont, J.P., Roget, S. and Adda, J. (1976) [Camembert aroma: identification of minor constituents.] *Le Lait*, **56**, 595–9.

Dumont, J.P., Adda, J. and Rousseaux, P. (1981) Exemple de variation de l'arôme à l'intérieur d'un même type de fromage: Le Comté. *Lebensm.-Wiss. u.-Technol.*, **14**, 198–202.

Dumont, J.P., Mourgues, R. and Adda, J. (1983) Potato-like off-flavour in smear-coated cheese: a defect induced by bacteria, in *Sensory Quality in Foods and Beverages* (eds A.A. Williams and R.K. Atkin), Ellis Horwood Ltd., Chichester, pp. 424–8.

Dunn, H.C. and Lindsay, R.C. (1985) Evaluation of the role of microbial Strecker-derived aroma compounds in unclean-type flavors of Cheddar cheese. *J. Dairy Sci.*, **68**, 2859–74.

Eckert, Th. von, Knieps, A. and Hoffmann, H. (1964) N-dimethylmethioninol – eine Substanz mit Camembert-Aroma. *Zeitschrift für Naturforschung*, **19b**, 1082–3.

El Soda, M., Korayem, M. and Ezzat, N. (1986) The esterolytic and lipolytic activities of the lactobacilli. III. Detection and characterization of the lipase system. *Milchwissenschaft*, **41**, 353–5.

Fahey, R.C. and Newton, G.L. (1983) Occurrence of low molecular weight thiols in biological systems, in *Functions of Glutathione: Biochemical, Physiological, Toxicological, and Clinical Aspects* (eds A. Larsson *et al.*), Raven Press, New York, pp. 251–260.

Fernándes, L. and Steele, J.L. (1993) Glutathione content of lactic acid bacteria. *J. Dairy Sci.*, **76**, 1233–42.

Forss, D.A. (1972) Odor and flavor compounds from lipids. *Prog. Chem. Fats Other Lipids*, **13**, 177–258.

Forss, D.A., Jacobsen, V.M. and Ramshaw, E.H. (1967) Concentration of volatile compounds from dilute aqueous solutions. *J. Agric. Food Chem.*, **15**, 1104–7.

Frutos, M. de, Sanz, J. and Martinez-Castro, I. (1991) Characterization of artisanal cheeses by GC and GC/MS analysis of their medium volatility (SDE) fraction. *J. Agric. Food Chem.*, **39**, 524–30.

Galesloot, Th. E. (1960) The oxidation–reduction potential of cheese. *Neth. Milk Dairy J.*, **14**, 111– 140.

Gallois, A. and Langlois, D. (1990) New results in the volatile odorous compounds of French cheeses. *Le Lait*, **70**, 89–106.

Gallois, A., Kergomard, A. and Adda, J. (1988) Study of the biosynthesis of 3-isopropyl-2-methoxypyrazine produced by *Pseudomonas monastaetrolens*. *Food Chem.*, **28**, 299–309.

Gmür, W., Bosset, J.-O. and Plattner, E. (1986) [Solubility of some important cheese constituents in supercritical carbon dioxide.] *Lebensm.-Wiss. u. -Technol.*, **19**, 419–25.

Gmür, W., Bosset, J.-O. and Plattner, E. (1987a) [Direct coupling of fluid extraction – capillary fluid chromatography. I. Theoretical optimisation of some important instrument parameters.] *J. Chromatogr.*, **388**, 143–50.

Gmür, W., Bosset, J.-O. and Plattner, E. (1987b) [Direct coupling of fluid extraction – capillary fluid chromatography. II. Production of a prototype and examples of its use.] *J. Chromatogr.*, **388**, 335–49.

Gmür, W., Bosset, J.-O. and Plattner, E. (1987c) [Direct coupling of fluid extraction – capillary fluid chromatography. III. Experimental optimisation of pressure and temperature programmes in fluid chromatography as applied to the analyis of dairy products.] *Mitt. Gebiete Lebensm. Hyg.*, **78**, 21–35.

Golovnya, R.V., Zhuravleva, I.L. and Kharatyan, S.G. (1969) Gas chromatographic analysis of amines in volatile substances of *Streptococcus lactis*. *J. Chromatogr.*, **44**, 262–8.

Gotti, R., Andrisano, V., Cavrini, V. and Bongini, A. (1994) Determination of glutathione in pharmaceuticals and cosmetics by HPLC with UV and fluorescence detection. *Chromatographia*, **39**, 23–8.

Green, M.L. and Manning, D.J. (1982) Development of texture and flavour in cheese and other fermented products. *J. Dairy Res.*, **49**, 737–48.

Griffith, R. and Hammond, E.G. (1989) Generation of Swiss cheese flavor compounds by the reaction of amino acids with carbonyl compounds. *J. Dairy Sci.*, **72**, 604–13.

Grosch, W. (1994) Determination of potent odorants in foods by aroma extract dilution analysis (AEDA) and calculation of odour activity values. *Flavour and Fragrance Journal*, **9**, 147–68.

Groux, M. and Moinas, M. (1974) La flaveur des fromages. II. Etude comparative de la fraction volatile neutre de divers fromages. Le Lait, 54, 44– 52.

Guichard, E., Berdagué, J.L., Grappin, R. and Fournier, N. (1987) Affinage et qualité du Gruyère de Comté. V. Influence de l'affinage sur la teneur en composés volatils. Le Lait, 67, 319–38.

Guth, H. and Grosch, W. (1994) Identification of the character impact odorants of stewed beef juice by instrumental analyses and sensory studies. J. Agric. Food Chem., 42, 3862–6.

Ha, J.K. and Lindsay, R.C. (1991a) Contribution of cow, sheep, and goat milks to characterizing branched-chain fatty acid and phenolic flavors in varietal cheeses. J. Dairy Sci., 74, 3267–74.

Ha, J.K. and Lindsay, R.C. (1991b) Volatile branched-chain fatty acids and phenolic compounds in aged Italian cheese flavors. J. Food Sci., 56, 1242–7, 1250.

Harper, W.J. (1959) Chemistry of cheese flavors. J. Dairy Sci., 42, 207–13.

Harper, W.J. and Kristoffersen, T. (1970) Biochemical aspects of flavor development in Cheddar cheese slurries. J. Agr. Food Chem., 18, 563–6.

Harper, W.J. and Wang, J.Y. (1980a) Amino acid catabolism in Cheddar cheese slurries. Formation of selected products from alanine. Milchwissenschaft, 35, 531–5.

Harper, W.J. and Wang, J.Y. (1980b) Amino acid catabolism in Cheddar cheese slurries. II. Evaluation of transamination. Milchwissenschaft, 35, 598–9.

Harper, W.J. and Wang, J.A. (1981) Amino acid catabolism in Cheddar cheese slurries. III. Selected products from glutamic acid. Milchwissenschaft, 36, 70–2.

Harper, W.J., Kristoffersen, T. and Wang, J.Y. (1978) Formation of free fatty acids during the ripening of fat modified cheese slurries. Milchwissenschaft, 33, 604–8.

Harper, W.J., Wang, J.Y. and Kristoffersen, T. (1979) Free fatty and amino acids in fat modified cheese. Milchwissenschaft, 34, 525–7.

Holden, J.T., Wildman, R.B. and Snell, E.E. (1951) Growth promotion by keto and hydroxy acids and its relation to vitamin B6. J. Biol. Chem., 191, 559–76.

Horwood, J.F. (1989) Headspace analysis of cheese. Aust. J. Dairy Technol., 44, 91–6.

Horwood, J.F. and Urbach, G. (1990) Correlation of flavour defects in Cheddar with headspace profile, in Posters and Brief Communications of the XXIII International Dairy Congress (ed. International Dairy Federation), Brussels, Vol. 1, p. 132.

Horwood, J.F., Lloyd, G.T., Ramshaw, E.H. and Stark, W. (1981) An off-flavour associated with the use of sorbic acid during Feta cheese maturation. Aust. J. Dairy Technol., 36, 38–40.

Horwood, J.F., Shanley, R.M. and Sutherland, B.J. (1994) Chemistry of cheese curd slurries: effect of oxygen on flavour and fatty acid production. Aust. J. Dairy Technol., 49, 63–9.

Imhof, R. and Bosset, J.-O. (1989) [Simple quantitative photometric determination of total carbonyls in biological media.] Mitt. Gebiete Lebensm. Hyg., 80, 409–19.

Imhof, R. and Bosset, J.O. (1994) Quantitative GC– MS analysis of volatile flavour compounds in pasteurized milk and fermented milk products applying a standard addition method. Lebensm.-Wiss. u. -Technol., 27, 265–9.

Imhof, R., Bosset, J.O. and Glättli, H. (1994a) Volatile organic aroma compounds produced by thermophilic and mesophilic mixed strain dairy starter cultures. Lebensm.-Wiss. u. -Technol., 27, 442–9.

Imhof, R., Gauch, R., Sieber, R. and Bosset, J.O. (1994b) Über einige flüchtige organische Verunreinigungen in Milch und Milchprodukten. Mitt. Gebiete Lebensm. Hyg., 85, 681–703.

Imhof, R., Glättli, H. and Bosset, J.O. (1995) Volatile organic aroma compounds produced by thermophilic and mesophilic single strain dairy starter cultures. Lebensm.-Wiss. u. -Technol., 28, 78–86.

Jarczynski, R. and Kiermeier, F. (1956) On the development of flavoring substances during the ripening of cheese. Proc XIVth Intern. Dairy Congress, Rome, 2, part 2, pp. 268–72.

Jocelyn, P.C. (1972) Biochemistry of the -SH Group. The Occurrence, Chemical Properties, Metabolism and Biological Function of Thiols and Disulphides, Academic Press Inc., London.

Karahadian, C., Josephson, D.B. and Lindsay, R.C. (1985) Contribution of Penicillium sp. to the flavors of Brie and Camembert cheese. J. Dairy Sci., 68, 1865–77.

Karl, V., Gutser, J., Dietrich, A., Maas, B. and Mosandl, A. (1994) Stereoisomeric flavour compounds LXVIII. 2-, 3-, and 4-alkyl-branched acids, Part 2: chirospecific analysis and sensory evaluation. Chirality, 6, 427–34.

Kepner, R.E., van Straten, S. and Weurman, C. (1969) Freeze concentration of volatile components in dilute aqueous solutions. J. Agric. Food Chem., 17, 1123–7.

Kiermeier, F. and Jarczynski, R. (1962) Quantitative determination of volatile sulphur compounds in soft cheeses. *Z. Lebensm.-Untersuch. u. -Forsch.*, **117**, 306–10, cited in *Dairy Sci. Abs.*, **25**, Abstract 297.

Kim, S.C. (1985) *Chemistry of sulfur compounds in cheese flavor and whey protein functionality*, PhD Thesis, University of Wisconsin, Madison.

Kim, S.C., Kim, M. and Olson, N.F. (1990) Interactive effect of H_2S production from cysteine and methanethiol production from methionine in milk-fat coated microcapsules containing *Brevibacterium linens*. *J. Dairy Sci.*, **57**, 579–85.

Kinsella, J.E. and Hwang, D.H. (1976) Enzymes of *Penicillium roqueforti* involved in the biosynthesis of cheese flavor. *CRC Crit. Rev. Food Sci. Nutr.*, **8**, 191–228.

Kitchen, B.J. (1985) Indigenous milk enzymes, in *Developments in Dairy Chemistry–3. Lactose and Minor Constituents* (ed. P.F. Fox), Elsevier Applied Science Publishers, London, New York, pp. 239–79.

Kowalewska, J., Zelazowska, H., Babuchowski, A., Hammond, E.G., Glatz, B.A. and Ross, F. (1985) Isolation of aroma-bearing materials from *Lactobacillus helveticus* culture and cheese. *J. Dairy Sci.*, **68**, 2167–71.

Kristoffersen, T. (1973) Biogenesis of cheese flavor. *J. Agr. Food Chem.*, **21**, 573–5.

Kristoffersen, T. (1985) Development of flavor in cheese. *Milchwissenschaft*, **40**, 197–9.

Kristoffersen, T. and Gould, I.A. (1959) Oxidation–reduction measurements on Cheddar cheese. *J. Dairy Sci.*, **42**, 901.

Kristoffersen, T., Stussi, D.B. and Gould, I.A. (1964b) Consumer packaged cheese. II. Chemical changes. *J. Dairy Sci.*, **47**, 743–7.

Kristoffersen, T., Gould, I.A. and Purvis, G.A. (1964a) Cheddar cheese flavor. III. Active sulfhydryl group production during ripening. *J. Dairy Sci.*, **47**, 599–603.

Lamparsky, D. and Klimes, I. (1981) Cheddar cheese flavour–its formation in the light of new analytical results, in *Flavour '81* (ed. P. Schreier), Walter de Gruyter, Berlin, pp. 557–77.

Law, B.A., Castanon, M.J. and Sharpe, M.E. (1976) The contribution of starter streptococci to flavour development in Cheddar cheese. *J. Dairy Res.*, **43**, 301–11.

Liardon, R., Bosset, J.-O. and Blanc, B. (1982) The aroma composition of Swiss Gruyère cheese. I. The alkaline volatile components. *Lebensm.-Wiss. und -Technol.*, **15**, 143–7.

Libbey, L.M. and Day, E.A. (1963) Methyl mercaptan as a component of Cheddar cheese. *J. Dairy Sci.*, **46**, 859–61.

Lin, Y.C., Kristoffersen, T. and Harper, W.J. (1979) Carbohydrate derived metabolic compounds in Cheddar cheese. *Milchwisenschaft*, **34**, 69–73.

Lindsay, R.C. (1994) Mechanisms for production of cheese flavor compounds. *Wisconsin Center for Dairy Research Annual Report July 1, 1993 to June 30, 1994*, pp. 74–5.

Lindsay, R.C. and Rippe, J.K. (1986) Enzymic generation of methanethiol to assist in the flavor development of Cheddar cheese and other foods, in *ACS Symposium Series No. 317 Biogeneration of Aromas* (eds T.H. Parliment and R. Croteau), American Chemical Society, pp. 286–308.

Lloyd, G.T. and Ramshaw, E.H. (1985) Objective assessment of flavour development during maturation of cheese, in *Specialty Cheeses for Australia Seminar II*, Australian Society of Dairy Technology, Melbourne, pp. 35–40.

Lopez, V. and Lindsay, R.C. (1993) Metabolic conjugates as precursors for characterizing flavor compounds in ruminant milks. *J. Agric. Food Chem.*, **41**, 446–54.

Maarse, H. and Belz, R. (1981) *Isolation, Separation and Identification of Volatile Compounds in Aroma Research*. Akademie-Verlag, Berlin.

Maarse, H., Visscher, C.A., Willemsens, L.C. and Boelens, M.H. (1989) *Volatile Compounds in Foods: Qualitative and Quantitative Data*, 6th ed, TNO-CIVO Food Analysis Institute, Zeist, The Netherlands.

Magak'yan, D.T., Zhuravleva, I.L., Dilanyan, Z.Kh. and Golovnya, R.V. (1976) [Gas chromatographic analysis of amines from volatile components of Chanakh pickled cheese.] *Prikladnaya Biokhimiya i Mikrobiologiya*, **12**, 253–8. (FSTA 76-10-P1805)

Manning, D.J. and Moore, C. (1979) Headspace analysis of hard cheeses. *J. Dairy Res.*, **46**, 539–45.

Manning, D.J. and Nursten, H.E. (1985) Flavour of milk and milk products, in *Developments in Dairy Chemistry–3 Lactose and Minor Constituents* (ed. P.F. Fox), Elsevier Applied Science Publishers, London, pp. 239–79.

Manning, D.J., Chapman, H.R. and Hosking, Z.D. (1976) The production of sulphur compounds in Cheddar cheese and their significance in flavour development. *J. Dairy Res.*, **43**, 313–20.

Martelli, A. (1989) [Volatile flavour components of Gorgonzola cheese.] *Revista della Società Italiana di Scienza del'Alimentazione*, **18**, 251–62.

Martínez-Castro, I., Sanz, J., Amigo, L., Ramos, M. and Martín-Alvarez, P. (1991) Volatile components of Manchego cheese, *J. Dairy Res.*, 58, 239–46.

McDonald, S.T. (1992) *Role of alpha-dicarbonyl compounds produced by lactic acid bacteria on the flavor and color of cheeses.* PhD Thesis, University of Wisconsin, Madison.

McSweeney, P.L.H., Fox, P.F., Lucey, J.A., Jordan, K.N. and Cogan, T.M. (1993) Contribution of the indigenous microflora to the maturation of Cheddar cheese. *Int. Dairy J.*, **3**, 613–34.

Meinhart, E. and Schreier, P. (1986) Study of flavour compounds from Parmigiano Reggiano cheese. *Milchwissenschaft*, **41**, 689–91.

Miller, A., III, Morgan, M.E. and Libbey, L.M. (1974) *Lactobacillus maltaromicus*, a new species producing a malty aroma. *Int. J. Syst. Bact.*, **24**, 346–54.

Moio, L., Langlois, D., Etievant, P.X. and Addeo, F. (1993) Powerful odorants in water buffalo and bovine Mozzarella cheese by use of extraction dilution sniffing analysis. *Ital. J. Food Sci.*, **5**, 227–37.

Moio, L., Semon, E. and Quere, J.L. Le (1994) 3-Hydroxy-5-methyl-2-hexanone, a new compound characterized by a melted cheese flavour in dairy products. *Ital. J. Food Sci.*, **6**, 441–7.

Morgan, M.E. (1976) The chemistry of some microbially induced flavor defects in milk and dairy foods. *Biotech. Bioeng.*, **18**, 954–65.

Morgan, S., O'Donnovan, C., Ross, R.P., Hill, C. and Fox, P.F. (1995) Significance of autolysis and bacteriocin-induced lysis of starter cultures in Cheddar cheese ripening, in *Proceedings of the 4th Cheese Symposium, 13–14th February, 1995, Fermoy* (eds T.M. Cogan, P.F. Fox and R.P. Ross), Teagasc, Co. Cork, pp. 51–60.

Moskowitz, G.J. and Noelck, S.S. (1987) Enzyme-modified cheese technology. *J. Dairy Sci.*, **70**, 1761–9.

Muller, C.J., Kepner, R.E. and Webb, A.D. (1971) Identification of 3-(methylthio)-propanol as an aroma constituent in Cabernet Sauvignon and Ruby Cabernet wines. *Am. J. Enol. Vitic.*, **22**, 156–60.

Ney, K.H. (1973) Technik der Aromauntersuchung. *Gordian*, **73**, 380–7.

Ney, K.H. (1981) Recent advances in cheese flavor research, in *The Quality of Foods and Beverages Vol. 1, Chemistry and Technology* (eds G. Charalambous and G. Iglett), Academic Press, New York, pp. 389–435.

Ney, K.H. (1985) [Flavour of Tilsit cheese.] *Fette Seifen Anstrichmittel*, **87**, 289–94.

Ney, K.H. and Wirotama, I.P.G. (1971) [Unsubstituted aliphatic monocarboxylic acids, alpha-keto-acids, and amines in Cheddar cheese aroma.] *Zeitschrift für Lebensmittel-Unter-suchung und -Forschung*, **146**, 337–43.

Ney, K.H. and Wirotama, I.P.G. (1972) [Investigation of the aroma of Edelpilzkäse, a German blue mould cheese]. *Zeitschrift für Lebensmittel-Untersuchung und -Forschung*, **149**, 275–9.

Ney, K.H. and Wirotama, I.P.G. (1973) [Unsubstituted aliphatic monocarboxylic acids and α-keto-acids in Camembert.] *Zeitschrift für Lebensmittel-Untersuchung und -Forschung*, **152**, 32–4.

Ney, K.H. and Wirotama, I.P.G. (1978) [Investigation of the aroma constituents of Fontina – an Italian cheese.] *Fette Seifen Anstrichmittel*, **80**, 249–51.

Okumura, J. and Kinsella, J.E. (1985) Methyl ketone formation by *Penicillium camemberti* in model systems. *J. Dairy Sci.*, **68**, 11–5.

Olson, N.F. (1994) The relationship of proteolytic patterns in cheese to the flavor and texture quality of Cheddar cheese. *Annual Report, Wisconsin Center for Dairy Research, June 30, 1994*, pp. 125–6.

Palo, V. and Hollá, L. (1982) GLC-spectrum of volatile substances in Czechoslovakian cheeses. *Brief Communications, XXI Int. Dairy Congr.*, p. 518.

Palo, V. and Hollá, L. (1983) A study of the GLC-spectrum of easily volatile substances of Czechoslovak cheeses with the application of the "Head space" method from cheese solution. *Pol'nohospodárstvo*, **29**, 1029–36.

Papoff, C.M., Delacroix-Buchet, A., Le Bars, D., Campus, R.L. and Vodret, A. (1995) Hydrolysis of bovine *β*-casein C by plasmin. *Ital. J. Food Sci.*, **7**, 157-68.

Parliment, T.H., Kolor, M.G. and Rizzo, D.J. (1982) Volatile components of Limburger cheese. *J. Agric. Food Chem.*, **30**, 1006–8.

Parodi, P.W. (1979) Stereospecific distribution of fatty acids in bovine milkfat triglycerides. *J. Dairy Res.*, **46**, 75–81.

Perret, G.R. (1978) *Volatile components of Cheddar cheese – characterisation, biochemistry of formation and flavour significance.* M Appl Sci. Thesis, Victoria Institute of Colleges, Melbourne, Australia. (Quoted in Urbach, 1993.)

Ponce-Trevino, R., Richter, R.L. and Dill, C.W. (1987) Influence of lactic acid bacteria on the production of volatile sulfur-containing compounds in Cheddar cheese slurries. *J. Dairy Sci.*, **70** (suppl. 1), 59.

Ponce-Trevino, R., Richter, R.L. and Dill, C.W. (1988) Observations on effects of oxidation – reduction potential on volatile sulfhydryl production in Cheddar cheese slurries. *J. Dairy Sci.*, **71** (suppl. 1), 278.

Prante, P.H. and Duus, D. (1986) [Catty flavour in cheese.] *Meieriposten*, **75**, 596–8.

Preininger, M. and Grosch, W. (1994) Evaluation of key odorants of the neutral volatiles of Emmentaler cheese by the calculation of odour activity values. *Lebensm.-Wiss. u. -Technol.*, **27**, 237–44.

Preininger, M., Rychlik, M. and Grosch, W. (1994) Potent odorants of neutral volatile fraction of Swiss cheese (Emmentaler), in *Trends in Flavour Research* (eds H. Maarse and D.G. van der Heij), Elsevier Science B.V., Amsterdam, London, New York, Tokyo, pp. 267–70.

Price, J.C. and Manning, D.J. (1983) A new technique for the headspace analysis of hard cheese. *J. Dairy Res.*, **50**, 381–5.

Ramshaw, E.H. (1985) Aspects of the flavour of phenol, methylphenol, and ethylphenol. *CSIRO Food Research Quarterly*, **45**, 20–2.

Ramshaw, E.H., Roberts, A.V., Mayes, J.J. and Urbach, G. (1990) Phenolic off-flavours in Cheddar cheese, in *Posters and Brief Communications XXIII International Dairy Congress, Montreal* (ed. International Dairy Federation), Brussels, Vol. 1, p. 147.

Reiter, B. and Sharpe, M.E. (1971) Relationship of the microflora to the flavour of Cheddar cheese. *J. Appl. Bact.*, **34**, 63–80.

Roberts, M., Bruinenberg, P.G., Limsowtin, G. and Wijesundera, R.C. (1995) Development of an asceptic cheese curd slurry system for cheese ripening studies. *Aust. J. Dairy Technol.*, **50**, 66–9.

Samples, D.R. (1985) *Some factors affecting the production of volatile sulfhydryl compounds in Cheddar cheese slurries.* PhD Thesis, Texas A&M University.

Schormüller, J. (1962) [New results in the biochemistry of cheese ripening.] *Nutritio et Dieta*, **3**, 68–86.

Schormüller, J. (1968) The chemistry and biochemistry of cheese ripening. *Adv. Food Res.*, **16**, 231–334.

Scott, T.W., Cook, L.J., Ferguson, K.A., McDonald, I.W., Buchanan, R.A. and Loftus Hills, G. (1970) Production of poly-unsaturated milk fat in domestic ruminants. *Aust. J. Sci.*, **32**, 291–3.

Sensidoni, A., Rondinini, G., Peressini, D., Maifreni, M. and Bortolomeazzi, R. (1994) Presence of an off-flavour associated with the use of sorbates in cheese and margarine. *Ital. J. Food Sci.*, **6**, 237–42.

Sieber, R., Bütikofer, U., Baumann, E. and Bosset, J.O. (1990) [The occurence of benzoic acid in cultured milk products and cheese.] *Mitt. Gebiete Lebensm. Hyg.*, **81**, 484–93.

Siek, T.J., Albin, I.A., Sather, L.A. and Lindsay, R.C. (1969) Taste thresholds of butter volatiles in deodorized butteroil medium. *J. Food Sci.*, **34**, 265–7.

Siek, T.J., Albin, I.A., Sather, L.A. and Lindsay, R.C. (1971) Comparison of flavor thresholds of aliphatic lactones with those of fatty acids, esters, aldehydes, alcohols, and ketones. *J. Dairy Sci.*, **54**, 1–4.

Singh, S. and Kristoffersen, T. (1970) Factors affecting flavor development in Cheddar cheese slurries. *J. Dairy Sci.*, **53**, 533–6.

Singh, S. and Kristoffersen, T. (1971) Influence of lactic cultures and curd milling acidity on flavor of Cheddar curd slurries. *J. Dairy Sci.*, **54**, 1589–94.

Singh, S. and Kristoffersen, T. (1972) Cheese flavor development using direct acidified curd. *J. Dairy Sci.*, **55**, 744–9.

Sloot, D. and Harkes, P.D. (1975) Volatile trace components in Gouda cheese. *J. Agric. Food Chem.*, **23**, 356–7.

Sloot, D. and Hofman, H.J. (1975) Alkylpyrazines in Emmental cheese. *J. Agric, Food Chem.*, **23**, 358.

Smith, P.W., Parks, O.W. and Schwartz, D.P. (1984) Characterization of male goat odors: 6-*trans*-nonenal. *J. Dairy Sci.*, **67**, 794–801.

Stadhouders, J. and Veringa, H.A. (1973) Fat hydrolysis by lactic acid bacteria in cheese. *Neth. Milk Dairy J.*, **27**, 77–91.

Stark, W. and Adda, J. (1972) Acides gras volatils du Livarot et du Pont-l'Évêque. *La technique laitière*, No. 746, pp. 15, 17.

Stark, W., Urbach, G., Hamilton, J.S. and Forss, D.A. (1973) Volatile compounds in butter oil. III. Recovery of added fatty acids and δ-lactones from volatile-free butter oil by cold-finger molecular distillation. *J. Dairy Res.*, **40**, 39–46.

Swaisgood, H.E. (1982) Chemistry of milk protein, in *Developments in Dairy Chemistry – 1 Proteins* (ed. P.F. Fox), Elsevier Applied Science Publishers, London, New York, pp. 1–59.

Swiatek, A. and Poznanski, S. (1959) Effect of common salt content upon hydrogen sulphide development in Trappist cheese. *Proc. 15th International Dairy Congress, London*, **3**, 1487–94.

Tanaka, H. and Obata, Y. (1969) Studies on the formation of the cheese-like flavor. *Agric. Biol. Chem.*, **33**, 147–50.

Thomas, T.D. (1987) Acetate production from lactate and citrate by non-starter bacteria in Cheddar cheese. *N.Z. J. Dairy Sci. Technol.*, **22**, 25–8.

Tsugo, T. and Matsuoka, H. (1962) The formation of volatile sulphur compounds during the ripening of the semi-soft white mould cheese. *Proc. XVIth International Dairy Congress, Copenhagen*, **B**, 385–94.

Urbach, G. (1987) Dynamic headspace gas chromatography of volatile compounds in milk. *J. Chromatogr.*, **404**, 163–74.

Urbach, G. (1990) Effect of feed on flavor in dairy foods. *J. Dairy Sci.*, **73**, 3639–50.

Urbach, G. (1993) Relations between cheese flavour and chemical composition. *Int. Dairy Journal*, **3**, 389–422.

Wiederholt, K.M. and Steele, J.L. (1994) Glutathione accumulation in lactococci. *J. Dairy Sci.*, **77**, 1183–8.

Wijesundera, C. and Urbach, G. (1993) Flavour of Cheddar cheese. Final report to the Dairy Research and Development Corporation, Project CSt66. DRDC, Glen Iris, Victoria, Australia.

Wilkinson, M.G. (1992) *Studies on the acceleration of Cheddar cheese ripening*, PhD Thesis, University College Cork, Ireland.

Wilkinson, M.G., Guinee, T.P., O'Callaghan, D.M. and Fox, P.F. (1994) Autolysis and proteolysis in different strains of starter bacteria during Cheddar cheese ripening. *J. Dairy Res.*, **61**, 249–62.

Wilson, R.D. (1989) Flavour volatiles from New Zealand milk fat. Paper presented at *Fats For The Future II*, Auckland, NZ, February 13–17, 1989, International Conference on Fats, Royal Society of New Zealand, Wellington, NZ.

Woo, A.H. and Lindsay, R.C. (1984) Concentration of major free fatty acids and flavor development in Italian cheese varieties. *J. Dairy Sci.*, **67**, 960–8.

Woo, A.H., Kolloge, S. and Lindsay, R.C. (1984) Quantification of major free fatty acids in several cheese varieties. *J. Dairy Sci.*, **67**, 874–8.

Wood, A.F., Aston, J.W. and Douglas, G.K. (1985) The determination of free amino acids in cheese by capillary column gas–liquid chromatography. *Aust. J. Dairy Technol.*, **40**, 166–9.

Yamamoto, K. (1976) Method for giving foodstuffs a flavour resembling that of a dairy product. British Patent 1423004.

Bibliography

Reviews

Adda, J. (1984) Formation de la flaveur, in *Le Fromage* (ed. A. Eck) Lavoisier, Paris, pp. 330–40.

Adda, J. (1986) Flavour of dairy products, in *Developments in Food Flavours* (eds G.G. Birch and M.G. Lindley), Elsevier Applied Science Publishers, London, pp. 151–72.

Adda, J. (1987) Mechanisms of flavour formation in cheeses, in *Milk T– T the Vital Force* (ed. XXII International Dairy Congress), D. Reidel Publishing Company, Dordrecht, pp. 169–77.

Bosset, J.O. and Gauch, R.(1988) Simple sample preparation for quantitative multiple-head-space determination of volatile components using adsorption cartridges. *J. Chromatogr.*, **456**, 417–20. (Emmental, Gruyère, Tilsit, Appenzell, Sbrinz, Raclette.)

Fox, P.F. and McSweeney, P.L.H. (1995) Chemistry, biochemistry and control of cheese flavour, in *Proceedings of the 4th Cheese Symposium, 13th–14th February, 1995, Fermoy* (eds. T.M. Cogan, P.F. Fox and R.P. Ross), Teagasc, Co. Cork, pp. 135–59.

Hammond, E.G. (1989) The flavors of dairy products, in *Flavor Chemistry of Lipid Foods* (eds D.B. Min and T.H. Smouse), American Oil Chemists' Society, pp. 222–36.

Hardy, J. and Adda, J. (1984) [Organoleptic properties of cheese.], in *Le Fromage* (ed. A. Eck), Diffusion Lavoisier, Paris, pp. 320–40.

Harper, W.J. and Kristoffersen, T. (1956) Biochemical aspects of cheese ripening. *J. Dairy Sci.*, **39**, 1173–5.

Imhof, R. and Bosset, J.O. (1994) Relationship between microorganisms and formation of aroma compounds in fermented dairy products (Review) *Z. Lebensm. Unters. Forsch.*, **198**, 267–76.

Law, B.A. (1982) Flavour compounds in cheese. *Perfumer & Flavorist*, **7**, 9–21.

Nakae, T. and Elliott, J.A. (1965) Production of volatile fatty acids by some lactic acid bacteria. II. Selective formation of volatile fatty acids by degradation of amino acids. *J. Dairy Sci.*, **48**, 293–9.

Reps, A., Hammond, E.G. and Glatz, B.A. (1987) Carbonyl compounds produced by the growth of *Lactobacillus bulgaricus*. *J. Dairy Sci.*, **70**, 559–62.

Roudot-Algaron, F., le Bars, D., Einhorn, J., Adda, J. and Grippon, J.C. (1993) Flavor constituents of aqueous fraction extracted from Comté cheese by liquid carbon dioxide. *J. Food Sci.*, **58**, 1005–9.

Seth, R.J. and Robinson, R.K. (1988) Factors contributing to the flavour characteristics of mould-ripened cheese, in *Developments in Food Microbiology – 4* (ed. R.K. Robinson), Elsevier Applied Science Publishers, London, pp. 23–46.

Swiatek, A. and Poznanski, S. (1959) Effect of common salt content upon hydrogen sulphide development in Trappist cheese. *15th International Dairy Congress, London*, **3**, 1487–94.

Yuguchi, H., Hiramatsu, A., Doi, K., Ida, Ch. and Okonogi, Sh. (1989) Studies on the flavour of yogurt fermented with *Bifidobacteria* – significance of volatile components and organic acids in the sensory acceptance of yogurt. *Jpn. J. Zootech. Sci.*, **60**, 734–41.

Cheddar

Arora, G., Cormier, F. and Lee, B. (1995) Analysis of odor-active volatiles in Cheddar cheese headspace by multidimetional GC/MS/sniffing. *J. Agric. Food Chem.* **43**, 748–52.

Brechany, E.Y., Christie, W.W. and Banks, J.M. (1993) Chemical analysis and olfactory perception of flavour volatiles in Cheddar cheese. *Int. Dairy J.*, **3**, 553–4.

Christensen, K.R. (1994) Studies on the volatile compounds responsible for Cheddar cheese aroma. *Diss. Abs. Int.*, **B55**, 1240.

Dacremont, C. and Vickers, Z. (1994) Concept matching technique for assessing importance of volatile compounds for Cheddar cheese aroma. *J. Food Sci.*, **59**, 981–5.

Hill, A.R. and Ferrier, L.K. (1989) Composition and quality of Cheddar cheese. *Modern Dairy*, **68**, 58–60.

Manning, D.J. (1974) Sulphur compounds in relation to Cheddar cheese flavour. *J. Dairy Res.*, **41**, 81–7.

Manning, D.J. (1979) Chemical production of essential flavour compounds. *J. Dairy Res.*, **46**, 531–7.

Manning, D.J. and Price, J.C. (1982) Effect of redox potential on the flavour of Cheddar cheese, in *XXI International Dairy Congress, Vol. 1, Book 1*, pp. 507–8.

O'Keeffe, R.B, Fox, P.F. and Daly, C. (1976) Contribution of rennet and starter proteases to proteolysis in Cheddar cheese. *J. Dairy Res.*, **43**, 97–107.

Sharpe, M.E. and Franklin, J.G. (1962) Production of hydrogen sulphide by lactobacilli with special reference to strains isolated from Cheddar cheese. *Proc. VIIIth International Congress of Microbiology, Canada*, **B.11.3**, p. 46.

Vanderweghe, P. and Reineccius, G.A. (1990) Comparison of flavor isolation techniques applied to Cheddar cheese. *J. Agric. Food Chem.*, **39**, 1549–52.

Grana

Piergiovanni, L. and Volonterio, G. (1977) Study into the substances responsible for the formation of the aroma of Grana cheese. I. Methyl ketones. *Industria del Latte*, **13**, 31–46.
Piergiovanni, L., Volonterio, G. and Conti, G. (1983) [Aroma producing substances in Grana cheese. II. Methods for study of alcohols.] *Industria del Latte*, **19**, 3–18. More than 40 primary alcohols were detected together with thiols and primary and secondary amines.

Parmesan

Barbieri, G., Bolzoni, L., Careri, M., Mangia, A., Parolari, G., Spagnoli, S. and Virgili, R. (1994) Study of the volatile fraction of Parmesan cheese. *J. Agric. Food Chem.*, **42**, 1170–6.
Careri, M., Manini, P., Spagnoli, S., Barbieri, G. and Bolzoni, L. (1994) Simultaneous distillation–extraction and dynamic headspace methods in the gas chromatographic analysis of Parmesan cheese volatiles. *Chromatographia*, **3**, 386–92.
Virgili,R., Parolari, G., Bolzoni, L., Barbieri, G., Mangia, A., Careri, M., Spagnoli, S., Panari, G. and Zannoni. M. (1994) Sensory–chemical relationships in Parmigiano-Reggiano cheese. *Lebensm.-Wiss. u. -Technol.*, **27**, 491–5.

Manchego

Amigo, L., Ramos, M., Martín-Alvarez, P.J., Martínez-Castro, I. and Sanz, J. (1990) [Development of volatile components during ripening of Manchego cheese], in *Brief Communications of the XXIII International Dairy Congress, Montreal, October 8–12, 1990, Vol. II*, International Dairy Federation, Brussels, p. 489.
Wirotama, I.P.G., Ney, K.H. and Freytag, W.G. (1973) [Investigation of the flavour of Manchego cheese.] *Zeitschrift für Lebensmittel-Untersuchung und -Forschung*, **153**, 78–82.

Gruyère

Bosset, J.O. and Liardon, R. (1984) The aroma composition of Swiss Gruyère cheese. III. Relative changes in the content of alkaline and neutral volatile components during ripening. *Lebensm. -Wiss. u. -Technol.*, **18**, 178–85.
Bosset, J.O., Collomb, M. and Sieber, R. (1993) The aroma composition of Swiss Gruyère cheese. IV. The acidic volatile components and their changes in content during ripening. *Lebensm. -Wiss. u. -Technol.*, **26**, 581–92.
Imhof, R., Isolini, D. and Bosset, J.O. (1990) Differences between homofermentative and heterofermentative lactobacilli with respect to the production of volatile flavour components in Swiss Gruyère cheese and in culture supernatants. *Lebensm. -Wiss. u. -Technol.*, **23**, 305–11.

Domiati

Collin, S., Osman, M., Delcambre, S. El-Zayat, A.I. and Dufour, J.-P. (1993) Investigation of volatile flavour compounds in fresh and ripened Domiati cheeses. *J. Agric. Food Chem.*, **41**, 1659–63.

Mozzarella

Moio, L., Dekimpe, J., Etievant, P.X. and Addeo, F. (1993) Volatile flavour compounds of water buffalo Mozzarella cheese. *Ital. J. Food Sci.*, **1**, 57–68.

Romano

Lee, K.-C. M., Shi, H., Huang, A.-S., Carlin, J.T., Ho, C.-T. and Chang, S.S. (1986) Production of Romano cheese flavor by enzymic modification of butterfat, in *ACS Symposium Series 317 Biogeneration of Aromas* (eds T. Parliment and R. Croteau), American Chemical Society, pp. 370–8.

Camembert

Moinas, M., Groux, M. and Horman, I. (1975) [The flavour of cheese. III. Minor constituents of Camembert aroma.] *Le Lait*, **55**, 414–17.

Goats' milk cheese

Vidal-Aragón, M.C., Sabio, E., González, J. and Mas, M. (1994) [Contribution to the study of volatile compounds in goat milk cheeses from Extremadura: effect of the season of manufacture]. *Alimentaria*, **31**, 25–9. [*Dairy Sci. Abs.* (1995) **57**, Abs. no. 3982.]

Various cheeses

Bosset, J.O. and Gauch, R. (1993) Comparison of the volatile flavour compounds of six European "AOC" cheeses by using a new dynamic headspace GC–MS method. *Int. Dairy J.*, **3**, 359–77.

Engels, W.J.M. and Visser, S. (1994) Isolation and comparative characterization of components that contribute to the flavour of different types of cheese. *Neth. Milk Dairy J.*, **48**, 127–40.

Oruhanbayala, Andoh, K. (1994) [Studies by headspace gas chromatography on flavor components in various natural cheese.] *Journal of Rakuno Gakuen University, Natural Science*, **19**, 407–21. [*Dairy Science Abstracts* (1995) **57**, Abs. no. 3976.]

Parma-type cheese

Rafecas, M., Boatella, J. and Torre, M.C. de la (1986) [Analysis of volatile compounds from a Parma-type cheese by adsorption onto activated charcoal.] *Revista de Agroquimica y Tecnologia de Alimentos*, **26**, 597–601. Mainly esters, secondary alcohols and alkan-2-ones.

Processed cheese

Jung, J.H. and Yu, J.H. (1988). [Studies on the flavour intensity and quality in processed cheese made from different amounts of Cheddar cheese.] *Korean J. Dairy Sci.*, **10**, 34–43.

Patents

Pittet, A.O., Muralidhara, R., Luccarelli, D. Jr, Miller, K.P. and Vock, M.H. (1986) Flavoring with gem dithioethers of phenylalkanes. US Patent 4585663. (Acetophenone dimethyl mercaptal has cheesy flavour.)

Yokoyama, H., Sawamura, N. and Motobayashi, N. (Fuji Oil Co. Ltd) (1993). Method of ripening cheese under high pressure. US Patent 5180596. (Shortens ripening period and reduces proliferation of contaminant organisms.)

Reduced-fat cheese

Banks, J.M., Brechany, E.Y. and Christie, W.W. (1989) The production of low fat Cheddar-type cheese. *J. Soc. Dairy Technol.*, **42**, 6–9.

Banks, J.M., Hunter, E.A. and Muir, D.D. (1994) Sensory properties of Cheddar cheese: effect of fat content on maturation. *Milchwissenschaft*, **49**, 8–12.
Brandsma, R.L., Mistry, V.V., Anderson, D.L. and Balwin, K.A. (1994) Reduced fat Cheddar cheese from condensed milk. 3. Accelerated ripening. *J.Dairy Sci.*, **77**, 897–906.

Nitrosamines

Smiechowska, M., Przybylowski, P. and Kowalski, B. (1994) Study of formation of volatile N-nitrosamines in Zulaw, Gouda and Edam cheese. *Polish Journal of Food and Nutrition Sciences*, **3**, 71–80. [*Dairy Sci. Abs.* (1995) **57**, Abstract no. 3969]

Accelerated ripening

Kim, S.C. and Olson, N.F. (1989) Production of methanethiol in milk fat-coated microcapsules containing *Brevibacterium linens* and methionine. *J. Dairy Res.*, **56**, 799–811.
Puchades, R., Lemieux, L. and Simard, R.E. (1989) Evolution of free amino acids during the ripening of Cheddar cheese containing added lactobacilli strains. *J. Food Sci.*, **54**, 885–8, 946.

Organoleptic testing

Roberts, A.K. and Vickers, Z.M. (1994) Cheddar cheese aging: changes in sensory attributes and consumer acceptance. *J. Food Sci.*, **59**, 328–34.

Physical properties

Boekel, M.A.J.S. van and Lindsay, R.C. (1992) Partition of cheese volatiles over vapor, fat and aqueous phases. *Neth. Milk Dairy J.*, **46**, 197–208.
Manning, D.J., Ridout, E.A., Price, J.C. and Gregory, R.J. (1983) Effect of reducing the block size on the flavour of Cheddar cheese. *J. Dairy Res.*, **50**, 527–34.

Statistical analysis

Banks, J.M., Brechany, E.Y., Christie, W.W., Hunter, E.A. and Muir, D.D. (1992) Volatile components of Cheddar cheese as indicator indices of cheese maturity, flavour and odour. *Food Research Applied Electrostatic Precipitation International*, **25**, 365–73.
Vangtal, A. and Hammond, E.G. (1986) Correlation of the flavor characteristics of Swiss-type cheeses with chemical parameters. *J. Dairy Sci.*, **69**, 2982–93.

9 Proteolytic systems of dairy lactic acid bacteria
F. MULHOLLAND

9.1 Introduction

Lactic acid bacteria (LAB), such as lactococci and lactobacilli, play an essential role in the manufacture of cultured dairy products such as cheese. They are primarily responsible for the acidification of the milk, the key stage in the manufacture of such products, and are also considered to be directly involved in the formation of the characteristic flavour notes formed in cheeses during ripening. Many of the enzymes found within or secreted by LAB are either directly or indirectly responsible for these processes. The economic importance of cultured dairy products, and manufacturers' demands for starter cultures that will produce a consistent product in the modern high throughput manufacturing facility, has led to in-depth investigation of these organisms, their physiology, biochemistry and genetics. Chapter 4 has already given an in-depth analysis of carbohydrate metabolism in LAB and this chapter will concentrate on the increase in knowledge being rapidly assembled in nitrogen metabolism, and the role played by the proteolytic system of LAB.

9.2 Milk as a growth medium

Lactococci and other lactic acid bacteria used as dairy starters are nutritionally fastidious, requiring exogenous sources of nucleotides, vitamins and amino acids to support growth. This amino nitrogen deficiency in LAB can be addressed by two routes:

1. Directly, by taking up amino acids or small peptides present in the extracellular medium into the cell.
2. Indirectly, if there are insufficient small nutrients, by hydrolysing proteins in the medium to amino acids and peptides of a transportable size.

Milk as a growth medium for starter culture has a limited supply of free amino acids and small peptides. Thomas and Mills (1981), and Law and Kolstad (1983), both reviewed data which demonstrated that milk cannot provide in the free form all the essential amino acids such as histidine, leucine, glutamic acid and methionine required for growth, and that lactococci rely on their proteolytic enzymes to hydrolyse the external milk proteins to supply these essential nutrients.

Milk has a number of proteins [total protein 3.0–3.5% (wt/wt)], of which the caseins represent about 80%. The four different types of casein found in milk, α_{s1}, α_{s2}, β and κ are organized into micelles to form soluble complexes. It has generally been accepted that the caseins are the major source of amino acids utilized by lactococci for growth in milk. Exterkate and de Veer (1987a) demonstrated that optimal growth of *Lactococcus lactis* subsp. *cremoris* HP required the presence of both β- and κ-casein in the medium and it was further speculated by Smid *et al.* (1991) that κ-casein hydrolysis may be a source of the essential amino acid histidine. It is perhaps significant that one of the early-formed peptides from proteinase hydrolysis, κ-casein 96-106, is histidine-rich.

9.3 Proteolysis

The function of proteolysis in dairy starter cultures is two-fold. Firstly, it relates to the requirement of the organisms to multiply in a low amino acid, high protein medium such as milk; proteolysis is also an essential part of the cheese ripening process whereby flavour notes are generated as a result of protein breakdown by enzymes, many of which are derived from the starter cultures.

9.3.1 Localization of the proteolytic system

From recent studies that will be discussed, it is now clear that the localization of the proteolytic system of lactococci is clearly differentiated into three main components;

- extracellular enzyme(s) capable of breaking down the milk proteins to peptides and amino acids;
- cell membrane-spanning transport systems capable of translocating both peptides and amino acids into the organism; and
- intracellular enzymes capable of breaking down the transported peptides to amino acids.

At one time it was thought that the extracellular proteinase was part of a complex extracellular proteolytic system with extracellular peptidases also involved in the breakdown of the milk proteins, following initial hydrolysis by the proteinase. To date, however, no strong evidence exists for extracellular peptidases and the current model of the lactococcal proteolytic system is shown in Figure 9.1. With this model, the only extracellular hydrolysis is due to the cell wall proteinase. Small peptides already present in the milk or derived from the proteolytic cleavage are then translocated inside the LAB by the oligopeptide transporter where they can be hydrolysed to their constituent amino acids by an array of intracellular peptidases.

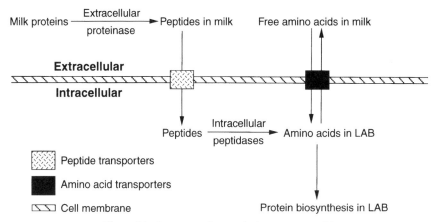

Figure 9.1   Lactococcal proteolytic system – model.

9.3.2   Extracellular proteolytic enzymes

(a) The cell wall proteinase.   Proteinase activity is essential in lactococci for rapid growth in milk. The need to understand the mechanisms behind the economically important traits of strain stability and media regulation of proteinase activity by medium composition has resulted in the cell wall proteinase probably being the most studied enzyme in lactococci. While other proteinases have been isolated and characterized (e.g. NisP and some intracellular proteinases), it has been shown by among others, Exterkate (1975) and Thomas *et al.* (1975), that most lactococcal proteinase activity is associated with the cell wall fraction. Hugenholtz *et al.* (1987) demonstrated by immunogold labelling that the proteinase was located at the outside of the cell wall. At one time it was thought that there were several proteinases in lactococci, based on biochemical characterization and specificity studies, but it is now clear there is only one main extracellular proteinase activity (Nissen-Meyer and Sletten, 1991).

Genetic variation within strains, however, plays an important role in the application of starter cultures in cheese manufacture. Complete nucleotide sequences have been determined from *Lactococcus lactis* Wg2 (Kok *et al.*, 1988a), SK11 (Vos *et al.*, 1989a), and NCDO763 (Kiwaki *et al.*, 1989). Two genetic variants of the proteinase, PI (from Wg2) and PIII (from SK11) have been extensively characterized. The genetic analysis shows a 98% homology between the two variants, differing in only 44 out of 1902 amino acids, with the SK11 enzyme also having an additional 60 amino acid duplication near the C-terminus. These differ in their caseinolytic activities, which are of significance in cheese ripening. Visser *et al.* (1986) earlier showed, before the genetic identities had been determined, that PIII is able to cleave both α- and β-casein, whereas PI from *Lactococcus lactis* subsp. *cremoris* Wg2 cleaves primarily β-casein with a specificity different from PIII. The enzymology of the lactococcal cell wall proteinase has been extensively reviewed recently by Exterkate (1995).

Whereas it was originally considered that only limited hydrolysis was achieved by the proteinase, necessitating the need for further hydrolysis of the products by extracellular peptidases, it has recently been shown by Juillard *et al.* (1995b) that the PI variant from *Lactococcus lactis* subsp. *cremoris* Wg2 can substantially hydrolyse β-casein into more than 100 different oligopeptides with one-fifth of them small enough to be taken up by the oligopeptide transport system. These peptides could potentially supply lactococci with all the amino acids required to sustain growth.

The genetic analyses confirm the extracellular location of the proteinase, which show the proteinase to be synthesized as the preproenzyme with a molecular mass of $\sim 200\,kDa$. At the N-terminus there is a 33-amino acid region with signal peptide characteristics. The mature proteinase sequence starts as an aspartic acid at position 188 of the predicted amino acid sequence, with amino acids 34–187 forming the pro-sequence. In certain regions the cell wall proteinase is similar to the serine proteinases of the subtilisin family, particularly in the regions that correspond to the catalytic site. In subtilisin the catalytic triad is formed by Asp_{32}, His_{64} and Ser_{221}; in the lactococcal proteinase, the catalytic triad is formed by Asp_{30}, His_{94} and Ser_{443}. Unlike subtilisin, which is secreted into the growth medium, the proteinase has a C-terminal anchor region which attaches the proteinase to the external side of the cell membrane. Deletion of the last 300 amino acids from the C-terminus results in extracellular release without affecting the caseinolytic specificity (Kok *et al.*, 1988b; Vos *et al.*, 1989b).

Immediately upstream of the gene encoding the proteinase is a highly conserved region encoding a protein involved in processing the proenzyme into the active enzyme. This gene, *prtM*, transcribes from the same promotor region as the proteinase, but in the opposite direction. The prtM protein is a lipoprotein which Haandrikman *et al.* (1989, 1991) and Vos *et al.* (1989b) showed to be involved in the activation of the proenzyme. Lactococci carrying the proteinase gene, but lacking the *prtM* gene were found to be phenotypically proteinase deficient, but were still expressing immunoreactive proteinase antigen.

(b) Proteinase stability and regulation of proteinase activity.   It was also shown several years ago by Lawrence *et al.* (1976) that the lactococcal proteinase activity is unstable, with proteinase positive strains (Prt$^+$) spontaneously converting to proteinase negative (Prt$^-$), the latter only able to grow to 10–25% of maximum cell density attained by the Prt$^+$ strain. Curing experiments by McKay and Baldwin (1974) and Gasson (1983) revealed that the proteinase gene was located on a plasmid. Loss of plasmids

resulted in loss of proteinase activity. Kok *et al.* (1985) showed that this activity could then be restored by introducing the proteinase-encoding recombinant plasmid, pGVK500, into the plasmid-free *Lactococcus lactis* MG1363.

Recently, Juillard *et al.* (1995a) demonstrated that growth of *Lactococcus lactis* in milk is biphasic for Prt$^+$ strains and monophasic for Prt$^-$ strains. The first phase for Prt$^+$ strains is identical to that found with the Prt$^-$ strains with no casein hydrolysis. It was postulated that this first phase utilizes peptides already found in milk (from the non-protein nitrogen fraction), which are transported into the organism via the oligopeptide transporter. Once this supply is exhausted, the Prt$^-$ strains are unable to grow further while the Prt$^+$ strains are able to go into a second growth phase using peptides (derived from the milk proteins) produced by the action of the proteinase as the source of amino nitrogen nutrients.

The proteinase is also the subject of regulation by the growth medium; Exterkate (1979) and Hugenholtz *et al.* (1984) demonstrated that more proteinase was produced when lactococci were grown in milk than in standard laboratory media. It was also reported by Laan *et al.* (1993) and Meijer *et al.* (1996) that this production was dependent on the nitrogen source and the growth phase, with the highest levels of activity found during growth in milk and the lowest in the peptide-rich laboratory medium, M17. The factors now considered to be responsible for the regulation of the proteinase are peptides present in the medium. Marugg *et al.* (1995) used reporter gene experiments to demonstrate that proteinase production in *Lactococcus lactis* SK11 could be regulated more than 10-fold. It was also demonstrated that by adding specific dipeptides to growth media, proteinase expression was reduced. Not all dipeptides tested caused this repression; leucylproline and prolylleucine were the effective peptides. It was also shown that the di/tripeptide transporter plays an important function in this process since this repression by peptides was not observed in mutants defective for the uptake of di/tripeptides.

9.3.3 *Amino acid and peptide transport systems*

As highlighted above, in the regulation of proteinase activity by peptides and in supplying oligopeptides nutrients for growth, the transport systems play a significant role in these processes. Any discussion on proteolysis in LAB necessarily requires an analysis of these systems, particularly the di/tripeptide transporter and the oligopeptide transporter which have been the subject of much recent research.

Kunji *et al.* (1993) demonstrated the two peptide transport systems existed in lactococci, a di/tripeptide transporter and an oligopeptide transporter. Using a mutant of *Lactococcus lactis* MG1363 that was resistant to the toxic analogues of alanine and alanine-containing di- and tripeptides, it was shown that the mutant was unable to transport alanine, dialanine and trialanine but was able to take up peptides larger than three amino acids.

At least three types of amino acid transporter systems have also been identified: (i) a proton motive force-driven transport for leucine, isoleucine, valine, alanine, glycine, serine, threonine and lysine; (ii) an antiport system for arginine and ornithine; and (iii) a phosphate bond-linked system for glutamic acid, glutamine, aspartic acid and asparagine. These have been reviewed by Konings *et al.* (1989). Although little has been referenced recently it is anticipated that the LAB amino acid transport systems will be the subject of further research in the near future, since their integrity and coordination are vital for efficient biomass production of starter cultures.

(a) Di/tripeptide transporter.   The di/tripeptide transporter (DtpT) in lactococci was first described by Smid *et al.* (1989a). Using peptidase-free lactococcal membrane vesicles, the specific uptake of Ala–Glu was shown to be driven by the electrical potential and the chemical gradient of protons across the membrane (a proton motive force-dependent system). The gene encoding the DtpT was fully characterized by Hagting *et al.* (1994), confirming it be the only di/tripeptide transport system in lactococci. The *DtpT* gene encodes a protein of 463 residues, which by hydropathicity profiling could form 12 membrane-spanning regions, indicative of its membrane location and function.

It was initially thought from work by Smid *et al.* (1989b) that the DtpT was essential for growth of lactococci using β-casein as the only amino nitrogen source. More recently, however, Kunji *et al.* (1995) and Juillard *et al.* (1995b), demonstrated that DtpT$^-$ mutants are able to grow on a β-casein medium equally well as the wild-type.

Smid and Konings (1990), demonstrated that the DtpT has a high affinity for proline-containing di- and tripeptides. They further showed that the addition of proline-containing dipeptides severely inhibited growth on casein-containing media, from which they concluded that normal growth required a balanced supply of different di- and tripeptides, which compete for the same DtpT system for entry into the cell. Another explanation of this finding was made by Laan *et al.* (1993), who proposed that the dipeptide, prolylleucine, was responsible for regulation of proteinase production, thus influencing cell growth. This has since been supported by the work of Marugg *et al.* (1995), which demonstrated that specific dipeptides, transported via the DtpT system, play an essential part in the regulation of proteinase production in *Lactococcus lactis*. Using reporter genes to monitor for proteinase production, the addition specific dipeptides, leucylproline and prolylleucine were shown to repress the expression of the *prtP* and *prtM* promotors. The significant role of the DtpT in this regulation was demonstrated by the inability of DtpT$^-$ mutants to show this regulatory action by these prolyl peptides.

(b) Oligopeptide transporter. Tynkkynen *et al.* (1993) have genetically characterized this system as an ABC (ATP-binding cassette) transporter. It consists of a two ATP-binding proteins, OppD and OppF, two integral membrane proteins, OppB and OppC, and a substrate binding protein, OppA.

Increasingly, the oligopeptide transporter (Opp) is considered the essential route of supplying lactococci with the peptides essential for growth in milk. Construction of mutants with a defective OppA system results in an inability to transport and utilize peptides larger than three amino acid residues and an inability to grow in milk. It has been suggested by Law (1978) and Rice *et al.* (1978) that the maximal size of peptide transportable into lactococci is 5–6 amino acids in length. In growth studies on chemically defined medium lacking leucine, and then supplemented with leucyl-containing peptides as the sole leucine source, Tynkkynen *et al.* (1993) showed that the Opp system could transport oligopeptides up to octapeptides. Further oligopeptide utilization studies in Prt^+ and Prt^- strains by Juillard *et al.* (1995a) and Opp deletion studies by Kunji *et al.* (1995) have demonstrated that Opp^- mutants were unable to sustain the first phase growth observed in Prt^- strains. This was considered to be due to the inability of the Opp^- mutants to take up oligopeptides already present in milk, which the Prt^- wild-type are able to utilize, leading to the conclusion that oligopeptides are the main source of nitrogen for *Lactococcus lactis* during growth in milk, and that the Opp system is the essential component for supplying these peptide nutrients.

9.3.4 Intracellular proteolytic enzyme: peptidases

For the LAB peptidases, a nomenclature system was devised by Tan *et al.* (1993a) to try to simplify and standardize the naming of peptidases from LAB caused by the sudden proliferation in these enzymes being characterized. The initial name is based on its biochemical activity and shortened to a three-letter code, e.g. glutamyl aminopeptidase, which is shortened to GAP. When the gene for the peptidase has been sequenced, the enzyme takes the pep prefix, followed by a letter to identify it; e.g. GAP is now known as pepA.

Table 9.1 shows a list of peptidases that have been purified and characterized from lactococci using this nomenclature. With the exceptions of a 36-kDa aminopeptidase isolated from the cell wall of *Streptococcus cremoris* AC1 by Geis *et al.* (1985), and Peptidase 53 characterized by Sahlstrøm *et al.* (1993), neither of which has gene information published, all lactococcal peptidases for which this genetic information is available are located intracellularly. It has in general been observed that a corresponding complement of peptidases has also been found in lactobacilli, although some exceptions have been found. Atlan *et al.* (1994) have characterized a cell

Table 9.1 Peptidases characterized from Lactococci

Peptidase	Abbreviated name	Molecular weight (kDa)	Enzyme type	Specificity	Location
Aminopeptidases					
Aminopeptidase N	PepN	95	Metallo	X⇓Y–Z………	Intracellular
Aminopeptidase C	PepC	50	Thiol	X⇓Y–Z………	Intracellular
Glutamyl aminopeptidase	PepA	38	Metallo	Glu(Asp)⇓Y–Z…	Intracellular/cell wall
Pyrrolidone carboxylyl peptidase	PCP	25	Thiol[a]	pGlu⇓Y–Z	Intracellular
Tripeptidase	PepT	46	Metallo	X⇓Y–Z	Intracellular
Dipeptidase	PepV	49	Metallo	Y⇓Z	Intracellular
Peptidase 53		53	Thiol	X⇓Y–Z…	Cell wall
Endopeptidases					
Neutral endopeptidase (NOP)	PepO	71.5	Metallo	W–X⇓Y–Z….	Intracellular
Oligopeptidase	PepF	70	Metallo	W–X⇓Y–Z….	Intracellular
Proline-specific peptidases					
X-Prolyl dipeptidyl aminopeptidase	PepX	88	Serine	X–Pro⇓Y–Z…	Intracellular
Prolidase	PRD	42	Metallo	X⇓Pro	Intracellular
Proline iminopeptidase	PIP	50	Metallo	Pro⇓X(–Y)	Intracellular
Aminopeptidase P	PepP	43	Metallo	X⇓Pro–Pro–Y–Z…	Intracellular

[a]From Mierau, 1996.

wall-located proline iminopeptidase from *Lactobacillus delbrueckii* subsp. *bulgaricus* CNRZ 397 for which no corresponding lactococcal enzyme has been identified.

The majority of specificity studies for these peptidases have been carried out using chromogenic substrates or peptides not normally associated with lactococcal physiology. With the possible exception of the dipeptidases and the tripeptidases, and some limited work on pepN and endopeptidases, little is known about the size of peptide that these peptidases can hydrolyse, and, with some exceptions, e.g. Baankreis *et al.* (1995), their actions on the important casein-derived peptides found in cheese are still largely unknown.

(a) Aminopeptidase N (PepN). PepN was initially isolated by Tan and Konings (1990), and is regarded as general aminopeptidase with a specificity for peptide substrates with N-terminal Lys, Arg and Leu. Table 9.2 shows the kinetic data obtained by Niven *et al.* (1995) showing the ability of PepN to hydrolyse lysine from the N-terminal of series of peptide substrates of increasing chain length (Lys–Phe–Gly$_n$, $n = 1$–8). Using V_{max}/K_m as a measure of the overall effectiveness it revealed that the optimal peptide substrate is Lys–Phe–(Gly)$_4$. This is largely in agreement with the size barrier for peptides entering the organism via the oligopeptide transporter. Analysis of the PepN gene by Strøman (1992), Tan *et al.* (1992b) and van Alen-Boerrigter *et al.* (1991), and immunohistochemical studies by Tan *et al.* (1992a) and Exterkate *et al.* (1992), agree that PepN is an intracellular enzyme.

Its role in amino acid nutrition and growth is not yet clear. Despite earlier reports that PepN$^-$ mutants do not affect growth in milk, Mierau (1996) has recently reported that a PepN$^-$ mutant does grow slightly, but significantly more poorly than the wild-type in milk. Baankreis (1992) had earlier shown that the intracellular extract of a PepN$^-$ mutant has an impaired ability to hydrolyse tetra-, penta- and hexapeptide substrates, but still retains strong di- and tripeptide hydrolysis capabilities. When grown on a selected medium, limited in the essential amino acid methionine except through a tetrapeptide, the PepN$^-$ mutant did show a reduced growth rate.

Table 9.2 The kinetic parameters of the hydrolysis of Lys–Phe–(Gly)$_n$ oligopeptides by pepN from *Lactococcus lactic* subsp. *cremoris* Wg2. (From Niven *et al.*, 1995.)

Substrate	K_m (mM)	V_{max} ($\mu mol \cdot min^{-1} \cdot U^{-1}$)	V_{max}/K_m
Lys–Phe–Gly	0.29	3.64	12.55
Lys–Phe–(Gly)$_2$	0.30	4.23	14.10
Lys–Phe–(Gly)$_4$	0.31	7.46	24.06
Lys–Phe–(Gly)$_6$	0.56	8.66	15.25
Lys–Phe–(Gly)$_8$	1.87	28.71	15.35

This suggests that the enzyme is involved in oligopeptide processing but that, under normal growth conditions in milk, enough essential nutrients can be provided by other proteolytic actions to largely compensate for the absence of this peptidase activity.

Comparison of pepN activities between different strains of lactococci has shown that significant differences do exist. Strøman (unpublished data), in an analysis of pepN activity within 34 different commercial strains, found over 50 times variation in the activity present in cell extracts. Whether or not this variation is due to increased or decreased expression, or due to a genetic variation in activities (or a combination of both) is not yet known, but some evidence does exist that the latter could play an essential role in the effectiveness of these enzymes during ripening.

While this enzyme may not be essential individually for growth on milk, its involvement in cheese-making may be significant. Its ability to hydrolyse peptides containing hydrophobic amino acids such as leucine, and to a lesser extent phenylalanine, have made it a likely candidate as a debittering enzyme. Tan *et al.* (1993b) demonstrated that the addition of PepN was able to reduce bitterness in a tryptic digest of β-casein. A debittering role in cheese was observed by Baankreis (1992) who showed that using increasing concentrations of pepN$^-$ mutants resulted in increased bitterness.

The role of pepN in accelerating or enhancing flavour development in cheese is debatable according to Christensen *et al.* (1995), who introduced a *Lactobacillus helveticus* CNRZ pepN gene into *Lactococcus lactis* subsp. *cremoris* SK11 with a resulting 100-fold increase in pepN activity in cell extracts relative to the wild-type. The resulting constructed strain was used to manufacture cheese and found to have 1000-fold higher pepN activity levels within the cheese than the control. No bitterness was observed in either the experimental or control cheeses and although free amino acid levels were high in the experimental cheese there were no sensory differences reported between that and the control.

(b) Aminopeptidase C (PepC).   A second general aminopeptidase (PepC) with a broader specificity range than PepN, and active on Lys-, Phe-, His-, Glu-, Leu-β-napthylamides, was characterized by Neviani *et al.* (1989). PepC is a cysteine-type proteinase, inhibited by iodoacetamide and *p*-chloromercuric benzoic acid. Again, analysis of the *PepC* gene by Chapot-Chartier *et al.* (1993), and immunohistochemistry studies by Tan *et al.* (1992a) show it to be an intracellular enzyme. No significant effect of deleting the pepC activity was reported in growth studies in milk by Mierau (1996). Significantly, however, in the same study, removing both pepN and pepC activity had a clear effect on growth rate, greater than that found for the pepN$^-$ mutant alone, suggesting that lack of general aminopeptidase activity in the pepC$^-$ mutant was largely compensated for by the pepN activity. No pepC$^-$ trials in cheese have been reported.

(c) Glutamyl aminopeptidase (PepA). A third aminopeptidase, PepA, with a specificity to N-terminal glutamic acid- and aspartic acid-containing peptides has also been described. This metallopeptidase was initially purified by Exterkate and de Veer (1987b) and reported to be membrane-bound or associated. From immunohistochemical studies by Baankreis (1992) PepA does appears to be cell wall/extracellularly located. The same enzyme with identical N-terminal sequence, however, has also been purified from intracellular extracts by Niven (1991) and Bacon et al. (1994). The *pepA* gene has recently been cloned and sequenced by I'Anson et al. (1995) and shows no signal or membrane anchoring sequences to support an extracellular location.

Glutamic acid is an essential amino acid for the growth of lactococci, with insufficient free levels in milk to support growth. The role of pepA in this process has yet to be determined. Growth in milk experiments using PepA⁻ mutants by I'Anson et al. (1995) show the enzyme to be non-essential, but required for optimal growth to be achieved.

PepA may also have a significant role in development of flavours in ripening cheese. Glutamate is a recognized flavour-enhancer, although its role in cheese flavour development is not clearly understood. Fractionation studies by Aston and Creamer (1986) demonstrated that glutamate is the most prevalent amino acid in the water-soluble fraction in mature Cheddar cheese. This fraction was shown by McGugan et al. (1979) to contain the components that make the greatest contribution to the intensity of the flavour. Cheese trials using *PepA⁻* starter strains have not yet been reported.

(d) Tripeptidase (PepT). A tripeptidase, pepT, has been purified by Bosman et al. (1990) and Bacon et al. (1993). Both show pepT to be an intracellular metallopeptidase, a dimer with an molecular weight of 105 000 Da. Mierau et al. (1994), cloned and sequenced the *pepT* gene, confirming its intracellular location. PepT will only cleave tripeptides, hydrolysing the N-terminal amino acid; it has a very broad range, able to hydrolyse acidic, basic and neutral amino acids from the N-terminus of tripeptides and, unusually, was also shown to hydrolyse N-terminal proline from the tripeptide, Pro–Gly–Gly. Growth studies by Mierau et al. (1994) in milk and in chemically defined media with casein as the sole source of essential amino acids showed no significant difference between wild-type and pepT⁻ mutants, suggesting that other peptidases within lactococci are able to compensate for the absence of this enzyme. No cheese trials with pepT⁻ mutants have been reported.

(e) Dipeptidase (PepV). Lactococcal dipeptidase activities have been purified and characterized by Hwang et al. (1981, 1982) and van Boven et al. (1988). Both found dipeptidase to be intracellular metallopeptidases, but

Table 9.3   Kinetic values obtained for Lactococcal peptidases

Peptidase	K_m range (mM)	V_{max} (μmol·min^{-1}·mg^{-1})	Reference
PepN	0.55	30	Tan and Konings (1990)
PepC	4.5	3.6	Neviani et al. (1989)
PepA	0.22–1.82	1.014–108.18	Bacon et al. (1994)
PepT	0.15–0.38	151	Bosman et al. (1990)
PepV	1.6–7.9	3700–13 000	van Boven et al. (1988)

also showed several differences between the two characterizations. In the former the dipeptidase had a molecular weight of 100 000 Da by gel filtration while the latter was found to be 49 000 Da by gel filtration and SDS–PAGE. The kinetics and substrate specificities were also significantly different, suggesting two different enzymes, although this has yet to be shown in any further studies.

The enzyme kinetics obtained for the dipeptidase characterized by van Boven et al. (1988) are unusual in comparison with other aminopeptidases of lactococci. Table 9.3 shows that the dipeptidase, while having a relatively low affinity for dipeptides, has a very much higher V_{max} than is found with other peptidases, whereby the dipeptidase is able very rapidly to hydrolyse its substrate. This observation has also been confirmed in the author's laboratory (S. Movahedi, personal communication). The physiological significance of this observation is unknown.

The gene encoding the dipeptidase, pepV, has been cloned and sequenced in Lactobacillus delbrueckii subsp. lactis DSM 7290 by Vongerichten et al. (1994) and shows considerable homology to a pepV gene characterized in lactococci (P. Strøman, personal communication). To date, however, no deletion mutants have been successfully constructed.

(f) Endopeptidases (PepO and PepF).   Several, apparently different, endo-peptidases have now been reported in lactococci; LEP-I by Yan et al. (1987a), a monomer with a molecular weight of 98 000 Da; LEP-II by Yan et al. (1987b); a dimer with an overall molecular weight of 80 000 Da, pepO by Tan et al. (1991), Pritchard et al. (1994) and Baankreis et al. (1995), a monomer with a molecular weight of 70 000 Da; an alkaline endopeptidase by Baankreis (1992), with an overall molecular weight of 180 000 Da; and pepF, an oligopeptidase by Monnet et al. (1994) a monomer with a molecular weight of 70 000 Da. All are reported to be metallopeptidases.

Significantly, only two different endopeptidase gene have been cloned and sequenced, confirming their separate identities. [PepO, by Mierau et al. (1993), and pepF by Monnet et al. (1994)]. In both cases, analysis of the gene suggests an intracellular location. Also, neither show any significant difference in milk growth experiments using pepO⁻ and pepF⁻ mutants compared with the wild-type.

The role for these intracellular endopeptidases in the hydrolysis of casein peptides for growth is debatable. PepF has not been shown to be active on peptides with less than seven amino acids and lactococci already have several general intracellular peptidases present (PepN, PepC, pepV, pepT), that are capable of hydrolysing the 2–6 amino acid peptides thought to be able to enter via the peptide transporters. PepO is active on pentapeptides (Met-enkephalin) and larger peptides. Indeed, it has been shown to hydrolyse substantially larger peptides such as glucagon (mol.wt 3483) and the β-chain of insulin (mol.wt 3496), but not on any of the whole caseins.

The role of endopeptidases in peptide conversion in cheese has been studied by Baankreis et al. (1995). They have shown that under conditions prevailing in cheese (pH 5.4, 800 mM NaCl and 13°C) the neutral endopeptidase (NOP) which is thought to be similar, but not completely identical to PepO, is active, and probably plays a crucial role in the degradation of an important bitter peptide found in cheese, β-casein 193–209, a peptide resistant to hydrolysis by the lactococcal cell wall proteinase. In the same study, under the same prevailing conditions, the alkaline endopeptidase described by Baankreis (1992) does not appear to play a significant role in the conversion of cheese peptides because of its low activity at the pH of cheese and the competition from the NOP and the cell wall proteinase.

(g) XPro dipeptidyl aminopeptidase (PepX).   The first lactococcal peptidase to be cloned and sequenced was the *pepX* gene by Mayo et al. (1991) and Nardi et al. (1991). X-Pro dipeptidyl aminopeptidase, PepX, that this gene encodes for was initially considered to be one of the most likely candidates as an essential enzyme associated with the casein hydrolysis due to the proline-rich nature of β-casein, comprising 17% of the amino acid.

The initial purification of PepX by KieferPartsch et al. (1989) gave a cell wall location of the enzyme. All other purifications of the enzyme [by Zevaco et al. (1990), Booth et al. (1990a) and Lloyd and Pritchard (1991)], however, indicated an intracellular location, a position supported by analysis of the *pepX* gene where no signal sequence or membrane anchor regions were found. PepX is inhibited by phenyl methyl sulphonyl fluoride and is therefore considered to be a serine-type peptidase.

Conflicting information exists on the role of this enzyme in growth of lactococci. Nardi et al. (1991) reported that a pepX⁻ mutant in *Lactococcus lactis* subsp. *lactis* NCDO763 was only able to grow in milk at 60% of rate of the wild-type, while Mayo et al. (1993) found the deletion did not affect the growth of lactococci in milk. Mierau (1996) also found no significant effect of deleting pepX on growth in milk. The use of pepX⁻ mutants in cheese by Baankreis (1992) showed that cheeses manufactured using increasing concentrations of the pepX⁻ strain in the starter culture exhibited decreasing organoleptic quality, but did not increase bitterness.

(h) Other proline-hydrolysing peptidases.   As noted above, proline is a significant constituent of β-casein. With the exception of pepX, however, compared with the other peptidases of lactococci little is known about the proline-specific peptidases of lactococci and they have been studied more in lactobacilli, e.g. the proline iminopeptidase from *Lactobacillus delbrueckii* subsp. *bulgaricus* CNRZ 397 by Atlan *et al.* (1994).

Booth *et al.* (1990b) have identified at least three other intracellular proline-specific peptidases in lactococci. Baankreis and Exterkate (1991) purified and characterized an iminopeptidase-like enzyme that hydrolyses di- and tripeptides containing proline or hydrophobic amino acids at the N-terminal. This is a metalloenzyme, a dimer with a molecular weight of 110 000 Da. No gene information is yet available and the role of the enzyme in amino acid metabolism within lactococci remains unclear. At the time of publication it was not clear how the substrate for this proline iminopeptidase (or proline dipeptidase (prolinase)) was formed within lactococci, leading to Baankreis and Exterkate (1991) speculating that an aminopeptidase P would need to be present in lactococci in order to produce the substrate for this enzyme.

Mars and Monnet (1995) have recently characterized an intracellular lactococcal aminopeptidase P that has a unique activity, hydrolysing peptides with the sequence X-Pro-Pro at the N-terminus. This is a metallopeptidase, a monomer with molecular weight of 43 000 Da. Again, no genetic data exist for this enzyme and consequently no deletion analysis data available.

Another proline-specific peptidase that has been characterized by Kaminogawa *et al.* (1984) and Booth *et al.* (1990c) is prolidase, an X-Pro specific dipeptidase. This is metallopeptidase with a native molecular weight of 42 000 Da. With the recent work by Marugg *et al.* (1995) demonstrating the significance of prolyl-containing dipeptides in the regulation of proteinase expression, the function and activity of the prolidase and the prolinase within lactococci needs further examination. To date, however, no further work on the prolidase has been reported.

(i) Multiple peptidase deletion studies.   Although it has been speculated for many years, the recent multiple peptidase deletion mutants studies by Mierau (1996) where the *pepX, pepO, pepT, pepC* and *pepN* genes have been inactivated, either singly or in combinations, provides the first direct evidence that these peptidases are involved in milk protein degradation for growth. Multiple mutations lead to slower growth rates in milk with a general trend being that growth rates decreased when more peptidases were

inactivated. The restricted growth of these multiple peptidase-deficient mutants in chemically defined media lacking leucine as a free amino acid, but allowing growth when supplemented with various leucyl-containing peptides, further demonstrated the involvement of these peptidases in peptide breakdown for growth.

Significantly, these deletion mutants have been constructed such that the final organism contains only lactococcal DNA, with no undesirable non-lactococcal selective marker genes present. These are considered 'food-grade' and will make a significant contribution to the understanding of the role the peptidases play in cheese ripening, although to date, no cheese experiments using these mutants have been reported.

9.4   Conclusions and topics for further research

An issue that has yet to be addressed properly is the significance of natural genetic variance of peptidases between different strains. The lactococcal cell wall-associated proteinase clearly has several natural genetic variants. These variants have some different cleavage properties on casein which are considered important in cheese manufacture. Whether or not genetic variation occurs in peptidases has not yet been examined. An example of this could be the ability of the lactococcal dipeptidase to cleave N-terminal glutamate dipeptides. The dipeptidase isolated from *Lactococcus lactis* subsp. *cremoris* Wg2 by van Boven *et al.* (1988) does not cleave Glu–Ala, while a similar dipeptidase from *Lactococcus lactis* subsp. *lactis* NCDO 712 is able to cleave this and other glutamate-containing dipeptides (S. Movahedi, personal communication). To address this issue, more substantial specificity studies, using standardized conditions, and in some cases, with more natural peptide substrates allowing comparisons of individual peptidase specificities between different (industrially important) strains are required. Further studies at the genetic level may then show the rationale for these specificity differences. Linked to this is the level of different peptidase activities found in individual strains. Little work has yet been done comparing the amount of activity in different strains, either in terms of actual amount of an enzyme or the specific activity found in the individual strains, although this variance has been demonstrated for both pepN and pepX (P. Strøman, personal communication).

A third area for future study is regulation of expression of the peptidases. The first paper on this subject has recently been published by Meijer *et al.* (1996) but requires further investigation.

Chapter 8 of this volume refers to recent literature on significant flavour compounds produced by LAB enzymes from amino acids, which are not therefore reviewed here. Finally, a review of the literature shows that research into the role of lipases and esterases from lactic acid bacteria in lipolysis is not advanced, with only limited knowledge and no clear understanding of their potential role in cheese ripening. Lawrence *et al.* (1976), Umemoto and Sato (1978), Kamaly *et al.* (1990) and Tsakalidou *et al.* (1992) have found lipase or esterase activities in lactococcal extracts and Holland and Coolbear (1996) have recently purified an intracellular esterase capable of hydrolysing tributyrin.

Acknowledgements

F.M. is funded by the Office of Science and Technology.

9.5 References

Aston, J.W. and Creamer, L.K. (1986) Contribution of the components of the water-soluble fraction to the flavour of Cheddar cheese. *N.Z. J. Dairy Sci. Technol.*, **21**, 229–48.

Atlan, D., Gilbert, C., Blanc, B. and Portalier, R. (1994) Cloning, sequencing and characterization of the *pepIP* gene encoding a proline iminopeptidase from *Lactobacillus delbrueckii* subsp. *bulgaricus* CNRZ 397. *Microbiology*, **140**, 527–35.

Baankreis, R. (1992) *The role of Lactococcal peptidases in cheese ripening*, PhD Thesis, University of Amsterdam, The Netherlands.

Baankreis, R. and Exterkate, F.A. (1991) Characterization of a peptidase from *Lactococcus lactis* subsp. *cremoris* HP that hydrolyses di- and tripeptides containing proline or hydrophobic residues as the aminoterminal amino acid. *Syst. Appl. Microbiol.*, **14**, 317–23.

Baankreis, R., van Schalkwijk, S., Alting A.C. and Exterkate, F.A. (1995) The occurrence of two intracellular oligoendopeptidases in *Lactococcus lactis* and their significance for peptide conversion in cheese. *Appl. Microbiol. Biotechnol.*, **44**, 386–92.

Bacon, C.L., Wilkinson, M., Jennings, P.V., Fhaolain, I.N. and O'Cuinn, G. (1993) Purification and characterization of an aminotripeptidase from cytoplasm of *Lactococcus lactis* subsp. *cremoris* AM2. *Int. Dairy J.*, **3**, 163–77.

Bacon, C.L., Jennings, P.V., Fhaolain, I.N. and O'Cuinn, G. (1994) Purification and characterization of an aminopeptidase A from cytoplasm of *Lactococcus lactis* subsp. *cremoris* AM2. *Int. Dairy J.*, **3**, 503–19.

Booth, M., Fhaolain, I.N., Jennings, P.V. and O'Cuinn, G. (1990a) Purification and characterization of a post-proline dipeptidyl aminopeptidase from *Streptococcus cremoris* AM2. *J. Dairy Res.*, **57**, 79–88.

Booth, M., Donnelly, W.J., Fhaolain, I.N., Jennings, P.V. and O'Cuinn, G. (1990b) Proline specific peptidases of *Streptococcus cremoris* AM2. *J. Dairy Res.*, **57**, 89–99.

Booth, M., Jennings, P.V., Fhaolain, I.N. and O'Cuinn, G. (1990c) Prolidase activity of *Lactococcus lactis* subsp. *cremoris* AM2: partial purification and characterization. *J. Dairy Res.*, **57**, 245–54.

Bosman, B.W., Tan, P.S.T. and Konings, W.N. (1990) Purification and characterization of a tripeptidase from *Lactococcus lactis* subsp. *cremoris* Wg2. *Appl. Environ. Microbiol.*, **56**, 1839–43.

Chapot-Chartier, M.-P., Nardi, M., Chopin, M.-C., Chopin, A. and Gripon, J.-C. (1993) Cloning and sequencing of pepC, a cysteine aminopeptidase from *Lactococcus lactis* subsp. *cremoris* AM2. *Appl. Environ. Microbiol.*, **59**, 330–3.

Christensen, J.E., Johnson, M.E. and Steele, J.L. (1995) Production of Cheddar cheese using a *Lactococcus lactis* subsp. *cremoris* SK11 derivative with enhanced aminopeptidase activity. *Int. Dairy J.*, **5**, 367–79.

Exterkate, F.A. (1975) An introductory study of the proteolytic system of *Streptococcus cremoris* strain HP. *Neth. Milk Dairy J.*, **29**, 303–18.

Exterkate, F.A. (1979) Accumulation of proteinase in the cell wall of *Streptococcus cremoris* strain AM_1, and its regulation of production. *Arch. Microbiol.*, **120**, 247–54.

Exterkate, F. A. (1995) The lactococcal cell envelope proteinases: differences, calcium-binding effects and role in cheese ripening. *Int. Dairy J.*, **5**, 995–1018.

Exterkate, F.A. and de Veer, G.J.C.M. (1987a) Optimal growth of *Streptococcus cremoris* HP in milk is related to β- and κ-casein degradation. *Appl. Microbiol. Biotechnol.*, **25**, 471–5.

Exterkate, F.A. and de Veer, G.J.C.M. (1987b) Purification and some properties of a membrane-bound aminopeptidase A from *Streptococcus cremoris*. *Appl. Environ. Microbiol.*, **59**, 3640–7.

Exterkate, F.A., de Jong, M., de Veer, G.J.C.M. and Baankreis, R. (1992) Location and characterisation of aminopeptidase N in *Lactococcus lactis* subsp. *cremoris* HP. *Appl. Microbiol. Biotechnol.*, **37**, 46–54.

Gasson, M.J. (1983) Plasmid complementation of *Streptococcus lactis* NCDO 712 and other lactic streptococci after protoplast-induced curing. *J. Bacteriol.*, **154**, 1–9.

Geis, A., Bockelmann, W. and Teuber, M. (1985) Simultaneous extraction and purification of a cell wall-associated peptidase and a β-casein specific protease from *Streptococcus cremoris* AC1. *Appl. Microbiol. Biotechnol.*, **23**, 79–84.

Hagting, A., Kunji, E.R.S., Leenhouts, K.J., Poolman, B. and Konings, W.N. (1994). The di- and tripeptide transport protein of *Lactococcus lactis*. *J. Biol. Chem.*, **269**, 11391–9.

Haandrikman, A.J., Kok, J.,Laan, H., Soemitro, S., Ledeboer, A.T., Konings, W.N., Kok, J. and Venema, G. (1989) Identification of a gene required for the maturation of an extracellular lactococcal serine proteinase. *J. Bacteriol.*, **171**, 2789–94.

Haandrikman, A.J., Meesters, R., Laan, H., Konings, W.N., Kok, J. and Venema, G. (1991) Processing of the lactococcal extracellular serine proteinase. *Appl. Environ. Microbiol.*, **57**, 1899–904.

Holland R. and Coolbear, T. (1996) Purification of tributyrin esterase from *Lactococcus lactis* subsp. *cremoris* E8. *J. Dairy Res.*, **63**, 131–40.

Hugenholtz, J., Exterkate, F.A., and Konings, W.N. (1984) The proteolytic systems of *Streptococcus cremoris*: an immunological analysis. *Appl. Environ. Microbiol.*, **48**, 1105–10.

Hugenholtz, J., van Sinderen, D., Kok, J. and Konings, W.N. (1987) Cell wall-associated proteases of *Streptococcus cremoris* Wg2. *Appl. Environ. Microbiol.*, **53**, 853–9.

Hwang, I.-K., Kaminogawa, S. and Yamauchi, K. (1981) Purification and properties of a dipeptidase from *Streptococcus cremoris* H61. *Agric. Biol. Chem.*, **45**, 159–65.

Hwang, I.-K., Kaminogawa, S. and Yamauchi, K. (1982) Kinetic properties of a dipeptidase from *Streptococcus cremoris*. *Agric. Biol. Chem.*, **46**, 3049–53.

I'Anson, K.J.A., Movahedi, S., Griffin, H,G., Gasson, M.J. and Mulholland F. (1995) A non-essential glutamyl aminopeptidase is required for optimal growth of *Lactococcus lactis* MG1363 in milk. *Microbiology*, **141**, 2873–81.

Juillard, V., Le Bars, D., Kunji, E.R.S., Konings, W.N., Gripon, J.-C. and Richard, J. (1995a) Oligopeptides are the main source of nitrogen for *Lactococcus lactis* during growth in milk. *Appl. Environ. Microbiol.*, **61**, 3024–30.

Juillard, V., Laan, H., Kunji, E.R.S., Jeronimus-Stratingh, C.M., Bruins, A.P. and Konings, W.N. (1995b). The extracellular P_1-type proteinase of *Lactococcus lactis* hydrolyzes β-casein into more than one hundred different oligopeptides. *J. Bateriol.*, **177**, 3472–8.

Kamaly, K.M., Takayama, K. and Marth, E.H. (1990) Acylglycerol acylhydrolase (lipase) activities of *Streptococcus cremoris*, and their mutants. *J. Dairy Sci.*, **73**, 280–90.

Kaminogawa, S., Azuma, N., Hwang, I.-K., Suzuki, Y. and Yamauchi, Y. (1984) Isolation and characterization of a prolidase from *Streptococcus cremoris* H61. *Agric. Biol. Chem.*, **48**, 3035–40.

Kiefer-Partsch, B., Bockelmann, W., Geis A. and Teuber, M. (1989) Purification of an X-prolyl-dipeptidyl aminopeptidase from the cell wall proteolytic system of *Lactococcus lactis* subsp. *cremoris*. *Appl. Microbiol. Biotechnol.*, **31**, 75–8.

Kiwaki, M., Ikemura, H., Shimizu-Kadota, M. and Hirashima, A. (1989). Molecular characterization of a cell wall-associated proteinase gene from *Streptococcus lactis* NCDO 763. *Mol. Microbiol.*, **3**, 359–69.

Kok, J., van Dijl, J.M., van der Vossen, J.M.B.M. and Venema, G. (1985) Cloning and expression of a *Streptococcus cremoris* Wg2 proteinase in *Bacillus subtilis* and *Streptococcus- lactis*. *Appl. Environ. Microbiology*, **50**, 94–101.

Kok, J., Leenhouts, K.J., Haandrikman, A.J. Ledeboer, A.M. and Venema, G. (1988a) Nucleotide sequence of the cell wall proteinase gene of *Streptococcus cremoris* Wg2. *Appl. Environ. Microbiol.*, **54**, 231–8.

Kok, J., Hill, D., Haandrikman, A.J., de Reuver, M.J.B., Laan, H. and Venema, G. (1988b) Deletion analysis of the proteinase gene of *Streptococcus cremoris* Wg2. *Appl. Environ. Microbiol.*, **54**, 239–44.

Konings, W.N., Poolman, B. and Driessen, A.J.M. (1989) Bioenergetics and solute transport in Lactococci. *CRC Crit. Rev. Microbiol.*, **16**, 419–76.

Kunji, E.R.S., Smid, E.J., Plapp, R., Poolman, B. and Konings, W.N. (1993) Di-tripeptides and oligopeptides are taken up via distinct transport mechanisms in *Lactococcus lactis. J. Bacteriol.*, **175**, 2052–59.

Kunji, E.R.S., Hagting, A., de Vries, C.J., Juillard, V., Haandrikman, A.J., Poolman, B. and Konings, W.N. (1995) Transport of β-casein-derived peptides by the oligopeptide transport system is a crucial step in the proteolytic pathway of *Lactococcus lactis. J. Biol. Chem.*, **270**, 1569–74.

Laan, H., Bolhuis, H., Poolman, B., Abee, T. and Konings, W.N. (1993) Regulation of proteinase synthesis in *Lactococcus lactis. Acta Biotechnol.*, **13**, 95–101.

Law B.A. (1978). Peptide utilization by group N streptococci. *J. Gen. Microbiol.*, **105**, 113–18.

Law, B.A. and Kolstad, J. (1983) Proteolytic systems in lactic acid bacteria. *Antoine van-Leeuwenhoek*, **49**, 225–45.

Lawrence, R.C., Fryer, T.F. and Reiter, B. (1976) Rapid method for the quantitative estimation of microbial lipases. *Nature*, **213**, 1264–5.

Lloyd, R.J. and Pritchard, G.G. (1991) Characterization of an X-prolyl dipeptidyl aminopeptidase from *Lactococcus lactis* subsp. *lactis* H1. *J. Gen. Microbiol.*, **137**, 49–55.

Mars, I. and Monnet, V. (1995) An aminopeptidase P from *Lactococcus lactis* with original specificity. *Biochim. Biophys. Acta*, **1243**, 209–15.

Marugg, J.D., Meijer, W., van Kranenburg, R., Laverman, P., Bruinenberg, P.G. and deVos, W.M. (1995) Medium-dependent regulation of proteinase gene expression in *Lactococcus lactis*: control of transcription initiation by specific dipeptides. *J. Bacteriol.*, **177**, 2982–9.

Mayo, B., Kok, J., Venema, K., Bockelmann, W., Teuber, M., Reinke, H. and Venema, G. (1991) Molecular cloning and sequence analysis of the X-prolyl dipeptidyl aminopeptidase gene from *Lactococcus lactis* subsp. *cremoris. Appl. Environ. Microbiol.*, **57**, 38–44.

Mayo, B., Kok, J., Bockelmann, W., Haandrikman, A., Leenhouts, K.J. and Venema, G. (1993) Effect of X-prolyl dipeptidyl aminopeptidase deficiency on *Lactococcus lactis. Appl. Environ. Microbiol.*, **59**, 2049–55.

McGugan, W.A., Emmons, D.B. and Larmond, E. (1979) Influence of volatile and non-volatile fractions on intensity of Cheddar cheese flavor. *J. Dairy Sci.*, **62**, 398–402.

McKay, L.L. and Baldwin, K.A. (1974) Simultaneous loss of proteinase and lactose-utilising enzyme activities in *Streptococcus lactis* and reversal of loss by transduction. *Appl. Microbiol.*, **28**, 342–46.

Meijer, W.M., Marugg, J.D. and Hugenholtz, J. (1996) Regulation of proteolytic enzyme activity in *Lactococcus lactis. Appl. Environ. Microbiol.*, **62**, 156–61.

Mierau, I. (1996) *Peptide degradation in* Lactococcus lactic in-vivo: *a first exploration*, PhD Thesis, University of Groningen, The Netherlands.

Mierau, I., Tan, P.S.T., Haandrikman, A.J., Kok, J., Leenhouts, K.J., Konings, W.N. and Venema, G. (1993) Cloning and sequencing of the gene for lactococcal endopeptidase, an enzyme with sequence similarity to mammalian enkephalinase *J. Bacteriol.*, **175**, 2087–96.

Mierau, I., Haandrikman, A.J., Velterop, O., Tan, P.S.T., Leenhouts, K.J., Konings, W.N., Venema, G. and Kok, J. (1994) Tripeptidase gene (*pepT*) of *Lactococcus lactis*: molecular cloning and nucleotide sequencing of *pepT* and construction of a chromosomal deletion mutant. *J. Bacteriol.*, **176**, 2854–61.

Monnet, V., Nardi, M., Chopin, A., Chopin, M.-C. and Gripon, J.-C. (1994) Biochemical and genetic characterization of PepF, an oligopeptidase from *Lactococcus lactis. J. Biol. Chem.*, **269**, 32070–6.

Nardi, M., Chopin, M.-C., Chopin, A., Cals, M.-M. and Gripon, J.-C. (1991) Cloning and DNA sequence analysis of an X-prolyl dipeptidyl amino-peptidase gene from *Lactococcus lactis* subsp. *lactis* NCDO763. *Appl. Environ. Microbiol.*, **57**, 45–50.

Neviani, E., Boquien, C.Y., Monnet, V, Phan Thanh, L. and Gripon, J.-C. (1989) Purification and characterization of an aminopeptidase from *Lactococcus lactis* subsp. *cremoris* AM2. *Appl. Environ. Microbiol.*, **55**, 2308–14.

Nissen-Meyer, J. and Sletten, K. (1991) Purification and characterization of the free form of the lactococcal extracellular proteinase and its autoproteolytic cleavage products. *J. Gen. Microbiol.*, **137**, 1611–18.

Niven G.W. (1991) Purification and characterization of aminopeptidase A from *Lactococcus lactis* subsp. *lactis* NCDO712. *J. Gen. Microbiol.*, **137**, 1207–12.

Niven, G.W., Holder, S.A. and Strøman, P. (1995) A study of the substrate specificity of aminopeptidase N from *Lactococcus lactis* subsp. *cremoris* Wg2. *Appl. Microbiol. Biotechnol.*, **44**, 100–5.

Pritchard, G.G., Freebairn, A.D. and Coolbear, T. (1994) Purification and characterization of an endopeptidase from *Lactococcus lactis* subsp. *cremoris* SK11. *Microbiology.*, **140**, 923–30.

Rice, G.H., Stewart, F.H.C., Hillier, A.J. and Jago, G.R. (1978) The uptake of amino acids and pepides by *Streptococcus lactis*. *J. Dairy Res.*, **45**, 93–107.

Sahlstrøm, S., Chrzanowska, J. and Sørhaug, T. (1993) Purification and characterization of a cell wall peptidase from *Lactococcus lactis* subsp. *cremoris* IMN-C12. *Appl. Environ. Microbiol.*, **59**, 3076–82.

Smid, E.J. and Konings, W.N. (1990) Relationship between utilization of proline and pro-line-containing peptides and growth of *Lactoccus lactis*. *J. Bacteriol.*, **172**, 5286–92.

Smid, E.J., Driessen, A.J.M. and Konings, W.N. (1989a) Mechanisms and energetics of dipeptide transport in membrane vesicles of *Lactococcus lactis*. *J. Bacteriol.*, **171**, 292–8.

Smid, E.J., Plapp, R. and Konings, W.N. (1989b) Peptide uptake is essential for growth of *Lactococcus lactis* in milk. *J. Bacteriol.*, **171**, 6135–40.

Smid, E.J., Poolman, B. and Konings, W.N. (1991) Casein utilisation by lactococci. *Appl. Environ. Microbiol.*, **57**, 2447–52.

Strøman, P. (1992) Sequence of a gene (lap) encoding a 95.3 kDa aminopeptidase from *Lactococcus lactis* subsp. *cremoris* Wg2. *Gene*, **113**, 107.

Tan, P.S.T. and Konings, W.N. (1990) Purification and characterization of an aminopeptidase from *Lactococcus lactis* subsp. *cremoris* Wg2. *Appl. Environ. Microbiol.*, **56**, 526–32.

Tan, P.S.T., Pos, K.M. and Konings, W.N. (1991) Purification and characterization of an endopeptidase from *Lactococcus lactis* subsp. *cremoris* Wg2. *Appl. Environ. Microbiol.*, **57**, 2593–9.

Tan, P.S.T., Chapot-Chartier, M.-P., Pos, K.M., Rousseau, M., Boquein, C.-Y., Gripon, J.-C. and Konings,W.N. (1992a) Localization of peptidases in Lactococci. *Appl. Environ. Microbiol.*, **58**, 285–90.

Tan, P.S.T., van Alen-Boerrigter, I.J., Poolman, B., Siezen, R.J., de Vos, W.M. and Konings, W.N. (1992b) Characterization of the *Lactococcus lactis* pepN gene encoding an aminopeptidase homologous to mammalian aminopeptidase N. *FEBS Lett.*, **306**, 9–16.

Tan, P.S.T., Poolman, B. and Konings, W.N. (1993a) Proteolytic enzymes of *Lactococcus lactis*. *J. Dairy Res.*, **60**, 269–86.

Tan, P.S.T., van Kessel, T.A.J.M., van de Veerdonk, F.L.M., Zuurendonk, P.F., Bruins, A.P. and Konings, W.N. (1993b) Degradation and debittering of a tryptic digest from *β*-casein by aminopeptidase N from *Lactococcus lactis* subsp. *cremoris* Wg2. *Appl. Environ. Microbiol.*, **59**, 1430–36.

Thomas, T.D. and Mills, O.E. (1981) Proteolytic enzymes of starter bacteria. *Neth. Milk Dairy J.*, **35**, 255–73.

Thomas, T.D., Jarvis, B.D.W. and Skipper, N.A. (1975) Localization of proteinase(s) near the cell surface of *Streptococcus lactis*. *J. Bacteriol.*, **118**, 329–33.

Tsakacidou, E., Zoidou, E. and Kalantzopoulos, G. (1992) Esterase activities of cell free extracts from strains of *Lactococcus lactis* subsp. *lactis* isolated from traditional Greek cheese. *J. Dairy Res.*, **59**, 111–13.

Tynkkynen, S., Buist, G., Kunji, E.R.S., Kok, J., Poolman, B., Venema, G. and Haandrikman, A. (1993) Genetic and biochemical characterisation of the oligopeptide transport system of *Lactococcus lactis*. *J. Bacteriol.*, **175**, 7523–32.

Umemoto, Y. and Sato, Y. (1978) Lipolysis by intracellular lipase of *Streptococcus lactis* against its neutral lipids obtained by growth at low temperature. *Agric. Biol. Chem.*, **42**, 221–5.

van Alen-Boerrigter, I.J., Baankreis, R. and de Vos, W.M. (1991) Characterization and overexpression of the *Lactococcus lactis* pepN gene and localization of its product, Aminopeptidase N. *Appl. Environ. Microbiol.*, **57**, 2555–61.

van Boven, A., Tan, P.S.T. and Konings, W.N. (1988) Purification and characterization of a dipeptidase from *Streptococcus cremoris* Wg2. *Appl. Environ. Microbiol.*, **54**, 43–9.

Visser, S., Exterkate, F.A., Slangen, C.J. and de Veer, G.J.C.M. (1986) Comparative study of action of cell wall proteinases from various strains of *Streptococcus cremoris* on bovine α_{s1}-β- and κ-casein. *Appl. Environ. Microbiol.*, **52**, 1162–66.

Vongerichten, K.F., Klein, J.R., Matern, H. and Plapp, R. (1994) Cloning and nucleotide sequence analysis of *pepV*, a carnosinase gene from *Lactobacillus delbrueckii* subsp. *lactis* DSM 7290, and partial characterization of the enzyme. *Microbiology*, **140**, 2591–600.

Vos, P., Simons, G., Siezen, R.J. and de Vos, W.M. (1989a) Primary structure and organization of the gene for a prokaryotic cell envelope-located proteinase. *J. Biol. Chem.*, **264**, 13579–85.

Vos, P., van Asseldonk, M., van Jeveren, F., Siezen, R.J., Simons, G. and de Vos, W.M. (1989b) A maturation protein is essential for the production of active forms of *Lactococcus lactis* SK11 serine proteinase located in or secreted from the cell envelope. *J. Bacteriol.*, **171**, 2795–802.

Yan, T.-R., Azuma, N., Kaminogawa, S. and Yamauchi, K. (1987a) Purification and characterization of a substrate-size-recognizing metalloendopeptidase from *Streptococcus cremoris* H61. *Appl. Environ.. Microbiol.*, **53**, 2296–302.

Yan, T.-R., Azuma, N., Kaminogawa, S. and Yamauchi, K. (1987b) Purification and characterization of a novel metalloendopeptidase from *Streptococcus cremoris* H61. *Eur. J. Biochem.*, **163**, 259–65.

Zevaco, C., Monnet, M. and Gripon, J.-C. (1990) Intracellular X-prolyl dipeptidyl peptidase from *Lactococcus lactis* subsp. *lactis*: purification and properties. *J. Appl. Bacteriol.*, **68**, 357–66.

10 Molecular genetics of dairy lactic acid bacteria
M.J. GASSON

10.1 Introduction

The genetics of lactic acid bacteria has seen very significant advances over the past decade, leading to a detailed molecular understanding of several key industrial properties and to the development of sophisticated technology for genetic manipulation. It is relevant to dairy interests that the most advanced species in terms of genetic understanding and technical sophistication is *Lactococcus lactis*, although other dairy species have also advanced rapidly. An introduction to the genetics of lactic acid bacteria with an emphasis on *L. lactis* is presented here. The reader is urged to consult the many excellent reviews that are available for a more complete treatment of individual topics.

10.2 Genetics of industrially relevant traits

Dairy fermentation processes rely on the properties of starter lactic acid bacteria, and the performance of these strains in turn depends on their genetic make-up. For several key properties it has been possible to isolate and characterize the genes involved and in some cases genetic manipulation has been used to seek their improvement and enhanced stability. A good example of practical improvement is the introduction of bacteriophage resistance by plasmid transfer. Plasmids in fact play an especially significant role in carrying genes for several important properties such as lactose catabolism and proteinase production as well as bacteriophage resistance (Gasson, 1993). The inherent instability of plasmid DNA has been addressed by genetic techniques, a notable example being the introduction of lactose catabolism and proteinase genes into the bacterial chromosome (Leenhouts *et al.*, 1991a; MacCormick *et al.*, 1995).

10.2.1 Lactose fermentation

The most fundamental role of lactic starter bacteria is their conversion of the milk sugar lactose into lactic acid which confers both flavour and preservation. The homofermentative dairy lactococci and *Lactobacillus casei* make use of an atypical catabolic pathway for lactose fermentation which

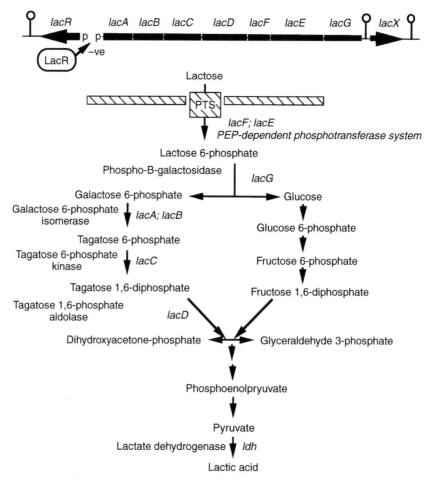

Figure 10.1 Upper diagram: organization of the lactose operon. Lower diagram: the lactose catabolic pathway.

starts with lactose uptake by the phosphoenolpyruvate (PEP)-dependent phosphotransferase system. As a consequence, intracellular lactose is phosphorylated and the substrate for further metabolism is lactose 6-phosphate rather than lactose. This is broken into glucose and galactose 6-phosphate and while the glucose can follow a normal glycolytic pathway the galactose 6-phosphate requires the dedicated tagatose 6-phosphate pathway for its further catabolism before rejoining the Embden–Meyerhoff pathway as triose phosphates (Figure 10.1).

It is very well established that the genes for lactose catabolism are encoded by plasmid DNA in *Lactococcus lactis* (McKay, 1982) and this helped with their cloning and sequence analysis (Maeda and Gasson, 1986;

deVos and Gasson, 1989; de Vos *et al.*, 1990; van Rooijen *et al.*, 1991; deVos and Simons, 1994). As also shown in Figure 10.1, genes encoding the lactose specific components of the phosphotransferase system, phospho-β-galactosidase and the enzymes of the tagatose 6-phosphate pathway are arranged in an operon that is subject to negative regulation by a repressor protein LacR. The latter is encode by a linked but divergently transcribed gene (van Rooijen and deVos, 1990; van Rooijen *et al.*, 1992).

The lactose operon has been stabilized by its integration into the lactococcal chromosome using *in vivo* genetic methods facilitating its exploitation in food compatible cloning strategies (MacCormick *et al.*, 1995) and the strong regulated lactose operon promoter has been used for the expression of heterologous genes (deVos and Simons, 1994). These aspects are considered in more detail below (section 10.3).

In other dairy lactic acid bacteria, including many lactobacilli and *Streptococcus thermophilus*, a more conventional route for lactose catabolism is followed with the use of an ATP-driven permease and β-galactosidase. In some cases these genes have also been cloned and characterized, notably in *Streptococcus thermophilus* (Mercenier *et al.*, 1994).

10.2.2   Casein degradation

Casein degradation plays a vital part in determining the properties of fermented dairy products and the process is, to a significant extent, controlled by the starter lactic acid bacteria. Most of these strains are nutritionally fastidious, requiring several free amino acids or small peptides to sustain their growth on a minimal defined medium. Since milk does not provide these amino acids it is essential for starter lactic acid bacteria growth that they degrade milk protein and to this end they possess a highly evolved and efficient proteolytic system. As with most lactic acid bacteria genetics, the characterization of the proteolytic system is most complete in the lactococci (see also, Chapter 9).

In *L. lactis* the initial stages of casein breakdown are controlled by a cell surface proteinase. This property involves plasmid genes which may be linked to or independent from those for lactose catabolism. The lactococcal proteinase genes from several *L.lactis* strains have been cloned and characterized (Kok and deVos, 1994). The lactococcal proteinase is a large molecule of over 200 kDa and its gene is a long open reading frame that can be translated into a polypeptide of 1902 amino acids in the case of *L. lactis* Wg2. The first 187 amino acids are removed in the mature proteinase as a result of the molecule's secretion, which involves cleavage of a 33-amino acid amino-terminal secretory leader and its subsequent activation by autocatalytic cleavage of the following 157-amino acid propeptide sequence. The proteinase is bound to the cell surface at its carboxy-terminus where a classical Gram-positive membrane anchor sequence can be identified (Kok *et al.*, 1988; Kok and deVos, 1994). The proteinase gene *prtP* is linked to a second divergently transcribed proteinase gene, *prtM*, which encodes a 33 kDa maturation enzyme (Haandrickman *et al.*, 1989; Vos *et al.*, 1989a). The latter is also cell surface-bound but in this case the maturation protein has an amino-terminal lipoprotein cleavage site between the leucine and cysteine residues at positions 21 and 22 respectively (Haandrickman *et al.*, 1991a). Deletion studies have confirmed that PrtM is essential for the proteinase to be active and it appears to play an as yet undefined role in the maturation of PrtP by autocatalytic cleavage (Haandrickman *et al.*, 1991b). The amino acid sequence of the lactococcal proteinase exhibits strong homology to others in the serine proteinases family, including *Bacillus* subtilisin. The catalytic domain is within the first 500 amino acids and the characteristic active site residues of serine proteinases are present as Asp_{30}, His_{94} and Ser_{433}, as is the essential 'oxyanion hole' residue Asn_{196} (Vos *et al.*, 1991; Siezen *et al.*, 1991).

Cloned and sequenced proteinase genes from several different strains with distinct casein cleavage specificities are available (Kok *et al.*, 1988; Vos *et al.*, 1989b; de Vos *et al.*, 1989). These have been exploited in the construction of a series of hybrid proteinase genes which have contributed to an appreciation of which parts of the molecule dictate specificity (Vos *et al.*, 1991). These studies have been extended by the construction of several site-directed mutants and molecular modelling studies (Siezen *et al.*, 1993). In this way the active site and substrate-binding regions of the proteinase have been confirmed experimentally and residues 747 and 748 in the extended lactococcal sequence have also been assigned a role in substrate binding. From a more applied standpoint this work provides an opportunity to manipulate flavour generation, as it is established that the cleavage specificity of some lactococcal proteinases generates proline-rich bitter peptides from β-casein, whereas others lack this property; therefore genetics offers the possibility of changing the casein cleavage properties of starter lactococci by plasmid transfer, proteinase gene cloning as well as by the use of hybrid or site-directed mutant genes.

In addition to the cell surface proteinase, starter bacteria have many peptidase enzymes, some of which are of significance for casein breakdown both for starter culture nutrition and in the creation of flavour. In recent years considerable effort has been devoted to the cloning and characterization of peptidase genes which now include endopeptidases, general amino-peptidases, proline-specific peptidases and oligopeptidases (Kok and deVos, 1994). This work has generally involved the initial purification of individual peptidase enzymes followed by the use of reversed genetics to isolate the equivalent gene. In some cases peptidase genes have been cloned by complementing *E. coli* mutations. The availability of cloned genes is of particular value in determining whether individual peptidases have a role in the support of starter growth and most importantly in the generation of

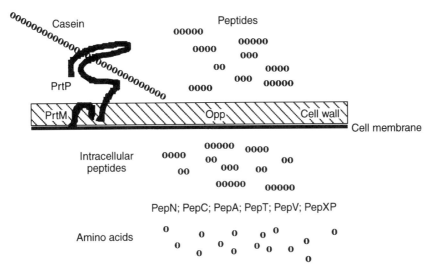

Figure 10.2   Genetically characterized components of the lactococcal proteolytic system.

mature cheese flavour. The most effective analysis has involved the sequential deletion of individual peptidase genes from *L. lactis* using 'silent' gene replacement techniques (see section 10.3). The constructed strains are invaluable for assessing the effect of peptidase loss on growth rate in milk and they can be used in experimental cheese-making trials to assess flavour generation. This strategy has been complemented by genetic engineering to increase the levels of specific peptidase enzymes. A summary of the cloned and characterized peptidase genes of *L. lactis* is given in Table 10.1. One of the more interesting outcomes of this genetic analysis is the observation that none of the peptidases identified to date has the features of a secreted or cell surface-presented enzyme. This is of particular significance as it suggests that, with the exception of the primary proteinase, the proteolytic process is intracellular (c.f. Chapter 9). A model of the proteolytic system of *L.lactis* is presented in Figure 10.2.

The process of peptide and amino acid transport into the cell has also been studied genetically (Kok and deVos, 1994). The best-characterized system is associated with the endopeptidase PepO. A spontaneous mutant of the laboratory *L. lactis* strain MG1614 was found to be defective in oligopeptide uptake and the equivalent genes were cloned by complementing this defect. The subsequent DNA sequence analysis revealed an operon of four genes *oppF*, *oppD*, *oppB* and *oppC*, followed by a second operon consisting of *oppA* and the endopeptidase gene *pepO*. Analysis of the amino acid sequences of proteins expressed by the *opp* genes revealed homology to the Opp proteins of *Bacillus subtilis* and *Salmonella typhimurium* and has led

Table 10.1 Cloned and characterized lactococcal peptidase genes

Gene	Enzyme	Class	Cleavage specificity	Monomer molecular weight weight (kDa)	Reference
pepA	aminopeptidase A	metallo	Asp(Glu)↑X–Y--	43	l'Anson *et al.*, 1995
pepC	aminopeptidase C	thiol	X↑Y–Z--	50	Chapot-Chartier *et al.*, 1992
pepN	aminopeptidase N	metallo	X↑Y–Z--	95	van Alen-Boerrigter *et al.*, 1991
pepXP	X-propyl dipeptidylaminopeptidase	serine	Pro↑Y–Z	90	Mayo *et al.*, 1991
pepV	dipeptidase	metallo	X↑Y	49	Unpublished
pepO	endopeptidase	neutral	---X↑Y---	70	Mierau *et al.*, 1993
pepT	tripeptidase	metallo	X↑YZ	52	Mierau *et al.*, 1994

to a model of the lactococcal oligopeptide uptake system. This involves initial substrate binding by the lipoprotein anchored OppA protein and transport involving two hydrophobic membrane-spanning proteins OppB and OppC and two ATP-binding proteins OppD and OppF (Tynkynnen *et al.*, 1993; Kok and deVos, 1994).

In addition to the homologous proteolytic machinery, the potential of introduced *Bacillus* neutral proteinase has been explored. For this the cloned gene has been introduced on a variety of expression cassettes and its influence on cheese maturation is being evaluated (van der Guchte *et al.*, 1991; Kok and deVos, 1994). The heterologous proteinase has a dramatic effect and the present challenge is to control its activity so as to prevent excessive ripening.

10.2.3 Bacteriophage resistance

One of the most serious problems of dairy fermentations is the susceptibility of starter cells to bacteriophages. The significance of the problem has led to a variety of culture management strategies that are designed to minimize the risk of vat failure due to bacteriophage. The analysis of bacteriophage-insensitive strains has led to the discovery and characterization of a large number of genetic systems that confer broad-spectrum resistance to bacteriophages (Hill, 1993; Klaenhammer and Fitzgerald, 1994). These systems have been divided into three categories based loosely on the mode of bacteriophage resistance that they confer.

1. Restriction/modification systems that protect bacteria from the introduction of foreign DNA are widespread and some of these protect lactic acid bacteria from bacteriophages (Hill *et al.*, 1989; Sanders and Schultz,1990; Hill, 1993; Klaenhammer and Fitzgerald, 1994). The bacteriophage DNA is recognized as foreign and degraded by a nuclease. The latter operates in tandem with another enzyme which modifies the bacterial DNA so that it is not subject to degradation. This process of self-protection limits the effectiveness of restriction/modification as a mechanism of bacteriophage defence because once a bacteriophage becomes modified it is no longer susceptible to restriction. Thus, lactococcal restriction/modification systems typically reduce the plaque-forming ability of bacteriophages by several orders of magnitude, but there is a significant risk that some modified bacteriophages will survive.
2. A distinct mechanism of resistance operates to prevent the introduction of bacteriophage DNA into the cell by inhibiting the adsorption and penetration of the bacteriophage. Where the mechanism has been studied, the presence of an atypical cell surface has been implicated in the resistance mechanism. Plasmid pSK112 causes the appearance of a galactosyl-containing lipoteichoic acid in the cell surface of resistant cells

which probably prevents the binding of bacteriophage to its receptor (Sijtsma *et al.*, 1988, 1990a,b). In the case of plasmid pCI528 a surface layer was demonstrated by electron microscopy and shown to consist of a hydrophilic polymer containing rhamnose and galactose (Lucey *et al.*, 1992). When this polymer was removed by treatment with alkali, bacteriophage was again able to attack the cells.

3. A third bacteriophage defence mechanism operates at a later stage after bacteriophage DNA has entered the cell. Several such systems have been discovered and they cause abortive infection of the bacteriophage (Hill, 1993; Klaenhammer and Fitzgerald, 1994). These systems usually provide a high-level resistance in which no plaques or very small plaques are formed following bacteriophage infection. In contrast to restriction/modification, any bacteriophage that survive the abortive infection mechanism remain susceptible in subsequent infections. Several genes for these Abi processes have been cloned and characterized and their DNA sequences have revealed the existence of distinct systems, such as *abiA* from pTR2030 (Hill *et al.*, 1990), *abiC* from pTN20 (Durmaz *et al.*, 1992) and *abi416* from pIL611 (Bidnenko *et al.*, 1993).

It has been noted that naturally occurring plasmids often carry genes for several bacteriophage resistance determinants. Thus, plasmid pTR2030 has genes for abortive infection and restriction/modification and plasmid pNP40 has genes for abortive infection and adsorption inhibition (Hill, 1993; Klaenhammer and Fitzgerald, 1994).

The use of conjugation to introduce bacteriophage resistance plasmids into dairy starter strains is now a well-established process in the industry and it has a proven track record of success. The availability of a growing array of individual bacteriophage resistance genes offers the future prospect of genetically engineered resistance to bacteriophages. As explained earlier, natural plasmids seem to have evolved to provide effective resistance through the concerted effect of multiple genes and it has been observed that single cloned genes can be less effective. It is also thought that the effective life of engineered resistance would be enhanced if resistance mechanisms were combined before their release into the industry, thereby minimizing the risk of bacteriophages surviving the resistance and evolving immunity to it.

In this regard it is noteworthy that there is evidence of bacteriophage evolution to overcome resistance mechanisms. An industrial bacteriophage isolate that is insensitive to the restriction system encoded by the bacteriophage resistance plasmid pTR2030 has been characterized. This appears to have been achieved by the bacteriophage acquiring the modification gene from pTR2030 and thereby allowing the bacteriophage DNA to be methylated and protected from the restriction endonuclease (Hill *et al.*, 1991). This event is important in demonstrating the potential of natural evolutionary processes to circumvent the efforts of the bacterial geneticist.

10.2.4 Starter cell lysis

One concern about highly bacteriophage-resistant cultures is that their capacity for flavour generation may be impaired. This could be associated with a resistance to lysis and this property has been targeted for genetic manipulation. It is conceivable that the release of a cocktail of intracellular starter enzymes is a significant step in flavour generation, and the development of controlled lysis of starter strains is a current biotechnological target.

Two molecular approaches are firstly to exploit the inherent autolysins of starter bacteria, and secondly to use cloned bacteriophage lysin genes. Strategies to control the expression of the autolysin or bacteriophage lysin genes and thereby starter cell lysis have been used. The original use of lysin genes for this purpose involved the lysin gene of bacteriophage ML3 and did not make use of controlled expression (Shearman *et al.*, 1992). This lysin gene was cloned together with its own constitutive bacteriophage promoter, and when introduced into *L. lactis* an autolytic phenotype was created whereby the culture grew normally through exponential phase but lysed early in the stationary phase (Figure 10.3). The inclusion of sucrose to provide osmotic buffering prevented lysis and facilitated culture maintenance. This property is probably caused by the fact that the lysin is active in degrading the cell wall from outside and when expressed intracellularly it has no lytic effect. As soon as one cell lyses by a natural process in stationary

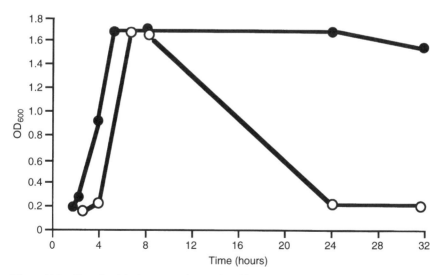

Figure 10.3 Growth of *L. lactis* carrying a plasmid vector that expresses the bacteriophage ML3 lysin gene. A control culture carrying plasmid pFI148 (●) grows normally and holds a high optical density in stationary phase, whereas the lysin-expressing strain pFI145 (○) grows normally to stationary phase, but then lyses.

phase a cascade effect is initiated in which more and more lysin is released as the cells start to lyse.

10.2.5   Antimicrobial peptides

The lactic acid bacteria produce a wide variety of antimicrobial peptides and some have potential in protecting foods and animal feed, including silage, from pathogenic and spoilage bacteria. Some of these peptides exhibit a broad-spectrum activity against Gram-positive bacteria and in the case of the lantibiotic nisin, activity has been extended to Gram-negative bacteria by its combination with chelating agents. Several antimicrobial peptides have been subjected to molecular genetic analysis and in some cases a good knowledge of their biosynthetic pathway has emerged. Although there is diversity in the structures of these compounds it is possible to identify two groups of related antimicrobial peptides that maintain well-conserved features across different lactic acid bacterial species. These are the small heat-stable bacteriocins and the lantibiotics (Klaenhammer, 1993; de Vuyst and Vandamme, 1994; Dodd and Gasson, 1994). An example of an antimicrobial peptide that does not fall into these groups is the large heat-labile bacteriocin, helvetican J produced by *Lactobacillus helveticus* (Joerger and Klaenhammer, 1990; Fremaux and Klaenhammer, 1994).

The small heat-stable proteins are unmodified peptides and include the lactococcins of *L. lactis* (Holo *et al.*, 1991; van Belkum *et al.*, 1991, 1992; Stoddard *et al.*, 1992; van Belkum, 1994), lactocin F of *Lactobacillus acidophilus* (Klaenhammer *et al.*, 1994), leucocin A-UAL of *Leuconostoc gelidum* (Hastings *et al.*, 1991; Stiles, 1994) and pediocin PA-1 of *Pediococcus acidilactici* (Marugg *et al.*, 1992; Ray, 1994). These molecules range in size from a 37-amino acid peptide in the case of leucocin A-UAL to a 57-amino acid peptide in the case of lactacin F. In all of the above examples genetic analysis has been undertaken and the DNA sequence of the structural gene that encodes the antimicrobial peptide and some of the associated genes have been determined. This has revealed conserved features, the most notable of which is the presence of an amino-terminal extension that is cleaved from the translated gene during antimicrobial peptide secretion. Although the various antimicrobial peptides have quite distinct structures, the amino-terminal leaders have well-conserved features of which the presence of two glycine residues immediately in front of the leader cleavage site is the most obvious. Genes involved in immunity have been identified and a particular feature of the biosynthesis of these heat-stable peptides is the involvement of ABC transport proteins. Thus, it seems that their biosynthesis involves a dedicated secretory process that depends on these ancillary proteins as well as the amino-terminal leader, the latter being cleaved during transport out of the cell.

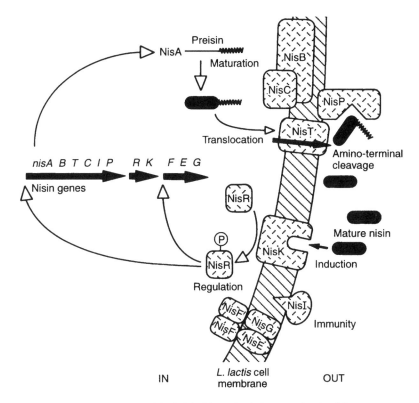

Figure 10.4 Model of nisin biosynthesis (see text for details).

Lantibiotics are a novel class of antimicrobial peptide that contain atypical amino acids and single sulphur lanthionine rings that are introduced into a gene-encoded precursor peptide by post-translational modification. The most commercially significant lantibiotic is nisin, produced by *L. Lactis*, but lantibiotics are produced by several different species of lactic acid bacteria. Other examples include lacticin 481 from *L. lactis* (Piard *et al.*, 1993; Piard, 1994), lactocin S from *Lactobacillus sake* (Nes *et al.*, 1994) and carnocin U149 from *Carnobacterium* (Stiles, 1994a). It is now established that nisin biosynthesis depends on a group of 11 chromosomally-located genes *nisA, B, T, C, I, P, R, K, F, E, G* that are arranged in a cluster (Rauch *et al.*, 1994). The mode of nisin biosynthesis and the role of the various *nis* gene products is depicted in Figure 10.4.

Lantibiotics, including nisin, are produced via a ribosomally synthesized precursor peptide with an amino-terminal extension (Buchman *et al.*, 1988; Kaletta and Entian,1989; Dodd *et al.*, 1990). The unusual amino acids dehydroalanine, dehydrobutyrine, lanthionine and β-methyllanthionine, are derived from the protein amino acids serine, threonine and cysteine. Serine and threonine are dehydrated to form dehydroalanine and dehydrobutyrine respectively and these derivatives are in turn condensed with cysteine to generate the lanthionine and β-methyllanthionine rings. The amino-terminal extension acts as a leader peptide that probably plays a role in the maturation reactions as well as secretion. During the latter process the leader is cleaved from the mature nisin molecule by a leader peptidase encoded by *nisP* (van der Meer *et al.*, 1993). Secretion involves an ABC transporter encoded by *nisT* (*Kuipers et al.*, 1993) and the maturation process probably involves the products of the *nisB* and *nisC* genes (Steen *et al.*, 1991; Engelke *et al.*,1992; Kuipers *et al.*, 1993). Immunity is conferred by the *nisI* gene (Kuipers *et al.*, 1993) with the possible involvement of the *nisFEG* operon. The expression of the nisin biosynthesis genes is positively controlled by a two-component regulator encoded by the *nisR* and *nisK* genes and nisin acts as the inducer for its own biosynthesis (Kuipers *et al.*, 1995a).

The knowledge gained from these molecular studies has been used successfully in the generation of lactococcal expression systems for the production of nisin variants with altered properties. These systems involve replacing the naturally occurring structural gene for precursor nisin (*nisA*) with a variant gene that has undergone site-directed mutagenesis (Dodd *et al.*, 1992, 1995, 1996; Kuipers *et al.*, 1992; Kuipers *et al.*, 1995b; Rollema *et al.*, 1995). Characterization and structural confirmation of the resulting nisin variants has generated useful functional data regarding the molecular requirements for an active nisin. In addition, this work has initiated a multidisciplinary study aimed at elucidation of the mechanism of action of nisin, at the molecular level. A protein engineering strategy of this type can be applied to the more practical objective of improving relevant properties of nisin, with potential to extend its commercial application as an antimicrobial agent. Examples of this might include provision of a broader inhibitory spectrum and enhanced specific activity against particular target organisms such as food pathogens and spoilage bacteria. In addition, the effectiveness of nisin might be improved by increasing its solubility and stability.

10.3   Gene cloning techniques

The introduction of DNA into lactic acid bacteria was revolutionized by the development of transformation by electroporation and this is now the universally used technique (Harlander, 1987; Holo and Nes, 1989; Gasson and Fitzgerald, 1994). The simplest approach to the maintenance of introduced genes is by their ligation to selectable plasmid vectors. A variety of plasmid replicons have been used in the construction of plasmid vectors for the lactic acid bacteria (deVos and Simons, 1994). These include those from the heterologous drug resistance plasmid pAMB1 (Simon and Chopin, 1988) and several homologous cryptic plasmids. The latter include plasmids that replicate by a rolling circle (RCR) mechanism such as pSH71 (Anderson and Gasson, 1985) and pWV01 (Kok et al., 1984) as well as theta replicating plasmids that use double-stranded DNA (O'Sullivan and Klaenhammer, 1993). The former have been very widely used and they have the advantage of replicating in a very wide range of bacteria including E. coli, conferring the features of a shuttle vector and simplifying cloning strategies. One disadvantage of these RCR vectors is that the single-stranded DNA seems to promote structural instability, making the cloning of some genes difficult. Both RCR and theta replicons have been used to generate vectors with a range of different copy numbers facilitating variation in the dosage of cloned genes.

The basic gene cloning vectors have been developed into a range of specialized vectors for promoter, terminator and secretory leader isolation, gene expression and secretion as well as cell surface anchoring (de Vos and Simons, 1994). For gene expression, promoter-active fragments derived by random selection from plasmids, the chromosome and bacteriophage genomes have been used. In addition, the controllable promoters from characterized genes have been exploited. One of the more widely used promoters is that derived from the lactose operon. This is negatively regulated by a linked, but divergently-transcribed repressor gene (see section 10.2 and Figure 10.1). The level of control is limited and significant expression remains in the uninduced state. Even so, the lactose promoter has been exploited successfully to control expression of the E. coli T7 RNA polymerase in a lactococcal version of one of the most effective of E. coli expression systems (Wells et al., 1993). Recently, the promoters of the nisin biosynthesis operon have been characterized and the major promoter upstream of the nisin structural gene nisA has been exploited for heterologous gene expression (W. deVos, personal communication). This promoter is subject to positive regulation by a classical two-component regulatory system that is activated by nisin (Kuipers et al., 1995a). For the expression of heterologous genes this offers complete control with no expression in the absence of nisin and product formation quantitatively related to inducer concentration through to a very high level.

An alternative to the use of plasmid gene cloning vectors is to integrate genes into the bacterial chromosome. This has the advantage of providing stable inheritance without the need to maintain a selection. Progressively more sophisticated approaches to chromosomal integration have been developed over the years (deVos and Simons, 1994; Gasson and Fitzgerald, 1994). Initially the process relied on suicide vectors that could not replicate in the target host species of lactic acid bacteria (Chopin et al., 1989;

Leenhouts *et al.*, 1989). The homologous DNA needed to direct integration, together with the gene to be integrated, were first cloned in *E. coli* and antibiotic selection was then used to isolate transformants in which chromosomal integration had been achieved. The process was dependent on relatively high rates of transformation as the frequency of recombination into the chromosome was itself relatively low. While much has been achieved in this way, more sophisticated delivery methods have been developed. These have been derived from an understanding of the replication of RCR replicons in which the origin of DNA replication is activated by a *trans*-acting protein RepA. One delivery vector system relies on a temperature-sensitive mutation in the *repA* gene which facilitates transformation and maintenance of the plasmid at a low permissive temperature (Maguin *et al.*, 1992). Following transformation with these so-called pG$^+$ host vectors the temperature is elevated, resulting in the inhibition of DNA replication and loss of the plasmid unless chromosomal integration takes place. This allows the introduction of a plasmid delivery vector before the selection for chromosomal integration and provides a more effective technique that is invaluable for strains with poor transformation efficiency.

Another strategy is to break the linkage of the origin of DNA replication and the *repA* gene. This has been achieved by the integration of the *repA* gene into the chromosome of strains of *E. coli*, *Bacillus subtilis* and *L. lactis* such that a plasmid construct with an origin of replication but no *repA* gene would be maintained (Leenhouts *et al.*, 1991b). Obviously in a normal host strain replication would be impossible and the vector would act as a suicide vector. In this case the replication machinery is derived from *Lactococcus* rather than *E. coli* and this provides a more food-compatible construct. By combining these two approaches a particularly effective system has evolved in which a delivery vector carrying no *repA* gene is paired with a helper plasmid carrying a temperature-sensitive *repA* gene. Both plasmids are maintained at a low permissive temperature, but elevation of the temperature results in loss of the helper plasmid and facilitates the selection of delivery vector integration into the chromosome.

Additional developments in chromosome manipulation include strategies to rescue adjacent genes by direct transformation of cleaved chromosomal DNA into *E. coli* (Godon *et al.*, 1994) and the exploitation of an IS element to facilitate random vector integration by transposition. The latter provides a powerful approach to gene inactivation and physical marking that is of especial value in the analysis of complex biological traits.

10.3.1 Food-compatible cloning systems

One of the most useful developments of chromosomal integration is silent integration of a cloned gene, deletion or other genetic manipulation in which no foreign DNA other than the intended genetic change remains. This is one example of a genetic technology designed to be acceptable for use in a food situation (deVos and Simons, 1994). One development of the chromosomal integration vectors is to include a β-galactosidase on the plasmid construct so the presence of the latter can readily be detected by the blue colour of colonies growing on X-gal-containing plates. The gene replacement vector can be integrated by single crossover recombination with selection for the vector's antibiotic resistance gene and subsequently its release by a second crossover recombination event can be followed by loss of the β-galactosidase gene.

Where gene expression is required, the site of chromosomal integration can be exploited to provide a resident promoter to facilitate transcription of the integrated DNA. A good example of this is the integration of a heterologous gene within a chromosomally located lactose operon in L. lactis. A bacteriophage lysin gene that kills Listeria was recently integrated in the lactococcal chromosome within the lacG gene, providing controlled expression at a level only slightly below that achieved with the same lactose operon promoter on a multicopy plasmid vector (Payne et al., 1996).

Where it is preferred to clone genes on a plasmid vector, as may be the case when copy number is a factor in achieving maximal expression, a food-compatible selection is required. There are several approaches that could be adopted. A bacteriocin or lantibiotic resistance, or immunity gene, would provide a selection that depends on the addition of the relevant antimicrobial agent to the growth medium. This has been achieved in the case of the lantibiotic nisin (Froseth and McKay, 1992). A more elegant strategy is to make use of a metabolic gene to provide a positive selection. One of the best and most effective examples of this involved the well-characterized lactococcal lactose operon. A background strain of L. lactis is used in which the lactose operon is integrated in the chromosome. The gene that codes for the Factor III component of the lactose uptake system lacF is deleted by the silent gene replacement technique described above. This lactose-negative phenotype can be converted to a lactose-positive phenotype when the chromosomal lacF deletion is complemented by a plasmid based copy of the same lacF gene. A food-compatible vector has this lacF gene under control of a constitutive lactococcal promoter and it will be effectively maintained simply by growth on lactose. Thus, such a vector is self-selecting, includes no foreign DNA, and is especially useful in milk fermentation where lactose is the sole energy source for starter growth (de Vos and Simons, 1994; MacCormick et al., 1995).

10.3.2 Natural gene transfer systems

While recombinant DNA technology provides a sophisticated method for genetic manipulation, there remains a place for the exploitation of natural gene transfer systems (Gasson and Fitzgerald, 1994). The most useful of these is the bacterial mating process of conjugation that is sometimes associated with the larger lactic acid bacteria plasmids. Indeed, one of the most successful uses of genetics to provide new strains for industrial exploitation is the development of bacteriophage-resistant derivatives by the introduction of self-transmissible bacteriophage resistance plasmids. One of the advantages of such natural gene transfer is avoidance of the need for regulatory approval that is a vital part of recombinant DNA technology. In a similar way the method may also be less likely to cause a negative consumer reaction.

While the transfer of bacteriophage resistance plasmids is of proven value, and similar transfer of the nisin conjugative transposon has been effective in the generation of exploitable new strains, conjugation is of limited value while it is solely associated with individual genetic systems. An important recent development was the discovery that the chromosome of *L. lactis* can be mobilized by conjugation under the control of a chromosomally located sex factor (Gasson *et al.*, 1995). This provides an important extension to natural gene transfer technology and one of the more exciting potential developments is its use to construct interstrain hybrids. This could be especially valuable in combining industrially relevant traits for which the genetic basis is ill-defined, or where it is the product of multiple genes. Hybrid construction is an important approach to industrial strain improvement in yeast and in future an analogous technology might be developed for the lactic acid bacteria.

10.4   Whole genome analysis

An important advance in the genetics of lactic acid bacteria is the development of physical and genetic chromosome maps (Gasson and Fitzgerald, 1994). The fact that these bacteria are nutritionally fastidious has limited the application of conventional genetic mapping which relies on the use of mutant strains, especially auxotrophic mutants. Recently, new approaches to the physical mapping of whole genomes have been developed. These involve the digestion of whole chromosomes with rare cutting restriction endonucleases that generate relatively few DNA fragments, even from such large target molecules. The most significant factor is the choice of an enzyme with a recognition site made up of bases that are less common in the target DNA, which for the AT-rich lactococci means GC-containing sites. Restriction endonucleases *Sma*I, *Not*I and *Apa*I have proved particularly effective (Le Bourgeois *et al.*, 1989).

The second important technological factor in physical mapping is the use of an electrophoretic technique known as pulsed field gel electrophoresis (PFGE). This allows the unusually large DNA fragments that rare cutting restriction endonucleases generate from entire chromosomes to be separated and their sizes can then be estimated. A variety of strategies can be used to order the fragments and solve a physical map of the chromosome. In the case of the lactococci, some innovative approaches involving integrative vectors were developed specifically to facilitate mapping (Le Bourgeois *et al.*, 1992a). Cloned genes and 'housekeeping genes' with well-conserved DNA sequences can be located on the physical map by hybridization or by the use of integrative vectors that introduce an extra rare cutting restriction site at the position of the gene. For two lactococcal strains, *L. lactis* MG1363 (Le Bourgeois *et al.*, 1995) and *L. lactis* IL1403 (Le Bourgeois *et al.*, 1992b), very detailed physical and genetic maps have been produced. These maps provide a blueprint for the further characterization of lactococcal genomes and the next step is to initiate a whole genome sequencing project for *L. lactis*.

10.5 Future prospects

It is clear that genetic technology and understanding of several key industrially relevant traits have advanced to the point where genetically manipulated strains of lactic acid bacteria with advantageous properties are likely to be available in the relative short term. Bacteriophage resistance by plasmid transfer is in place and a new generation of strains constructed by recombinant DNA technology are imminent. The proteolytic machinery is well characterized and strains have been manipulated to provide cell lysis. These various developments should lead on to strains with improved ripening capacity at some stage in the future. Antimicrobial production is well characterized and systems for the generation of variant molecules are available. This should lead to improvements in food safety and the prevention of spoilage by innovative means. One area that is just beginning to be explored is the molecular genetics of global regulation and the nature of lactic acid bacteria interaction with the environment. This holistic approach to their physiology and genetics may shed some light on the more difficult aspects of culture performance in an industrial setting. Certainly there are aspects of performance that will not be accessed by a simplistic 'single gene for a trait' approach and the interaction of multiple genes and the response of starter cells to their environment are key areas for future study.

Regulation of genetic technology and its acceptance by the consumer are also critical factors for the introduction of genetically manipulated starter cultures. The regulatory process is evolving as more products of gene technology become available and issues such as the desirability of labelling continue to be debated. It is encouraging that approval for food use has been granted to genetically manipulated yeasts. Many of the advances envisaged for lactic acid bacteria rely on rearrangements of genes from within the organism's pre-existing gene pool and there are some very elegant genetic strategies to provide acceptable food-compatible genetic technology that should win regulatory approval. The consumer issue is difficult to predict, but there is little doubt that genetic technology will increase in prominence, and as it becomes more commonplace it may become widely accepted. Certainly its take-up will depend on the generation of demonstrable benefit and to that extent its future may depend on the realization of its promise in terms of industrial and quality benefits.

10.6   References

Anderson, P.H. and Gasson, M.J. (1985) High copy number plasmid vectors for use in lactic streptococci. *FEMS Microbiol. Lett.* **30**, 193–6.

Bidnenko, E., Valyasevi, R., Cluzel, J.-P., Parreira, R., Hillier, A., Gautier, M., Anba, J., Ehrlich, D. and Chopin, M.-C. (1993) Amelioration de la resistance des lactocoques aux bacteriophages. *Le Lait*, **73**, 199–205.

Buchman, W.B., Banerjee, S. and Hansen, J.R. (1988) Structure, expression and evaluation of a gene encoding the precursor of nisin, a small protein antibiotic. *J. Biol. Chem.*, **263**, 16260–6.

Chapot-Chartier, M.-P., Nardi, M., Chopin, M.-C., Chopin, A. and Gripon, C. (1992) Cloning and sequence of pepC, a cysteine aminopeptidase gene from *Lactococcus lactis* subsp. *cremoris* AM2. *Appl. Environ. Microbiol.*, **59**, 330–3.

Chopin, M.-C., Chopin, A., Rouault, A and Galleron, N. (1989) Insertion and amplification of foreign genes in the *Lactococcus lactis* subsp. *lactis* chromosome. *Appl. Environ. Microbiol.*, **55**, 1769–74.

deVos, W.M. and Gasson, M.J. (1989) Structure and expression of the *Lactococcus lactis* gene for phospho-β-galactosidase (*lacG*) in *Escherichia coli* and *L. lactis. J. Gen. Microbiol.*, **135**, 1833–46.

deVos,W.M. and Simons, G.M. (1994) Gene cloning and expression systems in lactococci, in, *Genetics and Biotechnology of Lactic Acid Bacteria* (eds M.J. Gasson and W.M. deVos), Blackie Academic & Professional, Glasgow, pp. 52–105.

deVos, W.M., Boerrigter, L., van Rooijen, R.J., Reiche, B. and Hengstenberg, W. (1990) Characterization of the lactose specific enzymes of the phosphotransferase system in *Lactococcus lactis. J. Biol. Chem.*, **265**, 22554–60.

de Vuyst, L. and Vandamme, E.J. (1994) *Bacteriocins of Lactic Acid Bacteria*, Blackie Academic & Professional, Glasgow.

Dodd, H.M. and Gasson, M.J. (1994) Bacteriocins of lactic acid bacteria, in *Genetics and Biotechnology of Lactic Acid Bacteria* (eds M.J. Gasson and W.M. deVos), Blackie Academic & Professional, Glasgow, pp. 211–51.

Dodd, H.M., Horn, N. and Gasson, M.J. (1990) Analysis of the genetic determinant for the production of the peptide antibiotic nisin. *J. Gen. Microbiol.* **136**, 555–66.

Dodd, H.M., Horn, N., Hao, Z. and Gasson, M.J. (1992) A lactococcal expression system for engineered nisins. *Appl. Environ. Microbiol.*, **58**, 3683–93.

Dodd, H.M., Horn, N. and Gasson, M.J. (1995) A cassette vector for protein engineering the lantibiotic nisin. *Gene*, **162**, 163–4.

Dodd, H.M., Horn, N. and Gasson, M.J. (1996) Gene replacement strategy for engineering nisin. *Microbiology*, **142**, 47–55.

Durmaz, E., Higgins, D.L. and Klaenhammer, T.R. (1992) Molecular characterization of a second abortive phage resistance gene present in *Lactococcus lactis* subsp. *lactis* ME2. *J. Bacteriol.* **174**, 7463–9.

Engelke, G., Gutowski-Eckel, Z., Hammelmann, M and Entian, K.-D. (1992) Biosynthesis of the lantibiotic nisin: genomic organisation and membrane localization of the *nisB* protein. *Appl. Environ. Microbiol.*, **58**, 3740–3.

Fremaux, C. and Klaenhammer, T.R. (1994) Helveticin J, a large heat-labile bacteriocin from *Lactobacillus helveticus*, in *Bacteriocins of Lactic Acid Bacteria* (eds L. de Vuyst and E.J. Vandamme), Blackie Academic & Professional, Glasgow, pp. 397–418.

Froseth, B.R. and McKay, L.L. (1992) Development and application of pFM011 as a possible food-grade cloning vector. *J. Dairy Sci.*, **74**, 1445–53.

Gasson, M.J. (1993) Progress and potential in the biotechnology of lactic acid bacteria. *FEMS Microbiol. Lett.*, **12**, 3–20.

Gasson, M.J. and Fitzgerald, G.F. (1994) Gene transfer systems and transposition, in *Genetics and Biotechnology of Lactic Acid Bacteria* (eds M.J. Gasson and W.M. deVos), Blackie Academic & Professional, Glasgow, pp. 1–51.

Gasson, M.J., Godon, J.-J., Pillidge, C.J., Eaton, T.J., Jury, K. and Shearman, C.A. (1995) Characterisation and exploitation of conjugation in *Lactococcus lactis*. *Int. Dairy J.*, **4**, 757–62.

Godon, J.-J., Jury, K., Shearman, C.A. and Gasson, M.J. (1994) The *Lactococcus lactis* sex factor aggregation gene *cluA*. *Mol. Microbiol.*, **12**, 655–63.

Haandrickman, A.J., Kok, J., Laan, H., Soemitro, S., Ledeboer, A.M., Konings, W.N. and Venema, G. (1989) Identification of a gene required for the maturation of an extracellular serine proteinase. *J. Bacteriol.*, **171**, 2789–94.

Haandrickman, A.J., Kok, J. and Venema, G. (1991a) Lactococcal proteinase maturation protein PrtM is a lipoprotein. *J. Bacteriol.*, **173**, 4517–25.

Haandrickman, A.J., Meesters, R., Laan, H., Konings, W.N., Kok, J. and Venema, G. (1991b) Processing of the lactococcal extracellular serine proteinase. *Appl. Environ. Microbiol.*, **57**, 1899–904.

Harlander, S.K. (1987). Transformation of *Streptococcus lactis* by electroporation, in *Streptococcal Genetics* (eds J.J. Ferretti and R. Curtiss). American Society for Microbiology, Washington, DC, pp. 229–33.

Hastings, J.W., Sailer, M., Johnson, K., Roy, K.L., Vederas, J.C. and Stiles, M.E. (1991) Characterization of leucocin A-UAL 187 and cloning of the bacteriocin gene from *Leuconostoc gelidum*. *J. Bacteriol.*, **173**, 7491–500.

Hill, C. (1993) Bacteriophage and bacteriophage resistance in lactic acid bacteria. *FEMS Microbiol. Rev.* **12**, 87–108.

Hill, C., Pierce, K. and Klaenhammer, T.R. (1989) The conjugative plasmid pTR2030 encodes two defense mechanisms in lactococci, restriction modification (R^+/M^+) and abortive infection (Hsp^+). *Appl. Environ. Microbiol.*, **55**, 2416–19.

Hill, C., Miller, L.A. and Klaenhammer, T.R. (1990) Nucleotide sequence and distribution of the pTR2030 resistance determinant (*hsp*) which aborts bacteriophage infection in lactococci. *Appl. Environ. Microbiol.*, **56**, 2255–8.

Hill, C., Miller, L.A., and Klaenhammer, T.R. (1991) *In vivo* genetic exchange of a functional domain from a type IIA methylase between lactococcal plasmid pTR2030 and a virulent bacteriophage. *J. Bacteriol.*, **173**, 4363–70.

Holo, H. and Nes, I.F. (1989). High-frequency transformation, by electroporation, of *Lactococcus lactis* subsp. *cremoris* grown with glycine in osmotically stabilized media. *Appl. Environ. Microbiol.*, **55**, 3119–23.

Holo, H., Nillsen, O. and Nes, I.F. (1991) Lactococcin A, a new bacteriocin from *Lactococcus lactis* subsp. *cremoris*: isolation and characterization of the protein and its gene. *J. Bacteriol.*, **173**, 3879–87.

I'Anson, K.J.A., Movahedi, S., Griffin, H.G. and Gasson, M.J. (1995) A non-essential glutamyl aminopeptidase is required for optimal growth of *Lactococcus lactis* MG1363 in milk. *Microbiology*, **141**, 2873–81.

Joerger, M.C. and Klaenhammer, T.R. (1990) Cloning, expression and nucleotide sequence of the *Lactobacillus helveticus* 481 gene encoding the bacteriocin helveticin J. *J. Bacteriol.*, **171**, 6339–47.

Kaletta, C. and Entian, K.-D. (1989) Nisin, a peptide antibiotic, cloning and sequencing of the *nisA* gene and posttranslational processing of its peptide product. *J. Bacteriol.*, **171**, 1597–601.

Klaenhammer, T.R. (1993) Genetics of bacteriocins produced by lactic acid bacteria. *FEMS Microbiol. Rev.*, **12**, 87–108.

Klaenhammer, T.R. and Fitzgerald, G.F. (1994) Bacteriophages and bacteriophage resistance, in *Genetics and Biotechnology of Lactic Acid Bacteria* (eds M.J. Gasson and W.M. deVos), Blackie Academic & Professional, Glasgow, pp. 106–68.

Klaenhammer, T.R., Ahn, C. and Muriana, P.M. (1994) Lactacin F, a small heat-stable bacteriocin from *Lactobacillus johnsoni*, in *Bacteriocins of Lactic Acid Bacteria* (eds L. de Vuyst and E.J. Vandamme), Blackie Academic & Professional, Glasgow, pp. 377–96.

Kok, J. and deVos, W.M. (1994) The proteolytic system of lactic acid bacteria, in *Genetics and Biotechnology of Lactic Acid Bacteria* (eds M.J. Gasson and W.M. deVos), Blackie Academic & Professional, Glasgow, pp. 169–210.

Kok, J., van der Vosen, J.M.B.M. and Venema, G. (1984) Construction of plasmid cloning vectors for lactic streptococci which also replicate in *Bacillus subtilis* and *Escherichia coli*. *Appl. Environ. Microbiol.*, **50**, 94–101.

Kok, J., Leenhouts, K.J., Haandrikman, A.J., Ledeboer, A.M., and Venema, G. (1988) Nucleotide sequence of the cell wall proteinase of *Streptococcus cremoris* Wg2. *Appl. Environ. Microbiol.*, **54**, 231–8.

Kuipers, O.P., Rollema, H.S., Yap, W.M.G., Boot, H.J., Siezen, R.J. and deVos, W.M. (1992) Engineering dehydrated amino-acid residues in the antimicrobial peptide nisin. *J. Biol. Chem.*, **267**, 2430–4.

Kuipers, O.P., Beerthuyzen, M.M., Siezen, R.J. and deVos, W.M. (1993) Characterization of the nisin gene cluster *nisABTCIPR* of *Lactococcus lactis* and evidence for the involvement of the *nisA* and *nisI* genes in product immunity. *Eur. J. Biochem.*, **216**, 281–91.

Kuipers, O.P., Beerthuyzen, M.M., de Ruyter, P.G.G.A., Luesink. E.J. and deVos, W.M. (1995a) Autoregulation of nisin biosynthesis in *Lactococcus lactis* by signal transduction. *J. Biol. Chem..*, **45**, 27299–304.

Kuipers, O.P., Rollema, H.S., Beerthuyzen, M.M., Siezen, R.J. and deVos, W.M. (1995b) Protein engineering and biosynthesis of nisin and regulation of transcription of the structural *nisA* gene. *Int. Dairy J.*, **5**, 785–95.

Le Bourgeois, P., Mata, M and Ritzenthaler, P. (1989) Genome comparison of *Lactococcus* strains by pulsed-field gel electrophoresis. *FEMS Microbiol. Lett.*, **59**, 65–70.

Le Bourgeois, P. Lautier, M., Mata, M. and Ritzenthaler, P. (1992a) New tools for the physical and genetic mapping of *Lactococcus* strains. *Gene*, **111**, 109–14.

Le Bourgeois, P., Lautier, M., Mata, M. and Ritzenthaler, P. (1992b) Physical and genetic map of the chromosome of *Lactococcus lactis* subsp. *lactis* IL1403. *J. Bacteriol.*, **174**, 6752–62.

Le Bourgeois, P., Lautier, M., van den Berghe, Gasson, M.J. and Ritzenthaler, P. (1995) Physical and genetic map of the *Lactococcus lactis* subsp. *cremoris* MG1363 chromosome: comparison with that of *Lactococcus lactis* subsp. *lactis* IL1403 reveals a large genome inversion. *J. Bacteriol.*, **177**, 2840–50.

Leenhouts, K.J., Kok, J. and Venema, G. (1989) Campbell-like integration of hetereologous plasmid DNA into the chromosome of *Lactococcus lactis* subsp. *lactis*. *Appl. Environ. Microbiol.*, **55**, 394–400.

Leenhouts, J.K., Kok, J. and Venema, G. (1991a) Chromosomal stabilization of the proteinase genes in *Lactococcus lactis*. *Appl. Environ. Microbiol.*, **57**, 2548–75.

Leenhouts, J.K., Kok, J. and Venema, G. (1991b) Lactococcal plasmid pWVOI as an integration vector for lactococci. *Appl. Environ. Microbiol.*, **57**, 2562–7.

Lucey, M., Daly, C. and Fitzgerald, G.F. (1992) Cell surface characteristics of *Lactococcus lactis* harbouring pCI528, a 46-kb plasmid encoding inhibition of bacteriophage adsorption. *J. Gen. Microbiol.*, **138**, 2137–43.

MacCormick, C.A., Griffin, H.G. and Gasson, M.J. (1995) Construction of a food-grade host/vector system for *Lactococcus lactis* based on the lactose operon. *FEMS Microbiol. Lett.*, **127**, 105–9.

Maeda, S. and Gasson, M.J. (1986) Cloning, expression and location of the *Streptococcus lactis* gene for phospho-β-galactosidase. *J. Gen. Microbiol.*, **132**, 331–40.

Maguin, E., Duwat, P., Hege, T., Ehrlich, D. and Grus, A. (1992) New thermosensitive plasmid for gram-positive bacteria. *J. Bacteriol.*, **174**, 5633–8.

Marugg, J.D., Gonzalez, C.F., Kunka, B.S., Ledeboer, A.M., Pucci, M.J., Toonen, M.Y., Walker, S.A., Zoetmulder, L.C.M. and Vandenbergh, P.A. (1992) Cloning, expression and nucleotide sequence of genes involved in production of pediocin PA-1 from *Pediococcus acidilactici* PAC 1-0. *Appl. Environ. Microbiol.*, **58**, 2360–7.

Mayo, B., Kok, J., Venema, K., Bockelmann, W., Teuber, M., Reinke, H. and Venema, G. (1991) Molecular cloning and sequence analysis of the X-prolyl dipeptidyl aminopeptidase gene from *Lactococcus lactis* subsp. *cremoris*. *Appl Environ. Microbiol.*, **57**, 38–44.

McKay, L.L. (1982) Regulation of lactose metabolism in dairy streptococci, in *Developments in Food Microbiology-1*, Applied Science Publishers, London, pp. 153–82.

Mercenier, A., Pouwels, P.H. and Chassy, B.M. (1994) Genetic engineering of lactobacilli, leuconostocs and *Streptococcus thermophilus*, in *Genetics and Biotechnology of Lactic Acid Bacteria* (eds M.J. Gasson and W.M. deVos), Blackie Academic & Professional, Glasgow, pp. 252–93.

Mierau, I., Tann, P.S.T., Haandrikman, A.J., Mayo, B., Kok, J., Konings, W.N. and Venema, G. (1993) Cloning and sequencing of the gene for a lactococcal endopeptidase, an enzyme with sequence similarity to mammlian enkephalinase. *J. Bacteriol.*, **175**, 2087–96.

Mierau, I., Haandrikman, A.J., Velterop, O., Tan, P.S.T., Leenhouts, K.L., Konings, W.N., Venema, G. and Kok, J. (1994) Tripeptidase genes (pepT) of *Lactococcus lactis*: molecular cloning and nucleotide sequencing of pepT and construction of a chromosomal deletion mutant. *J. Bacteriol.*, **176**, 2854–61.

Nes, I.F., Mortvedt, C.I., Nissen-Meyer, J. and Skaugen, M. (1994) Lactocin S, a lanthionine-containing bacteriocin isolated from *Lactobacillus sake* L45, in *Genetics and Biotechnology of Lactic Acid Bacteria* (eds M.J. Gasson and W.M. deVos), Blackie Academic & Professional, Glasgow, pp. 435–49.

O'Sullivan, D.J. and Klaenhammer, T.R. (1993) High- and low-copy-number *Lactococcus* shuttle cloning vectors with features for clone screening. *Gene*, **137**, 227–31.

Payne, J., MacCormick, C.A., Griffin, H.G. and Gasson, M.J. (1996) Exploitation of a chromosomally integrated lactose operon for controlled expression in *Lactococcus lactis*. *FEMS Microbiol. Lett.*, **136**, 19–24.

Piard, J.-C. (1994) Lactacin 481, a lantibiotic produced by *Lactococcus lactis* subsp. *lactis* CNRZ 481, in *Genetics and Biotechnology of Lactic Acid Bacteria* (eds M.J. Gasson and W.M. deVos), Blackie Academic & Professional, Glasgow, pp. 251–71.

Piard, J.C., Kuipers, O.P., Rollema, H.S., Desmazeaud, M.J. and deVos, W.M. (1993) Structure, organization and expression of the *lct* gene for lacticin 481, a novel lantibiotic produced by *L.lactis* subsp. *lactis*. *J. Biol. Chem.*, **268**, 16361–8.

Rauch, P.J., Kuipers, O.P., Siezen, R.J., and deVos, W.M. (1994) Genetics and protein engineering of nisin, in *Bacteriocins of Lactic Acid Bacteria* (eds L. de Vuyst and E.J. Vandamme), Blackie Academic & Professional, Glasgow, pp. 223–49.

Ray, B. (1994) Pediocins of *Pediococcus* species, in *Bacteriocins of Lactic Acid Bacteria* (eds L. de Vuyst and E.J. Vandamme), Blackie Academic & Professional, Glasgow, pp. 465–96.

Rollema, H.S., Kuipers, O.P., Both, P., deVos, W.M. and Siezen, R.J. (1995) Improvement of solubility and stability of the antimicrobial peptide nisin by protein engineering. *Appl. Environ. Microbiol.*, **61**, 2873–8.

Sanders, M.E. and Schultz, J. (1990) Cloning of phage resistance genes from *Lactococcus lactis* ssp. *cremoris* KH. *J. Dairy Sci.*, **73**, 2044–53.

Shearman, C.A., Jury, K. and Gasson, M.J. (1992) Autolytic *Lactococcus lactis* expressing a lactococcal bacteriophage lysin gene. *Biotechnology*, **10**, 196–9.

Siezen, R.J., deVos, W.M., Leunissen, J.A.M. and Dijkstra, B.W. (1991) Homology modelling and protein engineering strategy of subtilases, the family of subtilisin-like proteinases. *Prot. Eng.*, **4**, 719–37.

Siezen, R.J., Bruinenberg, P.G., Vos, P., van Alen-Boerrigter, I.J., Nijhuis, M., Alting, A.C., Exterkate, F.A. and deVos, W.M. (1993) Engineering of the substrate binding region of the subtilisin-like, cell envelope proteinase of *Lactococcus lactis*. *Prot. Eng.*, **6**, 927–37.

Sijtsma, L., Sterkenburg, A. and Wouters, J.T.M. (1988) Properties of the cell walls of *Lactococcus lactis* subsp. *cremoris* SK110 and SK112 and their relation to bacteriophage resistance. *Appl. Environ. Microbiol.*, **54**, 2808–11.

Sijtsma, L., Jansen, N., Hazeleger, W.C., Wouters, J.T.M. and Hellingwerf, K.J. (1990a) Cell surface characteristics of bacteriophage-resistant *Lactococcus lactis* subsp. *cremoris* SK110 and its bacteriophage sensitive variant SK112 . *Appl. Environ. Microbiol.*, **56**, 3230–3.

Sijtsma, L., Hellingwerf, K.J. and Wouters, J.T.M. (1990b) Isolation and characterization of lipoteichoic acid, a cell envelope component involved in preventing phage adsorption from *Lactococcus lactis* subsp. *cremoris* SK110. *J. Bacteriol.*, **172**, 7126–30.

Simon, D. and Chopin, A. (1988) Construction of a vector plasmid family and its use for molecular cloning in *Streptococcus lactis*. *Biochimie*, **70**, 559–67.

Steen, M.T., Chung, Y.J. and Hansen, J.N. (1991) Characterization of the nisin gene as part of a polycistronic operon. *Appl. Environ. Microbiol.*, **57**, 1181–8.

Stiles, M.E. (1994a) Bacteriocins produced by *Carnobacterium* species. in *Bacteriocins of Lactic Acid Bacteria* (eds L. de Vuyst and E.J. Vandamme), Blackie Academic & Professional, Glasgow, pp. 451–60.

Stiles, M. (1994b) Bacteriocins produced by *Leuconostoc* species, in *Bacteriocins of Lactic Acid Bacteria* (eds L. de Vuyst and E.J. Vandamme), Blackie Academic & Professional, Glasgow, pp. 497–506.

Stoddard, G.W., Petzel., J.P., van Belkum, M.J., Kok, J. and McKay, L.L. (1992) Molecular analyses of the lactococcin A gene cluster from *Lactococcus lactis* subsp. *lactis* biovar. *diacetylactis* WM4. *Appl. Environ. Microbiol.*, **58**, 1952–61.

Tynkynnen, S., Buist, G., Kunji, E., Kok, J., Poolman, B., Venema, G. and Haandrickman, A. (1993) Genetic and biochemical characterization of the oligopeptide transport system of *Lactococcus lactis*. *J. Bacteriol.*, **175**, 7523–32.

van Belkum, M.J. (1994) Lactococcins, bacteriocins of *Lactococcus lactis*, in *Bacteriocins of Lactic Acid Bacteria* (eds L. de Vuyst and E.J. Vandamme), Blackie Academic & Professional, Glasgow, pp. 301–18.

van Belkum, M.J., Hayema, B.J., Jeeninga, R.E., Kok, J. and Venema, G. (1991) Organization and nucleotide sequences of two lactococcal bacteriocin operons. *Appl. Environ. Microbiol.*, **57**, 492–8.

van Belkum, M.J., Kok, J. and Venema, G. (1992) Cloning, sequencing and expression in *Escherichia coli* of *lcnB*, a third bacteriocin determinant from the lactococcal bacteriocin plasmid p9B4-6. *Appl. Environ. Microbiol.*, **58**, 572–7.

van der Guchte, M., Kodde, J., van der Vossen, J.M.B.M., Kok, J. and Venema, G. (1991) Heterologous gene expression in *Lactococcus lactis* subsp. *lactis*: synthesis, secretion and processing of the *Bacillus subtilis* neutral proteinase. *Appl. Environ. Microbiol.*, **56**, 2606–11.

Van der Meer, J.R., Polman, J., Beerthuyzen, M.M., Siezen, R.J., Kuipers, O.P. and de Vos, W.M. (1993) Characterization of the *Lactococcus lactis* Nisin A operon genes *nisP*, encoding a subtilisin-like serine protease involved in precursor processing, and *nisR*, encoding a regulatory protein involved in nisin biosynthesis. *J. Bacteriol.*, **175**, 2578–88.

van Alen-Boerrigter, I.J., Baankreis, R. and de Vos, W.M. (1991) Characterization and overexpression of the *Lactococcus lactis* pepN gene and localization of its product, aminopeptidase N. *Appl. Environ. Microbiol.*, **57**, 2555–61.

van Rooijen, R.J. and deVos, W.M. (1990) Molecular cloning, characterization and nucleotide sequence of *lacR*, a gene encoding the repressor of the lactose phosphotransferase system of *Lactococcus lactis*. *J. Biol. Chem.*, **265**, 18499–503.

van Rooijen, R.J., van Schalkwijk, A. and deVos, W.M. (1991) Molecular cloning, characterization and nucleotide sequence of the tagatose 6-phosphate gene cluster of the lactose operon of *Lactococcus lactis*. *J. Biol. Chem.*, **266**, 7176–81.

van Rooijen, R.J., Gasson, M.J. and deVos, W.M. (1992) Characterization of the promoter of the *Lactococcus lactis* lactose operon: contribution of flanking sequences and LacR repressor to its activity. *J. Bacteriol.*, **174**, 2273–80.

Vos, P., van Asseldonk, M., van Jeveren, F., Siezen, R.J., Simons, G. and deVos, W.M. (1989a) A maturation protein is essential for the production of active forms of *Lactococcus lactis* SKII serine proteinase located in or secreted from the cell envelope. *J. Bacteriol.*, **171**, 2795–802.

Vos, P., Simons, G., Siezen, R.J. and deVos, W.M. (1989b) Primary structure and organization of the gene for a proteolytic, cell envelope-located serine proteinase. *J. Biol. Chem.*, **264**, 13579–85.

Vos, P., Boerrigter, I.J., Buist, G., Haandrickman, A.J., Nijhuis, M., de Reuver, M.B., Siezen, R.J., Venema, G., deVos, W.M. and Kok, J. (1991) Engineering of the *Lactococcus lactis* proteinase by construction of hybrid enzymes. *Prot. Eng.*, **4**, 479–84.

Wells, J.M., Wilson, P.W., Norton, P.M., Gasson, M.J. and LePage, R.W.F. (1993) *Lactococcus lactis* high level expression of tetanus toxin fragment C and protection against lethal challenge. *Mol. Microbiol.*, **8**, 1155–62.

11 Sensory evaluation of dairy flavours
J. BAKKER

11.1 Introduction

It is believed that olfaction contributes the most crucial sensory information for assessing the quality and palatability of foods and is therefore of over-riding importance in determining the flavour of foods (Maruniak, 1988). The flavour of foods is usually defined as a combination of aroma and taste. Many dairy products have complex sensory properties, and it is often difficult to relate a particular sensory attribute with chemical measurements of flavour. For example, the development of the flavour and texture of cheese takes place over a long period of time, starting during cheese-making and continuing during the maturation period. Contributions to flavour development are made by the fermentation of the starter culture, and the enzymic modifications of the constituents in the milk, such as their activity towards lactose, lipids and proteins. Proteolysis influences the texture, but also leads to the formation of flavour peptides and free amino acids, which form the precursors for the development of aroma compounds. Lipolysis is important for the development of the typical flavour of mould-ripened cheeses, while the starter bacteria have a direct impact on the flavour of fresh cheeses such as cottage cheese. The complexity of the natural flavours in cheese has been reviewed recently (Bakker and Law, 1994).

Although considerable research has been directed towards the processes and components contributing to the sensory attributes, there are still many gaps in our understanding of the mechanisms which underlie the changes in such complicated products as cheese. For example, there is still an on-going scientific debate regarding the most important volatile components contributing to Cheddar cheese flavour (see Chapter 8). Thus, it would seem that sensory analysis, which has a number of useful analytical tools available, can contribute to our understanding of the sensory properties of dairy produce as well as add useful information regarding the effect of the various processes on these properties.

Changes in composition of foods may affect the release of flavours while eating, but also the formation of flavours, such as those formed during cheese maturation. Since consumers are becoming increasingly aware of the health risks associated with high fat intake, there is a development towards new foods containing less fat; such a change is expected to influence the release and perception of flavour. The aim of this chapter is to discuss the recent developments in sensory methodology, including the effects of the food matrix on flavour release and perception, in particular related to dairy products.

11.2 Sensory mechanisms

Sensory perception of a food is the result of information about a number of aspects of the food. The sensory properties of a food are conveyed to the brain by a number of senses; colour and surface structure are assessed by vision, structural information is collected by tactile and kinaesthetic senses combined with hearing when the food is eaten, volatile flavour compounds are perceived by the sense of smell, while the water-soluble stimuli are perceived by taste. The trigeminal nerve recognizes irritants such as capsaicin giving hot sensations. Astringency is believed to be more a sensation than a taste, thought to occur through the precipitation of combined salivary proteins and tannins, which do not usually occur in dairy produce. In addition, temperature also gives perceptual information about the food. It is generally accepted that we have four basic tastes; sweet, sour, bitter, with possibly a fifth taste, umami, best described as savoury. Flavour is usually defined as the perception of tastants and volatile aroma compounds during eating, while terms such as aroma, odour and smell are generally used to describe the sensations while sniffing the food. More detailed information about olfaction and taste can be found in the literature (Piggott, 1984).

The mechanisms of the sensory perception of both taste and volatile aroma compounds also forms an important area of research; the study of perception mechanisms of volatile compounds is currently a fast-moving research field. The volatiles are transported to the olfactory epithelium as part of the air flow, where the mechanisms of perception are believed to involve detection by protein binding (Buck and Axel, 1991). The perception is the result of the interpretation of the signals given to the brain. The mechanisms of olfaction are excellently reviewed elsewhere (Lancet, 1986) and are outside the scope of this chapter. An increased understanding of the mechanisms of perception will help to improve our understanding about how sensory data can be linked to chemical information regarding flavour.

11.3 Panels for sensory analysis

Sensory analysis can be used to investigate the properties of a food or to gain information about the human senses. Generally, two types of panel are used, a trained panel, usually specially selected for their sensory acuity, and a consumer panel. The trained panel is quite small, about 12 people, and is primarily used to give information about the food properties. The more detailed sensory analysis such as descriptive profiling or time–intensity measurements requires considerable resources from a taste panel. Hence, there has been a move towards recruiting panels which are paid specifically for these tasks. Depending on the scientific question the scientist is address-ing, either a trained panel or a consumer panel will be used. Typically, a trained panel is used to provide detailed analytical results consisting of both quantitative and qualitative descriptions of the sensory properties of a food sample. More detailed information regarding the panel recruitment, selec-tion, training, etc. can be found in textbooks, such as Moskowitz (1966). Consumer panels are generally used to collect hedonic, preference or acceptance information about the samples. There are also numerous studies regarding the reasons for consumers' food choice, their attitudes towards aspects of the foods as well as the influence of other factors such as mood, satiety and the label information. Discussion of these studies falls outside the scope of this chapter, and it is reviewed elsewhere (MacFie and Thomson, 1994). An overview of aspects of sensory perception of taste, smell, texture and colour as well as sensory methodology is given by Piggott (1984).

11.3.1 Quantitative descriptive analysis

The techniques used for sensory analyses have been changed radically over the last decades. However, at times sensory analysis has struggled to become a science, with its own procedures, data and methods of interpretation (Moskowitz, 1993). Moskowitz discusses the historic development of sen-sory analysis and also discusses the current debates in this research field. A major advance in sensory analysis towards becoming a formal science was the development of descriptive analysis, developed at the Arthur D. Little Company (Caul, 1957). They developed the flavour profile technique, which was the first method for describing the sensory characteristics and their magnitudes. This profiling technique provided data which were both quali-tative and quantitative, although statistical methods able to analyse such data sets still needed to be developed. This technique has evolved to quantitative descriptive analysis, able to describe accurately the sensory properties of food. Quantitative descriptive analysis allows the differences within a sample set to be described quantitatively, using a set of well-defined terms often specifically developed for that food, and it forms an important tool in the definition of the sensory characterisics of dairy products.

The development of descriptive profiling methods in sensory analysis has largely gone hand-in-hand with the development of computer hardware and software to collect the data sets. An overview of the type of hardware and software available describes how the introduction of the personal computer has revolutionized sensory data collection and analysis (Punter, 1994). The author discusses data collection, and how the development of the friendly interface between the assessor and the computer allows data collection from panellists after only minimal training. Information is also provided about the type of statistical packages developed for data analyses. Current packages for personal computers can perform analysis previously only possible on mainframe computers, thus making statistical processing of data more easily available to sensory scientists.

The data can also be collected using line scales on paper. The line scales are anchored on the left with 'none' of the attribute and on the right with 'maximum' of the attribute and are usually between 10 and 15 cm long. The attributes are generated in discussion with the panel, which has available a range representative of the samples to be assessed. Often, definitions of the terms are prepared, and even reference standards can be made available for training the assessors. A trial run of quantitative descriptive analysis can be done and statistical analysis of the data shows whether the panellists have received sufficient training on the use of these attributes.

The development of well-tested statistical techniques has contributed towards the establishment of quantitative descriptive analysis as a well accepted and widely used sensory technique, which has in turn contributed to the foundation of sensory analysis as a scientific discipline. An off-shoot of this technique – free choice profiling – allows the assessors to use their own terms to describe the differences between the samples, and was first used to examine a range of port wines (Willims and Langron, 1984). This technique also collects quantitative information of the intensity of attributes, without the need to train the assessors on the use of an agreed set of terms.

Descriptive analysis is a useful tool to describe the sensory differences between sample sets; its use has recently also been demonstrated in the identification of defects in milks (Lawless and Claasen, 1993). A range of descriptive terms on attribute scales was generated by a trained panel, which enabled significant discrimination between untreated, oxidized and rancid milks. The data showed good correlation with consumer acceptability judgements. Discriminant analysis showed that the descriptive profiles could be used to identify defects by their root causes. However, a better understanding regarding the basis of the sensory differences will also come from our increase in knowledge about the effect of the food matrix on the stimulus. In addition, individual differences between subjects regarding preferences, effect of external factors which may affect their choice of foods, as well as the perception mechanisms are all expected to contribute to the interpretation of sensory data.

11.3.2   *Time–intensity measurements*

One of the latest developments in sensory analysis is more extensive use of time–intensity measurements; the development and use of this method is reviewed by Cliff and Heymann, 1993. Time–intensity measurements are an extension of the scaling method in use for quantitative descriptive analysis and consist of the continuous scaling of perceived intensity of an attribute, until the perceived attribute intensity has dropped to zero or the recording period has stopped. Quantitative descriptive analysis gives information regarding the intensity of attributes only, whereas time–intensity measurements give, in addition to maximum intensity, information regarding the start time of the stimulus perception, the time needed to reach maximum intensity, total duration of the stimulus, and area under the curve. Other parameters such as rate of appearance, rate of extinction or drop in intensity relative to first introduction, can also be derived (Lee and Pangborn, 1986).

Before the development of computer software to collect data, panellists were asked to record their intensity perception on moving paper, or they were asked to record it at time intervals often given by a verbal clue. Nowadays, most time–intensity recordings are made by moving a mouse along a line scale on a computer screen. The line scale is often marked with none of the attribute on one end and maximum of the attribute on the other end. It is important that the panellists are well trained in the use of this technique, including the use of the mouse, and have been given precise instructions regarding placing the food in the mouth and the start of the recording procedure. It is equally important that the panellists are not aware that they are producing a curve, since that may lead to the reproduction of similar curves, which may not accurately reflect the perceived intensity of the attribute. Time–intensity measurements are quite time consuming, hence in our laboratory we first analyse the samples using qualitative descriptive analysis. From these results we select a subset of terms which were used to discriminate significantly between the samples.

There have been a number of new methods published on averaging the time–intensity information (Overbosch et al., 1986; Liu and MacFie, 1990; MacFie and Liu, 1992; Dijksterhuis and van den Broek, 1995) and analysing the data (Dijksterhuis et al., 1994). The enhanced resolution this method allows increases our understanding of the factors affecting the release of flavours from a food matrix. One question that remains to be answered is whether a change in fat content – such as a reduction in fat to produce a low-fat cheese – can affect the timing of flavour release, and hence produce an imbalance in flavours responsible for the lower acceptability of the sensory properties. However, there are still numerous questions about the use of this technique which remain unanswered (Lawless and Clark, 1992), such as the relationship between a continuous time–intensity profile and profiling information, the effect of using a continuous-rating method, and whether time–intensity methods are prone to any of the systematic response biases that are known to occur with other sensory methods. A number of other issues such as the possible changes of the sample and the effect on the assessment of the attribute need to be considered. For example, textural attributes may alter during eating as a result of the food being broken down to crumbs or being formed into a bolus. The textural attributes may also change as a result of temperature changes, such as the cheese sample being warmed up in the mouth. Another issue is possible perceptual changes during eating, for example volatile flavour perception is prone to adaptation, which means that when we are being subjected to a continuous stimulus, the perception of that stimulus is being suppressed. Thus, it would seem that we may be more sensitive to detecting changes in stimulus quality as a function of time than quantity as a function of time.

11.3.3   Experimental design and statistical analysis

Sample presentation is considered of particular importance to balance the effect of order of presentation and the carry-over effect of a preceding sample over a series of presentations of the same set of samples. Muir and Hunter (1991) reported a sensory evaluation study of Cheddar cheese to investigate order of tasting and carry over effects on the results. The authors found that the order of tasting introduces significant bias to the sensory evaluation of Cheddar cheese. The cheese tasted first is likely to have a marked positive bias when overall preference or acceptability is being judged. Experimental designs to overcome these effects have been published (MacFie et al., 1989). Aspects of experimental design as well as statistical methods for analysis of sensory data sets are excellently described elsewhere (O'Mahony, 1985; Piggott, 1986). The methodology for sensory analysis is reviewed by numerous authors, some describing mainly the product testing and sensory evaluation aspects (Moskowitz, 1996), while others consider the psychological basis of sensory evaluation (McBride and MacFie, 1990). When analysing sensory data it is important to evaluate the assessor performance, and to check for outliers, possibly due to specific anosmia, confusion about the attributes or reflecting individual differences. Sensory scientists are more interested in sample differences than human differences, and outliers can greatly affect the analysis. The better the overall performance of the panel – in other words the closer their agreement in qualitative and quantitative scoring – the more significant differences can be determined between the samples. Generally, methods for the analysis of multidimensional data sets (principal component analysis, generalized Procrustes analysis), such as obtained by quantitative descriptive analysis, reduce the data to a number of principal axes, on which the largest possible amount of variation is reflected. The first two axes account usually for a large percentage of the total variation, and form a two-dimensional plot on which the sample means are located. The attributes used to separate the samples on the axis can be used to interpret the differences between the samples.

Free choice profiling data are analysed using generalized Procrustes analysis (Gower, 1975; Arnold and Williams, 1986). This analysis assumes that the assessors perceive the samples in the same way in the sensory space, but that they may use different descriptors to assess the same attribute. This technique also allows for the use of different locations and proportions of the scale. Hence the analysis of data using generalized Procrustes analysis is less straightforward than principal component analysis. Examples of this technique used for Cheddar cheese, with further discussion regarding the advantages and drawbacks, are found in the literature (McEwan *et al.*, 1989; Muir *et al.*, 1995).

The incorporation of communication elements into experimental design is an issue gaining increasing support, as a product evaluated for its acceptability by the consumer is often not simply the physical formulation of the product alone (Moskowitz, 1993). This author gave the example of low-fat cheese, carrying the compelling message low-fat. In paired comparison tests, the consumers prefer the full-fat or modestly reduced-fat cheese, for texture and flavour. However, the attractiveness of low fat is significant, and can make a large difference in the marketability of an otherwise modestly acceptable product. The mixture of information and product leads to a different decision than would be obtained by the sensory evaluation of the product alone.

The detailed sensory data such as obtained using quantitative descriptive analysis can be used to link to consumer preference data in order to understand which attributes relate to like/dislike of the food. Such information has great value for product development. There are many research topics associated with the collection of consumer preference data, such as the effect of labelling, mood, information and segmentation of the population. This is discussed in a recent book (MacFie and Thomson, 1994), which also includes a chapter by MacFie and Greenhoff on preference mapping, a technique which relates the preference data to a multidimensional data set of attributes.

Further information regarding the physical and chemical properties of a food contributing to the sensory attributes of that food can be obtained by relating sensory data to chemical and physical measurements. Although instruments can measure many different food properties, these methods do not take into account the interactions between taste and smell, or the perceptor mechanisms in the olfactory system. Recent reviews describe how the various instrumental flavour data can be linked to sensory information (Piggott, 1990; Martens *et al.*, 1994). One common difficulty is the definition of sensory threshold values, which are often unreliable. The threshold value is the lowest concentration of a stimulus that can be perceived: it depends on the person, and is specific for both compound and food matrix. The odour unit – defined as the concentration of a flavour in a food divided by the threshold value – is often used. Scores lower than 1 are considered to be unimportant, while most importance is attributed to the higher scores. However, the fundamental shortcomings of this approach are the use of inaccurately defined threshold values, and the fact that no account is taken of interactions between taste and olfaction as well as between compounds and differences in intensity functions. For example, it is not known whether a number of sub-threshold concentrations of stimuli can contribute to perception.

There are a number of statistical methods available for linking compositional data to sensory information, such as single and multiple regressions, partial least squares, principal component regression and canonical correlation. The correlations between attributes and chemical or physical measurements need to be interpreted by the researcher, since no cause or mechanism can be determined in such a way. More information regarding the techniques can be found in O'Mahony (1985) and Piggott (1986). Piggott (1990) pointed out that all the methods currently used have strengths and weaknesses. A correlation does not mean that there is a causal link between sensory attribute and chemical or physical measurement. He also warns against drawing conclusions from chance correlations, in particular with using partial least squares analysis, and stresses the need to validate. The predictive ability of regression models should be evaluated by separating the data into a training set, from which the model is built, and a test set, used to evaluate the model.

One example of correlations between flavour characteristics and chemical measurements of flavour during Cheddar cheese maturation is the work by Barlow et al. (1989). They monitored the maturation of Cheddar cheeses using both sensory methods and gas-chromatographic analysis of volatile flavour compounds. Using regression analysis they found that the chemical constituents which related most strongly with the panels' Cheddar flavour score were the volatile sulphur compounds, in particular hydrogen sulphide, lactic acid, butanone, water-soluble nitrogen and acetic acid. They were able to make reasonable predictions of the 12-month Cheddar flavour score using these scores obtained after 3 months. They concluded that a number of groups of compounds provide correlations with Cheddar flavour scores, reinforcing Mulder's theory that Cheddar flavour results from a number of components when they are present at the correct concentrations and have the appropriate balance (Mulder, 1952).

11.4   Flavour release and perception

Predictions about the sensory properties of the food do not easily follow from knowledge of the aroma components. This can in part be attributed to gaps in our knowledge concerning the interactions of flavour compounds with major constituents in food such as fat, protein and carbohydrates, and the effect of food structure on the release of flavour during eating. Although flavour is defined as a combination of taste and odour sensations, in particular the release of volatile flavours is known to be influenced by the food matrix. Hence, there is a growing awareness that the flavours, in particular the volatile ones, of a food need to be available for perception during the period of eating. Flavour interactions with the food matrix and their effects on perception have been reviewed recently (Bakker, 1995).

The concentration of a stimulus consisting of a flavour compound in simple solution can be correlated with its perceived intensity (Stevens, 1957). However, if the carrier for the stimulus is a food, it is extremely difficult to define what fraction of the total flavour in the food is available for perception (Darling *et al.*, 1986). The possible interactions of the added flavour compound with the structure and the composition of the food may affect the release of flavour. In order to perceive the volatile odours of food before eating, the odours must travel through the nose to the olfactory epithelium to be detected by the receptor cells. When food is eaten, the volatile flavour molecules released from the food into the mouth pass back up through the nasopharynx and into the nose in the opposite direction to when they are sniffed (Lawless, 1991). Before a subject may perceive an aroma, a sufficiently high concentration of aroma molecules must be released from the food and stimulate the olfactory system. Hence, aroma perception during eating depends on the nature and concentration of volatiles present in the food, as well as their availability for perception as a result of interactions between the components in the food. Availability will be further influenced by the processes of eating, such as mastication, which breaks down the food structure; mixing the food with saliva and possible airflows due to breathing will also affect the release of the aroma compounds.

The partition coefficient of a food volatile at equilibrium is its concentration in the headspace divided by its concentration in the food phase at a defined temperature. The partition coefficient forms an important parameter in the mathematical modelling of flavour release (Overbosch *et al.*, 1991). However, it is believed that during eating, equilibrium partitioning between food and mouth space is not achieved. Since eating is a relatively rapid process, and most foods are swallowed within a minute, it is the initial rate of flavour release which is believed to be relevant for perception. Differences in the initial rate of release may affect the time required to reach the threshold concentration of a volatile; this may be determined by differences in the time–intensity measurements, in particular the time needed to start detecting the stimulus. The rate of release after perception of the threshold concentration is expected to influence the time required to reach maximum intensity, another parameter which can be determined using time–intensity analysis. Time–intensity measurements are expected to make a significant contribution to the research on the effect of the food matrix on flavour release. Mathematical modelling of the processes involved during eating will focus the research experiments towards the important factors affecting the release and perception of flavour.

11.4.1   Mastication

Chewing, one of the main activities used to break a food down, will also influence release and perception of flavour. Recently developed methodology using electromyography has made it possible to study chewing patterns (Brown, 1994). Electromyography monitors the interaction between food and consumer during mastication by recording the activity in the jaw-closing muscles. The activity differs for different foods and changes during mastication as the food is progressively broken down (Brown, 1996). Chewing behaviour for individual subjects can be determined from the electromyograms, and a numerical description for chewing behaviour consistent for each subject can be derived to express differences between subjects (Roberts and Vickers, 1994). Further developments in this research area may be the development of 'fingerprints' characteristic for textural properties, and lead to a better understanding of texture perception. A better understanding is expected also about the effect of the breakdown of food during eating on the release and perception of flavour, in particular when electromyography is combined with time–intensity measurements. Using both techniques, the observed differences in the time–intensity parameters can be related to the individual chewing parameters.

Cheddar cheeses with a range of textural characteristics were differentiated using electromyography, quantitative descriptive analysis and instrumental measures texture using an Instron (Jack *et al.*, 1993). A large percentage of the sensory differences was accounted for by the scores of soft, smooth, hard, coarse, sticky, mouthcoating, rubbery and chewy. The electromyography traces of masticatory muscle activity were unique for each subject for different samples, hence the prediction of texture perception from these data was also subject-dependent. Similarly, correlations between Instron measurements and electromyography were not consistent from one subject to another.

The complexiy of the interactions among food composition and structure, food breakdown during mastication and flavour release has been demonstrated in a study at the author's laboratory (W.E. Brown, D.J. Mela and N. Daget, personal communication). A series of chocolates with different fat contents and solid fat indices were eaten, and time–intensity measurements of chocolate aroma and sweetness during eating were recorded using trained assessors. Results showed that the mastication patterns, as revealed by electromyographs of the masticatory muscles, differed between chocolate samples and also between individuals. The temporal perception of aroma and tastant reflected differences between the chewing patterns for the two subjects. Clearly, flavour and tastant release were influenced by the composition and structure of the food matrix, but perception of the released compounds is modified by food breakdown pattern of individual consumers.

11.4.2   Sensors

There is a rapid current development of electronic sensors, built into instruments which are often referred to as 'electronic noses'. The concept of an electronic nose requires understanding of the biological olfactory system as well as the technology of volatile chemical sensing devices. A number of sensors with different specificities are built into a sensor array. Computer software is developed to process the responses of the sensors to a specific signal and compares the response to a library of responses stored for interpretation purposes. The final step will be to interpret this information in terms of sensory characteristics that the human nose would perceive. Thus, the electronic nose is capable of mimicking the human sense of olfaction or smell. An electronic analogue of the human nose is still a long way in the future.

Odour sensors have been built based on tin oxide gas sensors and on conducting polymer chemiresistors. In order to produce a successful sensor array, which can be used to detect differences between a large range of samples, the sensors in the array need to have a good specificity, high stability, rapid response, good reproducibility, and a small size. In addition, the sensors should display selectivity towards volatile aroma compounds which contribute significantly to the flavour characteristics of our foods. Much emphasis of the current research is focused on the development of conducting polymer chemiresistors, since the relative ease of producing them and the wide diversity of different polymers that can be produced may well hold the key to a successful electronic nose.

Currently these instruments give a rapid overall impression of the volatile compounds in the headspace of a sample, without any detailed information regarding the qualitaive or quantitative composition. The sensors of the array all respond to a greater or lesser extent to the volatile compounds to which they are exposed. With the help of the computer software this picture of responses is displayed graphically and can be matched with other stored sample traces. One practical application documented in the literature is the use of tin oxide sensors to monitor fish freshness. There would appear to be endless possible applications for sensors which could be designed to monitor specific compounds in process control, or to monitor specific markers of food spoilage, or to act as sensors to mark the optimum sensory properties of a mature Cheddar cheese. However, most of these applications are still futuristic, and the currently available equipment is continuously being improved. Validation of electronic noses with human perception panels is only one of the many research areas still open. A useful book edited by

Gardner and Bartlett (1992) with contributions of many experts working in this field gives an excellent overview of the state of the art in this exciting and fast-moving area of research.

11.4.3   Fat and flavour release

The influence of the physicochemical properties of both the flavour compound and the fat content has been demonstrated by the determination of air–oil partition coefficients by instrumental headspace measurements, and such information has recently been reviewed (Bakker, 1995). The partition coefficient of a flavour determines the equilibrium concentration of that flavour in the headspace, and is expected to relate to the perception of the volatiles from the food matrix (Overbosch et al., 1991), although the mechanical breakdown of the food during eating is considered to influence the release rate of volatiles from a food and hence also contribute to the perception (Bakker, 1995). This review concluded that fat greatly influences the partition coefficients determined under equilibrium conditions as well as dynamic measurements of flavour release. Besides the properties of the flavour compound itself, numerous factors of the fat were shown to influence the release of flavour, such as temperature, melting point of fat, and solid fat index. In addition, the composition of the emulsion, such as the fat content as well as emulsion structure, were shown to influence the release of flavour.

One investigation on partitioning of cheese volatiles concluded that reducing the fat content in a Cheddar cheese may also affect the concentration of sulphur-containing volatiles in the headspace (Boekel and Lindsay, 1992). A clear example of the effect of fat content on sensory perception is an evaluation of taste recognition threshold concentrations of styrene in oil-in-water emulsions and yoghurts with increasing fat content from 0.3 to 2.1 mg kg^{-1} (Linssen et al., 1993). Their results showed a linear increase of the flavour threshold with fat content. The authors attributed their finding to the very good solubility of styrene in fat, hence increasing the fat content of the emulsion led to an increasing amount of styrene in the fat phase. The calculated concentration in the aqueous phase of the emulsion required for the flavour threshold recognition was determined to be fairly constant and in agreement with the threshold concentration determined in water. Thus, even a small concentration of fat can influence the flavour threshold of a very hydrophobic compound.

The effect of emulsifier on the release of a flavour compound as determined by time–intensity measurements has been studied (Haring, 1990). The presence of an emulsifier in an oil gave rise to the formation of smaller oil droplets as a function of mastication time. Perceived intensity of the flavour compound was increased and longer lasting in the presence of emulsifier. The composition of fat has also been shown to influence the flavour perception (Lee, 1986). The release and perception of diacetyl was studied in a slow-melting saturated (stearin) fat with high solids content and a fast-melting unsaturated (olein) fat. Time–intensity measurements showed that stearin gave a slower flavour release rate and a lower flavour intensity than olein fat. Many other factors which may possibly influence the release of flavour from dairy foods still need to be investigated, such as type of fat, particle size of the fat droplets and distribution of fat in the food matrix.

11.4.4 Protein and flavour release

Research to date shows that proteins in model solutions can bind a number of flavour compounds to a greater or lesser extent (Bakker, 1995). Binding of flavours to proteins has implications for the perception of added flavour formulations. Selective binding of volatiles to proteins also has implications for the flavour formulation added to a protein-containing food. Available data indicate that volatiles are bound to proteins by hydrophobic interactions with discrete binding zones (Solms, 1986), thus probably limiting the availability of these compounds for sensory perception. The properties of the protein as well as the volatile compound determine the extent of binding. In addition, chemical reactions between proteins and volatiles can also occur. A number of sensory studies have been carried out, indicating a reduction of flavour intensity in the presence of protein. Flavour binding has been reviewed for whey proteins (Kinsella, 1989), while three reviews also consider the effect of these interactions on flavour perception (Solms, 1986; Kinsella, 1989; Overbosch et al., 1991). By implication, one would expect the progressive breakdown of proteins in cheese, as described in Chapter 1, to influence flavour binding, adding to the tendency of older cheese to be perceived as more intensely flavoured. Sensory studies carried out on simple model food solutions containing milk proteins indicate that interactions and possible reactions between flavour compounds and protein in aqueous solutions can reduce the perceived aroma intensity compared with water. Quantitative descriptive sensory analysis of vanillin, a flavour compound of importance for ice-creams, in whey protein concentrate and sodium caseinate solutions containing 2.5% sucrose to reduce the bitter taste of vanillin, showed a significant decrease in vanillin intensity with increasing concentration of whey protein concentrate from 0.125 to 0.5%, as compared with the reference containing vanillin and sucrose only (Hansen and Heinis, 1991). No such decrease was observed for sodium caseinate. The observed differences between the extent of flavour intensity loss for the two proteins was attributed to Schiff base formation or interaction between vanillin and the thiol group of cysteine in whey protein concentrate. Another example is the effect of benzaldehyde, (+)limonene and citral in whey protein concentrate and sodium caseinate solutions containing 2.5% sucrose on sensory assessment of intensity (Hansen and Heinis, 1992). The (+)limonene and benzaldehyde flavour intensity scores decreased as the whey protein concentrations increased, while for citral this trend was not statistically significant. The caseinate solutions showed smaller differences in intensity scores, which were often not significant. Possible explanations for these effects are non-polar interactions specifically with casein, and Schiff base formation with whey protein. Vanillin interaction has also been studied in milk protein isolates in sweetened drinks, looking at the impact of heat procesing and emulsifier on perceived vanillin flavour intensity (McNeil and Schmidt, 1993), with emulsifier and heating both affecting vanilla intensity.

The effect of the protein structure on binding still needs to be elucidated in order to understand the binding mechanisms. Little information is available regarding the binding properties of proteins when they are incorporated in food-like structures such as dairy foods, where fat content, the food structure and other flavour compounds can all influence flavour binding to proteins and the release of flavours during eating processes. Studies on the effect of mechanical action to mimic the conditions of food breakdown, mixing, stirring and dilution on the release of flavours are sparse. The effect of eating on the release and perception of flavour bound to proteins still needs to be studied. More investigations are required to elucidate the mechanisms influencing the release of flavour from food matrices. Sensory analysis is expected to be a useful tool in helping to elucidate the factors affecting the release of flavours from dairy foods.

11.4.5 Flavour release in the mouth

One of the earliest analyses of flavour retention in the mouth from standard solutions was done using an aqueous extraction of the mouth (Wilson and Ottley, 1981). These results indicated that flavours differ in the length of time they are retained in the mouth. Direct measurements of volatiles in breath has been used to detect residual industrial solvents in breath (Benoit et al., 1983; Hussein et al., 1983). To determine flavour released into the airspace during mastication, a direct method to determine the release of flavour components at the nose, breath-by-breath as a function of time, has been reported (Soeting and Heidema, 1988). These results showed variability in breathing patterns and total amounts of air expired, both within and between subjects. Total amounts of flavour compound (2-pentanone) expired over 2 minutes indicated a more than 30-fold difference between the 17 panel members. Despite this variability the authors were able to determine some major effects, such as a faster release of the flavour compound from water than from oil. Recently a method to measure volatile profiles during eating was developed (Linforth and Taylor, 1993). The authors concluded that the most successful method was trapping on Tenax followed by gas-chromatographic analysis. The authors reported different concentrations of menthone and menthol released in the headspace and in the nose space while eating extra-strong mints, sampled by collecting air from breath from the nose. Further developments of techniques to measure the release of flavours during eating will further our understanding about the factors affecting release, and will give us information about the volatile concentrations giving rise to sensory responses of flavour. Instrumental measurements of flavour release, without concentrating the flavours, would allow the collection of information more directly comparable with time–intensity measurements.

11.5 Cheese

One of the early studies on the characteristics of Cheddar cheese and their relation with acceptability was done by quantitaive descriptive analysis, also referred to as profiling, and free-choice profiling (McEwan *et al.*, 1989). In the comparison of the two techniques, seven varieties of Cheddar cheese were evaluated for odour, flavour and texture attributes. A common vocabulary was developed for this set (Table 11.1). The results obtained using the two techniques were similar and showed that strength of odour and flavour, and the texture terms 'rubbery' and 'grainy' were important attributes in separating the samples. Preference judgements obtained from a separate consumer group showed that the low-fat Cheddar was the least liked, while the vegetarian and the mature Cheddar were the most liked. The authors suggested that more research needs to be done towards developing mathematical models to relate quantitative descriptive analysis data to acceptability information from consumers.

The key attributes for the sensory classification of hard cheeses were determined by Muir *et al.* (1995). A panel of 16 assessors was used to describe the sensory properties of a range of hard cheeses encompassing the main types for sale in the United Kingdom. The sensory terms developed to discriminate between these cheeses is similar to the terms used for Cheddar cheese (McEwan *et al.*, 1989; Table 11.1). Despite the differences in methodology for these studies, similar terms were identified to allow discrimination between the samples. The important attributes to describe the differences between the samples were intensity, rancid and creamy for odour, cowy/unclean and creamy for flavour and rubbery, pasty and grainy for texture. The authors concluded that it will be feasible to construct a sensory key for cheese, based on selected characterisics. This definition of the sensory properties of cheese will enhance opportunities for trade and provide a useful tool for underpinning scientific and technological research.

Quantitative descriptive analysis has also been used to assess the sensory properties of Cheddar cheese during maturation, using a set of terms overlapping with the terms used for the studies described above, although it is not clear to which category some of the terms belong (Piggott and

Table 11.1 Vocabulary of sensory attributes developed for quantitative descriptive analysis of Cheddar cheese (McEwan *et al.*, 1989), a set of hard cheeses (Muir *et al.*, 1995) and a set of maturing Cheddar cheeses (Piggot and Mowat, 1991).

	Terms		
Category	Cheddar cheese	Hard cheese	Maturing Cheddar cheese
Odour	strength creamy/milky sour rindy mature	intensity creamy/milky sulphur/eggy fruity/sweet rancid	sour
Flavour	creamy/milky strength sour acid mature salty smoky rindy	creamy/milky intensity acid/sour animal/cowy/unclean salty fruity rancid sulphur/eggy bitter sweet	milky strength sour pungent salty cheesy rancid mouldy bitter sweet buttery processed maturity
Texture/mouthfeel	tongue-tingling soft-firm rubbery mouth coating grainless	pasty firm rubbery mouth coating grainy	moist soft to firm rubbery mouth coating grainy smooth crumbly chewy after-taste

Mowat, 1991). The mouthfeel, strength and aftertaste of the cheeses were largely dependent on the maturation time and conditions (temperature). Over time the milky/buttery flavour changes to a more sour, rancid and pungent characteristic. Textural changes occur during maturation, but these depend to a considerable extent on the production procedure and maturation conditions.

Forty-two cheeses selected to represent a wide range in milk source, country of origin, fat, moisture and microbial ripening properties were used to develop terminology to describe their flavour attributes (Heissener and Chambers, 1993). These authors defined a list of attributes, with definitions of the terms. They also provided references for each term, consisting of a food having a prevalence of the attribute, or a chemical compound.

Although the authors felt that the terms only apply to matured cheeses and that the number of terms can be reduced, it would seem that definitions combined with reference standards will allow panels to be trained on the same terms, thus allowing a comparison between data. A comparison between a Scottish and a Norwegian panel showed that training of the panels enhances the ability to discriminate between the individual sensory attributes (Hirst and Muir, 1994). The agreement on a standard vocabulary allowed a clear interpretation of the results, and showed discrimination between the samples using often the same attribute.

Sensory analysis has been used to study numerous aspects of cheese-making on the sensory properties of the cheese, for example the effect of storage temperature during maturation (Grazier et al., 1991), or type of salt (sodium chloride or mixture of sodium chloride and potassium chloride) used during processing (Reddy and Marth, 1994). Sensory studies are being extended to include the evaluation of attributes related to intended use of Cheddar cheese, referred to as the perception of appropriateness (Jack et al., 1994). In a range of Cheddar cheeses, textural characteristics and melting properties were identified as a major factor of appropriateness. All these studies add to our knowledge of the effect of raw material, processing and maturation on the sensory attributes of the end-product. However, despite the wealth of information available, our understanding of, for example, Cheddar flavour is still incomplete.

11.5.1 Low-fat cheese

A current interest in product development is the production of foods with a lower fat content, without the loss of sensory properties. Since fat is an important carrier of many hydrophobic flavours – many of which have a hydrophobic character – the change of fat is often assumed to affect the timing of flavour release, and thus the perception of flavour. A number of studies have been done to investigate the effect of composition differences on the temporal aspect of perception. Stampanoni and Noble (1991b) studied the influence of fat, acid and salt on the temporal perception of saltiness, firmness and sourness of cheese analogues. An increase in salt content and acid content both resulted in increased time–intensity parameters for saltiness and sourness respectively. Saltiness and sourness continued to be perceived after total breakdown of the cheese structure. An increase in fat (from 10% to 25%) resulted in a softer cheese, which was perceived to be more sour, but had no significant effect on the temporal aspect of perception.

A study of cheese analogues with varying amounts of fat (10, 17.5 and 25%), sodium chloride (0.5 and 2.0%) and citric acid (0.1 and 1.2%) showed that the texture properties were significantly affected by these changes; in particular lowering the fat content resulted in more springy and less cohesive samples (Stampanoni and Noble, 1991a). Sensory evaluation of reduced-fat cheeses has also been used to investigate the effect of processing variables on sensory quality (Drake *et al.*, 1995). Further research is needed to elucidate the factors affecting the release and perception of flavour from reduced-fat cheeses in order to manufacture low-fat cheese with acceptable sensory properties.

11.6   Yoghurt

An investigation of the descriptive analysis of strawberry and lemon yoghurt and consumer acceptability ratings showed that overall liking was based on the fruit flavour, sweetness, and sourness liking and on the sweetness/ sourness balance of the samples (Barnes *et al.*, 1991). One study reports the development of 32 terms to describe yoghurt samples; using analysis of the preliminary experiment, a subset of 12 terms was selected, and a confirmatory experiment was done using this reduced vocabulary (Hunter and Muir, 1993). The following terms were found to describe the sensory differences:

odour	sour, creamy, sweet
flavour	sour/acid, creamy, sweet, other
after-taste	sour/acid, chalky
texture	viscosity, chalky, serum separation

Studies have also been done on the effect of starter cultures on the sensory attributes determined by quantitative descriptive analysis (Ulberth and Kneifel, 1992; Rohm *et al.*, 1994). The effect of sucrose and fat content on the sensory attributes of strawberry yoghurt showed, as expected, that the ratings for sweetness and fattiness increased with increasing sucrose and fat contents (Tuorila *et al.*, 1993). Sucrose enhanced perceived fattiness and fat enhanced sweetness, while sourness was suppressed by both. Another study on effect of fat content on the timing of flavour perception was done with strawberry yoghurts (Tuorila *et al.*, 1995). The scoring of sweetness and sourness was not affected by fat level, and their data did not support the assumption that fat delays the perception of these attributes. However, it remains likely that fat content does affect the temporal perception of fat-soluble volatile flavour compounds.

11.7   Ice-cream

A recent study on ice-creams prepared with vanilla flavours and different matrix compositions was done using quantitative descriptive analysis (King, 1994). The results showed that the ice-cream matrix has a significant effect on the perception of flavour. A modification of the fat distribution in the finished product also influences flavour release.

11.8 Conclusions

Sensory analysis can play a useful role in determining the sensory attributes of dairy foods, in particular since the important flavour compounds contributing to sensory properties of some dairy products, such as Cheddar cheese, are still ill-defined. Sensory analysis has been used successfully to investigate the sensory attributes of dairy foods. Efforts have also been made to develop well-defined terms to describe the differences between cheeses, and a number of investigations have been done to establish the effects of various processing and maturation changes on the sensory properties of cheese. Scientific information illustrates that interactions of flavour compounds with components of the food matrix do affect the release and perception of flavour. However, our understanding of the mechanisms underlying these interactions is far from complete. Further development of methods to measure flavour release during eating, and of linking such information to time–intensity measurements, will provide information about flavour release and perception in human subjects.

11.9 References

Arnold, G.M. and Williams, A.A. (1986) The use of generalised Procrustes techniques in sensory analysis, in *Statistical Procedures in Food Research* (ed. J.R. Piggott), Elsevier Applied Science, London, New York, pp. 233–53.

Bakker, J. (1995) Flavor interactions with the food matrix and their effects on perception, in *Ingredient Interactions, Effects of Food Quality* (ed. A.G. Gaonkar), Marcel Dekker, Inc., New York, pp. 411–39.

Bakker, J. and Law, B. (1994) Cheese flavour, in *Understanding Natural Flavors* (eds J.R. Piggott and A. Paterson), Blackie Academic & Professional, Glasgow, pp. 283–7.

Barlow, I., Lloyd, G.T., Ramshaw, E.H., Miller, A.J., McVabe, G.P. and McCabe, L. (1989) Correlations and changes in flavour and chemical parameters of Cheddar cheeses during maturation. *Aus. J. Dairy Technol.*, **May**, 7–18.

Barnes, D.L., Harper, S.J., Bodyfelt, F.W. and McDaniel, M.R. (1991) Correlation of descriptive and consumer panel flavour ratings for commercial prestirred strawberry and lemon yoghurts. *J. Dairy Sci.*, **74**, 2089–99.

Benoit, W.R., Lovett, A.M., Nacson, S. and Ngo, A. (1983) Breath analysis by atmospheric pressure ionisation mass spectrometry. *Anal. Chem.*, **55**, 805–7.

Boekel, van M.A.J.S. and Lindsay, R.C. (1992) Partition of cheese volatiles over vapour, fat and aqueous phases. *Neth. Milk Dairy J.*, **46**, 197–208.

Brown, W.E. (1994) Development of a method to investigate differences in chewing behaviour in humans. I. Use of electromyography in measuring chewing. *J. Texture Studies*, **23**, 1–16.

Brown, W.E. (1996) Measurement of perceived texture in foods – the use of mastication analysis, in *Characterisation of Foods. Emerging Methods* (ed. A. Gaonkar), Elsevier Applied Science, London, New York, pp. 309–27.

Buck, L. and Axel, R. (1991) A novel multigene family may encode odorant receptors: a molecular basis for odor recognition. *Cell* **65**, 175–87.

Caul, J.F. (1957) The profile method of flavor analysis. *Adv. Food Res.*, **7**, 1–40.

Cliff, M. and Heymann, H. (1993) Development and use of time–intensity methodology for sensory evaluation: a review. *Food Res. Int.*, **26**, 375–85.

Darling, D.F., Williams, D. and Yendle, P. (1986) Physico-chemical interactions in aroma transport processes from solution, in *Interactions of Food Components* (eds G.G. Birch and M.G. Lindley), Elsevier Applied Science, London, New York, pp. 165–88.

Dijksterhuis, G. and van den Broek, E. (1995) Matching the shape of time–intensity curves. *J. Sensory Studies*, **10**, 149–61.

Dijksterhuis, G., Flipsen, M. and Punter, P. (1994) Principal component analysis of T–I curves: three methods compared. *Food Quality and Preference*, **5**, 121–7.

Drake, M.A., Herrett, W., Boylston, T.D. and Swanson, B.G. (1995) Sensory evaluation of reduced fat cheeses. *J. Food Sci.*, **60**, 898–901, 905.

Gardner, J.W. and Bartlett, P.N. (eds) (1992) *Sensors and Sensory Systems for an Electronic Nose*, Kluwer Academic Publishers, Dordrecht.

Gower, J.C. (1975) Generalized procrustes analysis. *Psychometrika*, **40**, 33–51.

Grazier, C.L., Bodyfelt, F.W., McDaniel, M.R. and Torres, J.A. (1991) Temperature effects on the development of Cheddar cheese flavor and aroma. *J. Dairy Sci.*, **74**, 3656–68.

Hansen, A.P. and Heinis, J.J. (1991) Decrease of vanillin perception in the presence of casein and whey proteins. *J. Dairy Sci.*, **74**, 2936–40.

Hansen, A.P. and Heinis, J.J. (1992) Benzaldehyde, citral, and δ-limonene flavour perception in the presence of casein and whey proteins. *J. Dairy Sci.*, **75**, 1211–15.

Haring, P.G.M. (1990) Flavour release: from product to perception, in *Flavour Science and Technology* (eds Y. Bessiere and A.F. Thomas), John Wiley & Sons, Chichester, pp. 351–4.

Heissener, D.M. and Chambers, E. (1993) Determination of the sensory flavour attributes of aged natural cheese. *J. Sensory Studies*, **8**, 121–32.

Hirst, D.H. and Muir, D.D. (1994) Definition of sensory properties of hard cheese: a collaborative study between Scottish and Norwegian panels. *Int. Dairy J.*, **4**, 743–61.

Hunter, E.A. and Muir, D.D. (1993) Sensory properties of fermented milks: objective reduction of an extensive sensory vocabulary. *J. Sensory Studies*, **8**, 213–27.

Hussein, M.M., Kachikian, R. and Pidel, A.R. (1983) Analysis for flavor residuals in the mouth by gas chromatography. *J. Food Sci.*, **48**, 1884–5.

Jack, F.R., Piggott, J.R. and Paterson, A. (1993) Relationships between electromyography, sensory and instrumental measures of Cheddar cheese texture. *J. Food Sci.*, **58**, 1313–17.

Jack, F.R., Piggott, J.R. and Paterson, A. (1994) Use and appropriateness in cheese choice, and an evaluation of attributes influencing appropriateness. *J. Food Qual. Pref.*, **5**, 281–90.

King, B.M. (1994) Sensory profiling of vanilla ice-cream: flavour and base interactions. *Lebens.-Wiss. u. Technol.*, **27**, 450–6.

Kinsella, J.E. (1989) Flavor perception and binding to food components, in *Flavor Chemistry of Lipid Foods* (eds D.B. Min and T.H. Smouse), American Oil Chemists' Society, Illinois, pp. 376–403.

Lancet, D. (1986) Vertebrate olfactory reception. *Annu. Rev. Neurosci.*, **9**, 329–55.

Lawless, H. (1991) The sense of smell in food quality and sensory evaluation. *J. Food Quality*, **14**, 33–60.

Lawless, H.T. and Claassen, M.R. (1993) Validity of descriptive and defect oriented terminology systems for sensory analysis of fluid milk. *J. Food Sci.*, **58**, 108–12, 119.

Lawless, H.T. and Clark, C.C. (1992) Psychological biases in time–intensity scaling. *Food Technol.*, **46**, 81–90.

Lee, W.E. III (1986) A suggested instrumental technique for studying dynamic flavour release from food products. *J. Food Sci.*, **51**, 249–50.

Lee, W.E. III and Pangborn, R.M. (1986) Time–intensity: the temporal aspects of sensory perception. *Food Technol.*, **40**, 71–82.

Linforth, R.S.T. and Taylor, A.J. (1993) Measurement of flavour release in the mouth. *Food Chem.*, **48**, 115–20.

Linssen, J.P.H., Janssens, A.L.G.M., Reitsma, H.C.E., Bredie, W.L.P. and Roozen, J.P. (1993) Taste recognition threshold concentrations of styrene in oil-in-water emulsions and yoghurts. *J. Sci. Food Agric.*, **61**, 457–62.

Liu, Y.H. and MacFie, H.J.H. (1990) Methods for averaging time–intensity curves. *Chemical Senses*, **15**, 471–84.

MacFie, H.J.H. and Liu, Y.H. (1992) Development in the analysis of time–intensity curves. *Food Technol.*, **46**, 92–7.

MacFie, H.J.H. and Thomson, D.M.H. (eds) (1994) *Measurement of Food Preferences*, Blackie Academic & Professional, Glasgow.

MacFie, H.J.H., Bratchell, N., Greenhoff, K. and Vallis, L.V. (1989) Designs to balance the effect of order of presentation and first-order carry-over effects in Hall tests. *J. Sensory Studies*, **4**, 129–48.

Martens, M., Risvik, E. and Martens, E. (1994) Matching sensory and instrumental analysis, in *Understanding Natural Flavors* (eds J.R. Piggott and A. Patterson), Blackie Academic & Professional, Glasgow, pp. 60–76.

Maruniak, J.A. (1988) The sense of smell, in *Sensory Analysis of Foods* (ed. J.R. Piggott), Elsevier Science Publishers Ltd, pp. 25–68.

McBride, R.L. and MacFie, H.J.H. (eds) (1990) *Psychological Basis of Sensory Evaluation*, Elsevier Applied Science, London.

McEwan, J.A., Moore, J.D. and Colwill, J.S. (1989) The sensory characteristics of Cheddar cheese and their relationship with acceptability. *J. Soc. Dairy Technol.*, **42**, 112–17.

McNeil, V.L. and Schmidt, K.A. (1993) Vanillin interaction with milk protein isolates in sweetened drinks. *J. Food Sci.*, **58**, 1142–7.

Moskowitz, H.R. (1993) Sensory analysis procedures and viewpoints: intellectual history, current debates and future outlooks. *J. Sensory Studies*, **8**, 241–56.

Moskowitz, H.R. (ed.) (1996) *Product Testing and Sensory Evaluation of Foods, Marketing and R & D Approaches*, Food and Nutrition Press, Inc., Westport, Connecticut, USA.

Muir, D.D. and Hunter, E.A. (1991) Sensory evaluation of Cheddar cheese: order of tasting and carry over effects. *Food Qual. Pref.*, **3**, 141–5.

Muir, D.D., Hunter, E.A., Banks, J.M. and Horne, D.S. (1995) Sensory properties of hard cheese: identification of key attributes. *Int. Dairy J.*, **5**, 157–77.

Mulder, H. (1952) Taste and flavour forming substances in cheese. *Neth. Milk Dairy J.*, **6**, 157–68.

O'Mahony, M. (ed.) (1985) *Sensory Evaluation of Food: Statistical Methods and Procedures*, Marcel Dekker, Inc., New York.

Overbosch, P., van den Enden, J.C. and Keur, B.M. (1986) An improved method for measuring perceived intensity/time relationships in human taste and smell. *Chemical Senses*, **11**, 331–8.

Overbosch, P., Afterof, W.G.M. and Haring, P.M.G. (1991) Flavor release in the mouth. *Food Rev. Int.*, **7**, 137–84.

Piggott, J.R. (ed.) (1984) *Sensory Analysis of Foods*, Elsevier Science Publishers, London.

Piggott, J.R. (ed.) (1986) *Statistical Procedures in Food Research*, Elsevier Applied Science, London.

Piggott, J.R. (1990) Relating sensory and chemical data to understand flavour. *J. Sensory Studies*, **4**, 261–72.

Piggott, J.R. and Mowat, R.G. (1991) Sensory aspects of maturation of Cheddar cheese by descriptive analysis. *J. Sensory Studies*, **6**, 49–62.

Punter, P.H. (1994) Software for data collection and processing, in *Understanding Natural Flavors* (eds J.R. Piggott and A. Paterson), Blackie Academic & Professional, Glasgow, pp. 97–112.

Reddy, K.A. and Marth, E.H. (1994) Sensory evaluation of Cheddar cheese made with sodium chloride or mixtures of sodium and potassium chloride. *J. Sensory Studies*, **9**, 187–204.

Roberts, A.K. and Vickers, Z.M. (1994) A comparison of trained and untrained judges' evaluation of sensory attribute intensity and liking of Cheddar cheeses. *J. Sensory Studies*, **9**, 1–20.

Rohm, H., Kovac, A. Kneifel, W. (1994) Effects of starter cultures on sensory properties of set-style yoghurt determined by quantitative descriptive analysis. *J. Sensory Studies*, **9**, 171–86.

Soeting, W.J. and Heidema, J. (1988) A mass spectrometric method for measuring flavour concentration/time profiles in human breath. *Chemical Senses*, **13**, 607–17.

Solms, J. (1986) Interactions of non-volatile and volatile substances in foods, in *Interactions of Food Components* (eds G.G. Birch and M.G. Lindley), Elsevier Applied Science Publishers, London, pp. 189–210.

Stampanoni, C.R. and Noble, A.C. (1991a) The influence of fat, acid, and salt on the temporal perception of firmness, sourness and saltiness of cheese analogs. *J. Text. Studies*, **22**, 381–92.

Stampanoni, C.R. and Noble, A.C. (1991b) The influence of fat, acid and salt on selected taste and texture attributes of cheese analogs: a scalar study. *J. Text. Studies*, **22**, 367–80.

Stevens, S.S. (1957) On the psychological law. *Psychological Reviews*, **64**, 153–81.

Tuorila, H., Sommerdahl, C., Hyvonen, L., Leporanta, K. and Merimaa, P. (1993) Sensory attributes and acceptance of sucrose and fat in strawberry yoghurts. *Int. J. Food Sci. Technol.*, **28**, 359–69.

Tuorila, H., Sommerdahl, C. and Hyvonen, L. (1995) Does fat affect the timing of flavour perception? A case study with yoghurt. *Food Qual. Pref.*, **6**, 55–8.

Ulberth, F. and Kneifel, W. (1992) Aroma profiles and sensory properties of yoghurt and yoghurt related products. II. Classification of started cultures by means of cluster analysis. *Milchwissenschaft*, **47**, 432–4.

Williams, A.A. and Langron, S.P. (1984) The use of free choice profiling for the evaluation of commercial ports. *J. Sci. Food Agric.*, **35**, 558–68.

Wilson, H.K. and Ottley, L.W. (1981) The use of a transportable mass spectrometer for the direct measurement of industrial solvents in breath. *Biomed. Mass Spectrometry*, **8**, 606–10.

Index